RFID AT ULTRA AND SUPER HIGH FREQUENCIES

Identification Radiofréquence et Cartes à Puces – Description, 2nd edition,
Dunod, France, 2001

Applications en Identification Radiofréquence et Cartes à Puce sans Contact,
Dunod, France, 2003

Identification Radiofréquence et Cartes à Puces – Applications, Dunod, France, 2004

Multiplexed Networks for Embedded Systems, John Wiley & Sons, Ltd, Chichester, 2007

RFID and Contactless Smart Card Applications, John Wiley & Sons, Ltd, Chichester, 2007

Illustrations in the text: Alain and Ursula Bouteveille-Sanders

RFID AT ULTRA AND SUPER HIGH FREQUENCIES

Theory and application

Dominique Paret
Consultant – Senior Technical Expert, DP-Consulting

Translated by: Roderick Riesco, MA, Member of the Institute of Translation and Interpreting, UK

A John Wiley and Sons, Ltd, Publication

Originally published in the French language by Dunod as *Identification et traçabilitié en UHF–SHF,
RFID de Super Hautes Fréquences* by Dominique Paret. © 2005 Dunod.

This edition © 2009, John Wiley & Sons, Ltd

Registered office
John Wiley & Sons Ltd, The Atrium, Southern Gate, Chichester, West Sussex, PO19 8SQ, United Kingdom

For details of our global editorial offices, for customer services and for information about how to apply for
permission to reuse the copyright material in this book please see our website at www.wiley.com.

Library of Congress Cataloguing-in-Publication Data

Paret, Dominique.
  RFID at ultra and super high frequencies : theory and application /
Dominique Paret.
      p. cm.
  Includes bibliographical references and index.
  ISBN 978-0-470-03414-9 (cloth)
  1. Radio frequency identification systems.   2.   Radio frequency–Identification.
  3. Wireless communication systems.   I.   Title.
  TK6553.P32 2009
  621.384'151–dc22

                                                      2009031446

A catalogue record for this book is available from the British Library.

ISBN: 978-0-470-03414-9 (H/B)

Set in 10/12pt, Times Roman by Thomson Digital, Noida, India
Printed in Great Britain by CPI Antony Rowe, Chippenham, Wiltshire

Contents

About the Author

Dominique Paret

Consultant – Senior Technical Expert, DP-Consulting

Dominique graduated in Electronic Engineering from Bréguet/ESIEE and holds a DEA Sciences from Paris VI University. He started his carrier in 1968 as TV Group Leader at Consumer Application Labs in RTC–Compelec (where he created several patents in the analogue and digital fields).

Before creating DP-Consulting (training and consulting services) in May 2006, he spent 40 years with Philips Semiconductors/NXP France as Technical Support Manager in "Innovation and Emerging Business". More recently he has worked mainly on two leading edge subjects:

- **new concepts for automotive applications** such as CAN, LIN, very high speed bus, real-time triggered concepts – FlexRay, Safe by Wire, SBC, FailSafe SBC systems;
- **identification** including RFID (from 125 kHz up to 5.8 GHz, UWB), contact and contactless smart cards, e_government applications (e_passport, e_visas, e_ID cards), NFC (near field communications), biometrics, geo-localization and cryptographic systems.

Dominique Paret is also an official National Representative Delegate in several standardization organizations such as the French national body "AFNOR", International "ISO" working groups and ECMA, for contactless smart cards and RFID (proximity ISO 14 443, vicinity 15 693, RFID 18 000 – x, EPC, Biometrics, ICAO) and other consortia for the automotive sector covering LIN, CAN, FlexRay and TMP systems. The author also acts as one of the contactless/RFID technical experts in COFRAC (*Comité Français d'Accréditation*).

Additionally, Dominique Paret teaches industrial LAN, RFID, smart cards and digital TV techniques and technologies as "Senior Lecturer" in several Electronic Engineering High Schools in France (ESIEE, ESEO, ESISAR, ENSEA, ISEN, ESIGETEL, ESAIP). He occasionally teaches at F'Satie Pretoria University, South Africa, and INPT in Rabat, Morocco.

Dominique Paret has written many technical books on I2C, CAN, FlexRay protocols and applications, embedded networks, RFID, NFC and color display resolution, in French, English, Spanish and Korean (published respectively by Dunod, John Wiley, Paraninfo, and Arcon Publishers).

In 2007, Dominique Paret co-founded FILRFID (*Fédération des Industriels et Editeurs de Logiciels pour RFID*). In FILRFID the main mission and task is to establish relationships between industries, universities, engineers, high schools, professional academies, industrial and national research laboratories, national research agencies/ANR, and all French government ment departments (*Ministère de l'Enseignement et de la Recherche et Ministère de l'Industrie/ DGE*), in order to help and promote RFID activities.

Dominique Paret is also an active member of EESTEL (*Experts Européens en Systèmes de Transactions Electroniques*) and NFClub.

2009, France

DP-Consulting

dp-formations & dp-services

10, rue Georges Langrognet
- 92190 - Meudon -
France

phone: + 33 (0)9.75.72.07.39
e-mail: dp-consulting@orange.fr

Preface

Contactless identification has now become a mature technology worldwide, and it is only in certain areas of application, such as RFID at UHF and SHF, that some finishing touches are still required. Having worked in this field for many years, I felt the need to draw up a 'report' on this area of UHF and SHF. Up to the present, very little basic or practical technical information or training has been available for engineers, technicians and students. I hope that this book will go some way to making up for this shortage and will be the most comprehensive guide to this mysterious field of RFID at UHF and SHF available at a given date. Consequently, the book is not intended to be encyclopaedic, but rather to be a solid and thorough technical introduction to the subject. It is thorough in the sense that all the 'real' aspects of these contactless applications (principles, technologies, components, standards, regulations, applications, safety, etc.) are dealt with in detail.

Furthermore, in order not to intimidate the reader with the theoretical proofs that are required for an understanding of the devices used, I have paid considerable attention to the educational aspect of the text and have ensured that the relationship between theory and technology, finance and other aspects, is clear to the reader at all times.

The Structure of the Book and How to Use It

To help the reader to comprehend this subject of RFID at UHF and SHF, this book has been structured in five main sections.

Part One, *Chapters 1 to 3*. This provides an overall introduction to the subject, the physical principles used in RFID and, above all, the terminology and definitions required for a clear understanding of the language used by users in this field. Then, to whet your appetite, the last chapter of this part gives a brief overview of the applications and markets involved and the main players (but not the only ones) in these markets.

Part Two, *Chapters 4 to 10*. This has been designed to impart the minimum (and unavoidable) technical and theoretical knowledge to ensure that you will avoid making too many unnecessary mistakes when working in this highly specialist field of RFID. For some readers, this part will be a recapitulation of their earlier studies, but, as I have learnt from experience, it will also be a source of long-term support for many other readers who are newcomers to this field. Please do not be put off by all the equations. They may look complicated, but really they are not so difficult. I have also done my best to make them understandable to all readers – even those who have a few gaps in their mathematical and scientific education. In short: exercise your brain and you will be well rewarded (see below)!

In this second part, *Chapter 4* summarizes (again with reference to RFID applications at UHF and SHF) the theory of the propagation of RF waves and the calculation of the electromagnetic field components and transferred power. This is followed by *Chapters 5 to 9* – the heart of the book – which form a consistent and uniform whole. This is the crux of the matter. These chapters are essential and indispensable, as you will see. This set of chapters will show you how to deal with a range of subjects, including the principles of forward and return links, theoretical and actual operating distances, principles, techniques, problems of back scattering, environmental effects, etc., and will draw clear dividing lines between the subjects. You should always remember that this is essentially a single theme. When this has been assimilated, *Chapter 10* of this part summarizes all of the above and provides a well-deserved reward in the form of concrete examples, which are highly detailed, numbered, quantified, analysed, etc., together with overall link budgets for both remotely powered and battery-assisted systems.

Part Three, *Chapters 11 to 14*. This section gets complicated – for good reasons! When the physical and mathematical theory has been learnt, it is time to move on to technical design and the concepts of values, forms and duration, etc., of bits, and also to the method of shaping the waves used for RF transmission and quantifying the many effects on the radiated spectra. As you will have realized, this part also forms a single entity – although it is set out in four large chapters.

Having reached this point, you will have all the elements of RFID at UHF and SHF at your fingertips, and you may think it is time to close the book. But no. There is more work to be done. There is a vast amount of material in the fourth section, which is obviously essential.

Part Four, *Chapters 15 to 17*. This includes two large chapters that describe everything directly or indirectly related (principally) to the physical parts of the standards (e.g. ISO, human exposure) for UHF and SHF RFID, on the one hand, and to worldwide and local regulations on the other hand. These chapters form one of the most important sources for the feasible (and unfeasible) industrial applications of RFID at these frequencies. They should therefore be read with great care, if you wish to avoid unpleasant surprises concerning applications and the legal and even criminal aspects of this field.

Part Five, *Chapters 18 to 20*. This is the part of the book that deals with specific and technological aspects of applications. Since the technology is constantly developing, it will provide some highly representative examples of tag and base station design, at a given date, with details of components and electronic subunits.

However, if you still have even the shadow of a doubt about any of these matters, you are always welcome to contact me by post or e-mail[*] with your questions.

In these few words I have summed up the content of this book and the best way of assimilating it. Meanwhile, I hope you gain both pleasure and profit from your reading; above all, enjoy it, because this book was written not for myself but for you! Seeing the size of the book, I am obviously doing you a great favour!

---

[*] dp-consulting@orange.fr

To complete this preface, I would point out that I have written two other books, under the titles *Identification Radiofréquence et Cartes à Puces – Description* (2nd edition, Dunod, France), denoted in the text as Reference 1, and *Identification Radiofréquence et Cartes à Puces – Applications* (Dunod, France; English translation, *RFID and Contactless Smart Cards Applications*, John Wiley & Sons, Ltd, 2007), denoted in the text as Reference 2, which supplement this book by providing more specific information on applications and details of their application, and which should meet the requirements of most users of RFID at LF and HF.

Since this field is constantly developing, I am aware that I will have to update the content of this book in three or four years' time, but at least the foundations and principles have been put in place.

I hope you enjoy reading the book. You are always welcome to send me your constructive comments and remarks about the content and structure of the book.

Acknowledgements

The UHF and SHF field of radio frequency identification (RFID) is very active and many highly skilled people are working in this area. In the course of my work, I have met many of them on numerous occasions, so it is very difficult for me to acknowledge all of them individually. At the risk of considerable unfairness, therefore, I should like to offer my special thanks to colleagues in the generic Contactless Identification teams at NXP Semiconductors, based at Hamburg, Germany and Gratkorn, near Graz in Austria, with whom I have had the pleasure of working in this area for many years. They include Michael Jerne, Reinhard Meindl, Franz Amtmann, Hubert Watzinger, Roland Brandl, Bernhard Grüber, Peter Raggam and Joseph Preishüber, now with CISC, and also many academics, lecturers, researchers and friends, especially:

- Christian Ripoll of ESIEE in Paris and François de Dieuleveult of the CEA, for their kind contributions to the chapter on base station architecture,
- Smail Tedjini, Philippe Marcel and Christophe Chantepy of ESISAR at Valence,
- Mohamed Latrach and Patrick Plainchault of ESEO at Angers,

with whom I have had the pleasure of teaching in this area of application for many years.

Then there are my professional colleagues ('friends and competitors', but friends above all), whom I see regularly at standardization meetings at the AFNOR (CN 31) and ISO (SC 31): they will know who they are. I would like to thank them for their comments and observations about the technical content and structure of this book, and for their positive attitude and good fellowship, which have done so much to advance this field of RFID.

Similarly, I offer my sincere thanks to Manuela Philipsen and Martin Bührlen of NXP Semiconductors for the many documents and photographs that they have kindly supplied, to Sylvie Bourgeois for being undaunted enough to carry out a thorough check on the consistency and typography of the many equations in this book, and finally to Sophie Gilet for her kind assistance with the digital section of the book.

I would also like to thank Mr Rod Riesco for the high quality of the translation of the original work.

To conclude on a less cheerful note, I wish to dedicate this book to the memory of a dear friend of many years, Alain Berthon, from Texas Instruments, who died, much too young, in June 2005, and who, during his years at AFNOR and ISO, was one of the leading contributors to the birth and development of RFID applications. People may work in competing companies in the field, but nevertheless they still respect each other and develop deep friendships.

Dominique Paret
France, 2009

Note to Readers

Dear Readers,

Please take notice – this is really a warning!

Sadly for you, I have been known in the profession for many years (if you love your work, you don't notice the time passing) as a technically precise and painstaking author. Indeed, every single line in this text has been considered, reconsidered, pulled to pieces, exhaustively examined, and confirmed – so for your part, you will be expected to study this book word by word, line by line.

After these rather ominous words, for light relief you should know that many of my friends and readers, knowing that I have word-processed the manuscript, have told me, 'Dominique, you ought to let your readers have the text in .doc format, so they can view it in double spacing and read everything that's written between the lines of the published text!' Unfortunately, that won't work. In fact, believe it or not, everything is contained right within the lines; if I have omitted anything, it is because the context does not require it. Because of this, I have provided the notes above to help you understand the why and the how of the content, structure and presentation of this book. First of all, though, a brief remark for educational purposes.

This book contains a large number of equations. I have left them in the text to satisfy all those who wish to gain a comprehensive knowledge of RFID at UHF and SHF. I have provided detailed notes on each one (they are not just thrown in!). If you wish, you can skip the proofs, but you should at least bear in mind the initial assumptions and the end results.

And now – enjoy your reading!

Very important note

The attention of readers, future designers, manufacturers and users is drawn to the important fact that, in order to ensure accurate coverage of the field of contactless technology and RFID, this book describes very many patented technical principles (bit coding, modulation techniques, collision management, devices, etc.), which are the subject of licensing and associated rights and which have already been published externally in official professional texts and communications or at public conferences and seminars, etc., but whose use is strictly subject to the current law in terms of licensing rights, royalties, etc.

You should note especially that the use of standards such as the ISO standards does not mean that the content of the above lines can be disregarded.

For your information, here are some extracts from Annex A (regulatory) of ISO – *Reference to patent rights*:

The International Organization for Standardization (ISO) [and/or] International Electro-technical Commission (IEC) draws attention to the fact that it is claimed that compliance with this International Standard may involve the use of a patent concerning (..subject matter..) given in (..subclause..). The ISO [and/or] IEC take[s] no position concerning the evidence, validity and scope of this patent right. **The holder of this patent right has assured the ISO [and/or] IEC that he is willing to negotiate licenses under reasonable and non-discriminatory terms and conditions with applicants throughout the world.** *In this respect, the statement of the holder of this patent right is registered with the ISO [and/or] IEC. Information may be obtained from: [. . .name of holder of patent right. . .] [. . .address. . .] Attention is drawn to the possibility that some of the elements of this International Standard may be the subject of patent rights other than those identified above. ISO [and/or] IEC shall not be held responsible for identifying any or all such patent rights.*

Part One

RFID: General Features, Basic Principles and Market

This first section will provide a quick introduction to radio frequency identification (RFID) with an orientation towards RFID at ultra high frequency (UHF) and super high frequency (SHF). This section will introduce a lot of vocabulary and numerous definitions of terms, concepts and principles relating to frequencies, operating modes, and the like. I have therefore divided it into three chapters, as follows:

- some introductory words, definitions and vocabulary;
- a description of the general operating principles of the 'base station–tag' pair;
- and finally the market and fields of application for contactless and RFID technology.

Note

I would ask the reader to be as careful as possible when using these terms, which are frequently used unwisely, owing to ignorance, abuse of language, journalistic distortions, supposedly technical articles that are excessively or badly popularized, etc., often causing great confusion. RFID professionals, working in the most reputable organizations concerned with ISO standardization, have courageously and painstakingly compiled and edited a definitive 'RFID Vocabulary' (ISO 19762) to help users to understand each other more easily – so, use it if you can!

1

Introduction, Definitions and Vocabulary

In view of the imminent development of very many applications in the field of identification, systems of traceability and logistical monitoring, etc., using radio frequency identification (RFID), and having already written two books about some of these systems, operating mainly at frequencies below 135 kHz (mainly at 125 kHz) and 13.56 MHz (see References 1 and 2 mentioned in the Preface), I should now like to present this book, which is specifically concerned with RFID devices operating at ultra high frequency (UHF) and super high frequency (SHF). For various reasons (I will provide details of these subsequently), I preferred to wait until now to deal with this subject on a separate basis.

For the present, and for several years to come, this book will offer a wide-ranging theoretical, technical, technological and applications-related overview of RFID systems operating at UHF and SHF. I have also dealt in considerable detail with the themes of international standards (ISO, ETSI, FCC, etc.), the current regulations, the aspects of human exposure, etc., which cannot be ignored when working in this field.

The generic terms in the title of this book – identification, contactless tags and devices operating at UHF and SHF – cover a variety of different and controversial fields and subject areas. For example, while the words 'identification' and 'tracking' in relation to products will please a manufacturer wishing to monitor his output, the same words may also arouse fears of erosion of civil liberties and privacy. Similarly, the term 'tag' also suggests major benefits and ease of use for fast check-out procedures at large stores, flexibility for stocktaking, convenience for industrial and domestic supplies, opportunities for better protection against counterfeiting and the black market; however, the horrors of even more subtle market research may now lurk behind the purchase of any item!

That, then, is a very brief introduction to this vast, fascinating and contradictory world into which this book is designed to take you. It will deal with technical matters only, 'from theory to practice', as they say (see also References 1 and 2 mentioned in the Preface).

RFID at Ultra and Super High Frequencies: Theory and Application   Dominique Paret
© 2009 John Wiley & Sons, Ltd

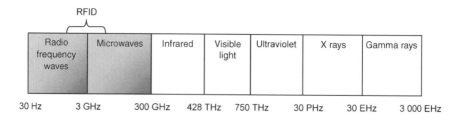

Figure 1.1   Electromagnetic spectrum of radio frequencies

1.1 To Understand Radio Frequency, We Must Know about Frequencies and Their Classification

We have arrived in the world of radio frequency identification (RFID). Let us start with frequencies, their definitions and their classifications.

1.1.1 General Classification of Radio Frequencies

To avoid any misunderstanding of terminology, and for simple practical reasons, the frequencies have been classified according to their values and/or related wavelengths. Figure 1.1 shows a summary of the international classification of frequencies.

As we all know, an electromagnetic wave is identified by its frequency of oscillation f (and/or its period $T = 1/f$) and its related wavelength λ. The relation between f and λ in air or in a vacuum is well known (see Table 1.1):

$$\lambda = cT = c/f$$

Table 1.1   Frequencies and wavelengths used at radio frequency (RF)

Band number	Abbreviation of band name	Frequency bands (the upper and lower limits are exclusive)	Metric names	Metric abbreviations of the band	Wavelengths λ (the upper and lower limits are exclusive)
−1	ELF	0.03 to 0.3 Hz	Gigametre	Gm	1 to 10 Gm
0	ELF	0.3 to 3 Hz	Hectomegametre	hMm	100 to 1000 Mm
1	ELF	3 to 30 Hz	Decamegametre	daMm	10 to 100 Mm
2	ELF	30 to 300 Hz	Megametre	Mm	1 to 10 Mm
3	ULF	300 to 3000 Hz	Hectokilometre	hkm	100 to 1000 m
4	VLF	3 to 30 kHz	Myriametre	Mam	10 to 100 km
5	*LF*	*30 to 300 kHz*	*Kilometre*	*km*	*1 to 10 km*
6	MF	300 to 3000 kHz	Hectometre	hm	100 to 1000 m
7	*HF*	*3 to 30 MHz*	*Decametre*	*dam*	*10 to 100 m*
8	VHF	30 to 300 MHz	Metre	M	1 to 10 m
9	*UHF*	*300 to 3000 MHz*	*Decimetre*	*dm*	*10 to 100 cm*
10	*SHF*	*3 to 30 GHz*	*Centimetre*	*cm*	*1 to 10 cm*

The band number N is the value of the exponent (0.3×10^{N} to 3×10^{N} Hz).
The term ELF denotes the set of bands from −1 to 2.

where c is the speed of light, i.e. the velocity of propagation of light in a vacuum (or in air), λ is in metres and f is in hertz. Thus

$$\lambda = \frac{3 \times 10^8}{f}$$

Please note that I have shown the frequency bands used in RFID, i.e. LF, HF (VHF), UHF and SHF, in bold italics in Table 1.1.

1.1.1.1 Radio Frequencies Accepted and/or Authorized for RFID

Figure 1.2 and Table 1.2 show the ranges of radio frequencies, out of the frequency bands indicated above, that are accepted by national and international regulatory bodies for RFID applications, together with their relative positions in the RF spectrum.

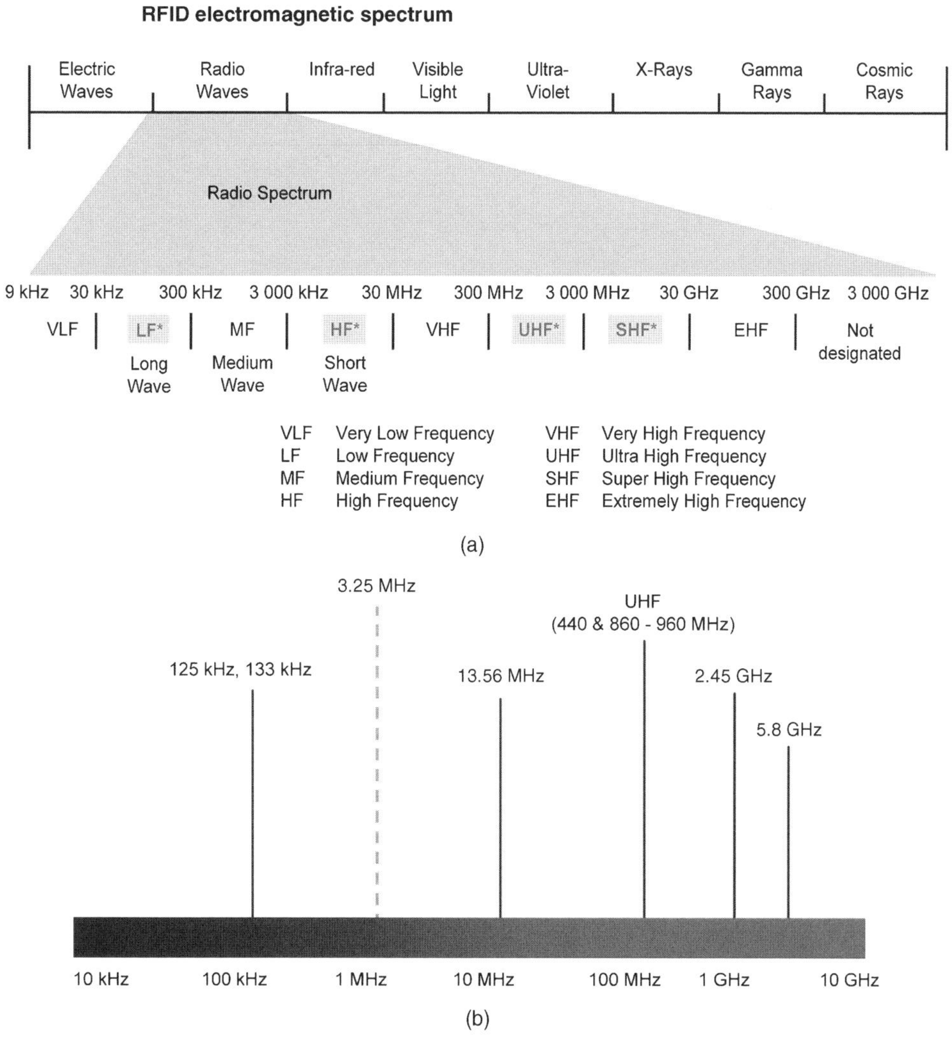

Figure 1.2 (a) RF and RFID electromagnetic spectrum. (b) Frequencies authorized/accepted for RFID applications

Table 1.2

Radio frequency bands		Radio frequencies accepted and/or authorized for RFID	
From 30 to 300 kHz	LF	Low frequencies	<135 kHz
From 3 to 30 MHz	HF	High frequencies	13.56 MHz
From 300 to 3000 MHz	UHF	Ultra high frequencies	433 and from 860 to 960 MHz
			2.45 GHz
From 3 to 30 GHz	SHF	Super high frequencies	5.8 GHz

Notes
Because of the closeness of its value and its physical properties, the 2.45 GHz frequency at the top of the UHF band is very often included in SHF band – and I will do this as well.

In my first two books (References 1 and 2 mentioned in the Preface), I have provided full details of applications using LF and HF; the present book will only be concerned with UHF and SHF applications.

For the time being, we have finished with the definitions of frequencies. Let us now look at the way they are used.

1.2 RFID: Who Uses It and What For?

Until recent times, the identification of objects and persons was practically always based on paper, card and other media, using written or printed codes and data processing, requiring either contact with the identifier (for writing) or direct visibility of it (for reading). For some years now, as the performance of radio frequency links and the associated electronic components has improved, research and development has been directed towards the possibility of replacing and enhancing the former identification methods with methods called *radio frequency identification* (RFID) or *contactless identification*. These new methods have been restricted for a long time by the impossibility of providing a remote power supply to the identifier, mainly because of its power consumption, which required the provision of a local power source (a battery or accumulator). However, with the tremendous advances in integrated circuit technology over the last decade, the dream of true contactless systems (using no batteries) has become a reality.

In Chapter 3, I will consider what this is used for (i.e. the applications) and who benefits (i.e. the market).

1.3 History

Identification has existed in numerous forms for many years. Looking at just the last few decades, it is clear that there has been an explosion in the growth of labelling for identifying many articles, with the appearance of paper bar code labels and their readers. At the same time, many industrial experiments and applications using electronic tags were developed, mainly concerned with monitoring industrial processes and animal identification (for sheep, pigs, cattle, horses, domestic pets, etc.), using implants, or monitoring individuals by access control systems (for buildings, transport, etc.). This concept of electronic tagging is now moving out of its industrial setting towards the field of high-level identification systems incorporating all the provision for confidentiality and secrecy that may be required.

These specialist subjects have been studied for many years, and we can now say that their application on a massive scale is imminent, if not already present. In addition to the widespread and well-known systems of electronic immobilizer devices for motor vehicles (more than 700 million fitted over a few years) operating on this secure contactless principle and transport smart cards (close to 2 billion operating at present, mainly in Asia, Europe and South America), there are also many experiments, pilot studies and major projects currently under way in the field of electronic tags.

Clearly, this has been a great stimulus to users of 'contact' devices, particularly users of tracking systems.

1.4 Radio Frequency (or Contactless) Identification and Its Range of Applications

Here is a brief outline of the range of applications of contactless electronic identification, or RFID, at the present time.

1.4.1 Contactless RFID

The term *radio frequency identification* (RFID) denotes any identification system operating by means of radio frequency waves. The term *contactless* is also frequently used, but it does not specify the kind of transmission (RF, IR, etc.). The field of contactless identification can be broken down into various subfields, the main ones being as follows.

Optical Vision
This requires the presence of a detector operating with direct vision of the identifier, using either the human eye, a reading device (laser, etc.) or a CCD (charge coupled device) camera. The most widespread example is that of standard printed labels, or bar code and 2D (two-dimensional) code labels.

The greatest problem (if it really is a problem) of these systems is the fact that direct vision is essential for reading and that it is dependent on the cleanness of the label (which may be stained or torn, for example). A second problem generally arises from the impossibility of updating the labels in a straightforward way – other than by simply replacing the labels. Nevertheless, we must not forget their greatest advantage, namely their very low cost!

LF and HF Link
By using radio frequency communication between the identifiers and the readers, it is possible to read at longer distances (not dependent on the resolution of the human eye or of an optoelectronic reader), without the need for direct physical optical vision of the identifier. This also makes it possible to propose 'bulk' reading systems, which can handle a large number of identifiers simultaneously in the radio frequency field of the reader, without the need to view them optically. With the aid of electronic systems, it is also possible to provide protection,

security, etc., for the information on or in the identifier. This often leads to the development of identifiers, which are frequently referred to as *intelligent bar codes* or *intelligent labelling*; these will be described more fully below.

As I have mentioned, the frequencies used under the name of 'LF and HF radio frequencies' extend from a few kilohertz to several tens of megahertz.

Links at Ultra High Frequencies (UHF) and Super High Frequencies (SHF)

The carrier frequencies for the operation of these devices at UHF are 433 and 860/960 MHz. For SHF, the most common applications are at 2.45–5.8 and sometimes 24 GHz. Although some of these identifiers have incorporated batteries (in which case they are called *battery assisted*, see below), these systems are to be classed as 'passive' devices, since they do not transmit any electromagnetic radiation (the fact that they have incorporated batteries does not in any way make them 'active'!). In fact, as will be described in detail in this book, these identifiers modulate their degrees or rates of reflection of the incident radiation (known as the 'mirror' effect) in order to enable the transmitter to interpret them. The transmitting source, which also receives the reflected radiation, can then interpret the modulation created by the identifier.

Because of the level of these carrier frequencies, high communication speeds are possible, resulting in short transaction times, of the order of tens of milliseconds. This also makes it possible to identify rapidly moving objects 'on the fly' (as in the case of trains, or cars at motorway toll booths, for example, bearing in mind that a speed of 10 m/s is equivalent to 36 km/h, or alternatively 144 km/h is equivalent to 40 m/s and therefore a vehicle covers 1 m in 25 ms).

The well-known technical problem of UHF and SHF applications is their very modest (or downright poor) ability to pass through most liquids and the human body (which is 80% water!), as well as their generally rather directional propagation – although this may sometimes be an advantage!

Another problem of links operating at these frequencies is the limitation on remote power supply to the identifier, since the ability to use a small antenna, because of the wavelengths associated with the frequencies concerned, means that energy recovery is rather limited, and therefore it is sometimes necessary to use a local power supply.

Infrared Links

Like super high frequency links, optical links of the infrared type (with wavelengths of around 800 nm) are often used in support of contactless identification devices (at motorway tolls, for example) to provide a higher data rate and greater directionality of the communication beam. In this case also, operating distances are generally high and the transponders are often supplied autonomously.

1.5 The Concept of Contactless Communication

I will now provide a brief interpretation of the concept of contactless communication. This will include a basic introduction to:

- the concept of contactless communication distances;
- the concept of power supply and power supply mode;

- communication and the communication model (ISO/OSI);
- the concept of the operating mode.

1.5.1 The Concept of Contactless Communication Distances

Please note that the term 'distance' refers to the total/maximum distance that can be covered by a transmission, while 'range' refers to the interval between two distances or limits.

Since this book is concerned with RFID and 'contactless' applications, it will now be useful to define the mechanical concept implied by 'contactless', in other words the concept of communication distances and consequently the applications envisaged by users of 'contactless' technology. First of all, however, in order to avoid needless debate, you should know that, surprising though it may seem, there is no ISO standard that defines the operating distance (in terms of metres) of RFID devices in the strict sense of the term.

Very Short Distance, i.e. from Contactless to Contact!
Surprisingly, there are many 'contactless' applications in which the operating distance between the base station and the identifier must be, or can be, virtually zero (in 'touch' systems); the essential point is that electrical insulation must be present for the purposes of the application.

Short Distance
'Short distance' applications (such as those operating at 13.56 MHz according to ISO 10536) generally operate over distances of the order of a few millimetres or tens of millimetres. These are contactless applications that 'make contact'.

Proximity
The same applies to the concept of 'proximity' contactless systems (e.g. ISO 14443), which use distances of the order of tens of centimetres. This family of applications includes contactless smart cards requiring 'voluntary action' for their presentation for applications in banking, payment, transport, access control, etc.

Vicinity
The same advantages and drawbacks are found in 'vicinity' contactless systems (see, for example, ISO 15693/ISO 18000-x). The distances required are of the order of 50 cm to 1 m and support 'hands-free' applications including access control, baggage recognition and monitoring at airports, movement of trolleys, etc.

Long Distance
This term is generally used for applications operating over distances of the order of 1 to 5–10 m. Examples are applications at the gates of super- and hypermarkets or reading from pallets. Beyond these distances, we speak of 'very long distance'.

Very Long Distance
In very long distance applications (over more than tens or even hundreds of metres) we tend to leave the field of remotely powered tags or transponders, which will be described

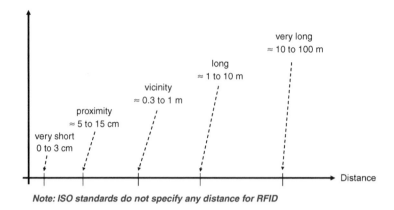

Figure 1.3   Operating distances in RFID

subsequently, and enter the world of systems using radio frequency links in which the identifiers have their own incorporated power supplies and operate on conventional 'radio' principles, using transmitters and receivers on each side (i.e. on the fixed and remote elements).

This type of system, known as 'active', is outside the scope of this book, and you should consult the excellent monograph by François de Dieuleveult, *Electronique Appliquée aux Hautes Fréquences* (Dunod, Paris, 2007), for assistance in developing projects of this type.

Figure 1.3 summarizes these classes of operating distances.

1.6 The Elements, Terms and Vocabulary of RFID

Figure 1.4 and Table 1.3 show the various elements of a contactless application, in the form of a block diagram and the layers of the OSI/ISO model.

Here is a rapid survey of the building blocks of an RFID system.

Remote Element
Let us start with the remote element, which has a memory (WORM, E2PROM, FLASH, etc.) for storing the data forming part of the application concerned and which provides control of the communication and finally the part providing RF transmission.

Table 1.3

	Remote element	Fixed element
Layer 7	Application	Application
Layer 2	Communication protocol	Communication protocol
Layer 1	Analogue part	Analogue part
	Antenna	Antenna
Medium	Electromagnetic radiation	Air coupling

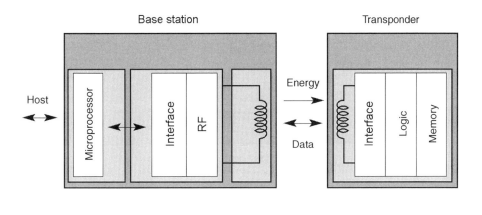

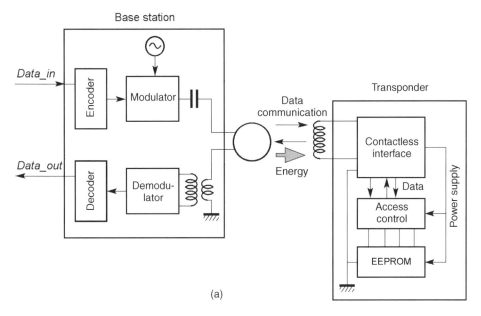

(a)

Figure 1.4   (a) Block diagram of the elements of a contactless application. (b) The OSI structure of an RFID application

Medium
The communication medium between the antennae of the remote element and the fixed part is usually air. The electromagnetic RF radiation carries the data.

Fixed Element
The fixed element comprises an analogue part, used for transmitting and receiving RF signals, the circuits for managing the protocol for communication with the identifier, the communication management system (collision management, authentication, encryption, etc.) and finally an interface for dialogue with the host system.

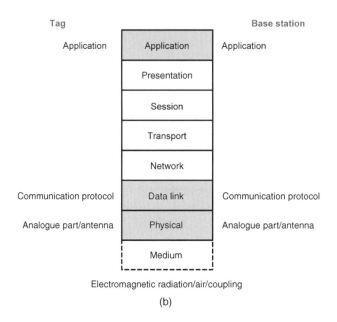

Figure 1.4   (*Continued*)

Host System
To complete this brief survey of an RFID system, we have the host system, which manages the application at the highest level.

Now that the general picture is in place, we can go on to examine each of these blocks more closely, but first of all it is time to learn some vocabulary, to avoid any misunderstandings.

1.7 Vocabulary: The Many Terms Used for the Elements of RFID

Instead of attaching specific names to the 'remote elements' and 'fixed elements' in the preceding section, I have rather avoided the issue! In fact, my profession is extravagant with its naming of the various elements used in contactless applications. It is therefore necessary to carry out a little pruning on this large stock of names given to the main components of contactless RFID applications.

1.7.1 Remote Element

Let us start with the remote element. We often encounter the following terms.

Identifier
I have chosen to use this term so far in my introduction to the subject, because it is sufficiently descriptive ... for now.

Tag

From the noun 'tag' meaning a label, and the verb 'to tag', i.e. to check or mark.

Pit

A more powerful term. PIT stands for *programmable identification tag*, which is used to identify an 'object'. Strictly speaking, a 'tag' is not (re)programmable – only a PIT is!

Data Carrier

This generic term (not restricted to contactless systems) denotes the carrier in the sense of the element that includes or contains the data, which is obviously the tag or the PIT.

Label, Smart Label

These are other names for 'tag'.

Transponder

The technology used for contactless applications is practically always based on an electronic device composed of an interrogating transmitter (fixed element) and responder (remote element), i.e. a transmitter/responder pair, abbreviated to 'transponder', meaning an object that can respond to commands sent by a 'transmitter' using a radio signal. This is one of the generic terms that will be used frequently in this book to avoid repetition.

ICC, PICC, VICC

You should note that ISO standards relating to contactless smart card applications often refer to the terms ICC, PICC and VICC, which stand for *integrated circuit card*, *proximity integrated circuit card* and *vicinity integrated circuit card* respectively. Although smart card applications appear to be outside the scope of this book, it is always possible that UHF or SHF tags may be produced in smart card format one day . . . or maybe smart cards operating at UHF and SHF!

Now let us look at the other end of the system.

1.7.2 Fixed Element

The fixed element, or what is called the fixed element (in fact, it may be a handheld or pistol-type reader), can also have many different names.

Base Station

The term 'base station' denotes the (generally fixed) command unit, which can request the reading, writing, management, etc., of the tag, using radio frequency communication.

Reader

This frequently used (and overused) term is illogical since, in most applications, the 'reader' can also transmit writing commands to the tag. Can we really call this unit a 'reader' if it can also 'write'? Preferably, we should restrict this term to its original meaning, to avoid confusion between pure readers, which can only read, and those that can both read and write. Thanks in advance!

Interrogator

Frequently used in the USA, this term, recognized by the ISO, is rather closer to the reality, since the base station does indeed give orders and/or commands to the transponder to tell it what to do. When the transponder is requested to dump or return its contents, then clearly this could be considered as an interrogation from the viewpoint of the base station, but when the base station sends data to be written to the transponder, is the base station really an 'interrogator'?

Initiator
Some radio frequency communication systems such as NFC (near field communication) use the term 'initiator', since there is always one element or base station that initializes the communication.

Now let us look at some other terms.

Modem
If we respect the derivation of the term 'modem' from 'MODulator/DEModulator', we should use it only to denote the electronic components responsible for modulating and demodulating the signals of transponders and base stations. The term should be avoided if possible in order to avoid confusion.

CD, PCD, VCD, etc
These terms, also used by the ISO, are some of the least misleading terms, since they denote coupling devices, proximity coupling devices and vicinity coupling devices, in other words the elements for coupling (in which direction(s)?) between the transponder(s) and base station(s).

In short, to avoid any confusion, I shall use the generic terms 'tag' or 'transponder' and 'base station' as far as possible, as in my opinion they are the most accurate (or least misleading!) descriptive names for these elements. Now let us look at the operating principle of this system.

1.8 Appendix: Units and Constants

Quantity	Symbol	Unit	Dimension
Current density	J	Ampere per square metre	$A\,m^{-2}$
Electric field strength	E	Volts per metre	$V\,m^{-1}$
Electric displacement	D	Coulomb per square metre	$C\,m^{-2}$
Conductivity	σ	Siemens per metre	$S\,m^{-1}$
Frequency	f	Hertz	Hz
Magnetic field	H	Ampere per metre	$A\,m^{-1}$
Magnetic induction (flux density)	B	Tesla ($V\,s\,m^{-2}$)	T
Density	ρ	Kilogram per cubic metre	$kg\,m^{-3}$
Permeability	μ	Henry per metre	$H\,m^{-1}$
Permittivity	ε	Farad per metre	$F\,m^{-1}$
Power flux density	S	Watt per square metre	$W\,m^{-2}$
Specific absorption rate	SAR	Watt per kilogram	$W\,kg^{-1}$
Wavelength	λ	Metre	m
Temperature	T	Kelvin	K

Physical constant	Symbol	Value	Dimension
Speed of light	c	2.997×10^8	$m\,s^{-1}$
Permittivity of free space	$\varepsilon\varepsilon_0$	$10^{-9}/36\pi = 8.854 \times 10^{-12}$	$F\,m^{-1}$
Permeability of vacuum/air	μ_0	$4\pi \times 10^{-7}$	$H\,m^{-1}$
Impedance of free space	Z_0	120π (or 377)	Ω

2

General Operating Principles of the Base Station–Tag Pair

We have now completed our brief survey of the components of an RFID system and the vocabulary used in this field. This second chapter is divided into four main sections, namely:

- energy transfer and communication modes;
- data communications;
- the concept of operating modes;
- a brief introduction to more specific problems of long distance applications.

This chapter has the twofold purpose of resolving many general queries about RFID and providing some small but important details of the general technical principles used daily in the RFID system. You should be aware that this is only an 'appetizer': we will examine the operation of these devices more thoroughly from Chapter 4 onwards, with special reference to UHF and SHF, and throughout the rest of the book; in other words, the main course is yet to come!

2.1 Energy Transfer and Communication Modes

Under this heading I will detail the differences between the provision of energy, the power supply to the electronic part of the transponder and finally the communication mode between the base station and transponder. There is often a confusion between these concepts and terms are often applied wrongly, mainly due to incorrect use of language. I would ask you to make the effort to use the 'true' terms wherever possible and to promote their use.

Let me come back for a moment to that word 'true' in the previous paragraph. I do not claim to be in sole possession of the truth or to be the fount of all knowledge – but sometimes it is useful to call a spade a spade. I will therefore attempt to clarify the main terms, on the basis of those listed in the ISO 19762 family of standards, 'Information Technology AIDC Techniques – Harmonized Vocabulary, Part 3 – Information Technology, AIDC Techniques – Harmonized Vocabulary – Radio Frequency Identification (RFID)', compiled by my friend Craig Harmon

RFID at Ultra and Super High Frequencies: Theory and Application   Dominique Paret
© 2009 John Wiley & Sons, Ltd

and other courageous colleagues on the ISO SC31, WG4 and SG3 committees, and especially those on the AFNOR CN31 committee; believe me, the successful outcome of this mission required both perseverance and self-denial.

Note

If you still have any concerns about these matters, please avoid any further philosophical agony by summoning up all your courage and joining the specialist RFID Vocabulary committees of the ISO!

Let us start by examining some general features of the energy transfer modes, where these are used, and the communication modes between the base station and the tag. First of all I will make the distinction between the transfer of energy from the base station to the tag and the exchange of data (forward and return links); I will then go on to examine their possible combinations.

Here is a brief survey of the different possible modes of energy transfer from the base station to the tag, for the remote power supply of the tag where this is possible.

Nonsimultaneous: Energy Transfer and Communication in Different Phases

In this first case (Figure 2.1(a)), the RF radiation propagated from the base station to the tag has the sole purpose of supplying energy to the tag in order to charge the 'power supply capacitance', which is incorporated in the tag so as to provide a sufficient supply to the whole tag to ensure its correct operation. After this power supply phase, the tag can receive commands from the base station and return data to it. The cycle is then repeated, with more energy supplied to the tag in order to continue the communication, and so on. Clearly, this is time consuming and can sometimes be inflexible.

Simultaneous: Energy and Communication in One Exchange

In this second case (Figure 2.1(b)), because of the principles and types of modulation used, the radiation from the base station can provide the energy supply and data exchange simultaneously during the exchange between the base station and the tag.

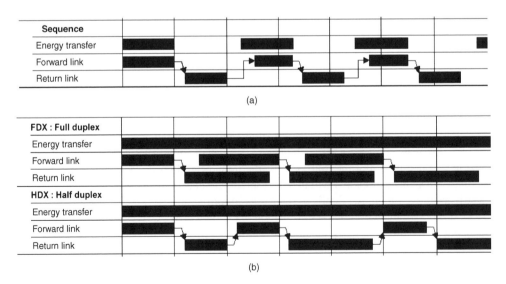

Figure 2.1   (a) Nonsimultaneous energy transfer mode. (b) Simultaneous energy transfer mode

Note that the great majority of tags available on the market operate on this principle.

I will continue with my explanations, assuming that the problem of the internal power supply of the tag has been overcome. We can re-examine this matter in detail subsequently, but let us start by simply looking at the principle of communication between the base station and the tag.

2.2 Forward Link and Return Link

Let us start by considering the exchanges between base stations and tags. They are of two kinds, and will be defined for the purposes of this book strictly as follows:

- 'from the base station to the tag', called the 'forward link';
- 'from the tag to the base station', called the 'return link'.

Also, to avoid any misunderstandings, we will assume that the tag, regardless of any on-board intelligence it may have, operates only in response to commands from the base station.

2.2.1 The Forward Link, from the Base Station to the Tag

The purpose of the forward link (from the base station to the tag) is:

- to transport energy, if possible, to the tag, so that the tag can carry out its tasks;
- to act as the support for data transmission from the base station to the tag;
- in the case of systems operating in *passive* mode (see the definition below), to provide a physical support for the communication from the tag to the base station during the return link phase of communication from the tag to the base station.

The forward link is theoretically provided by a device (the base station) that emits radiation at radio frequency. The device is therefore provided with a *transmitter*. Because of the presence of this transmitter, the forward link is described as *active*. The base station also incorporates a *receiver*. The base station is therefore a *transceiver*.

In the forward direction, the base station must enable the tag to send signals to it by using a digital (binary) code, a communication protocol and a carrier frequency modulation system, which has very little or no effect on the quality of any remote power supply. For this purpose we can use frequency modulation techniques such as FSK, or any of the numerous amplitude modulation methods such as ASK 100% and ASK x% (Figure 2.2); these will be described in detail in Chapter 12. In most of the regular commercial RFID applications, the transferred energy is sufficient for the remote power supply of the tag and also for the provision of the forward link from the base station to the transponder.

Let us start by examining this potential energy supply.

Energy Supplied, and the Concept of the Power Supply to the Tag

Except in the case of *chipless* tags (without integrated circuits – see Chapter 18), operating mainly according to the principles of wave propagation in surface wave filters, and disregarding for now the principles that will be adopted subsequently for the actual establishment of

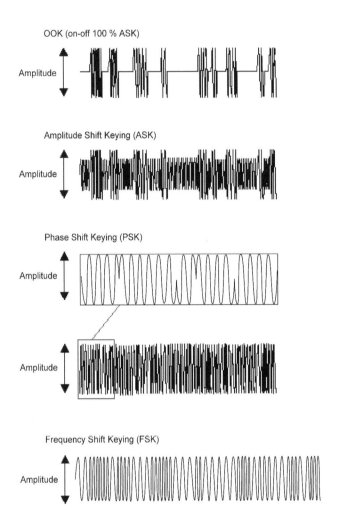

OOK (on-off 100 % ASK)

Amplitude

Amplitude Shift Keying (ASK)

Amplitude

Phase Shift Keying (PSK)

Amplitude

Amplitude

Frequency Shift Keying (FSK)

Amplitude

Figure 2.2   Examples of types of modulation

communications between the base station and the tag, the electronics of the integrated circuit incorporated in the tag must be supplied with power in order for the tag to operate correctly and carry out its communication tasks. Whatever some people might believe, or write in the press, this local power supply is not provided by the Holy Ghost!

Let us spend a little time on this matter. First of all, how does the tag draw on the energy it needs to operate? There are two possible cases:

- either the energy transmitted by the radiation from the base station (simultaneously or otherwise, see the preceding paragraphs) provides the power required by the tag or
- the energy transmitted by the radiation from the base station cannot provide a remote power supply to the tag – so something must be done to overcome this.

FINAL
DRAFT

INTERNATIONAL
STANDARD

**ISO/IEC
FDIS
19762-3**

ISO/IEC JTC 1
Secretariat: **ANSI**
Voting begins on: 2004-12-03
Voting terminates on: 2005-02-03

**Information technology — Automatic
identification and data capture (AIDC)
techniques — Harmonized vocabulary —**

**Part 3:
Radio frequency identification (RFID)**

*Technologies de l'information — Techniques d'identification
automatique et de capture de données (AIDC) — Vocabulaire
harmonisé —*

Partie 3: Identification par radiofréquence (RFID)

**05.05.50
passive tag**
RFID device which reflects and modulates a carrier signal received from an interrogator

**05.04.01
active tag**
RFID device having the ability of producing a radio signal

**05.04.13
interrogator**
fixed or mobile data capture and identification device using a radio frequency **electromagnetic field** to stimulate and effect a modulated data response from a **transponder** or group of transponders present in the **interrogation zone**

Exemple (ISO 18 000 – 4)

This part of ISO/IEC 18000 contains two modes. The first is a passive tag operating as an interrogator talks first while the second in a battery assisted tag operating as a tag talks first. The detailed technical differences between the modes are shown in the parameter tables.

Figure 2.3 RFID vocabulary, harmonized according to the ISO

To avoid serious confusion in the terminology, after lengthy debate the two cases described above are now officially described at the ISO by the terms 'remotely powered or batteryless' and 'battery assisted' respectively (Figure 2.3).

Remotely Powered or Batteryless Tags
For many reasons, including cost, weight, volume, size, etc., nearly all the latest RFID applications require tags that have no additional specific power sources on board. To enable the tags to operate correctly, therefore, the key technical requirement is to power them at a distance – hence the term 'remotely powered' – by using the magnetic/electromagnetic field in which it is located. The electrical energy (continuous power) required for this purpose must be supplied to the tags by making use of the energy contained and transported by the radiated

electromagnetic field and transmitted by the RF waves from the base station. Clearly, this means that the energy received and supplied must be at least sufficient to allow the tag to operate correctly across the whole range of its applications. In this case, we can say that a remote power supply has been provided to the tag, and thus the tag is 'remotely powered' (see Figure 2.4(a)).

At any given time, and depending on the technology, the frequency, the emitted power level according to the local radio frequency regulations and current law, etc., sufficient energy can generally be recovered to power the chip (the integrated circuit) on the tag within operating distances varying from several tens of centimetres to several metres. Clearly, this gives rise to technical and technological problems, in respect of low consumption, (very) specialized electronic circuitry, tight technical constraints, etc., but there is no such thing as a free lunch! The indisputable benefit is that we end up with an autonomous element of small size (no bulky or heavy battery) requiring no specialized maintenance.

Battery Assisted Tags, i.e. Tags Powered by Local Batteries

For many reasons (such as the desired operating distance, the technology used and the current regulations), the energy transmitted by the radiation emitted by the base station may be insufficient to power the tag remotely. For example, if we have a tag with a power consumption of 50 µW, working at a frequency of 2.45 GHz and an operating distance of 10 m, then, simply because of the attenuation of about 60 dB due to the medium (air), we would need 50 W of power emitted from the base station in order to power the tag remotely. A detailed proof of this is provided in Chapter 6. Unfortunately, the regulatory authorities limit the maximum emitted power to 25 mW, 500 mW and 4 W in this frequency band. In these conditions, then, we have no chance of making the tag work in a remotely powered mode!

To resolve this problem, we must provide an independent local energy source on the tag, in the form of a battery or accumulator, or the like, which is called an *on-board power supply*. In this case, we say that the tag is *battery assisted* (see Figure 2.4(a)). This also offers greater possibilities for operation, especially in terms of the communication distance. Indeed, since the base station no longer has to provide the energy for a remote power supply to the tag, the communication distances between the base station and tag can be greater (from about 15 to 100 m). On the other hand, the endurance of a tag is directly dependent on its power life, in other words the life of its battery or accumulator, which may or may not be rechargeable, and its power consumption. To overcome some of these problems, standby systems are usually incorporated in tags to prolong the battery life and consequently the life of the tag.

Leaving aside the problems of dimensions (volume, thickness, weight, etc.), mechanics (quality of contact, vibration, etc.), electrical factors (accidental reversal of battery polarities, etc.), economics (cost, etc.), the main problem of such a unit is that of the endurance (the life of the power supply element). To overcome this, let us consider the following two examples.

Without a Recharging Device

In this case, we have a battery whose endurance is conventionally defined as a number of coulombs; in other words, $q = it$, in A s. When the power consumption of the device (consumption of the tag + wake-up mode + mark-space ratio) is known, all the questions are answered... except for the problem of the need to change the battery from time to time, the consequent risks of data loss from the tag memory, unfortunate reversals of battery polarity, etc. In other words – the usual problems!

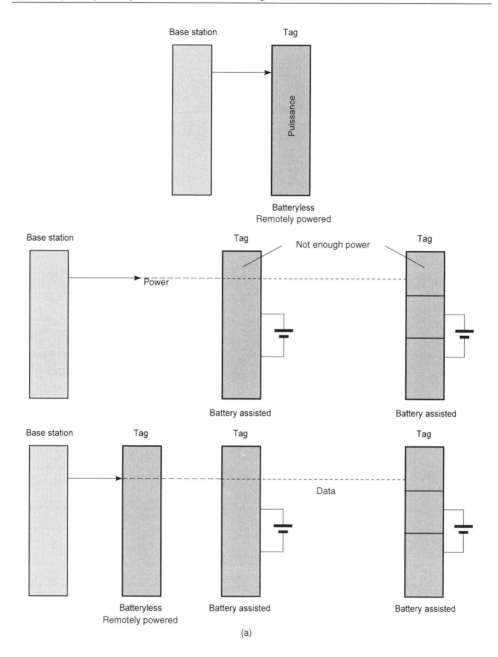

Figure 2.4 (a) A 'remotely powered' or 'batteryless' tag, 'battery assisted' tags and the 'forward link'. (b) Principle of load modulation. (c) Passive, active, remotely powered and battery-assisted tags. (d) Summary of the performance of passive, active, remotely powered and battery-assisted tags

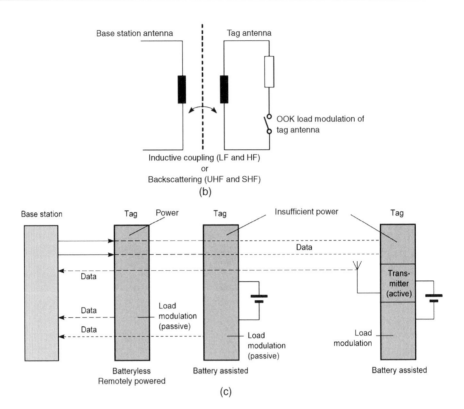

Figure 2.4   *(Continued)*

With a Recharging Device
Here, we have an accumulator on board the transponder, together with its appropriate charging and discharging circuit (often far from simple), and the associated problems of endurance (because, believe it or not, even the best accumulators will not last for ever) and replacement as for the batteries mentioned above. In spite of the problems, these systems with on-board chargers have some special advantages in terms of applications. For example, they can operate in a self-contained way at long distance, simply by using the power stored on board, and can recharge their accumulators when they are close to the base station, using the (electro)magnetic field produced by the base station. To sum up, then, remotely powered devices are always to be preferred wherever they can be used.

Important note
Quite frequently, and incorrectly, remotely powered tags are called 'passive', and battery-assisted tags are called 'active', which is meaningless! (See the explanations below for further details). Once again, please try to use the correct term!

2.2.2 The Return Link, from the Tag to the Base Station

Let me begin once more by defining some of the terms used in relation to the return link and the tags involved with it. Two little words, 'passive' and 'active', have created a vast amount of complication and confusion in the terminology that appears from time to time in the popular science press and even more often in the mass media, and is then passed on among consultants (whose knowledge is all too frequently general rather than specialized) and end users not fully up to date with the technology. I have heard all kinds of variations on these themes. Without claiming to be in possession of the truth, I can say that certain highly skilled people at the ISO, who give everything its proper name, have carefully defined a number of terms relating to this whole area (in the ISO family of standards 19762, Part C, which is specifically concerned with 'Information Technology: Automatic Identification and Data Capture (AIDC) Techniques: Harmonized Vocabulary: Radio Frequency Identification (RFID)'). In short, this is what you should understand by the terms 'passive' and 'active' – and take care not to confuse the power supply system/energy transfer and the forward and return link principles! See Figure 2.3 again.

Active and Passive Tags
Regardless of its type of power supply (remotely powered or battery assisted), the tag must be provided with a communication device to provide the return link from the tag to the base station, called the *return channel*. This can operate in different ways, according to the principles followed (Figure 2.4(a)).

Passive Tag
In this case, the tag uses the forward link radiation (generally unmodulated) supplied by the base station, and sends reply signals to the base station in the *half* or *full duplex* mode (see below, generally in the half duplex mode) in a 'passive' way, in other words without using any on-board transmitter function. The tag is then said to be 'passive'. On the other hand, there is

no reason why it should not be remotely powered by the incident radiation (thus being batteryless) or, alternatively, battery assisted!

Active Tag
In this case, for various reasons (the distance may be too long or the base station demodulators may not be sensitive enough, etc.), the tag cannot send signals to the base station without having a true transmitter on board – so we have to install one! Because of the presence of the transmitter, the return link is 'active', regardless of the type of power supply. The tag is then said to be 'active'.

To sum up, regardless of how the tag is powered, 'passive' means that the return link – from the tag to the base station – is provided without the aid of an RF transmitter. On the other hand, and once again regardless of the how the tag is powered, if the tag has a transmitter on board to respond to the base station it is called 'active'.

Therefore, do not confuse the type of power supply with the communication principle! A battery has never been active in its life! Maybe it gives a kind of life to something, but that is another story.

> **Note**
> You should know that 'semi-active' and 'semi-passive' tags do not exist in this series. Such language is nonsensical! How many times have I heard these meaningless terms? If they are supposed to denote tags, which provide their return links in a passive way (without transmitters) and are fitted with batteries, then they are simply 'battery-assisted passive tags' – nothing more.

The Secrets of the Transponder
Whether tags are passive or active, if their internal electronic architecture is such that they can only respond to commands sent by the base station, they are simple *responders*, and the system consisting of a couple, the base station (*transmitter*) on the one hand and the tag on the other hand, then operates in what is called a *transponder* mode of communication.

Having sorted out these matters of vocabulary, let us return to the actual technology and see how passive and active tags operate.

'Passive' Return Link Tags
These are the most commonly encountered types. In terms of numbers of tags, they represent more than 95% of applications.

In the case of passive return links from the tags to the base station, you should always remember that, regardless of its power supply (i.e. whether it is remotely powered or battery assisted), the tag has no transmitter on board, but it must be an element that can send signals to the base station at some stage. For this purpose, in order to provide the return link, the base station supplies a physical medium in the form of an unmodulated 'carrier' frequency, which is constantly maintained, and the tag sends signals to the base station by modulating the electrical characteristics of this carrier. For this purpose, two techniques that are very similar in physical terms, being based on the same principle of 'modulation of the impedance (in terms of resistance and reactance) of the tag antenna load' – also known as *load modulation* – are commonly used in most of the commercially available tags. Since the physical effects of this 'load modulation' vary according to the frequencies used (see Chapters 6 and 8), in RFID we normally speak of:

- 'magnetic coupling' at LF and HF;
- 'retroreflection' or 'reradiation' or *back scattering* of the incident radiation at UHF and SHF.

The Load Modulation Technique
When the base station supplies an unmodulated carrier to the tag to provide a physical medium for the return link, it allows the tag to operate entirely in its own way, according to its specific operating mode. The tag, operating at its own rate, modulates all or part of the equivalent electrical load that it represents at the terminals of its antenna (Figure 2.4(b)) by *on–off keying* (OOK) according to a specific binary coding. By doing so:

- at the RFID low frequencies of below 135 kHz and at the RFID high frequency of 13.56 MHz, its modulation of the load that it represents (by varying the resistance or capacitance) modifies the power consumption that it represents in the magnetic field and, because of the magnetic coupling present between the tag and the base station, tends to modify, via the existing magnetic coupling, the current flowing in the antenna circuit of the base station;
- at the RFID ultra high frequencies of 433, 860–960 MHz and 2.45 GHz, and at the super high frequency of 5.8 GHz, its modulation of the load impedance of its receiving antenna, in the case of propagating radiation, has the end result of modifying the proportion of radiation reflected (reradiated) towards the base station (the *back scattering* method, which will be explained in detail in this book), by causing a variation in the apparent radar cross-section, as will be explained in detail in Chapters 8 and 9.

In both of these cases, the tag does not have any transmitter function, properly speaking. The system operates in *transmitter* mode (base station) and *responder* mode (the same) and the whole assembly operates in *transponder* mode. The return link is thus based on the principle of a passive return link from the tag element.

'Active' Return Link Tags
In this case, there are various possibilities for the tag response phase:

- As above, during the return link phase, the base station supplies a carrier that can act as a medium for the return communication. Unfortunately, the signal level that would be returned by the tag (by back scattering, for example) if the return link were passive would be too weak to be processed directly by the base station. To overcome this problem, and again regardless of the power supply mode of the tag (remotely powered or battery assisted), the tag is provided with a low-power local transmitter to 'boost' the signals returned to the base station,
- In some systems, during the return link phase, the base station ceases to supply a carrier that could be used as the medium for the return signals. In this case, the tag sends signals to the base station by emitting and using, for example, either the same carrier frequency or a different frequency to provide the return link. In this case, since it is possible to have two independent carriers for the forward and return link channels, the exchange between the base

station and the tag can be carried out easily in either half duplex or full duplex mode. The exchanges can also be markedly faster ($>1 \, \text{M bit s}^{-1}$).

In both of the above examples (there are many other possibilities), the return link from the tag to the base station thus operates in the transmitter mode. The return link is therefore based on a principle of 'active operation', which is why the tag is called 'active'. Clearly, in order to achieve this, each tag must incorporate, as stated above, both a reception function (to detect the forward link from the base station) and an RF transmitter (to provide the return link). In fact, the devices described above commonly operate in the same way as a conventional set of radio receivers and transmitters. What we have here is something like a 'communicating object' – you could call it a 'mini-Walkie-Talkie'. Clearly, these forms of operation are more complicated and expensive, and require more energy at the tag, which in some cases at least requires the provision of a battery on the tag; however, please avoid the confusion of cause and effect that we often find in tabloid journalism or among the uneducated!

Another fact to consider is that, even if a tag is 'active' because it has an on-board transmitter, it does not necessarily operate as a true transmitter/receiver (transceiver). For example, it may well operate simply as a responder throughout its life, only responding to commands and forming a transponder system in combination with the base station. Anyone who still harbours doubts on this point should consider the operating modes of the new near field communication (NFC) devices now being implemented in mobile phones. However, before finishing with this subject, I must point out that hybrids such as 'battery-assisted passive' and 'active remotely powered' tags are quite feasible. I assure you that they actually exist. If you don't believe this, please see Chapter 10 for the details of battery-assisted passive tags.

Finally, you should now be fully equipped to avoid incorrect terminology, and your reward will be Figure 2.4(c) and (d), which summarize the 'passive', 'active', 'remotely powered', 'battery-assisted' and other functions.

Minimum Value of Back Scatter Coupling and the Duality of Energy Transfer and Communication

This requires further explanation. There must be a minimum level of magnetic coupling between the tag and the base station antenna if the tag's load modulation/variation is to be significant and detectable by the base station circuits dedicated for this purpose. The same applies in UHF and SHF, since there must be a significant variation in the back scattered signal in order to make it detectable by the receiver part of the base station. To make everything quite clear, then, we must distinguish carefully between the concept of the minimum quantity of energy to be transferred to the tag to remotely power it and the concept of the minimum return signal required to establish the return link so that the application runs correctly.

As you will have noticed, there is a close relationship between the transport of energy and the establishment of the forward and return links. The keystone of this whole structure is the relationship that must exist between the base station and the tag to ensure that everything operates correctly. Briefly, when setting up a tag-based system we must remember that energy is nothing without communication – otherwise we will simply have produced a remote power supply device!

As mentioned previously, for any given economical system, the first minimum level of coupling to be established is the one that provides sufficient voltage to power the tag remotely, with allowance for the tag's own load and power consumption. If these are known (tag manufacturers' specifications are expected to state the necessary voltage and current, and therefore the power consumption, of their product), then we can calculate in advance the minimum values of the parameters required to ensure that the application operates correctly.

The Maximum Level of Coupling/Reflection

Now that we are certain that the system can operate correctly in terms of energy and communication, let us move on to a topic that is frequently ignored or overlooked by many users. There are two related, but fundamentally different, problems that require us to consider the maximum level of the coupling/reflection coefficient:

- the first concerns tags that must be able to operate simultaneously at long or short distances from the base station antenna;
- the second concerns the safety of the operation of tags that are designed to operate nearly always at the same distance from the base station antenna.

In the first case, the designers generally do their utmost to optimize the long distance operation of the system, but must also ensure that it operates well when the tag is brought very close to the base station antenna and is kept there for a certain time (known as the exposure time). This is because the tag is subjected to new constraints (an intense field) when the distance decreases, and this must be taken into account. This dual operation in long and short distance modes is a serious matter and must be carefully designed for. Designers and users often simply estimate the minimum 'coupling' or 'back scattering/reflection' values according to the operating frequencies, since these factors contribute to the determination of the maximum operating distance of their system and pay little attention to what may happen if the transponder is close to the base station.

Let us now examine the second case in detail. Sometimes there is no need to worry about safety in operation, because the tags and base station(s) are designed to operate at a more or less fixed distance (such as about 8/10/15 cm for contactless cards or 1 to 2.5 m for reading from pallets), but it is possible that a tag or tags may accidentally be placed very close to the base station for a long time. This will become clearer if we consider the example of monitoring packages moving in bulk on a continuously moving conveyor, where the electronic tags (incorporated in paper labels) are applied to one face of each box and their physical position in space cannot be completely known at any instant. The base stations must have a high power level in order to read the most distant tags. Now suppose that, following a breakdown or emergency stoppage of the moving belt, conveyor, etc., the box stays in a fixed position facing the base station antenna. According to 'Murphy's law', it will inevitably be the box whose tag is closest to the base station antenna that is subjected to the most powerful field. I can assure you that, in some cases, I have actually observed blue smoke rising, not from the integrated circuits in the tags but from the paper label, which was unable to dissipate the heat and withstand the temperature. Several beautifully designed packages turned black as a result.

In other words, these two problems require us to pay attention to the maximum value of the coupling coefficient that the system can withstand, and its consequences.

Effects of the Maximum Value of Coupling/Reflection

The integrated circuit located in the tag has a number of intrinsic physical limitations in respect of current, voltage and power. Let us consider the effects on these if the coupling is too strong.

Current

The induced current supplied through the tag antenna may exceed the permitted and/or authorized maximum current for the input of the integrated circuit.

Voltage

It is also possible that the induced or received voltage may continue to rise because of the magnetic or electromagnetic fields (according to the frequencies used). In this case, we need to know whether or not the integrated circuit is capable of controlling the voltage at its terminals and can continue to operate. This is often directly related to the problem of maximum power input, which will now be examined, and which includes both of the preceding problems.

Power

Because of the combined effects of the voltage that must be controlled and a high current level to be limited, we must:

- either ensure that the tag continues to operate and must (or can) discharge excess power or
- ensure that the tag continues to operate by providing power limiting/regulation circuits or
- make the component cease to operate if a certain threshold is reached, by switching to a 'nondestructive' mode using, for example, an 'auto-shutdown' circuit.

Temperature

Most tags operate over ambient temperature ranges known as 'domestic' (0 to 70 °C) or 'industrial' (−40 to +85 °C). Clearly, tags are mainly required to dissipate energy when they are close to an antenna that is producing a strong field. All the unnecessary heat has to be disposed of, as I have pointed out above. For this purpose, we must know the thermal properties of the tag, especially its overall thermal resistance, in order to be able to estimate the maximum temperature of the chip and avoid destroying it (although this happens very rarely), but above all in order to ensure that this high temperature of the tag does not start a fire in the paper, cardboard or plastic to which it is attached. Surprisingly, this is the most dangerous point!

Very often, users who have no suitable solutions to hand (such as field measurement devices on board the tag) design base stations that start by analysing and processing tags located at a short distance, by emitting a weak magnetic field, after which these tags go into 'shutdown' mode. When the base stations cannot detect any more elements located at a short distance, they emit a stronger magnetic field to operate in the long distance mode.

Other systems operate with base stations in the auto-adaptive power emission mode, but in this case what happens when two tags are present at the same time, one very close to the antenna and the other far away? A good question!

2.3 Data Communications

This second part of the chapter is solely concerned with the way that communication is organized between the base station and tag(s) and vice versa, and the way that data are exchanged.

Communication is all very well, but what should we communicate? How should we communicate? And so on. First of all, a few simple words are needed about what the concept of communication means.

2.3.1 Communication Model

The general term 'communication' normally implies access to a medium, messages to be transmitted, techniques and problems relating to the methods and problems of transmission, collision management, reception, security, confidentiality of carriage, validity, etc., of a message, and, frequently, the way in which the message and everything relating to it is to be used.

Generally, we can analyse the composition and architecture of any communication and then establish its structure or model. Some years ago, in order to ensure functional compatibility between nonuniform equipment (in response to a variety of manufacturers, functionalities and equipment), the members of the International Standardization Organization (ISO) worked on this problem and developed a set of standards designed to define the interconnection and behaviour of each piece of equipment in relation to its functional environment. This open and general-purpose picture, or model, of communication is called 'OSI' (open systems interconnection) (see also Chapter 15).

The ISO/OSI Communication Model

ISO document 7498, published in November 1984, 'Information Processing Systems – OSI – Open Systems Interconnection – Basic Reference Model', commonly called the ISO/OSI model, defines a model that describes the set of tasks and formalizes the flow of exchanges between these, especially for communicating systems. This standard makes it possible to develop standardized systems and/or networks into which a variety of heterogeneous systems can subsequently be integrated; these standardized systems are therefore called, by definition, 'open systems'.

To allow this heterogeneity, the OSI model is limited to a specification of the functions to be provided by each open system and the protocols to be implemented between these systems, but carefully avoids stipulating any particular embodiment of these functions in the systems themselves. Thus the model specifies the overall behaviour of open systems in their exchanges with other open systems, but not their internal operation. For this purpose, each open system is considered to be an abstract machine called the 'open system model'. The ISO/OSI standards precisely define the operation (internal and external) of these open system models in the execution of the functions of the network that they form. A system designed according to this structure is called 'open' because it can be interconnected without difficulty (or almost! – Author's note): hence its name open systems interconnection (OSI). The highly detailed ISO/OSI model described in the standard is composed of seven layers, one on top of another, and numbered 1 to 7, and provides a clear description of a complex communication system (see Chapter 15 for more details).

Very often, it is unnecessary to specify and/or fulfil all the conditions described in all the layers when establishing a communication system. In this case, we speak of a 'compressed' model. For further information, you should turn to Chapter 15 where I detail the relationships between the ISO/OSI standards and RFID contactless applications.

2.3.2 Communication Modes

Accessing the Physical Layer/Medium (Air)

Access to the medium, usually air in our case, is always a tricky problem when more than one element may attempt to access it at once, something that occurs frequently in RFID when tags send simultaneous responses (the 'collision' phase). Before discussing this subject, and without going too deeply into the science of communication, let us briefly review some terms and definitions adapted to RFID applications that are used when describing modes of access to a medium.

Carrier Sense Multiple Access (CSMA)

The CSMA (carrier sense multiple access) protocol is a multiple-access technique in which stations wishing to produce or transmit data restrict their use of the common communications resource by means of a solution based on the presence or absence of another user's carrier. In RFID, this enables the base station (or the tag) to sense (*carrier sense*) whether another base station (or tag) is using the transmission channel, before it uses the channel itself. This principle is used, for example, in devices of the LBT (listen before talk) type used for RFID base stations in Europe (see Chapter 13).

Carrier Sense Multiple Access with Collision Detection (CSMA/CD)

The CSMA/CD (carrier sense multiple access with collision detection) protocol enables the base station (or the tag) to sense whether another base station (or tag) is using the communication channel before it transmits, and detects the presence of collisions if this is the case. This principle is also widely used in RFID in base stations operating in an LBT mode and managing collisions due to the simultaneous presence of numerous tags in the electromagnetic field.

Code Division Multiple Access (CDMA)

CDMA is a different technique in which each transmission is split into packets and a unique code is assigned to each packet. All the encoded packets are then combined mathematically into one signal, and when the signal is received each receiver extracts only the packets that relate to it. Multiple access in a simple transmission channel depends solely on the use of independently encoded modulations.

Time Distributed Medium Access (TDMA)

The aim of the method known as time distributed medium access (TDMA) is to provide time distributed access to the medium for each participant, thus creating a structure that avoids collisions: it is therefore also known as *collision avoidance*. With this method, it is also easy to design real-time systems, since the latencies are known and the bandwidth is known.

Flexible Time Division Multiple Access (FTDMA)

The flexible time division multiple access (FTDMA) method provides access to the medium by using a principle of *mini-slotting*. During this mini-slotting, access to the medium is allowed dynamically according to the priority assigned to the participants having data to transmit, using a hierarchy related to the value of the 'unique ID (UID)' contained in the message header. The purpose of this principle is to provide communication that is limited in respect of time and bandwidth.

2.4 The Principle of Communication

I will now describe how RFID communication can be initiated, and then we shall consider how it takes place.

2.4.1 Method of Initiating the Exchange/Communication Between the Base Station and the Tag

Here is another problem that gives rise to much debate. To put it simply, which party initiates the communication? The base station? Or the tag?

When remotely powered passive tags are used, there can be no communication without the remote power supply to the tag, in other words without the initial presence of the radio frequency carrier. The base station must therefore always be the first to start, by sending its carrier. Everybody agrees that this is the case, at least! However, from this point onwards, there are two different approaches to the actual initiation (or 'triggering') of the communication itself.

Tag Talks First (TTF) and Answer to Reset (ATR)
This first approach relates to the fact that, as soon as a tag enters the active volume (space) of the base station and is powered (i.e. it is remotely powered or permanently powered, or its power supply is woken up, etc.), following its internal reset, it immediately starts to communicate without any other preliminaries, by signalling its presence: this is the reason for the term TTF (tag talks first) or 'talk after reset'. The presence of the carrier radiation has automatically generated what is known as a request or invitation to dialogue, and this type of tag (TTF) entering the field then responds to this request of the answer to reset (ATR) type. This triggering option (TTF) works very well if we are sure that there will never be more than one tag at a time in the base station's coverage area – if not, we must look out for serious problems due to potential conflicts of signals from the many tags present.

Reader (Interrogators) Talks First (RTF or ITF) and Answer to Request (ATQ)
In order to avoid, mitigate or at least be able to manage the signal collision problems mentioned above, we must have tags that are polite enough to follow the rule 'listen before you speak!' When they enter the area of influence of the base station, the tags, being remotely powered, carry out their internal resets and immediately switch to a special logical state (often called 'ready'), in which they must wait patiently for a special command from the base station (a request command or possibly the presence of an answer to request (ATQ)) so that they can respond to it and inform it of their presence; consequently these systems are described as 'reader talks first (RTF)' or 'talk after request'. Theoretically, by comparison with TTF tags, RTF tags have an additional logic circuitry on board, which enables them to interpret the request command and, on paper, all things being equal, we would expect them to be slightly more expensive than TTF tags.

Coexistence of TTF and RTF
Clearly, it is extremely difficult to know what problems of coexistence may arise when both RTF and TTF tags are present simultaneously in the electromagnetic field, e.g. when the base station has already started communicating on its own initiative with an RTF tag and a TTF tag sneaks into the field and immediately starts to signal its presence. At the present time, it is impossible to say that this will never arise. Following this line of argument, TTF tags should

only be used in applications in which it is certain that they will never leave their operating location and contaminate other sites in which RTF tags may be present (e.g. where components are identified on a production line and the transponders are reused at the end of the line).

In conclusion, you should know that, in order to avoid such problems, some countries (in the Far East, for example) do not permit the use of tag talk first (TTF) transponders and that, in order to avoid these concerns, the ISO has only standardized RTF or ITF contactless smart cards (ISO 14443 and 15693) and tags (ISO 18000-x family of standards) for automatic item management.

I will consider these subjects in more detail, together with *listen before talk* (LBT) and *tag only talk after listening* (TOTAL) in Part Four of this book.

2.4.2 Controlling the Exchange

Now that you know which party can trigger, or start, the exchange, let us move on to the structure of the exchange (see also Figure 2.1(b) at the beginning of this chapter).

Half Duplex

The mode called 'half duplex' (often abbreviated to HDX) is an 'alternating' communication mode in which the forward and return data links are not simultaneous and in which the forward and return messages cannot, therefore, overlap each other.

Full Duplex

In this operating mode, called 'full duplex' (often abbreviated to FDX), the data exchanges in the forward and return links take place simultaneously. In the case of contactless and RFID applications using only one carrier frequency, there are various possible solutions, although these are rarely used at present. For example, there may be a forward link using a frequency modulated carrier frequency and a simultaneous return link using load modulation on the same carrier frequency. Many other communication principles for these full duplex links can, or could, be used.

The advantage of this exchange mode is that faster transaction times can be achieved (if the transaction times in the half duplex mode are not adequate), making it possible to meet the requirements of certain applications, at the cost of an increased complexity of the base station electronics, which must handle the forward and return communication protocols simultaneously in real time, together with the transmission errors that are always possible.

Interlaced Half Duplex

This mode is a derivative of the one described above. In this case, the transmissions are carried out in the full duplex mode in the base station (which can transmit and receive at the same time) and in the half duplex mode only in the tag.

Summary with Reference to RFID

In currently used RFID devices, the data exchange mode used most commonly (in 95% of systems, if not 100%) is half duplex, and thus the base station and tag(s) communicate alternately, in time segments, and not simultaneously, as would be the case in the full duplex mode. Note that some types of bit coding and/or carrier frequency modulation and/or combinations of these (see Chapter 12) may or may not permit communication in the full duplex mode.

Message to be Transmitted
The term 'message' includes the description, the meaning, the basis of the content, its form, its format, its origin, its destination and everything relating to these.

Transmission
Transmission relates to everything concerning the envelope, the protection of the transfer, the quality and reliability of the transfer, its speed, its net transfer rate, the global absence of errors and, if errors occur, their detection, signalling and recovery from them.

Use of the Message
The use of the message relates to the application itself and depends on its specific nature in terms of the user's requirements according to action initiated by it, the content of incoming messages, etc.

2.5 The Concept of Operating Modes

The following text will discuss what is meant by the term 'tag operating mode' and the proposed applications. Once again, tags can be classified into types according to their operating and communication modes.

Read-Only
This mode is simply a matter of reading the content of the tag, assuming that it contains something that someone has written to it. We must distinguish two different cases at this point:

- the tag may have been written to in advance by the manufacturer of the component or
- chip (with a unique number, number, special content, etc.) or
- the tag may have been delivered in a blank state by the manufacturer and may have
- been written to once only (WORM, write once, read multiple) by the user according to
- his requirements. When this has been done, the tag becomes 'read-only'.

Multiple Read/Write
In this operating mode, the tag can be reused or rewritten. Since there is no special protection, it can be used, for example, on production lines for monitoring operations carried out on products during their manufacture. Here again, we must draw a few distinctions.

Programmable/Read Mode (MTP, Multitimes Programmable)
In this mode, a given memory area of the transponder can be rewritten only a limited number of times (a hundred times, for example), allowing certain modifications of the content such as updates and versions of data, but the main purpose of the tag subsequently is to be read.

Read/Write Mode (R/W)
In this mode, there are no restrictions on the multiple rewriting of the tag, with no limit on the number of rewrites (subject only to the technological limits of the chip as regards the possibilities of writing to the E2PROM, which can be done from about 100 to 500 000 times).

Protected Read and Write
This relates to a higher level in the structure of the operation between the base station and tag. This 'protected' read and write mode covers a wide range of applications and requires a certain

amount of further explanation. Note that this is concerned with the reading of all or part of the content of the protected transponder under certain conditions (password, etc.) and the modification of the content under the same conditions.

Protected Read

This operating mode is the standard example of applications in which data classed as 'secret' are to be found in a specific location (memory area) for which the right of access is only granted if identity is proved to the tag in some way, e.g. by using a password, a hardware procedure (special timing), etc.

Write-Protected

This operating mode is often, but not always, combined with read-protected, and is used to safeguard the writing of the content of the memory:

- by creating areas (all or some) of the memory operating in the OTP (one time programmable) mode or
- by permanently locking a memory area to prevent any future writing, after a number of voluntary modifications of the content (an example is the monitoring of the manufacture of a product along the production line with rewriting allowed along the line, but with write-locking at the last stage of manufacture of the product, where the process is considered to be complete) or
- by allowing access to rewrite some or all of the memory subject to a specified condition or conditions, and subject to the entry of x more or less complex passwords.

If the memory has been write-protected, it operates as a PROM.

Secure Read and Write

This operating mode does not use any specific form of encryption, but uses conventional security methods (such as partner authentication or rolling codes) to provide more secure establishment of the communication between the base station and the tag. I will indicate the problems relating to these systems (such as potential problems of code synchronization). For more information, you should read or reread my previous books.

Encrypted Read and Write

The content of the communication between the tag and the base station can also be encrypted in order to foil anyone spying on the data exchanges. This is generally done by using crypto-graphic algorithms, which may be proprietary or standardized types using private keys (DES, triple DES, AES, for example) or public keys (RSA, ECC, for example).

Reading and Writing a Single Tag Present in the Electromagnetic Field

Only one tag is present in the operating field of the base station and communication control is a simple matter.

Reading and Writing Multiple Tags in the Electromagnetic Field

It is also possible that more than one tag will be present simultaneously (deliberately or accidentally) in the volume (x dm^3 or y m^3) in which the operation of the electromagnetic field produced by the base station takes place.

To be workable, the system must be designed so that it can identify all the tags present in the field of the base station and the base station must be able to communicate individually with some or all of them (e.g. with a subset of tags having one or more features in common). The problems of collision that may arise and the collision management procedures to be used (wrongly called 'anti-collision' systems, since collisions do indeed occur!) are described below (in Chapter 15; if necessary, see Reference 1 mentioned in the Preface as well). Note that the number of tags present simultaneously may be known or completely unknown and that the system must be able to manage all these cases.

General Note About Reading and Writing Tags
Nothing in the above text implies that the writing distances and/or speeds are identical to the reading distances or speeds of the tags. Some sets of applications can be satisfied with short distances and slow write speeds but require long distances and high speeds for the read phase – or vice versa.

2.6 General Operating Problems in Data Transmission
2.6.1 Application Specific

Except in some very special cases, the technical options to be chosen will depend on the technical constraints, environment, regulations, etc., arising from the topology of the application.

Topology of the Application
The topology of the application is very important for the choices that must be made. We must always ask at least the following questions:

- Is the base station/tag system mechanically fixed or mobile? If it is mobile, then what is the speed (some examples: production line conveyor belt: less than $1\,\mathrm{m\,s^{-1}}$; a running human: 5 to $10\,\mathrm{m\,s^{-1}}$; bicycle: $20\,\mathrm{m\,s^{-1}}$; train: 40 to $50\,\mathrm{m\,s^{-1}}$).
- Will one or more tags be present at any one time in the magnetic field?
- Does each tag contain only one application (simple identification of an object) or multiple applications?
- Is the transaction time important or irrelevant to the application?

Tags Moving Relative to the Base Station
In many applications it is necessary to deal with relative movement, which may be fast or slow, deliberate or accidental, between the base stations and the tags. Some examples are transport smart cards, data capture on the fly, monitoring components on production lines, pallets on forklifts and numbers on marathon runners.

  To ensure that the communication process is successful, we must consider the factors noted in the preceding paragraphs and we must also ensure that the tag remains in the electromagnetic field of the base station for at least the whole of the transaction time and that the minimum energy and coupling or back scattering required for the application are maintained, regardless of the positions and possible relative movements of the tag(s) with respect to the base station. When considering the time for which the tag(s) remains in the magnetic or electromagnetic field, we must allow for:

- the speed of the relative movement of the tag(s) with respect to the base station (remember, $l = v\,t$);
- the wake-up time, plus the reset time of the tag(s) when it (they) returns to the field;

- the collision management time, if more than one tag is present in the field at one time;
- the gross and net speed of the data transmitted from the tag to the base station and vice versa;
- the volume of data to be exchanged between the base station and each of the tags;
- and, finally, the positions or orientations of the tags in the field, which in some applications can be totally random (as in the case of access control badges or items in a supermarket trolley, for example), thus possibly requiring additional time for the operation of special multiple antenna devices, which may or may not be time-division multiplexed, rotating field devices, etc., which will be described more fully in Chapter 19 and whose effectiveness has been proven on many occasions.

Let us consider one of these points in detail.

Transaction Time

The time taken for a transaction depends on numerous parameters, as follows:

- the wake-up time, plus the reset time required by the tag to prepare itself for operation after it has entered the action space of the electromagnetic field (this time is generally several milliseconds or thereabouts);
- the time for managing possible collisions (collision management procedure) due to the simultaneous presence of more than one tag (when there is more than one label on more than one package on a pallet, for example), together with the time required for sequences of authentication, encryption and decryption if required (several milliseconds or thereabouts);
- the sum of the turn-round times for transmission/reception/transmission, etc., required for the correct progress of communications between base stations and tags;
- the time for accessing the correct application, where more than one application may be handled by a single tag and the correct application(s) must be selected before an operating session is commenced;
- the amount of data to be exchanged for the purposes of the application and their management (reading, writing to E2PROMs, calculations using data present in tags, etc.);
- the gross speed of the exchange, including, of course, the protocol structure of the communication (the header, prologue, epilogue, CRC, etc., of the communication protocol);
- the period for which the tag is physically present in the electromagnetic field, which may or may not depend on its relative movement in time with respect to the base station.

Examples of values of all these elements are given in Table 2.1.

Table 2.1

Function	Time taken to perform the function
Reset	About 5 ms
Answer to request and collision management sequence	3 ms
Selection	2 ms
Authentication (if required)	2 ms
Reading 16 bytes	2.5 ms
Reading 16 bytes + rereading for checking	9 ms
Total (approximate)	25 ms

This example shows that several moving tags can be processed quite easily. For movement at a fast walking speed of $6\,km\,h^{-1}$ ($=1.66\,m\,s^{-1}$), a field area with a length of 4 cm is sufficient for the transaction!

There is generally a fairly close relationship between the carrier frequency, the bit rate, the communication protocol and the maximum overall transaction time. To give you an idea of the figures, here are some orders of magnitude of overall transaction time:

- about 50 ms for a car immobilizer;
- about 25 to 30 ms for contactless smart card applications;
- about 10 to 20 ms per item for a system for identifying items on a pallet.

This may seem strange. These values are practically the same, but the reasons are different. For your information, car immobilizers operate at 125 kHz, smart cards operate at 13.56 MHz and pallet systems operate at UHF. I only mention this to underline the fact that you should not let yourself be influenced by coincidences that may mislead many users in their choice of system.

All the factors described above must of course be selected and/or determined in full compliance with current local regulations concerning emission levels and electromagnetic pollution, while meeting the requirements of the desired scope of operation.

2.6.2 Memory Size

Depending on the planned applications, it may be necessary to use data memories (EPROM/E2PROM/Flash E2PROM/FRAM) of varying sizes (from a few bits to several kilobytes or even megabytes). Regardless of this, it is necessary to know if, at a given instant, the application only uses a small amount of data at once or if all of the total memory content, or large parts of it (bytes, blocks, pages), have to be dumped sometimes (this is why the maximum memory size is important).

Since the tag is accessed via a series link, the time taken to 'dump' the memory must be taken into account. Also, for some types of memory (such as E2PROM), the write time is longer than the read time, and, furthermore, when we write to the memory we generally reread what has just been written (a 'READ after WRITE') to ensure that we have indeed written what was intended and that the tag did not leave the electromagnetic field during the theoretical write time. Clearly, all these time intervals contribute to the minimum transaction time, and therefore affect the choice of bit rate.

2.6.3 Bit Rates: Chosen or Necessary

If you can do a lot, you can do a little. Why not use high bit rates if you can get them? In principle, this is correct. However, why would you drive a sports car or a 4×4 in London if a smaller car was just as satisfactory? This shows you why the debate about kilobits per second is quite illusory. Should we choose speeds of several kilobits or several hundred kilobits per second? Once again, the question is: What is the useful transaction time that is really necessary for the application? And what are the possibilities offered by local regulations and their wavebands? Other questions are: What are the time losses due to the application topology (signal reflection, 'dense' environment, etc.; see Chapters 12 and 16) and the protocol used (effects of an excessively large header, prologue and epilogue, maximum time in the field, collision management, selection, transmission security protection, transmission secrecy, etc.)? All of these factors will raise the gross communication speed. It is up to you to do the maths!

2.7 More Specific Problems Relating to 'Long Distance' RFID Systems

To define our terms at the outset and avoid any confusion, I repeat that 'long distance' signifies the use of systems whose useful operating distance is, in standard conditions (without any clever tricks), between 1 and 10 metres, and that the space occupied by the radiated electromagnetic field is large. This concept of long distance systems implies a high probability of the simultaneous presence of many tags in the space in which the electromagnetic field is created, making it necessary to consider the phenomena of collisions and collision management, possible 'weak collisions' of signals, 'masking' effects and so on.

In view of this, while everything stated in the earlier paragraphs holds true for most contactless/RFID applications, we will have to consider some new problems that are specific to long distance applications when working with such systems.

2.7.1 The Distinctive Features of Long Distance Applications

To begin with, what is this strange-sounding field of applications? It includes the monitoring of baggage, parcels, boxes, pallets, containers, lorries (often moving), etc., or, more succinctly, bulky items whose physical volumes may be as much as several cubic metres, as well as markets that consume vast quantities of labels, e.g. labels for supermarket and hypermarket products. In other words, we are dealing with logistics.

A common feature of these markets is that, usually, the tag is, or becomes, due to its application, a consumable product that is thrown away after its momentary multiple uses (several read/write operations during its short active life). For a product to be considered 'consumable', it must be cheap and its functionality must not make it expensive either.

2.7.2 Key Parameters of Long Distance Systems

At the technical level, these systems can be defined by the following parameters:

- for the tag:
  - variable memory capacity;
  - small mechanical dimensions (including the tag antenna), where the chosen carrier frequency must require only a small and inexpensive antenna, and therefore the carrier frequency must generally be as high as possible, without increasing the complexity of the product;
- for the system:
  - reduced security for communication and access;
  - long reading distances and, if necessary, long writing distances for the memory;
  - large read/write volume (in m^3) (see below);
  - bulk reading;
  - a known transaction time, because of the large number of possible collisions and the need to provide a fast collision management mechanism;
  - compliance with the radio frequency pollution regulations.

Let us take a closer look at all these points.

Bulk Reading

The problem is that we must provide satisfactory operation over the desired (long) distances, while also reading all the tags (100%, not 99.9999%!) regardless of their relative angular positions with respect to the base stations and other tags – and guarantee this level of performance! (Saying this is easy – doing it is much harder and signing the deal is another story altogether!) Thus we need to read the tags, not in a particular orientation or in a specified plane, but in three-dimensional space! For this purpose, we use several base stations and/or several antennas for each base station at each site when working at UHF and SHF, combined with circular or elliptical polarization. I will discuss these vital points in detail throughout this book.

2.7.3 Energy Transfer and Data Exchange

In order to provide the required long distance performance for 'long distance' applications, we must consider many other problems.

One of the first things to do when determining the physical characteristics of the system is to examine the way in which the radiated power spectrum is to be optimized. Given that a high power level must be radiated to provide successful long distance transmission and remote power supply, we must be careful, on the one hand, not to exceed the maximum power permitted by local regulations and, on the other hand, to keep the sideband levels as low as possible, since the templates (or 'masks') permitted by the regulators are usually very low.

To achieve these differences between the carrier and sideband strength (of the order of 50 to 60 dB – see below), we must pay particular attention to the choice of bit coding and to the specification of the type of modulation associated with the carrier frequency. Furthermore, regardless of the carrier modulation used subsequently for transmitting data from the base station to the tag, we must remember that it will generate sidebands close to the carrier frequency, which are always more or less proportional to the level of the radiated carrier.

Notes
In long distance applications, where the detection of the weak response from the tag is concerned, base stations generally use digital signal processing methods (DSP), and therefore the weak data signal received from the tag is not a hindrance.

To summarize, the physical parameters of the base station (i.e. the bit coding and modulation type; for details, see Chapters 11 and 12) must be chosen and optimized in such a way that the energy transmission is optimized and the radiated sidebands are as weak as possible with respect to the carrier, and in such a way as to provide the best match between the energy contribution to provide the best possible remote power supply, the communication bit rate, the type of modulation and the structure of the radiated spectrum.

As I have pointed out, and explained in detail, in my earlier books, the chosen solution must be balanced overall and not orientated towards the optimization of just one or two of the parameters.

This concludes the introduction and the guide to the specific terminology of this branch of identification represented by contactless applications, of which RFID is one example. Before moving on to the purely technical section, I shall finish this first part by giving you an idea of the market for these applications and the leading players.

3

The Market and Applications for Contactless Technology

This short chapter will provide a brief survey of the market, the distribution of applications, the frequencies generally chosen and used – and much more besides!

3.1 The Market for Contactless Technology and RFID

The market for contactless identification is indisputably one of the rare electronics markets in which major expansion is expected over the next twenty years. Its rate of growth is remarkably high. This market is currently divided into a number of main segments, which I will rapidly survey, in no particular order, since each has its own distinctive features.

3.1.1 Industrial RFID

This was one of the first segments of this market to appear. The numbers of transponders used for each application are generally in the mid-range (5000, 10 000 or 100 000 per year), since they are specific to particular projects (such as 'closed' applications in which the tags are often re-used), which are themselves linked to various industrial processes. These systems mainly operate at 125 kHz, but sometimes at 13.56 MHz, and are often preferred by small to medium sized businesses.

3.1.2 Motor Vehicle Immobilizers and 'Hands Free' Devices

The market for immobilizers (devices preventing the starting of motor vehicles, see Figure 3.1) is currently one of the largest segments in the contactless identification field (in terms of quantity). Close to a billion of these items (cumulated quantity) are now in use, and are working correctly – otherwise we would know about it!

RFID at Ultra and Super High Frequencies: Theory and Application   Dominique Paret
© 2009 John Wiley & Sons, Ltd

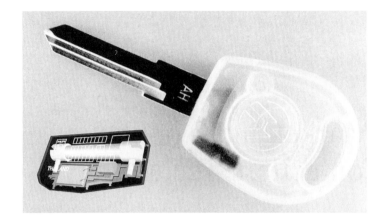

Figure 3.1   Immobilizer for a motor vehicle

3.1.3 Contactless Smart Cards

This segment is growing day by day. The applications are generally of the multiple-application type, the leading one being the transport sector (AFC, automatic fare collection), with close to 2 billion cards in use (cumulated amount at 1 January 2009), while the runner-up, in the last stages of development, is the contactless credit card, identity cards, passports and their electronic e-visas, and electronic wallets. These cards form part of the family known as 'voluntary validation action' proximity cards.

3.1.4 'Intelligent' or 'Communicating' Tagging, Labelling and Traceability

This segment is expected to outdo all the others in terms of quantity over the next few years. The main applications using RFID electronic labels will be of the 'bar code with added functionality' type and will be disposable. Here, the base unit is a million (or millions) of tags per day, i.e. several billion items per year! The main subsegments in this sector are those relating to monitoring applications for baggage, product distribution chains (SCM, supply chain management), items, packages and their delivery, assets, etc., and finally the location of items.

One very large subsegment consists of applications of the EAS (electronic article surveillance) type, in other words anti-theft systems (gates) for CDs, clothes, etc., widely used at the exits from large stores.

Figure 3.2(a) and (b) provides a global indication of these segments in terms of the number of items.

3.2 Applications for Tags

To avoid too much abstraction, and to provide you with a little diversion, here are some examples of existing applications of contactless identification using tags, listed according to their usual operating frequencies (although these do not necessarily limit future developments). At the present stage, I would ask you to forgive any exceptions that may arise.

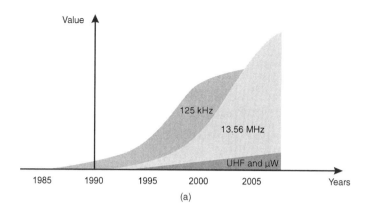

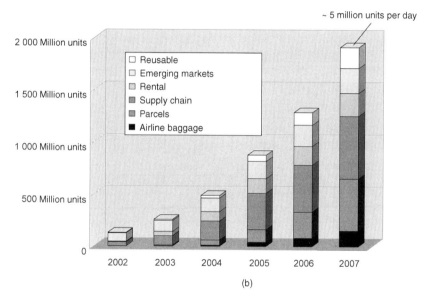

Figure 3.2 (a) RFID in terms of value. (b) The 'smart label' segment of the RFID market, in terms of quantity

Figure 3.3, kindly provided by Bernard Jeanne Beylot[1] shows, at a given date, in one page, the range of common RFID segments and applications.

3.2.1 Frequencies Below 135 kHz

Motor Vehicle Applications

- Immobilizers (start prevention systems for motor vehicles);
- immobilizers + remote control (remote keyless entry, RKE);

---

[1] Bernard Jeanne-Beylot, Consultant at CURITIBA, is a member of CN 31 of AFNOR and SC 31 of ISO, and organizes numerous conferences on the theme of RFID applications.

Identification of:	Persons						Products		Animals		Vehicles
Application types	Identity cards, passports, visas, driving licences	Electronic wallet	Checking transport documents, ticketing and leisure	Time clocks & personnel management	Access control for vehicles & pedestrians	Industrial identification	Identification and tracking in logistics	Other applications	Pets	Livestock	Immobilisers for cars, lorries, motor cycles
	Open and/or closed loop	Closed loop	Open and/or closed loop	Open and/or closed loop	Open and/or closed loop	Closed loop	Open and/or closed loop	Open and/or closed loop	Open loop	Open loop	Closed loop
Most commonly used frequencies	13.56 MHz	13.56 MHz	13.56 MHz	125 kHz 13.56 MHz	125 kHz 13.56 MHz	125 kHz 13.56 MHz	13.56 MHz UHF* 2.45 GHz*	125 kHz 13.56 MHz UHF* 2.45 GHz*	134.2 kHz	134.2 kHz	125 kHz
Standards & regulations (physical layer & air protocol)	(ISO 10536) ISO 14443 ISO 15693	(ISO 10536) ISO 14443	(ISO 10536) ISO 14443 ISO 15693	No standard, or ISO 14443 ISO 15693	No standard, or (ISO 10 536) ISO 14 443 ISO 15693	No standard, or ISO 15693 en attendant ISO 18000-2 < 135 kHz ISO 18000-3 / 13.56 MHz	No standard, or ISO 15693 en attendant ISO 18000-2 < 135 kHz ISO 18000-3 / 13.56 MHz ISO 18000-4 / 2.45 GHz ISO 18000-6 / UHF	No standard, or ISO 15693 en attendant ISO 18000-2 < 135 kHz ISO 18000-3 / 13.56 MHz ISO 18000-4 / 2.45 GHz	ISO 11784 (data) ISO 11785 (com) ISO 14223-X X=1 radio interface X=2 code structure X=3 application	ISO 11784 (data) ISO 11785 (com) ISO 14223-X X=1 radio interface X=2 code structure X=3 application	No standard, and/or proprietary system
Examples of applications used in France	Under test CIEC (citizen's electronic card)		Remote ticketing Calypso system for RATP & SNCF		Building entry control for postmen Vigik system for French Post Office	Industrial process monitoring, tool control, after-sales service	Tracking of products, packages, containers and pallets in the logistics flow	Identification of trees and street furniture, product authentication, maintenance	Electronic identification of domestic pets using implanted glasstags	Electronic identification of livestock using radio tags (ear tags, Bolus, etc.)	Transponders in car keys and/or contactless keys: PSA, Renault, VAG, Ford

* UHF and 2.45 GHz frequencies, depending on current statutory limits for France and Europe

Figure 3.3   Application segments of the RFID market. Reproduced by permission of B. J. Beylot

- passive keyless entry (PKE) and passive keyless start (PKS);
- industrial anti-intruder and access control systems (building security, etc.);
- automatic motorway tolls;
- loyalty cards, maintenance cards, service cards, vehicle service record, etc.;
- weight measurement (loaded, empty, etc.) for lorry loading.

Industrial Applications

- All forms of identification and tracking;
- alarm and theft prevention systems for shops, EAS;
- animal identification (at 134.2 kHz) + animal feed monitoring + freezer line;
- industrial laundry, cleaning of hotel linen and work clothing;
- tree monitoring in Paris (health, watering);
- on-board monitoring and management of production and supply lines;
- household and industrial waste collection and sorting;
- access control (buildings, airport zones, high-security areas);
- libraries (book return/borrowing, shelving, stocktaking);
- pallet monitoring (in shops, during transport, etc.);
- identification of gas cylinders for public use (butane, propane) and industrial use (medical);
- monitoring of pictures, art objects (furniture), pairs of new skis, hiring, etc.;
- monitoring documents in a company, monitoring trolleys (in factories, hospitals, etc.);
- baggage sorting and monitoring (at airports, etc.);
- monitoring 20-foot containers;
- source coding (source labelling);
- source coding (supermarkets);
- prevention of counterfeiting and black markets in luxury products (clothes, perfume, watches, etc.), casino chips, etc.

3.2.2 At 13.56 MHz

This frequency, located in one of the ISM (instrumentation, scientific and medical) reserved bands, is commonly used for the following applications.

Contactless Smart Cards
Figure 3.4 shows the best-known applications based on credit cards of the blue card, visa and other types, and, in general, all cards relating to electronic funds transfer, ticketing, transport, etc.

Electronic Wallets
These applications include, for example, retail purchasing (e.g. using the electronic purse application), prepayment (car parking, etc.), pay per view TV, etc.

Applications in Telecommunications
Cards used in public and private payphones and those fitted in some GSM phones for access rights and storing certain personal data.

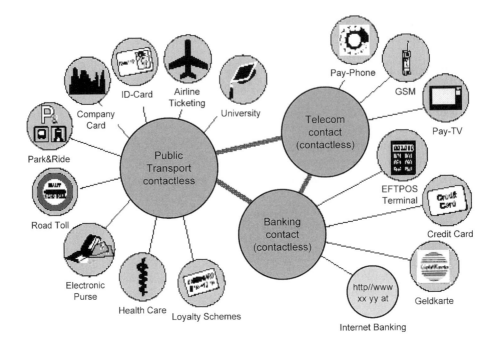

Figure 3.4   Examples of applications of contactless smart cards at 13.56 MHz

Transport
These are often multiservice or multiapplication cards, and form part of the large group of applications known as AFC (automatic fare collection). Some examples are:

- bus and urban transport systems;
- automatic air ticket check-in;
- air ticket reservation;
- park and ride;
- road tolls;
- prepayment at ski resorts and access to facilities (ski lifts, etc.).

Access Control

- Access of all kinds;
- secure access control (in nuclear power stations);
- management of working hours.

Applications in the Community

- Schools (access to canteens, sports halls, etc.);
- elderly persons (assistance to shopkeepers);
- leisure (libraries, swimming pools, tennis courts and other sports facilities);
- hotel reservation;
- public transport in general.

Monitoring and Tracking

- Parcel monitoring (DHL, Federal Express, TNT, UPS, or United Parcel Services, Chrono Post, mail-order companies, etc.);
- assistance in sorting and routing post, fleet management (monitoring hired vehicles, postal vehicles, company lorries, etc.), airline baggage control;
- document monitoring, stocktaking and shelving in libraries;
- item identification for large stores (ILM, item level management);
- source coding/labelling;
- EAS (electronic alarm surveillance) anti-theft devices at the exits from shops and supermarkets flow control (reception, despatch, stocktaking, order preparation, automatic sorting, etc.);
- CD, DVD rental and other rental applications;
- timing for sports events.

Personal Data and Official Documents

- National identity card/passport (including JPEG 2000 compressed photographs, etc.);
- driving licence;
- car registration papers;
- health cards;
- company cards;
- loyalty cards (customer loyalty).

3.2.3 At 443 and 860 to 960 MHz

- Remote control central locking systems for motor vehicles – RKE;
- wheel identification and location for tyre pressure monitoring systems (TPMS).

Identification and Tracking

- Alarm and theft prevention systems for shops, EAS;
- supply chain management;
- long distance identification;
- identification of parcels, pallets, containers, etc.;
- pallet monitoring (in shops, during transport, etc.);
- baggage sorting and monitoring (at airports, etc.) (Figure 3.5);
- monitoring 20-foot containers;
- source coding (source labelling);
- source coding (supermarkets).

3.2.4 At 2.45 and 5.8 GHz

- Remote payment of road tolls;
- labels (mainly in the USA and Japan, for reasons of authorized emitted power – see Chapter 15);
- real-time location (RTL) systems (for monitoring lorry fleets, for example).

Figure 3.5   Examples of RFID applications at UHF (doc from NXP)

3.2.5 Infrared

- Contactless travel passes;
- remote payment of road tolls.

3.3 Operators and Participants in the Market

Participants in this market cover a range of different types, classes and sectors. There are very few companies that cover the whole of a system, since each class of company has its own specialization.

To develop a working application, it must first be defined by the end clients or the decision makers for the project. We then need chips produced by integrated circuit manufacturers (chip producers), and these chips must be placed on a substrate including the antenna (by inlet manufacturers). Then everything has to be encapsulated in a special package (e.g. by card producers or manufacturers of paper or card labels), after which they are customized, and so on. Finally, a base station must be produced and incorporated in a whole system (the host system).

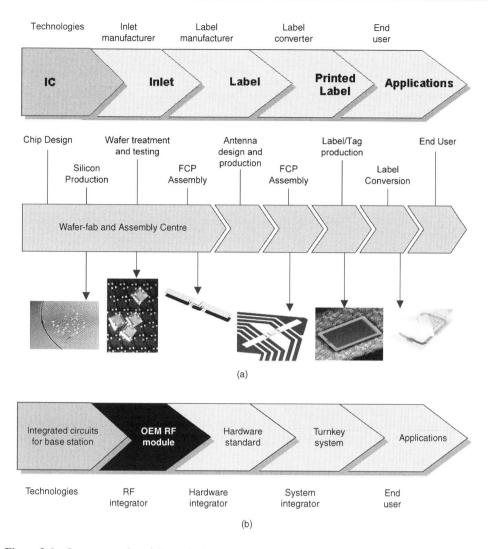

Figure 3.6 Operators and participants in the market (according to a document produced by NXP/Philips Semiconductors): (a) for transponders; (b) for base stations

Figure 3.6 is a schematic illustration of the sequence of operators and participants in this market.

3.3.1 Main Manufacturers of Transponders and Hardware

Manufacturers of Silicon Chips for Transponders and Base Stations

This is a tricky group to define, because, strictly speaking, we should make a distinction between companies that manufacture their own chips and sell them through their catalogues, 'fabless' companies (not having their own chip production facility) that develop integrated circuits for tags using computer-aided design and send these to the manufacturer to create a product for their catalogues, and finally companies that commission engineering companies to design, with or without their participation, an ASIC (application specific IC), which may or may not be freely available in the market.

However, many well-known major component manufacturers (chip producers) are included among the possible suppliers of components (chips) for contactless applications. The main ones are as follows, strictly in alphabetical order.

Integrated Circuits for Tags (Chip Producers)
Atmel, Hitachi, Infineon, MicroChip, µ-Electronique Marin, NEC, NXP Semiconductors, Sony, ST MicroElectronics, Texas Instruments/Tiris and many others.

Integrated Circuits for Base Stations
Atmel, Elmos, Infineon, Mietec, Nedap, NXP, ST MicroElectronic, etc., and numerous proprietary ASICs.

Antennas
Many antenna manufacturers offer, in addition to their standard ranges, all kinds of unusual custom-designed antenna coils using various technologies including copper films, printed circuits, screen printing, conductive ink deposition, gravure printing, etc. They include Cleo, FCI, Micro spire, Metget, Vogt and others.

Card Tag Manufacturers
When the chips are ready, they have to be encapsulated or packaged with their antennas in some kind of casing, in the form of inlets or inlays, key fobs, watches, ISO format smart cards, underground train tickets, paper labels, etc., so that they can be used.

Here again, there is a wide range of possible suppliers. The main ones, in alphabetical order, are: ASK, Balogh, Denso, Feig, Freudenberg, GemAlto, IER, Isra, Nedap, Oberthur Card System, Orga, Pav Card, Paragon, Pygmalyon, Rafsec, Nagra ID, Sagem, Sihl, Sokymat and Tagsys.

Systems Integrators
These companies offer all or part of a system on a 'turnkey' basis, from the elementary transponder to the complete final application, and including the design and production of the base station, the host and the application-specific software.

Design Houses
These firms generally provide possible solutions 'on paper' and file patents relating specifically to the planned applications. They usually provide consultancy services.

3.3.2 Decision Makers

This category includes all client organizations and companies that specify and choose the systems and components that they wish to use, but do not purchase them directly. Some examples are:

- transport companies and organizations;
- airlines (Lufthansa, British Airways, American Airlines, Air France, IATA, etc.);

- banking and interbank groups (Visa, etc.);
- major motor vehicle manufacturers;
- governments (Ministry of Industry and Commerce, Ministry of the Interior, Ministry of Health, Ministry of Agriculture);
- public services (Post Office);
- major distribution companies (Carrefour, DHL, Federal Express, TNT, etc.).

3.3.3 Users

Everybody – including animals and objects!

Part Two

Wave Propagation: Principles, Theories... and the Reality

The second, and very long, part of this book presents and describes the specific technical principles and theories used in contactless RFID applications operating at UHF and SHF. Together with Part Four, which is concerned with standards and regulations, it provides the essential grounding for this book. Therefore, please prepare yourself for a detailed perusal of this absolutely fundamental section.

I can assure you that I have not been deliberately long-winded; I have only introduced material that is strictly necessary. Unfortunately, 'strictly necessary' means 'rather dense and bulky' when it comes to providing a thorough grounding in RFID at UHF and SHF. Furthermore, having been frequently annoyed in the past by technical and scientific works full of formulae apparently conjured up out of thin air, and phrases along the lines of '... after simplification, and allowing for the usual approximations...', I have adopted a policy, for educational, professional and industrial purposes, of detailing the actual sources of the equations, formulae, etc., and pointing out exactly where simplifications have been made and their real consequences in terms of engineering. My excessive thoroughness is easily justified. When you work on your own RFID applications at UHF and SHF, you will soon discover that your systems hardly ever correspond to the theoretical cases described in the literature, and you will need to revisit the basic equations, without any approximation, in order to adapt them to your specific problems. At this point, you will be delighted to recall the hard work you put in, for the best of reasons, to gain all this useful knowledge. Thus, you have been warned....

As for this second part of the book, it consists of two major sections:

- The first of these, Chapter 4, provides a detailed and specific statement of the general principles of propagation and calculation of the strength and power of radiated electromagnetic fields, and of radiation resistance.
- The second section, which is essential for a proper understanding of RFID at UHF and SHF, consists of Chapters 5 to 10, dealing with wave propagation and its consequences, as follows:
  • First of all, in Chapter 5, I describe in detail the conventional phenomena of radio frequency wave propagation in free space.
  • Chapter 6 will deal with the recovery of power from the terminals of the receiving antenna and the estimation of operating ranges.
  • Chapter 7 will then detail the problems of wave propagation in 'less free' space, in other words how to handle everyday reality.
  • In Chapter 8 we will examine the theories and applications relating to reflection and back scattering of these waves.
  • In Chapter 9, we will look at back scattering techniques, their problems and their use.
  • Finally, with the aid of a comprehensive range of examples, Chapter 10 will demonstrate how these properties are widely used in UHF and SHF RFID techniques, which is the main purpose of this book.

Notation Conventions

A large number of equations will be used in the following chapters. In order to simplify the presentation of the text, I have made use of the following conventions (except, as usual, in a few specific cases which are difficult to deal with in any other way).

	Examples
A constant quantity is shown in either upper or lower case, in ordinary roman type.	Speed c
A variable quantity is shown in either upper or lower case, in *italics* (given that this is the electrical field that depends on time, distance, etc., and is therefore a variable quantity).	Electrical field E
A vector is shown in **bold roman upper case**.	Poynting vector $\mathbf{V}$
A scalar product (also known as a dot product) is represented by a **bold raised full stop** $\boldsymbol{\cdot}$.	
The symbol of the vector product (also known as the cross product) of two vectors is represented by the sign $\times$ in **bold type**.	$\mathbf{S} = \mathbf{E} \times \mathbf{H}$
The modulus of a vector is represented by two vertical bars, $\mid \mid$.	$\mid\mathbf{S}\mid$
The mean value of a variable is represented by $\langle \ \rangle$.	$\langle\mid\mathbf{E}\mid\rangle$ means 'the mean value of the modulus of the vector $\mathbf{E}$'

4

Some Essential Theory

UHF and SHF RFID applications cover a wide range and use many different methods and properties. The physical phenomena in this field are also numerous and vary according to the technologies that are used. I will provide details of these later on. First of all, let us briefly describe the physical concept of an electromagnetic wave.

4.1 The Phenomenon of Propagation and Radiation

If we think of UHF and SHF RFID systems as a mountain range, we are now at its foot! The first thing to bear in mind is that the operation of all these systems is based on the study of the phenomena of propagation of radio frequency waves in an environment that is far from ideal, allowing for their reflection, absorption, etc. – which are never simple problems.

The aim of this book is not to add to the numerous existing treatises on these matters, but to provide a detailed understanding of the specific problems of UHF and SHF RFID. Therefore, here is a brief refresher course on the phenomenon of propagation in the most conventional case, that of the radiation of a 'Hertzian dipole'.

Using this simple example, I will then go on to examine some real-world uses of RFID.

4.2 The Hertzian Dipole

By definition, a Hertzian dipole consists of a wire conductor whose length l and diameter D are very small with respect to the emitted wavelength λ. Assuming that a quantity of electricity $q(t) = Q_0 \sin \omega t$ flows in the antenna wire at each instant, the current flowing in the Hertzian dipole is

$$i(t) = \mathrm{d}q/\mathrm{d}t = Q_0 \omega \cos \omega t$$
$$i(t) = I_0 \sin \omega t$$

where I_0 is the peak amplitude of the current.

RFID at Ultra and Super High Frequencies: Theory and Application   Dominique Paret
© 2009 John Wiley & Sons, Ltd

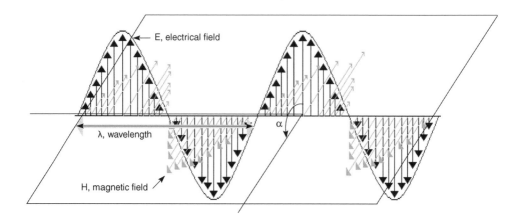

Figure 4.1   Example of representation of an electromagnetic wave

At a distance r from an element of length dl of the antenna in question, this current produces a magnetic field element of the form dH (Laplace's law):

$$dH = Idl \sin\frac{\alpha}{r^2}$$

This assumes that the 'effect' produced at a distance appears to be instantaneous, in macroscopic terms, with respect to its 'cause'. However, this is not so. In fact, to put it briefly, if a conductor (antenna) is raised to a certain potential, it creates around itself an electrical field E whose appearance is not instantaneous. It actually develops in air at the speed of light, denoted by c.

When a high-frequency alternating voltage is applied to the antenna, this produces an oscillation (or wave) of the electrical field, which also moves at a velocity of approximately c (depending on the medium that it passes through).

The resulting magnetic field H depends on the electrical field E. Consequently, there is also an oscillation of the magnetic field H, which moves in the same way. The set of the two fields, electrical E and magnetic H, is propagated at the speed of light. The combination of these two oscillations forms what is commonly called an 'electromagnetic wave', which is illustrated in Figure 4.1 (at a given instant).

I will show you later that, at long distance only, the magnetic field lies in a plane perpendicular to the electrical field.

4.2.1 Propagation Equation

The demonstration and mathematical solution of the propagation phenomenon and the values of the associated electrical and magnetic fields require the use of Maxwell's equations, Green's function, Bessel functions, the Legendre polynomial, the Lorentz and Hertz potentials, etc., and lie outside the scope of this book. Readers may consult any of the specialized scientific and theoretical books on these subjects,[1] and if this is reminiscent of their earlier education, I can at least offer them a few sweeteners to make the medicine of 'Laplace', 'd'Alembert',

---

[1] *Transmission and Propagation of Electromagnetic Waves*, K. F. Sander and G. A. L. Reed, Cambridge University Press, 1978, and *Fundamentals of Electromagnetic Waves*, R. Sheshadri, Addison Wesley.

'divergence', 'rotational', etc., rather easier to assimilate. In other words, plenty of general knowledge.

All we need to know is that Maxwell and Hertz did indeed formulate some general equations that govern these fields. These equations indicate that, in a vacuum, the fields conform to an equation of the following type, between the partial second-order derivatives of the two variables r (distance) and t (time):

$$\varepsilon_0 \mu_0 \frac{\partial^2 E_r}{\partial t^2} = \frac{\partial^2 E_r}{\partial r^2}$$

where $\mu_0 = 4\pi \times 10^{-7}\,\mathrm{H\,m^{-1}}$ and $\varepsilon_0 = 1/(36\pi) \times 10^{-9}\,\mathrm{F\,m^{-1}}$ are, respectively, the permeability and permittivity of a vacuum.

The general space–time solution of this equation (which is therefore a function of the two variables t and r) is of the following type:

$$E(t,r) = f\left(t - \frac{r}{\gamma}\right) = E_0 \sin\left[2\pi\left(\frac{t}{T} - \frac{r}{\lambda}\right)\right]$$

where

$$\gamma = \frac{1}{\sqrt{\varepsilon_0 \mu_0}}\,\mathrm{m\,s^{-1}}$$

is the velocity of propagation of the wave in a vacuum.

This type of equation shows that, at a point P of the abscissa space r, at the instant t, the value $E(r)$ is equal to the value that $E(t, r)$ had at the origin, at a time r/γ before the instant in question. This mathematical property thus indicates that the signal $E(t, r)$ is propagated in the direction of the r, with the velocity γ.

Given that, on the one hand, $\omega = 2\pi f = 2\pi/T$ and, on the other hand,

$$\lambda = cT = \frac{c}{f} = \frac{c}{\omega/(2\pi)}, \quad \text{i.e.}\,\frac{1}{\lambda} = \frac{\omega}{2\pi c}$$

we can write

$$E(t,r) = E_0 \sin\left[2\pi\left(\frac{t}{T} - \frac{r}{\lambda}\right)\right]$$

$$E(t,r) = E_0 \sin\left(\frac{2\pi t}{T} - \frac{2\pi r}{\lambda}\right)$$

$$E(t,r) = E_0 \sin\left(\omega t - \frac{r\omega}{c}\right)$$

$$E(t,r) = E_0 \sin\left[\omega\left(t - \frac{r}{c}\right)\right]$$

or, alternatively, by making $k = \omega/c = 2\pi/\lambda$ (sometimes denoted by β), which is called the 'wave number', we obtain

$$E(t,r) = E_0 \sin(\omega t - kr)$$

which is a function of t and r, at the frequency f or the pulsation $\omega =$ constant. The quantity kr represents the 'propagation phase shift'.

4.2.2 Equation of the Radiated Electrical Field E and Magnetic Field H

Referring to Figure 4.2, in the case of a wire antenna lying along the z axis, having a length l that is small with respect to the wavelength λ, called a *Hertzian dipole*, isolated in an unlimited space, where the radiation is produced by a current $i(t) = I_0 \sin \omega t$ of peak amplitude I_0 flowing

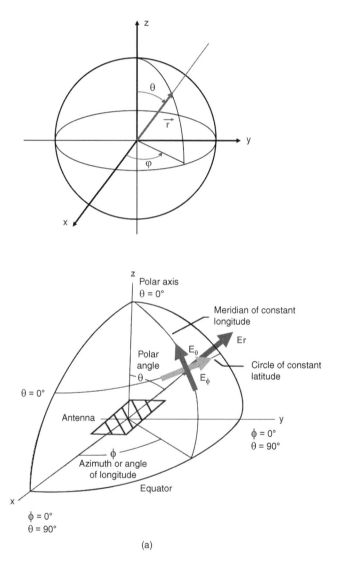

(a)

Figure 4.2   (a) Definition of the polar coordinates of the electrical field vector **E**. (b) Representation of the **E** and **H** vectors in polar coordinates

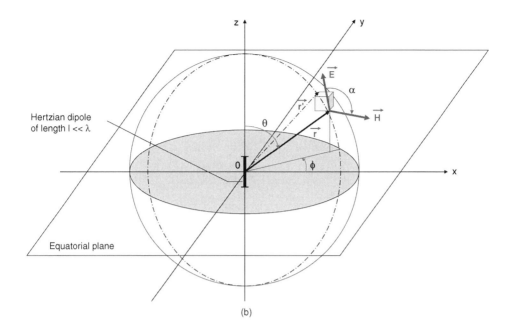

Figure 4.2   *(Continued)*

through the wire, we can show, in polar coordinates (in other words as a function of the distance r and of the angles θ, φ), that the relations yielding the components of the radiated fields – at a given instant – at a measurement point located at a distance r from the source O are those shown in Figure 4.2(b).

Regardless of the distance r, at a given point they form a set of two groups of three equations that are functions of r, θ and φ:

- for the magnetic field H:

$$H_r(t,r) = 0$$

$$H_\theta(t,r) = 0$$

$$H_\phi(t,r) = -\left[\frac{I_0 l \sin\theta}{4\pi} \left(\frac{1}{(jr)^2} + \frac{k}{jr} \right) \right] e^{j(\omega t - kr)}$$

- for the electrical field E:

$$E_r(t,r) = -\left[\frac{I_0 l \cos\theta}{2\pi\varepsilon_0\omega} \left(\frac{1}{(jr)^3} + \frac{k}{(jr)^2} \right) \right] e^{j(\omega t - kr)}$$

$$E_\theta(t,r) = -\left[\frac{I_0 l \sin\theta}{4\pi\varepsilon_0\omega} \left(\frac{1}{(jr)^3} + \frac{k}{(jr)^2} + \frac{k^2}{jr} \right) \right] e^{j(\omega t - kr)}$$

$$E_\varphi(t,r) = 0$$

Important

Because of the presence of j and the e^j, all these expressions represent values of complex numbers (remember that $\rho e^{j\varphi} = \rho(\cos\varphi + j\sin\varphi)$); see also Appendix 2 at the end of this chapter. We have also assumed that:

ε_0 = permittivity of the medium
c = speed (velocity of propagation in the medium in question)
$\omega = 2\pi f$
$\lambda = c/f$
$k = \dfrac{\omega}{c} = \dfrac{2\pi}{\lambda}$ = wave number

---

Those wishing to investigate this more closely should consult Appendix 1 at the end of the chapter.

Notes

1. Because of the symmetry of revolution about the z axis due to the position of the dipole on the z axis, all of the above equations are independent of the variable φ.
2. Because the values of H_r, H_θ and E_φ are zero, the scalar product $(\mathbf{E} \cdot \mathbf{H})$ of the vectors $\mathbf{E}$ and $\mathbf{H}$ is equal to 0. The vectors $\mathbf{E}$ and $\mathbf{H}$ are therefore orthogonal to each other, but not necessarily to r.

Important

Throughout the subsequent chapters I will use the value I_0 and will therefore be dealing with 'peak' values in all cases.

4.3 Classification of Fields and Regions of Space

As shown in the previous section, these basic equations depend on three main parameters:

- The time variable, t, as follows:

$$e^{j(\omega t - kr)} = e^{j\omega t}e^{-jkr}$$
$$e^{j(\omega t - kr)} = (\cos\omega t + j\sin\omega t)e^{-jkr}$$
$$e^{j(\omega t - kr)} = \frac{\cos\omega t + j\sin\omega t}{e^{jkr}} = \text{a sinusoidal function decreasing with distance}$$

- The variable r, according to the terms in $1/r$, $1/r^2$ and $1/r^3$. In cases where one or more of these terms can be disregarded by comparison with the others, the equations can be written in a simpler form.
- The product kr, defined as

$$kr = \frac{2\pi}{\lambda}r$$

4.3.1 Regions of Space

In the case of a very small antenna (i.e. where the sphere equivalent to the radiation source would have a very small diameter D), as a first approximation, it is customary to define three regions of space, namely the near, mid and far fields, according to the possible values of r with respect to the wavelength λ at which the system operates, in other words according to whether r is small or large with respect to $1/k = \lambda/(2\pi)$. For the sake of simplicity, starting from this value (as shown in Figure 4.3), I will demonstrate in a few paragraphs that two of the main parameters determining the phenomena of wave propagation (the impedance of air and the phase relationship between the vectors **E** and **H**) tend towards constant and stable values.

Near Field: $r \ll \lambda/(2\pi)$

In this region of space (in reality, $r < 0.63\sqrt{(D^3/\lambda)}r$), in other words in the immediate environment of the antenna, the values of the terms in $1/r^2$ or $1/r^3$ are much greater than those of the terms in $1/r$. This area, commonly called the Rayleigh region, is one of transmission by a strong transformer effect (where the received power P_r is practically equal to the emitted power P_e, unlike the situation in free space, where $P_r \ll P_e$).

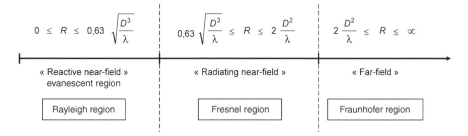

R = radius of the sphere
D = diameter of the antenna (a sphere of dimeter D including the antenna)
λ = wavelength associated with the frequency

(a)

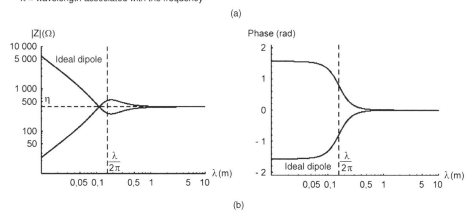

(b)

Figure 4.3 (a) Definition of the near and far fields. (b) Variation of the impedance of air and the phase E/H as a function of the near field ($r < \lambda/2\pi$) and the far field ($r > \lambda/2\pi$)

At a point very close to the antenna, the electrical field is produced by the antenna potential: this is the near field. At this distance, 'quasi-standing wave' conditions are present, and magnetic or electrical coupling can be established between a transponder and the source.

In my previous books (References 1 and 2 mentioned in the Preface), these matters were not dealt with in detail, since those books were mainly concerned with near field applications.

In this area, the radiated power flux density, which will be explained in detail a little further on (it is proportional to the scalar product of the moduli of **E** and **H**), decreases very rapidly, because it contains terms in $1/r^5$. As I will show subsequently, in this 'near field' region the value of the Poynting vector (the vector product **E** × **H**) can be imaginary, in which case no 'real' power is transmitted (only reactive power is transmitted), or it can be real, in which case power is transmitted.

Mid Field: r of the Same Order as λ

In this region of space (actually in the range from $0.63\sqrt{(D^3/\lambda)}$ and $2(D^2/\lambda)$), all the terms in $1/r$, $1/r^2$ and $1/r^3$ are also retained. Here, we find a radiating near field in which the Fresnel component is present when the main dimension of the antenna D is large with respect to the transmission wavelength λ but the distance r is of the same order as λ.

Far Field: r ≫ λ/(2π)

In this area (beyond $2(D^2/\lambda)$), also known as the Fraunhofer region, the terms in $1/r^2$ and $1/r^3$ become negligible, and the values of the fields E and H decrease according to $1/r$. This begins to occur at a distance r of approximately $\lambda/(2\pi)$ (a relation determined by the equation governing the real or imaginary value of the Poynting vector), since beyond this distance the field is due to a 'former' potential of the antenna. Thus we have left a quasi-standing wave region and entered a region of wave propagation and radiated field conditions. There is no longer any magnetic coupling effect.

In this region of space, as I shall show you, it is customary to define the radiation pattern of the antenna (which is omnidirectional for a very small loop antenna). Also, as I shall show you, in an electromagnetic field of an antenna where the predominant field components are those representing energy propagation, the angular distribution of the field is essentially independent of the distance from the transmitting antenna.

In these regions, the distribution is not affected by the structure of the antenna and the wave is propagated in the form of a plane wave. In the 'far field' region, the value of the Poynting vector (the vector product **E** × **H**) is real and power can be transferred.

4.3.2 Some Remarks on RFID Applications

Following these general statements – which are true at a given time for remotely powered systems, for existing semiconductor technology and for the current regulations – it will be useful to be aware of the 'near' or 'far' field regions in which a planned RFID application is to operate. Here, then, are some tables summarizing the relations between frequencies and the associated distances $\lambda/(2\pi)$, assuming a speed of light c of $3 \times 10^8 \, \mathrm{m\,s^{-1}}$.

LF and HF

At LF and HF (see Table 4.1), the desired and/or possible operating distances for RFID applications are always much smaller than the value of $\lambda/(2\pi)$. In these applications, therefore,

Table 4.1

	Frequency (MHz)	Wavelength (m)	$\lambda/(2\pi)$ (m)
LF	0.125	2400	382
HF	13.56	22.12	3.52

the tags operate in the 'near field' and the applications are primarily based on the principles of inductive loops and inductive magnetic coupling.

UHF and SHF

At UHF and SHF (see Table 4.2), the desired and/or possible operating distances for RFID applications are always much greater than the value of $\lambda/(2\pi)$. In these applications, therefore, the tags operate in the 'far field', and must therefore use principles of coupling with the base station other than the magnetic coupling principle, e.g. the phenomenon of wave propagation and reflection.

Table 4.2

	Frequency (MHz)	Wavelength (cm)	$\lambda/(2\pi)$ (cm)
UHF	433	69.2	11.02
	866	34.6	5.51
	915	32.8	5.22
SHF	2450	12.2	1.94
	5890	5.1	0.8

Summary
Where

$$\lambda = \frac{c}{f} = \frac{3 \times 10^8}{f}$$

the **near field (magnetic coupling)** $< \lambda/(2\pi) <$ **far field (wave propagation)** (see Table 4.3).

A Note on the Characteristics of SHF Waves

Some electromagnetic waves on the UHF–SHF boundary (mainly around 2.45 GHz) are strongly absorbed by water and human tissue. In a moist medium, therefore, the attenuation of microwaves may be very considerable. UHF (ultra high frequencies), used particularly in the USA and Australia for RFID applications in line with existing RF regulations, are not absorbed, or only slightly absorbed, by water molecules. Because of the small wavelengths used at these

Table 4.3

Frequency, f	Wavelength, λ	RFID operation in
150 kHz	2 km	Near field
10 MHz	30 m	Near field
900 MHz	33 cm	Far field
3000 MHz $=$ 3 GHz	10 cm	Far field

frequencies (at 2.45 GHz, $\lambda = 12$ cm), a smaller surface area is needed for the antenna, thus providing a high degree of freedom in the design of tags.

As will be shown in Chapter 13, the use of spread spectrum techniques such as DSSS and FHSS, when feasible, can often provide a high apparent immunity to noise and interference at these frequencies, which is why they are often used in industrial environments.

4.4 RFID Applications Using UHF and SHF, i.e. Far Field Applications

4.4.1 Values of E and H in the Far Field

As mentioned above, UHF and SHF RFID applications operate in the far field in almost all cases. In these conditions, the value of r is large and the effect of the term $1/r$ is predominant over the terms in $1/r^2$ and $1/r^3$. The equations for E and H can therefore be simplified, so the general far field equations become

$$E_\theta(t, r) = -\left(j \frac{I_0 l \sin\theta \, k^2}{4\pi\varepsilon_0\omega} \frac{}{r} \right) e^{j(\omega t - kr)}$$

$$H_\phi(t, r) = -\left(j \frac{I_0 l \sin\theta \, k}{4\pi} \frac{}{r} \right) e^{j(\omega t - kr)}$$

since the other four terms E_r, E_φ, H_r and H_θ present in the equations are negligible. Since the radial component E_r of the electrical field is zero, the wave is said to be 'transverse electromagnetic' (TEM) (Figure 4.4).

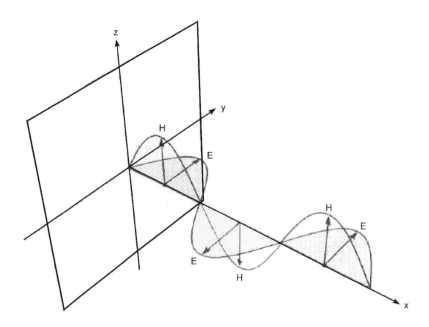

Figure 4.4   Far field, plane wave

Notes
Because of the symmetry of revolution about the z axis due to the position of the dipole on the z axis itself, for the same r and the same θ the two equations above are entirely independent of the variable φ. Because the values of E_r, E_φ, H_r, H_θ are negligible, the scalar product $(\mathbf{E} \cdot \mathbf{H})$ is equal to 0. The vectors $\mathbf{E}$ and $\mathbf{H}$ are therefore orthogonal to each other. Also, because E_r is now equal to 0 as well, each of the planes of the vectors $\mathbf{E}$ and $\mathbf{H}$ must also be perpendicular to the vector $\mathbf{r}$. The propagation of the radiated wave is therefore orthogonal to the plane of the vectors $\mathbf{E}$ and $\mathbf{H}$.

Finally, if we replace k with $2\pi/\lambda$, we have

$$E_\theta(t, r) = -\left(j\frac{\pi I_0 l \sin\theta}{\lambda^2 \varepsilon_0 \omega}\frac{1}{r}\right)e^{j(\omega t - kr)}$$

$$H_\phi(t, r) = -\left(j\frac{I_0 l \sin\theta}{2\lambda}\frac{1}{r}\right)e^{j(\omega t - kr)}$$

and $E_\theta(t, r)$ and $H_\varphi(t, r)$ are orthogonal to each other. Replacing ω with its value $\omega = 2\pi f = 2\pi/T$, and given that $\lambda = cT$, then $\omega = 2\pi c/\lambda$. Given also that

$$\varepsilon_0 = \frac{1}{36\pi} \times 10^{-9}\,\mathrm{F\,m^{-1}}$$

and $c = 3 \times 10^8\,\mathrm{m\,s^{-1}}$, we have

$$E_\theta(t, r) = -\left(j\frac{60\pi I_0 l \sin\theta}{\lambda}\frac{1}{r}\right)e^{j(\omega t - kr)}$$

$$H_\phi(t, r) = -\left(j\frac{I_0 l \sin\theta}{2\lambda}\frac{1}{r}\right)e^{j(\omega t - kr)}$$

which are expressions of the complex variable whose moduli are

$$|E_\theta| = \frac{60\pi I_0 l \sin\theta}{\lambda r}$$

$$|H_\varphi| = \frac{I_0 l \sin\theta}{2\lambda r}$$

Additionally, in order to simplify the calculations and the forms of the equations, this book will mainly be concerned with the equatorial plane of the antenna (defined for $\theta = \pi/2$ [$\sin\theta = 1$]), in other words the plane perpendicular to the axis of the dipole. On these assumptions (far field [$r \gg \lambda/(2\pi)$] and equatorial plane), the equations for the moduli (and thus for the peak values, because of the I_0) of the fields E and H are considerably simplified, as follows:

$$|E_\theta| = \frac{60\pi I_0 l}{\lambda r}$$

$$|H_\varphi| = \frac{I_0 l}{2\lambda r}$$

4.4.2 Relations between E and H, Impedance of Vacuum (or Air)

Again on the above assumptions, for the far field, we can calculate the ratio E_θ/H_φ:

$$\frac{E_\theta}{H_\phi} = \frac{\left(j\frac{I_0 l \sin\theta}{4\pi\varepsilon_0\omega}\frac{k^2}{r}\right)e^{j(\omega t - kr)}}{\left(j\frac{I_0 l \sin\theta}{4\pi}\frac{k}{r}\right)e^{j(\omega t - kr)}}$$

After simplification of the numerator and denominator, regardless of the value of θ, we have

$$\frac{E_\theta}{H_\varphi} = \frac{k}{\varepsilon_0\omega}$$

regardless of the values of t and r, where ε_0 is the permittivity of the medium and c is the speed (velocity of propagation). Thus,

$$k = \frac{\omega}{c} = \frac{2\pi}{\lambda}$$

and therefore

$$\frac{E_\theta}{H_\varphi} = \frac{1}{\varepsilon_0 c}$$

Because of the presence of the parameters ε_0 and c in this equation, the value of this ratio is constant and represents what is called the impedance of the vacuum (or of air), denoted Z_0. Thus,

$$Z_0 = \frac{1}{\varepsilon_0 c}\Omega$$

and, therefore, finally

$$E_\theta = Z_0 H_\varphi \, V \, m^{-1}$$

where Z_0 is the characteristic impedance of the vacuum (or air) in Ω, E_θ is the electrical field strength in $V\,m^{-1}$ and H_φ is the magnetic field strength in $A\,m^{-1}$.

Given also that the relation between the propagation speed (velocity) c, the permeability of the vacuum μ_0 and the permittivity of the vacuum ε_0 is

$$\varepsilon_0\mu_0 c^2 = 1$$

therefore

$$c = \frac{1}{\sqrt{\varepsilon_0\mu_0}}$$

Replacing c by its value in $Z_0 = 1/(\varepsilon_0 c) = \sqrt{\varepsilon_0\mu_0}/\varepsilon_0$, we have

$$Z_0 = \sqrt{\frac{\mu_0}{\varepsilon_0}}$$

and therefore, with $\mu_0 = 4\pi \times 10^{-7}\,\mathrm{H\,m^{-1}}$ and $\varepsilon_0 = 1/(36\pi \times 10^{-9})\,\mathrm{F\,m^{-1}}$, we have

$$Z_0 = \sqrt{\frac{4\pi \times 10^{-7}}{1/(36\pi \times 10^{-9})}} = 120\,\pi\,\Omega = 376.939\,\Omega = 377\,\Omega$$

Note

This result was quite predictable, since, by using the formulae obtained at the end of the preceding paragraphs,

$$|E_\theta| = \frac{60\pi I_0 l}{\lambda r} \text{ in the equatorial plane}$$

$$|H_\varphi| = \frac{I_0 l}{2\lambda r} \text{ in the equatorial plane}$$

it would have been easy to determine the ratio (E_θ/H_φ) directly and immediately find

$$Z_0 = E_\theta/H_\varphi = 120\pi\,\Omega = 377\,\Omega$$

Why be straightforward when you can be complicated? This method would have been less interesting ... and you would never have known that Z_0 was also equal to $\sqrt{(\mu_0/\varepsilon_0)}$. As they say, no pain, no gain!

Other Common Forms of the Relation Between E and H

Very frequently, for practical reasons, the relation between the moduli of the vectors **H** and **E** is expressed, not in absolute values but in the form of relative values, in other words in dB, dBm or dBµ. Given that

$$H = E/Z_0\ \mathrm{A\,m^{-1}}$$

we can write, in dB,

$$20\log H = 20\log(E/Z_0)$$
$$H(\text{in dB A m}^{-1}) = E(\text{in dB V m}^{-1}) - 20\log Z_0$$
$$H(\text{in dB A m}^{-1}) = E(\text{in dB V m}^{-1}) - 20\log(377)$$
$$= E(\text{in dB V m}^{-1}) - 40 - (20 \times 0.576)$$

and thus, finally, in the far field and in the equatorial plane:

$$H(\text{in dB A m}^{-1}) = E(\text{in dB V m}^{-1}) - 51.5\,\text{dB}$$

This relation describing the change from E to H – reflecting the influence of the impedance of a vacuum on the propagation of the signal – is well known and widely used, e.g. as a simple way of transposing the radiated values when determining conformity with ETSI radiation standards. (If necessary, you should consult my earlier book, Reference 1 mentioned in the Preface.)

Given the absolute values of the radiated fields, their equivalent values expressed in dBµ A m^{-1} for H and in dBµ V m^{-1} for E are very commonly used, as follows:

$$H \text{ (in dBμ A m}^{-1}) = 20 \log \frac{H \text{ (in A m}^{-1})}{1 \times 10^{-6}}$$

$$H \text{ (in dBμ A m}^{-1}) = 20 \log [H(\text{in A m}^{-1}) \times 10^6] = H(\text{in dB A m}^{-1}) + 120 \text{ dB}$$

and therefore

$$H(\text{in dB A m}^{-1}) = H(\text{in dBμ A m}^{-1}) - 120 \text{ dB}$$

By similar reasoning, we would have found values in dBμ V m^{-1} for E as follows:

$$E(\text{in dB V m}^{-1}) = E(\text{in dBμ V m}^{-1}) - 120 \text{ dB}$$

Therefore, once again

$$H(\text{in dBμ A m}^{-1}) = E(\text{in dBμ V m}^{-1}) - 51.5 \text{ dB}$$

4.4.3 Electromagnetic Power Density, Power and Energy

Poynting Vector
When describing propagation phenomena, we usually define a vector **S**, called the *Poynting vector*, by the following vector relation:

$$\mathbf{S} = (\mathbf{E} \times \mathbf{H})$$

an equation in which $\times$ represents the mathematical operation of finding the 'vector product'. By definition, regardless of the distance (far or near field), this vector is always perpendicular to the plane of the vectors **E** and **H**, since it is determined by the vector product (**E**×**H**) and its direction indicates the direction of radiation of the wave. This vector is also called the *radiant vector* (Figure 4.5).

The value of the modulus of the vector **S**, denoted $|S| = s_{\text{peak}}$, is therefore

$$s_{\text{peak}} = |\mathbf{S}| = |E| \, |H| \sin \alpha$$

where α is, by definition, the oriented angle between the vectors **E** and **H**, from **E** towards **H**; s_{peak} represents the maximum elongation of the value of the modulus of the vector **S**, since its equation allows for the values of **E** and **H** directly related to I_0, which is the maximum (peak) value of the current flowing in the dipole.

The physical magnitude associated with the value of s_{peak} is a power density per m^2 or, stated more precisely, a 'power flux density' in watts$_{\text{peak}}$ per m^2 (W$_{\text{peak}}$ m^{-2}), since |E| is expressed in V$_{\text{peak}}$ m^{-1} and |H| is expressed in A$_{\text{peak}}$ m^{-1}, i.e. the scalar product |E| |H| sin α in V A$_{\text{peak}}$ m^{-2} = W$_{\text{peak}}$ m^{-2}. Thus the value $|S| = s_{\text{peak}}$, the modulus of the Poynting vector **S**, represents a power flux density, in W$_{\text{peak}}$ m^{-2}.

Having defined the vector **S** and given its modulus a physical meaning, I will now go on to consider a curious entity, which is the value of the 'flux of this vector'. You may well wonder what this is.

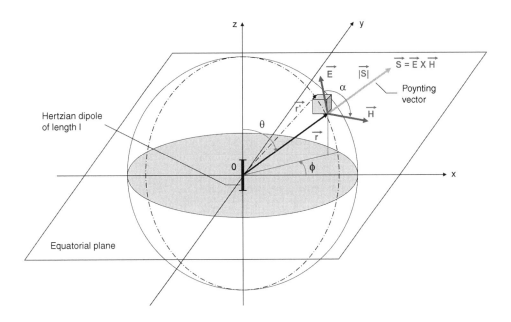

Figure 4.5 The Poynting vector in polar coordinates

The Flux of the Poynting Vector or the Flux of the Vector Quantity S across a Surface
The concept of the flux of a vector is related to the idea of a volumetric flow rate. To demonstrate
this by analogy (Figure 4.6), imagine a net stretched across a river, perpendicularly to the
velocity vector **v** of the current (expressed in $m\,s^{-1}$), which is assumed to be constant over
the whole surface $\Delta\sigma$ of one mesh of the net (expressed in m^2).

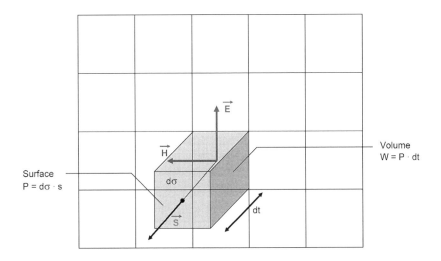

Figure 4.6 Flux of the vector of radiated electromagnetic power flux density

The quantity of water, i.e. the volume in litres passing through one mesh in the time interval Δt (in seconds) is proportional to:

- the modulus $|v|$ of the velocity of the current;
- the surface area of the mesh $\Delta\sigma$;
- the time interval Δt.

This quantity is denoted by τ (tau), where $\tau = |v|\, \Delta\sigma\, \Delta t$, and is therefore expressed in m s^{-1} · m^2 s = m^3). Therefore τ represents the volume of water contained in a parallelogram with a base of $\Delta\sigma$ and length $l = |v|\, \Delta t$.

This is simple enough, so far. Now it gets complicated.

The 'flux of the velocity vector $\mathbf{v}$' of the water through the mesh with an area of $\Delta\sigma$, for a quantity $\Delta\phi$, is then defined as the volume of water passing through the surface of the mesh per unit of time ($\Delta t = 1$ unit of time), i.e. $\Delta\phi = |v|\, \Delta\sigma$, a value expressed in m s^{-1} · m^2 s = m^3 s^{-1}. It is therefore a volume flow rate per second or, more simply, a volume flow rate, or just the 'flow rate' (per unit of time, of course).

Notes

If the mesh is not perpendicular to the flow of the river, then less water will pass through the equivalent surface and the flux will be reduced to $\Delta\phi = |v|\, \Delta\sigma \cos\theta$. In order to know which side of the mesh the water is flowing from, the surface must be orientated, thus becoming a vector $\Delta\sigma$.

The total flux, ϕ_{total}, i.e. the global flow rate of the river – averaged over all the meshes of the net, regardless of the position of each mesh with respect to the current – is usually written as

$$\phi_{\text{total}} = \iint |v|\,\mathrm{d}\sigma$$

Let us return to our own (electromagnetic) field. Forget the river, however tempting it looks, and go back to the Poynting vector $\mathbf{S}$ (power flux density). $\mathbf{S}$ represents the equivalent of the velocity vector described in the preceding analogy. Let us assume that the net is stretched perpendicularly to the Poynting vector $\mathbf{S}$ (which is expressed in W_{peak} m^{-2}) and that this vector is constant over the surface $\Delta\sigma$ of each mesh (in m^2). A quantity x (expressed in units y) passing through a mesh in the time interval Δt (in seconds) will be proportional to:

- the modulus $s_{\text{peak}} = |\mathbf{S}|$ of the power flux density;
- the surface area of the mesh $\Delta\sigma$;
- the time interval Δt.

This quantity is denoted by $\tau = s_{\text{peak}}\, \Delta\sigma\, \Delta t$ and is thus expressed as

$$(W_{\text{peak}}\, m^{-2} \cdot m^2\, s) = W_{\text{peak}} s = J_{\text{peak}} = \text{joule}_{\text{peak}}$$

Therefore τ represents the peak electromagnetic energy (due to the two vectors $\mathbf{E}$ and $\mathbf{H}$ from which the Poynting vector $\mathbf{S}$ originated) contained in the parallelogram with a base of $\Delta\sigma$ and a length of $s_{\text{peak}}\, \Delta t$.

Similarly, if we use $\Delta\phi$ to denote the flux of the Poynting vector $\mathbf{S}$ (power flux density) through a mesh with an area of $\Delta\sigma$, then the quantity $\Delta\phi = s_{\text{peak}}\, \Delta\sigma$, a value expressed in W_{peak} m^{-2} · m^2 = W_{peak} passing through the surface of the mesh per unit of time ($\Delta t = 1$ unit of time). The flux $\Delta\phi$ of the Poynting vector thus represents a rate of flow of electromagnetic energy (J_{peak}) per second, which of course is much better known under the name of radiated electromagnetic POWER (peak):

$$\phi_{\text{total}} = \iint s_{\text{peak}} d\sigma$$

a double integral calculated along the surface $\Sigma = $ peak radiated electromagnetic power (in W_{peak}). The flux $\Delta\phi$ of the Poynting vector represents a radiated electromagnetic power. Its value is obviously expressed in W_{peak}.

Note that, as mentioned above, the direction of the Poynting vector indicates the direction of propagation, and therefore the electromagnetic energy also flows in the direction of propagation. It is as inevitable as the flow of the aforementioned river.

The Poynting Theorem

To complete this lengthy explanation, which is essential for an understanding of the subsequent text, I will now generalize these concepts of power flux density and radiated electromagnetic power and energy by stating the Poynting theorem, which has been a hidden presence throughout the previous paragraphs.

At any point of a volume v of an isotropic linear medium, the elementary electromagnetic energy volume density w can be written thus:

$$\frac{\partial W}{\partial v} = w \, \text{J m}^3$$

This elementary volume density is equal to the sum of the electrostatic energy $\frac{1}{2}\varepsilon E^2$ and electromagnetic energy $\frac{1}{2}\mu H^2$ in the presence of

$$w = w_e + w_m$$

$$w = \frac{1}{2}\varepsilon E^2 + \frac{1}{2}\mu H^2$$

Clearly, the total energy W contained in the volume v is found by integration.

Given that $dW = w \, dv$ and $dW = w_e \, dv + w_m \, dv$, we find

$$W = \iiint w \, dv = \iiint (w_e + w_m) dv = W_e + W_m \, \text{J}$$

In the case of plane waves, as mentioned above, the mathematical relations associated with Maxwell's equations can be used to define an equality that is called the Poynting theorem, which is stated below.

The Poynting theorem

The instantaneous electromagnetic power (i.e. the variation of the sum of the electromagnetic energy W_m and the electrostatic energy W_e during the time interval dt – in fact, the decrease of energy, hence the '$-$' sign in the equation – passing through a closed surface Σ, i.e. contained in a volume dv) is equal to the flux of the Poynting vector $\mathbf{S} = \mathbf{E}_z \times \mathbf{H}_y$ passing through this closed surface, which limits this volume, or, in the form of an equation,

$$\text{Electromagnetic power} = -\frac{\partial W}{\partial t} = \iiint \frac{\partial}{\partial t}(w_m + w_e)\partial v = -\iint \frac{E_z H_y}{4\pi} \partial\sigma$$

> **Note**
> The presence of the divisor 4π in the denominator of the double integral is due to the fact that the integration according to $d\sigma$ is calculated over the whole surface Σ of any sphere of radius r whose surface area is $4\pi r^2$.

Summary

- The vector product $\mathbf{S} = \mathbf{E} \times \mathbf{H}$, called the Poynting vector, defines a power flux density s (expressed in $\mathrm{W\,m^{-2}}$) at one point in space.
- The derivative of the electromagnetic energy W (in J) contained in a volume with respect to time is the radiated electromagnetic power P (in W).
- The Poynting theorem states that the latter quantity (radiated electromagnetic power P) is also equal to the flux of the Poynting vector $\mathbf{S}$, where s is the power flux density.

To put this in another way:

$$\text{Electromagnetic power} = \text{flux of the Poynting vector}$$

$$= \iint \text{power flux density} \cdot d\sigma \, (\text{integral taken over the closed surface})$$

$$= \iiint \frac{dW}{dt} dv \, (\text{integral taken over the volume}$$

$$\text{contained by the above closed surface})$$

> **Note**
> I have been careful to use the term power 'flux density' throughout the preceding paragraphs. The simple term power 'density' is commonly used for the sake of simplicity. However, although the full term might appear rather cumbersome, I shall continue to use it, because it clearly indicates the difference between the values of s (power flux density) and P (radiated electromagnetic power), which are all too frequently confused by many users.

4.4.4 Radiated 'Instantaneous Electromagnetic' Power

I showed in the preceding paragraph that the flux of the Poynting vector S passing through this surface represents the instantaneous radiated 'electromagnetic' power passing through a closed surface (i.e. a volume). I shall now calculate its actual value. This is a three-stage process:

1. Now that we know the full form of the equations for the vectors $\mathbf{E}$ and $\mathbf{H}$, we can calculate the Poynting vector $\mathbf{S}$:

$$\mathbf{S} = \mathbf{E} \times \mathbf{H}$$

2. From this we can deduce the peak value of its associated modulus $s_{\text{peak}} = |S|$.
3. We can then deduce its mean value, denoted $\langle s_{\text{peak}} \rangle$.

4. Finally, we can calculate the mean flux $\langle \Delta\phi_{peak} \rangle$ of $\langle s_{peak} \rangle$, whose value is in fact the value of the total mean radiated power, $\langle P_{peak\ radiated} \rangle$, within this volume:

$$\langle P_{peak\ radiated} \rangle = \iint \langle s_{peak} \rangle d\sigma, \text{ a double integral calculated along the surface } \Sigma$$

where Σ is a closed surface of any size and shape that completely surrounds the transmitting antenna and $d\sigma$ is an element of the surface Σ.

Note

The calculation of the flux of the Poynting vector **S** is only useful if the surface in question is an element of a sphere of any radius r, which then only requires the calculation of the radial component S_r of the Poynting vector. Therefore, since

$$H_\theta = E_\varphi = 0$$

we find that

$$S = S_r = E_\theta H_\varphi$$

Following this highly theoretical work, we can move on at last to some practicalities, by carrying out a specific calculation of the kind used daily in RFID applications operating at UHF and SHF.

Electromagnetic Power Radiated in the Far Field by a Hertzian Dipole

To calculate the electromagnetic power radiated in the far field by a Hertzian dipole, we must first calculate the mean value of the Poynting vector.

Mean Value, $\langle |S_r| \rangle$ ($\langle s_{r\ peak} \rangle$), of the Poynting Vector in the Far Field

In the context of the UHF and SHF RFID applications with which we are concerned, and therefore still with reference to the far field ($r \gg \lambda/(2\pi)$, $1/r^2$ and $1/r^3 \lll 1/r$), since $H_\theta = E_\varphi = 0$, as I have demonstrated, the equations for the E and H fields presented above are simplified and can be written thus:

$$E_\theta(t, r) = j\left(\frac{I_0 l}{4\pi\varepsilon_0\omega} \frac{k^2}{r}\right) \sin\theta e^{j(\omega t - kr)}$$

$$H_\phi(t, r) = j\left(\frac{I_0 l}{4\pi} \frac{k}{r}\right) \sin\theta e^{j(\omega t - kr)}$$

The above equations indicate that these quantities $E_\theta(t, r)$ and $H_\varphi(t, r)$ depend on the complex variable $j(\omega t - kr)$ and therefore represent exponentially decreasing sinusoidal functions. They also show that, in these far field conditions, the vectors $E_\theta(t, r)$ and $H_\varphi(t, r)$ are orthogonal to each other.

We know that the Poynting vector is

$$S = S_r = E_\theta \times H_\varphi$$

and that its modulus s (power flux density) is therefore

$$s_r = |S_r| = |E|\,|H|\sin \alpha$$

The vectors $E_\theta(t, r)$ and $H_\varphi(t, r)$ are therefore orthogonal to each other ($\alpha = 90°$, $\sin \alpha = 1$):

$$s_r = |S_r| = |E|\,|H|\sin \alpha = |E|\,|H|$$

Important note
You should note that the value of s_r above is equal to the scalar product of the moduli of E and H, which, for the whole of this chapter so far, have been used to represent the maximum (peak) amplitudes, not the r.m.s. amplitudes, of the E and V fields. Therefore s_r represents the maximum (peak) amplitude of the Poynting vector.

At this stage, in the present case, if we wish to be very precise we must write

$$s_{r\text{peak}} = |S_r| = |E_\theta|\,|H_\varphi|$$

If we take $H_\varphi^*(t, r)$ as the 'conjugate' value of $H_\varphi(t, r)$, i.e.

$$H_\phi^*(t, r) = \frac{I_0 l}{4\pi}\frac{-jk}{r}\sin\theta e^{-j(\omega t - kr)}$$

and using $\langle \dots \rangle$ to denote the 'mean value of $\dots$' of a variable, then Appendix 3 of this chapter (to be read very carefully) provides a detailed proof that the mean value of a product of two sinusoidal functions (in phase) is equal to half the scalar product of the value of the moduli (i.e. of the maximum peak values) of the two functions, or to half the real value of the scalar product of one function by the conjugate of the other; in other words,

$$\langle s_{\text{peak}} \rangle = \text{mean of the modulus of } |S_r|$$

$$|S_r| = \langle s_{\text{peak}} \rangle = \tfrac{1}{2}|E|\,|H|$$

expressed in peak mean watts per m^2 (W$_{\text{peak mean}}$ m^{-2}) and $|E|$ and $|H|$ expressed in peak values:

$$|S_r| = \langle s_{\text{peak}} \rangle = \tfrac{1}{2}\,\text{Re}\,(E_\theta H_\phi^*) = \tfrac{1}{2}\,\text{Re}\,(E_\theta^* H_\phi)$$

where Re signifies 'real part of $\dots$'.

Note
As indicated in Appendix 3 of this chapter, the presence of the factor $\frac{1}{2} = (1/\sqrt{2} \times 1/\sqrt{2})$ provides a way of moving from the peak values of the moduli of the vectors $\mathbf{E}$ and $\mathbf{H}$ to the maximum (peak) mean value of the vector $\mathbf{S_r}$; $\langle s_{\text{peak}} \rangle$ is a mean, permanent, nonfluctuating value (continuous), which is expressed in W m^{-2}. For the sake of clarity, and to underline its physical significance, I will show its value in W$_{\text{mean peak}}$ m^{-2}, to emphasize that it represents the highest value of the mean of the modulus of the Poynting vector.

Now let us calculate the value of the following scalar product:

$$E_\theta H_\phi^* = \left(j\frac{I_0 lk^2 \sin\theta}{4\pi\varepsilon_0\omega r}e^{j(\omega t - kr)}\right)\left(-j\frac{I_0 lk \sin\theta}{4\pi r}e^{-j(\omega t - kr)}\right)$$

$$E_\theta H_\phi^* = \left[\left(\frac{1}{16}\frac{k^3}{\pi^2\varepsilon_0\omega}\right)\frac{I_0^2 l^2 \sin^2\theta}{r^2}\right]e^0$$

Replacing k by its value ω/c, the product $\varepsilon_0 c$ by $1/Z_0$ and finally ω by $2\pi f$ in the first member of the above equation, and given that $e^0 = 1$, we obtain

$$E_\theta H_\phi^* = \left(\frac{1}{16}\frac{Z_0 \times 4f^2}{c^2}\right)\frac{I_0^2 l^2 \sin^2\theta}{r^2}$$

Note that this expression is entirely 'real' and not 'imaginary' in nature.

Now we only need to multiply this result by $1/2$ to obtain the literal expression of the mean (peak) power flux density $\langle s_{peak}\rangle$; the radiated peak mean power flux density of a Hertzian dipole is therefore

$$\langle s_{peak}\rangle = \langle |S_r|\rangle = {}^1/_2\,\mathrm{Re}(E_\theta H_\phi^*)$$

$$\langle s_{peak}\rangle = \frac{Z_0}{8c^2}I_0^2 l^2 f^2 \frac{\sin^2\theta}{r^2}W_{peak\ mean}m^{-2}$$

The equation $\langle s_{peak}\rangle = f(r, \theta)$ for a point in space located at a distance r from the radiation source is frequently represented in polar coordinates.

The graph in Figure 4.7 shows the variations of $\langle s_{peak}\rangle$. This figure indicates:

- that the radiation pattern of the radiated maximum mean power flux density $\langle s_{peak}\rangle$ is not identical in all directions, and thus it is not in any way 'isotropic';

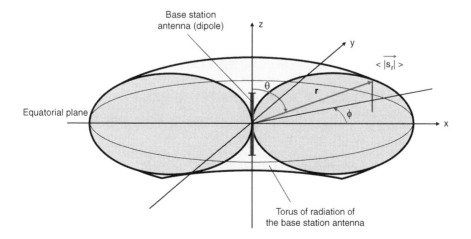

Figure 4.7 Representation of the theoretical diagram of power flux density radiation of a Hertzian dipole

- that it is zero on the principal axis of the dipole;
- that it is maximal in the equatorial plane perpendicular to the plane of the dipole;
- that this antenna is therefore 'directional' and has a gain in the equatorial plane (the method of calculating its value will be described in Chapter 6).

Now let us replace Z_0 by its numerical value in the above equation; given that $\lambda = cT = c/f$, we find

$$\langle s_{\text{peak}} \rangle = \frac{15\pi}{\lambda^2} I_0^2 l^2 \frac{\sin^2 \theta}{r^2} \, \text{W}_{\text{peak mean}} \text{m}^{-2}$$

This equation clearly shows that, in the far field, for a given frequency, the power flux density is inversely proportional to the square of the distance.

Note

We could have found these values directly from the equations presented previously:

$$E_\theta = \frac{60\pi I_0 l}{\lambda r} \sin \theta$$

$$H_\varphi = \frac{I_0 l}{2\lambda r} \sin \theta$$

and therefore

$$\langle s_{\text{peak}} \rangle = \langle |S_r| \rangle = {}^1\!/_2 |E| \, |H|$$

which is expressed in mean (peak) watts per m^2 thus:

$$\langle s_{\text{peak}} \rangle = \langle |S_r| \rangle = \frac{15\pi}{\lambda^2} I_0^2 l^2 \frac{\sin^2 \theta}{r^2} \, \text{W}_{\text{mean peak}} \text{m}^{-2}$$

Simple is best, sometimes.

Example

Let us see what happens in the equatorial plane. In this case, $\theta = \pi/2$; therefore $\sin^2 \theta = 1$:

$$\langle s_{\text{max dipole}} \rangle = \langle |S_r| \rangle = 15\pi \left(\frac{I_0 l}{\lambda r} \right)^2 \text{W}_{\text{mean peak}} \text{m}^{-2}$$

Total Mean Radiated Electromagnetic Power in the Far Field

Now that we have calculated the peak mean value of the mean power flux density $\langle s_{\text{peak}} \rangle$, we can calculate the total peak mean power $\langle P_{\text{peak}} \rangle$ radiated in the far field. I gave you the general formula for this above. Applying it to our specific case of mean values, we find

$$\langle P_{\text{peak}} \rangle = \iint \langle s_{\text{peak}} \rangle \, d\sigma$$

a double integral calculated along the surface Σ expressed in $W_{\text{mean peak}}$, where $\langle s_{\text{peak}} \rangle$ is the peak mean power flux density (in $W\,m^{-2}$), Σ is a closed surface of any size and shape completely surrounding the transmission antenna and $d\sigma$ is an element of the surface Σ.

The peak total mean power passing through a sphere having a centre O and any radius r (i.e. the flux [of the mean value] of the Poynting vector over this surface) is equal to the power radiated by the Hertzian dipole:

$$\langle P_{\text{peak}} \rangle = \iint \langle s_{\text{peak}} \rangle d\sigma$$

$$\langle P_{\text{peak}} \rangle = \iint \frac{Z_0}{8c^2} I_0^2 l^2 f^2 \frac{\sin^2 \theta}{r^2} d\sigma$$

Given that, in polar coordinates, the surface element $d\sigma$ of a sphere is written as follows (Figure 4.8):

$$d\sigma = (r d\theta)(r \sin \theta \, d\varphi) = (r^2 \sin \theta \, d\theta \, d\varphi)$$

we find

$$\langle P_{\text{peak}} \rangle = \iint \frac{Z_0}{8c^2} I_0^2 l^2 f^2 \frac{\sin^2 \theta}{r^2} (r^2 \sin \theta \, d\theta) d\varphi$$

$$\langle P_{\text{peak}} \rangle = \frac{Z_0}{8c^2} I_0^2 l^2 f^2 \iint (\sin^2 \theta \sin \theta) d\theta \, d\varphi$$

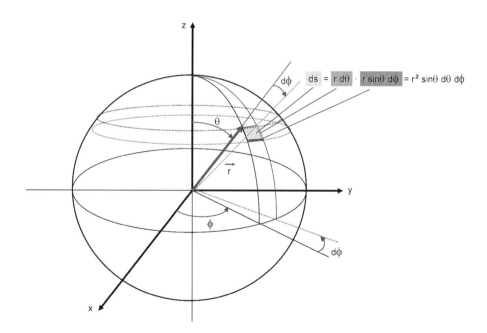

Figure 4.8 Surface element in polar coordinates

$$\langle P_{\text{peak}} \rangle = \frac{Z_0}{8c^2} I_0^2 l^2 f^2 \iint (\sin^3 \theta) d\theta \, d\varphi$$

Recalling the trigonometric identity:

$$\sin 3\theta = 3 \sin \theta - 4 \sin^3 \theta$$

and therefore that

$$\sin^3 \theta = \tfrac{1}{4}(3 \sin \theta - \sin 3\theta)$$

then

$$\int \sin^3 \theta \, d\theta = \frac{1}{4} \int (3 \sin \theta - \sin 3\theta) d\theta$$

$$\int \sin^3 \theta \, d\theta = -\tfrac{3}{4} \cos \theta + \tfrac{1}{12} \cos 3\theta$$

Therefore, calculating the value of the first integral for θ varying from 0 to 2π in the vertical plane (Figure 4.8):

$$\int \sin^3 \theta \, d\theta = (-\tfrac{3}{4} \cos \pi + \tfrac{1}{12} \cos 3\pi) - (-\tfrac{3}{4} \cos 0 + \tfrac{1}{12} \cos 0)$$

$$\int \sin^3 \theta \, d\theta = (\tfrac{3}{4} - \tfrac{1}{12}) - (-\tfrac{3}{4} + \tfrac{1}{12}) = \tfrac{8}{6}$$

Then, using the procedure for the second integration for φ varying from 0 to 2π in the horizontal plane (Figure 4.8),

$$\int \sin^3 \theta \, d\varphi = 2 \times \frac{8\pi}{6} = \frac{8\pi}{3}$$

we finally obtain

$$\int (\sin^3 \theta) d \, \theta d\varphi = \frac{8\pi}{3}$$

and, therefore, in conclusion,

$$\langle P_{\text{peak}} \rangle = P_{\text{mean tot peak (dipole)}} = \frac{\pi Z_0}{3c^2} I_0^2 l^2 f^2 \, W_{\text{mean peak}}$$

which is a value independent of r or, alternatively, replacing Z_0 with its value:

$$P_{\text{mean tot peak (dipole)}} = \frac{1}{12\pi\varepsilon_0 c^3} I_0^2 l^2 \omega^2 \, W_{\text{mean peak}}$$

If we replace the literal values of Z_0 and c by their respective numerical values, the above equation can also be written as

$$P_{\text{mean tot peak (dipole)}} = 40\left(\frac{\pi I_0 l}{\lambda}\right)^2 W_{\text{mean peak}}$$

Important notes
You should note that the (peak) total mean power radiated is independent of r and that it is therefore conservative. Note also the presence of I_0 – the peak (maximum) amplitude of the current – in this equation, but because we have correctly used the factor $^1/_2$ in the expression of $\langle s_{\text{peak}}\rangle$, this means that the total (peak) mean power radiated is indeed expressed in $W_{\text{mean peak}}$, i.e. the maximum (peak) value of the mean value radiated! I agree that this is complicated, but it is well worth making the effort to understand it!

Example

$$I_0 = 0.1\,A_{\text{peak}}$$
$$\lambda = 0.3\,m\ (f = 1\,GHz)$$
$$l = 0.01\,m\ (l \ll \lambda \rightarrow \text{conditions of the Hertzian dipole})$$

$$P_{\text{mean tot peak (dipole)}} = 40\left(\frac{\pi I_0 l}{\lambda}\right)^2$$

$$P_{\text{mean tot peak (dipole)}} = 40\left(\frac{3.14 \times 0.1 \times 0.01}{0.3}\right)^2 = 4.38\,mW_{\text{mean peak}}$$

Radiation Resistance of the Hertzian Dipole

The power that we found in the preceding section, i.e.

$$P_{\text{mean tot peak (dipole)}} = 40\left(\frac{\pi I_0 l}{\lambda}\right)^2 W_{\text{mean peak}}$$

is commonly considered to generally resemble power dissipated in an 'equivalent' resistance – called the radiation resistance – created by a mean continuous current (DC) or an r.m.s. alternating current (AC), such that $P_{\text{rms}} = RI_{\text{mean}}^2 = RI_{\text{rms}}^2$. Unfortunately, though, the term I_0 in the formula is actually a peak current, which complicates matters. To make this combination more uniform and ensure that the two equations are comparable, the term I_0 must be converted to I_{rms}, which is easily done.
Since $I_0 = \sqrt{2}I_{\text{rms}}$, it follows that

$$P_{\text{rms}} = \tfrac{1}{2}RI_0^2$$

and if we identify the first and the last equation, this means that

$$40\left(\frac{\pi l}{\lambda}\right)^2 = \frac{1}{2}R$$

and therefore, in conclusion, for a Hertzian dipole, and only if the length l is very small with respect to λ,

$$R_{\text{radiation}} = 80\left(\frac{\pi l}{\lambda}\right)^2 = f(\lambda)\Omega$$

$$R_{\text{radiation}} = 789.6\left(\frac{l}{\lambda}\right)^2 = f(\lambda)\Omega$$

Disregarding ohmic losses, this radiation resistance represents:

- in transmission, the equivalent load resistance for the source;
- in reception, the equivalent internal resistance created by the receiving antenna.

This formula is only valid in the case of the Hertzian dipole where the length l is very small with respect to λ; it gives completely erroneous results if, for example, $l = \lambda/4$ or $l = \lambda/2$. I will show you later how to calculate the radiation value for any value of l.

Important notes
1. There is no direct relationship between radiation resistance, ohmic resistance and antenna impedance!
2. Numerous confusions in the expression of formulae and their indices, etc., are encountered in the technical literature because $P_{\text{mean tot peak (dipole)}}$ is treated as equal to $P_{\text{rms}} = RI_{\text{rms}}^2$; in other words, the maximum (peak) value of the radiated mean power (which is essentially a permanent, continuous component) is changed to an equivalent alternating signal having the same dissipated power in watts.

Amplitude of the Electrical Field $E_{mean\ peak}$ of a Hertzian Dipole in the Equatorial Plane in the far Field

For this Hertzian dipole, in the far field ($r \gg \lambda/(2\pi)$), the electrical field E_θ observable in the equatorial plane ($\sin\theta = 1$) is expressed as a function of the total mean power radiated by the Hertzian dipole. Given that

$$E_\theta = Z_0 H\varphi$$
$$E_\theta = Z_0 \frac{I_0 l\, jk}{4\pi\, r} e^{j(\omega t - kr)}$$

and replacing k with its value $2\pi/\lambda$ where $\lambda = c/f$, we obtain

$$E_\theta = jZ_0 \frac{I_0 lf}{2cr} e^{j(\omega t - kr)}$$

We demonstrated previously that

$$P_{\text{mean tot peak}} = \frac{\pi Z_0}{3c^2} I_0^2 l^2 f^2 \text{ expressed in } W_{\text{mean peak}}$$

or alternatively

$$\sqrt{P_{\text{mean tot peak}}} = \sqrt{\frac{\pi Z_0}{3c^2}} I_0 lf$$

If we remove the value of $I_0 lf$ from this equation and introduce it into the last equation for E, we obtain

$$E_\theta = jZ_0 \frac{1}{2c\sqrt{\pi Z_0/(3c^2)}} \frac{\sqrt{P_{\text{mean tot peak}}}}{r} e^{j(\omega t - kr)}$$

$$E_\theta = j \frac{\sqrt{Z_0^2}}{\sqrt{4\pi Z_0/3}} \frac{\sqrt{P_{\text{mean tot peak}}}}{r} e^{j(\omega t - kr)}$$

Let us assume that

$$|E_\theta| = E_{\text{peak}} = j\sqrt{\frac{3Z_0}{4\pi}} \frac{\sqrt{P_{\text{moy tot peak}}}}{r}$$

$$E_{\text{peak}} = 9.487 \frac{\sqrt{P_{\text{mean tot peak}}}}{r}$$

where E_{peak} is expressed in $V_{\text{peak}} \, m^{-1}$ and $P_{\text{mean tot peak}}$ in $W_{\text{mean peak}}$.

Note

I have already shown that, in the same far field conditions,

$$E_\theta = \frac{60\pi I_0 l}{\lambda r} \quad \text{in the equatorial plane}$$

and

$$P_{\text{mean tot radiated}} = 40 \left(\frac{\pi I_0 l}{\lambda}\right)^2 \quad W_{\text{mean peak}}$$

Let us extract the value of the product $(I_0 l)$ from $P_{\text{mean tot radiated}}$. This gives

$$I_0 l = \frac{\lambda}{\pi} \sqrt{\frac{P}{40}}$$

and transfer this value into the equation for E_θ. We obtain

$$E_\theta = \frac{60\sqrt{P/40}}{r}$$

$$E_\theta = \frac{3\sqrt{10} \times \sqrt{P}}{r}$$

i.e. for a Hertzian dipole,

$$E_{\text{peak}} = 9.487 \frac{\sqrt{P_{\text{mean tot peak}}}}{r}$$

where E_{peak} is expressed in $V_{\text{peak}}\,m^{-1}$ and $P_{\text{mean tot peak}}$ in $W_{\text{mean peak}}$.

There is a very quick way of finding this value, but if we had not gone through the previous procedure, you would never have known that E_m was equal to

$$E_m = |E_\theta| = \sqrt{\frac{3Z_0}{4\pi}}\frac{\sqrt{P}}{r}$$

Once more: no pain, no gain!

Example 1

By way of example, still using a Hertzian dipole, let us assume that at a given date the current local regulations specify that we must not exceed 0.5 W EIRP at 2.45 GHz for outdoor applications (see Chapters 16 and 17 for further details). Let us calculate, for example, the maximum electrical field strength E_{rms} that this produces at 10 m:

$$E_{\text{rms}} = \frac{E_{\text{peak}}}{\sqrt{2}} = \frac{9.487}{\sqrt{2}}\frac{\sqrt{P_{\text{mean radiated rms}}}}{r}$$

$$E_m = 6.71\frac{\sqrt{0.5}}{10} = 474\,\text{mV}\,\text{m}^{-1}\,\text{rms}$$

Given that

$$E_m(\text{in dBmV}\cdot\text{m}^{-1}) = 20\log\frac{E\text{ in }V\cdot\text{m}^{-1}}{1\times 10^{-6}}$$

we find

$$E_m(\text{in dB\textmu V m}^{-1}) = 20\,([\log 4.74 + \log 10^{-1}] + \log 10^6)$$

$$E_m(\text{in dB\textmu V m}^{-1}) = 20(0.676 + 5) = +113.52\,\text{dB\textmu V m}^{-1}$$

and given that, additionally,

$$H(\text{in dB\textmu A m}^{-1}) = E(\text{in dB\textmu V m}^{-1}) - 51.5\,\text{dB}$$

$$H(\text{in dB\textmu A m}^{-1}) = 113.52 - 51.5 = 62.02\,\text{dB\textmu A m}^{-1}$$

Example 2

By way of example, let us calculate the maximum equivalent power (EIRP) corresponding to a radiated magnetic field of $+42\,\text{dB\textmu A m}^{-1}$. Given that H in dB A m$^{-1} = E$ in dB V m$^{-1} - 51.5$ dB, and therefore that $42\,\text{dB\textmu A m}^{-1}$ are equivalent to $42 + 51.5 = 93.5\,\text{dB\textmu V m}^{-1}$,

$$E_m(\text{in dB\textmu V m}^{-1}) = 93.5 = 20(4 + 0.676)$$

$$E_m(\text{in dB\textmu V m}^{-1}) = 20\,(\log 10\,000 + \log 4.74) = 20\log(47\,400)$$

i.e.

$$E_{m\,\text{max}} \text{ at } 10\,\text{m} = 47.4\,\text{mV}$$

$$E_m = 6.71\frac{\sqrt{P}_{\text{mean radiated rms}}}{r}$$

$$E_m = 6.71\frac{\sqrt{P}}{10}$$

$$\left(\frac{47.4\,\text{mV} \times 10}{6.71}\right)^2 = P = 4.999\,\text{mW}_{\text{EIRP rms max}} = 5\,\text{mW}_{\text{EIRP rms max}}$$

4.5 The Hertzian Dipole and a Dipole of any Length, λ/n and $\lambda/2$

Throughout this chapter I have mentioned the Hertzian dipole – a wire-shaped element whose length is very small with respect to the wavelength of the operating frequency used – through which a current flows. In the conventional context of RFID applications, this is often an unlikely situation. In reality, we are often concerned with the production of an antenna of finite dimensions whose length l is not in any way negligible with respect to the wavelength. For example:

- the antenna known as a '$\lambda/2$ dipole', which is symmetrical and supplied at its centre (Figure 4.9(a));
- the asymmetric $\lambda/4$ antenna, with one end connected to earth (Figure 4.9(b)).

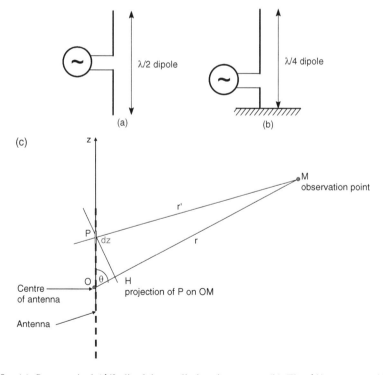

Figure 4.9 (a) Symmetrical '$\lambda/2$ dipole' supplied at its centre. (b) The $\lambda/4$ antenna with one end connected to earth. (c) Field radiated at one point

Theoretically, then, all the calculations I have set out for you since the beginning of this book need to be redone! You will be delighted to hear this, of course! However, let me calm things down by informing you that we can avoid this problem by considering the antenna of this new dipole of any length as a sequence of juxtaposed small elementary dipoles. Now we are home and dry, or nearly.

To evaluate the global radiation of this antenna in a given direction we must therefore take into account the propagation of the signal along the wire, meaning that the currents are not necessarily in phase from one point of the antenna to the next, and we must then also add up the contributions of each elementary dipole, allowing the phase shifts due to the differences in the paths of the radiated waves. To do this, let us look at an antenna of length l, laid along the vertical axis Z, and call its centre O: this point will then be assumed to be the origin (Figure 4.9(c)).

At a point P on the antenna, on one side z of the origin O, an element of length dz (a small elementary dipole) has a current flowing through it with the following equation:

$$I(z,t) = I_0(z)e^{jw_t}$$

This elementary dipole dz contributes to the propagation of the signal at any point in space, as explained above, and the conventional relation demonstrated previously will give the values of the fields E and H created at a distance r by the elementary dipole of length dz placed on z.

For example, the electrical field $E_\theta(t, r)$ will be as follows:

$$E_\theta(t,r) = -j\left[\frac{I_0 l \sin\theta}{4\pi\varepsilon_0\omega}\left(\frac{1}{r^3} + \frac{jk}{r^2} - \frac{k^2}{r}\right)\right]e^{j(\omega t - kr)}$$

Furthermore, at an observation point M located in the direction of the angle θ, the received electromagnetic field will be a function of the phase shift between the wave emitted by the point O and that emitted by the point P. The value of this phase shift OH can easily be calculated by projecting the point P on to the axis OM (the point H), which brings in the speed of propagation of the wave (c) and the value of cos θ.

In order to find the total values of the fields E and H, we must then add (in the sense of finding the 'integral sum') from $-l/2$ to $+l/2$ along the Z axis all the values representing the contributions of each of the elementary dipoles dz; in the case of a far field application, this gives

$$E_\theta(t,r) = \left[-j\frac{I_0 l \sin\theta}{4\pi\varepsilon_0\omega}\left(-\frac{k^2}{r}\right)\right]e^{j(\omega t - kr)}\int_{-1/2}^{+1/2} I_0(z)e^{jkz\cos\theta}dz$$

This equation can be solved only if we know the function $I_0(z)$ describing the distribution of the current I_0 along the antenna placed on the axis Z.

4.5.1 λ/n Dipole Antenna

One particularly interesting case is that of an antenna whose length l is finite and which is supplied with alternating current at its centre. This is because, in this case, a system of standing waves is established in the antenna and the currents vary from one point to another, but sooner or later they must form a current node at each end. The antenna will therefore have n preferred finite operating modes, where $n = 1, 2, 3, 4$, etc., and its length will be an integer multiple of the half-wavelength.

In the general case of a symmetrical dipole antenna whose length l is known, the distribution of current over the whole length of the antenna is therefore described as follows:

$$I_0(z) = I_0 \sin n\pi \left(\frac{z}{l} + \frac{1}{2}\right)$$

and we can then calculate the integral

$$\int_{-l/2}^{+l/2} I_0(z) e^{jkz \cos \theta} dz$$

Using a double integration by parts (see the previously cited References 1 and 2 mentioned in the Preface if necessary), we obtain the following exact result:

$$E_\theta(t, r) = \left[j \frac{I_0 Z_0}{2\pi} \frac{F(n, \theta)}{r}\right] e^{j(\omega t - kr)}$$

in which equation

$$F(n, \theta) = \frac{\cos\left[(n\pi/2)\cos\theta\right]}{\sin\theta} \text{ if } n \text{ is odd}$$

$$F(n, \theta) = \frac{\sin\left[(n\pi/2)\cos\theta\right]}{\sin\theta} \text{ if } n \text{ is even}$$

We can then calculate the field $H_\varphi(t, r)$. Remembering that $E_\theta(t, r) = Z_0 H_\varphi(t, r)$, we find

$$H_\phi(t, r) = \left[j \frac{I_0}{2\pi} \frac{F(n, \theta)}{r}\right] e^{j(\omega t - kr)}$$

Mean Value $\langle |S_r| \rangle$ of the Poynting Vector in the Far Field for a λ/n Dipole
Let us again calculate the effective mean value of the Poynting vector, in other words the radiated electromagnetic power flux density, for this case.

In UHF and SHF RFID applications, and therefore assuming that we are still in the far field ($r \gg \lambda/(2\pi)$, $1/r^2$ and $1/r^3 \lll 1/r$), as I have shown several times, the equations of the E and H fields can be simplified as follows:

$$E_\theta(t, r) = \left[j \frac{I_0 Z_0}{2\pi} \frac{F(n, \theta)}{r}\right] e^{j(\omega t - kr)}$$

$$H_\phi(t, r) = \left[j \frac{I_0}{2\pi} \frac{F(n, \theta)}{r}\right] e^{j(\omega t - kr)}$$

$E_\theta(t, r)$ and $H_\varphi(t, r)$ are values of the complex variable $j(\omega t - kr)$, and therefore represent sinusoidal functions. They also show that, in far field conditions, the vectors $\mathbf{E}_\theta$ and $\mathbf{H}_\varphi$ are perpendicular (orthogonal) to each other. It follows that for $\alpha = 90°$, $\sin \alpha = 1$, the value of $s = |S| =$ power flux density of the modulus of the vector $\mathbf{S}$ is equal to

$$s = |S| = |E| |H| \sin \alpha = |E| |H|$$

Given that $\langle \ldots \rangle$ denotes 'mean value of $\ldots$' of a variable, the mean value $\langle s \rangle$ of $\langle |S| \rangle$ will be as follows (see Appendices 1 and 2 of this chapter):

$$\langle |S| \rangle = \langle s \rangle = \tfrac{1}{2}|E|\,|H|$$

which is expressed in $W_{\text{mean peak}}\,m^{-2}$ ($|E|$ and $|H|$ are still expressed in peak values). Assuming that $H_{\varphi}^{*}(t, r)$, the 'conjugated' value of $H_{\varphi}(t)$, is

$$H_{\phi}^{*}(t, r) = \left[-j\,\frac{I_0}{2\pi}\,\frac{F(n, \theta)}{r}\right]e^{-j(\omega t - kr)}$$

the preceding equation can be written thus (see Appendix 3 at the end of the chapter):

$$\langle |S_r| \rangle = \langle s \rangle = \tfrac{1}{2}\,\mathrm{Re}(E_\theta H_\phi^{*}) = \tfrac{1}{2}\mathrm{Re}(E_\theta^{*}H_\phi)$$

Given the above equations, the scalar product $(E_\theta H_\phi^{*})$ is

$$E_\theta(t, r)H_\phi^{*}(t, r) = \left[\frac{I_0^2 Z_0}{4\pi^2}\left(\frac{F(n, \theta)}{r}\right)^2\right]$$

Note that this expression is entirely real, not imaginary.

Finally, the literal expression of the mean power flux density of a dipole of any length becomes:

$$\langle |S_r| \rangle = \langle s \rangle = \tfrac{1}{2}\mathrm{Re}(E_\theta H_\phi^{*}) = \frac{Z_0}{8\pi^2}I_0^2\,\frac{F^2(n, \theta)}{r^2}\quad \text{in } W_{\text{mean peak}}m^{-2}$$

4.5.2 $\lambda/2$ Dipole Antenna

A particularly interesting case is that of the antenna, widely used in UHF and SHF RFID, whose fundamental vibration mode corresponds to $n = 1$, producing a single alternation of current along the antenna, with the power supply at its centre. This mode is call the *half-wave mode*, which is why we speak of a $\lambda/2$ antenna.

Mean Value $\langle |S_r| \rangle$ of the Poynting Vector in the Far Field for a $\lambda/2$ Dipole

For a $\lambda/2$ dipole antenna, in other words one with $n = 1$, the peak mean power flux density is

$$\langle s \rangle = \langle |S_r| \rangle = \tfrac{1}{2}\mathrm{Re}(E_\theta H_\phi^{*})$$

$$\langle s \rangle = \frac{Z_0}{8\pi^2}I_0^2\,\frac{\cos^2[(\pi/2)\cos\theta]/\sin^2\theta}{r^2}\,W_{\text{peak mean}}m^{-2}$$

The curves in Figure 4.10 show the variations of $\langle s \rangle = f(\theta)$.

This figure shows that the effective mean radiated power flux density $\langle s \rangle$ is not in any way uniform, and that it is, in particular, zero in the axis of the dipole and maximal on the equatorial plane perpendicular to the plane of the dipole. Therefore, the radiation is not identical in all

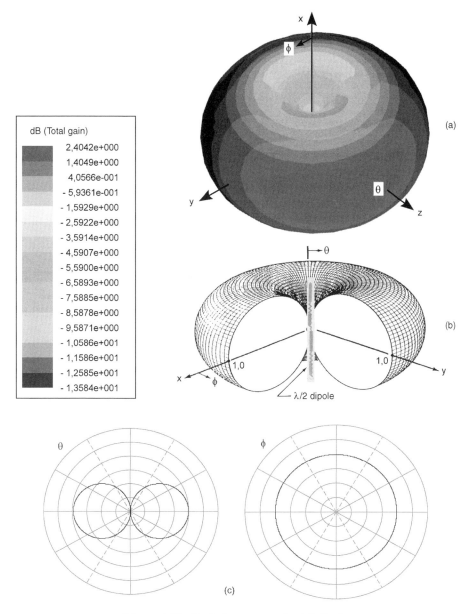

dB (Total gain)

	2,4042e+000
	1,4049e+000
	4,0566e-001
	- 5,9361e-001
	- 1,5929e+000
	- 2,5922e+000
	- 3,5914e+000
	- 4,5907e+000
	- 5,5900e+000
	- 6,5893e+000
	- 7,5885e+000
	- 8,5878e+000
	- 9,5871e+000
	- 1,0586e+001
	- 1,1586e+001
	- 1,2585e+001
	- 1,3584e+001

Figure 4.10   Radiation pattern of a $\lambda/2$ dipole

directions: in other words it is not 'isotropic'. This antenna is therefore *directional* and has a gain in the equatorial plane.

Now let us replace Z_0 with its numerical value 120π in the above equation:

$$\langle s \rangle = \frac{15}{\pi} I_0^2 \frac{\cos^2[(\pi/2)\cos\theta]/\sin^2\theta}{r^2} W_{\text{peak mean}} m^{-2}$$

This function has a maximum value, $s_{\text{mean peak max}}$, when $\theta = \pi/2$, i.e. in the direction of the equatorial plane:

$$s_{\text{mean peak max}} = \langle s \rangle = \frac{15}{\pi}\left(\frac{I_0}{r}\right)^2 \text{W}_{\text{mean peak}}\, \text{m}^{-2} \text{ for a } \lambda/2 \text{ dipole}$$

Peak Total Mean Electromagnetic Power Radiated in the Far Field by a λ/2 Dipole
With reference to a λ/2 dipole again, now that we have calculated the mean power flux density, $\langle s \rangle$, we can calculate the total mean power $\langle P \rangle$ radiated in the far field. I gave you the general formula for this above. Applying it to our specific case of mean values, we find

$$\langle P \rangle = \iint \langle s \rangle \, d\sigma$$

a double integral calculated along the surface Σ expressed in watts effective, where $\langle s \rangle$ is the mean power density (in W m^{-2}), Σ is a closed surface of any size and shape completely surrounding the transmitting antenna and $d\sigma$ is a surface element of Σ.

The peak total mean power passing through a sphere having a centre O and any radius r (i.e. the flux [of the mean value] of the Poynting vector over this surface) is equal to the power radiated by the dipole:

$$\langle P \rangle = \iint \frac{15}{\pi} I_0^2 \frac{\cos^2[(\pi/2)\cos\theta]/\sin^2\theta}{r^2} \, d\sigma$$

In polar coordinates, the surface element $d\sigma$ of a sphere is written as follows:

$$d\sigma = (r d\theta)\,(r \sin\theta \, d\varphi) = (r^2 \sin\theta \, d\theta d\varphi)$$

and we find

$$\langle P \rangle = \iint \frac{15}{\pi} I_0^2 \frac{\cos^2[(\pi/2)\cos\theta]/\sin^2\theta}{r^2} (r^2 \sin\theta \, d\theta) d\varphi$$

$$\langle P \rangle = \iint \frac{15}{\pi} I_0^2 \frac{\cos^2[(\pi/2)\cos\theta]}{\sin\theta} \, d\theta d\varphi$$

$$\langle P \rangle = \frac{15}{\pi} I_0^2 \iint \frac{\cos^2[(\pi/2)\cos\theta]}{\sin\theta} \, d\theta d\varphi$$

The literal value of the first integral $\int \cos 2[(\pi/2)\cos\theta]/\sin\theta \, d\theta$ is difficult to calculate (double integration by parts) and is beyond the scope of this book. Since, in our case, we need to find the finite value of this when θ varies from 0 to π, the easiest way is to estimate its value by a graphic/numeric evaluation method, which gives us (just for this once I am omitting the proof, but please trust me!)

$$\int_0^\pi \frac{\cos^2[(\pi/2)\cos\theta]}{\sin\theta} \, d\theta = 1.2188$$

As regards the second phase of the integration, which now becomes $\int 1.2188 \, d\varphi$ for φ varying from 0 to 2π, this is clearly as follows:

$$1.2188 \int_0^{2\pi} d\phi = 1.2188(2\pi - 0) = 7.658$$

and therefore, in conclusion,

$$P_{\text{mean tot radiated (dipole } \lambda/2)} = \frac{15}{\pi} I_0^2 \times 7.658 \, W_{\text{mean peak}}$$

Or, alternatively,

$$P_{\text{mean tot radiated (dipole } \lambda/2)} = 36.564 \, I_0^2 \, W_{\text{mean peak}}$$

Far Electrical Field Radiated by a λ/2 Dipole
I have just shown, in a general way, that, for a dipole, the electrical and magnetic fields are

$$E_\theta(t, r) = \left[j \frac{I_0 Z_0}{2\pi} \frac{F(n, \theta)}{r} \right] e^{j(\omega t - kr)}$$

$$H_\phi(t, r) = \left[j \frac{I_0}{2\pi} \frac{F(n, \theta)}{r} \right] e^{j(\omega t - kr)}$$

If the mode n is equal to 1, i.e. in the presence of a $\lambda/2$ dipole, we will use

$$F(n, \theta) = \frac{\cos\left[(n\pi/2)\cos\theta\right]}{\sin\theta} \text{ since } n \text{ is odd}$$

Also, in the equatorial plane $\theta = \pi/2$, we find that $F(n, \theta) = 1$, and at an instant t the modulus of $E_\theta(t, r)$ is then written as

$$E_{\pi/2}(r) = \frac{I_0 Z_0}{2\pi} \frac{1}{r}$$

Now let us replace Z_0 with its value (120π), and, extracting I_0 from the power formula $P_{\text{mean tot radiated (peak) } \lambda/2 \text{ dipole}}$, we obtain

$$E_{\pi/2}(r) = \frac{60}{r} \sqrt{\frac{P}{36.564}}$$

or, alternatively, for a $\lambda/2$ dipole:

$$E_{\pi/2}(r) = 9.923 \frac{\sqrt{P}}{r}$$

an equation where P is expressed as a peak mean value.

> **Important note**
> This value is independent of the operating frequency. A last comment: the values of 9.487 are for the case of a Hertzian dipole ($l \ll \lambda$), or 9.923 in the case of a dipole of length $l = \lambda/2$.

Radiation Resistance of the $\lambda/2$ Dipole
As mentioned above, the foregoing equation

$$P_{\text{mean tot radiated (dipole } \lambda/2)} = 36.564\, I_0^2 \, W_{\text{mean peak}}$$

is usually considered to resemble its general form

$$P_{\text{eff}} = RI_{\text{rms}}^2$$

but, unfortunately, the term I_0 that it includes is actually a peak current. To regain the uniformity and ensure that the two equations are comparable, the term I_0 must be converted to I_{rms}, which is easily done.

Since $I_0 = \sqrt{2} I_{\text{0rms}}$, it follows that $P_{\text{rms}} = \frac{1}{2} RI_0^2$. Then, if we identify the first and the last equation, this means that $36.564 = \frac{1}{2} R$. Thus

$$R_{\text{radiation}} = 36.564 \times 2 \, \Omega$$

and therefore, to conclude, in the case of the $\lambda/2$ dipole:

$$R_{\text{radiation}} = 73.128 \, \Omega$$

> **Very important**
> This result is very important, because it indicates that the radiation resistance of a $\lambda/2$ dipole is completely independent of the wavelength λ of the operating frequency used, since the length of the dipole antenna is $\lambda/2$ – and each frequency has its own $\lambda/2$ dipole.

4.6 List of the Main Formulae in this Chapter

$$\varepsilon_0 \mu_0 c^2 = 1$$

where

c = propagation speed (velocity) = $3 \times 10^8 \, \text{m s}^{-1}$
μ_0 = permeability of a vacuum = $4\pi \times 10^{-7} \, \text{H m}^{-1}$
ε_0 = permittivity of a vacuum = $\dfrac{1}{36\pi} \times 10^{-9} \, \text{F m}^{-1}$
$\omega = 2\pi f$
$\lambda = c/f$
$k = \omega/c = 2\pi/\lambda$ = wave number

Regardless of the distance r, at the point in question, we have a set of two groups of three equations that are functions of r, θ and φ:

- for the magnetic field H:

$$H_r(t,r) = 0$$

$$H_\theta(t,r) = 0$$

$$H_\phi(t,r) = \left[\frac{I_0 l \sin\theta}{4\pi}\left(\frac{1}{r^2} + \frac{jk}{r}\right)\right] e^{j(\omega t - kr)}$$

- for the electrical field E:

$$E_r(t,r) = \left[-j\frac{I_0 l \cos\theta}{2\pi\varepsilon_0\omega}\left(\frac{1}{r^3} + \frac{jk}{r^2}\right)\right] e^{j(\omega t - kr)}$$

$$E_\theta(t,r) = \left[-j\frac{I_0 l \sin\theta}{4\pi\varepsilon_0\omega}\left(\frac{1}{r^3} + \frac{jk}{r^2} - \frac{k^2}{r}\right)\right] e^{j(\omega t - kr)}$$

$$E_\varphi(t,r) = 0$$

Far fields

$$E_\theta(t,r) = \left(j\frac{\pi I_0 l \sin\theta}{\lambda^2 \varepsilon_0\omega}\frac{1}{r}\right) e^{j(\omega t - kr)}$$

$$H_\phi(t,r) = \left(j\frac{I_0 l \sin\theta}{2\lambda}\frac{1}{r}\right) e^{j(\omega t - kr)}$$

and $E_\theta(t,r)$ and $H_\varphi(t,r)$ are orthogonal to each other.
 Far field E and H fields ($r \gg \lambda/(2\pi)$)

$$|E_\theta| = \frac{60\pi I_0 l \sin\theta}{\lambda r}$$

$$|H_\varphi| = \frac{I_0 l \sin\theta}{2\lambda r}$$

Moduli of the far field E and H fields ($r \gg \lambda/(2\pi)$), and equatorial plane

$$|E_\theta| = \frac{60\pi I_0 l}{\lambda r}$$

$$|H_\varphi| = \frac{I_0 l}{2\lambda r}$$

Impedance of a vacuum

$$Z_0 = \frac{1}{\varepsilon_0 c}$$

$$Z_0 = \sqrt{\mu_0 \varepsilon_0} = 120\pi = 377\Omega$$

Relation between the electrical and magnetic fields

$$E_\theta = Z_0 H_\varphi$$

$$H \text{ (in dB}\mu \text{ A m}^{-1}) = E \text{ (in dB}\mu \text{ V m}^{-1}) - 51.5 \text{ dB}$$

4.7 Appendix 1: Brief Notes on Maxwell's Equations

More formally expressed, where $\mathbf{E}$ denotes the electrical field, $\vec{\mathbf{B}}$ the magnetic induction, $\vec{\mathbf{D}}$ the electrical excitation, $\vec{\mathbf{H}}$ the magnetic field, $\vec{\mathbf{F}}$ the Lorentz force, $\vec{v}$ the velocity, $\vec{\mathbf{A}}$ the vector potential, $\vec{\mathbf{J}}$ the current density, and based on the laws of Faraday, Ampère and Gauss,

$$\vec{\nabla} \times \vec{\mathbf{E}} = -j\omega\vec{\mathbf{B}} = -j\omega\mu_r\mu_0\vec{\mathbf{H}} \text{ (Faraday's law)}$$

$$\vec{\nabla} \times \vec{\mathbf{H}} = j\omega\vec{\mathbf{D}} + \vec{\mathbf{J}} = j\omega\varepsilon_0\varepsilon_r\vec{\mathbf{E}} + \vec{\mathbf{J}} \text{ (Ampère's law)}$$

$$\vec{\nabla} \cdot \vec{\mathbf{D}} = \rho \text{ (Gauss's law)}$$

$$\vec{\nabla} \cdot \vec{\mathbf{B}} = 0$$

$$\vec{\mathbf{F}} = -q(\vec{\mathbf{E}} + \vec{v} \times \vec{\mathbf{B}})$$

$$\vec{\nabla} \times \vec{\mathbf{A}} = \vec{\mathbf{B}}$$

$$\oint_C \vec{\mathbf{E}} \cdot d\vec{\mathbf{S}} = \frac{q}{\varepsilon_0\varepsilon_r} = \Phi_E$$

$$\oint_C \vec{\mathbf{B}} \cdot d\vec{\mathbf{S}} = 0 \oint_C \vec{\mathbf{B}} \cdot d\vec{\mathbf{S}} = \Phi_B$$

$$\oint \vec{\mathbf{E}} \cdot d\vec{\mathbf{l}} = -\mu_0\mu_r\frac{d\Phi_B}{dt}$$

$$\oint \vec{\mathbf{B}} \cdot d\vec{\mathbf{l}} = -\mu_0\mu_r\varepsilon_0\varepsilon_r\left(\frac{d\Phi_E}{dt} + i\right)$$

The set of equations that make up what are usually called Maxwell's equations are as follows:

$$\text{rot } H = J + \frac{\partial D}{\partial t}$$

$$\text{rot } E = -\frac{\partial B}{\partial t}$$

$$\text{div } D = \rho$$

$$\text{div } B = 0$$

Let $\beta = k$, the wave number:

$$\beta = \frac{2\pi}{\lambda} = \frac{\omega}{c}$$

A common way of presenting the solutions of these equations is as follows (in spherical coordinates):

$$
\begin{cases}
\vec{\mathbf{H}}_r = -\frac{I\,dl}{4\pi}\beta^2 2\cos\theta\left[\frac{1}{(j\beta r)^2} + \frac{1}{(j\beta r)^3}\right]e^{-j\beta r}\,\vec{r} \\[3mm]
\vec{\mathbf{H}}_\theta = -\frac{I\,dl}{4\pi}\beta^2 2\sin\theta\left[\frac{1}{j\beta r} + \frac{1}{(j\beta r)^2} + \frac{1}{(j\beta r)^3}\right]e^{-j\beta r}\,\vec{\theta} \\[3mm]
\vec{\mathbf{E}}_\varphi = -\frac{I\,dl}{4\pi}\beta^2 2\sin\theta\left[\frac{1}{(j\beta r)} + \frac{1}{(j\beta r)^2}\right]e^{-j\beta r}\,\vec{\varphi}
\end{cases}
$$

The value of the vector $\vec{\mathbf{H}}$ is

$$\vec{\mathbf{H}} = -\frac{I\,dl}{4\pi}\beta^2 2\cos\theta\left[\frac{1}{(j\beta r)^2} + \frac{1}{(j\beta r)^3}\right]e^{-j\beta r\vec{r}} - \frac{I\,dl}{4\pi}\beta^2 2\sin\theta\left[\frac{1}{j\beta r} + \frac{1}{(j\beta r)^2} + \frac{1}{(j\beta r)^3}\right]e^{-j\beta r\vec{\theta}}$$

The case of Near Field RFID Applications

Whatever the operating frequencies of RFID systems may be – LF, HF, UHF or SHF – if the product $\beta r \ll 1$ (i.e. $r \ll \lambda/(2\pi)$), only the terms in $1/(\beta r)^3$ are preponderant and therefore $e^{j\beta r}$ is substantially equal to $e^0 = 1$. The approximation of the equation above is called the *near field approximation*. In this case, the value of the vector $\vec{\mathbf{H}}$ becomes

$$\vec{\mathbf{H}} = j\frac{I\,dl}{4\pi\beta \cdot r^3}(2\cos\theta \cdot \vec{r} + \sin\theta \cdot \vec{\theta})$$

The magnetic field H then has the following properties:

- It is not propagated: the energy is stored, not radiated.
- It can be considered to be quasi-stationary.
- It is practically decoupled from the electrical field.
- It decreases with the cube of distance.

The magnetic field produced in air at the point P by a current I flowing through the element of length dl of a wire-like conductor having a closed profile Γ can be written as follows (this is the

Biot-Savart law):

$$\vec{\mathbf{H}} = \frac{I}{4\pi} \oint_\Gamma \frac{d\vec{\mathbf{l}} \times \vec{\mathbf{r}}}{r^3}$$

This equation is the basis for RFID operating at 125 kHz and 13.56 MHz, and has been described fully in my previous books (References 1 and 2 mentioned in the Preface); for your information, some 'near field' applications in UHF are also under development, under the name of *item management* (short distance) as opposed to *supply chain management* applications (long distance).

4.8 Appendix 2: Brief Notes on Complex Numbers

Let z be a complex number: $z = (a + jb)$. Its modulus will then be

$$\rho = \sqrt{a^2 + b^2}$$

and the tangent of its argument φ is $\tan \varphi = b/a$. This complex number can be written either in its trigonometric form or in its exponential form:

$$z = (a + jb) = \rho\,(\cos \varphi + j \sin \varphi) = \rho e^{j\varphi}$$

Example

Any signal of the form $y = a \cos(\omega t + \varphi)$ can be written in complex notation:

$$y = \frac{a}{2}\left(e^{j(\omega t + \phi)} + e^{-j(\omega t + \phi)}\right)$$

since

$$y = a/2\left\{[\cos(\omega t + \varphi) + j \sin(\omega t + \varphi)] + [\cos(\omega t + \varphi) - j \sin(\omega t + \varphi)]\right\}$$

By definition, the following value is called the *conjugate* of z and is denoted z^*:

$$z^* = (a + jb)^* = (a - jb)$$
$$(a + jb)^* = (a - jb)$$
$$\rho^* = \sqrt{a^2 + b^2} = \rho$$
$$\tan\varphi = -b/a$$
$$\varphi^* = -\varphi$$
$$(a + jb)^* = \rho[\cos(-\varphi) + j \sin(-\varphi)]$$
$$(a + jb)^* = \rho(\cos\varphi - j \sin\varphi)$$
$$(a + jb)^* = \rho e^{-j\varphi}$$

Question

Is the conjugate of a product of two complex numbers the product of the conjugates of each of them? As the equality below shows, the answer is yes:

(a) On the one hand:

$$
\begin{aligned}
[(a+jb)\,(c+jd)]* &= [(ac+jad+jbc+j^2bd]* \\
&= [(ac-bd)+j(ad+bc)]* \\
&= [(ac-bd)-j(ad+bc)]
\end{aligned}
$$

(b) On the other hand:

$$
\begin{aligned}
(a+jb)*(c+jd)* &= (a-jb)\,(c-jd) \\
&= (ac-jad-jbc+j^2bd) \\
&= [(ac-bd)-j(ad+bc)]
\end{aligned}
$$

(c) Therefore:

$$
[(a+jb)\,(c+jd)] = (a+jb)*(c+jd)*
$$

4.9 Appendix 3: Brief Notes on Powers Expressed as Complex Numbers

Take the sinusoidal values

$$
u(t) = U\sqrt{2}\cos \omega t
$$

and

$$
i(t) = I\sqrt{2}\cos(\omega t - \varphi)
$$

where U and I express their effective values. We can say that the instantaneous power is

$$
p(t) = u(t)i(t) = U\sqrt{2}\cos \omega t\, I\sqrt{2}\cos (\omega t - \varphi)
$$

After the usual trigonometrical manipulations, we obtain

$$
p(t) = 2UI\left[\frac{\cos (2\omega t - \varphi) + \cos \varphi}{2}\right]
$$

$$
p = UI[\cos (2\omega t - \varphi) + \cos \varphi]
$$

$$
p = UI\cos \varphi + UI\cos (2\omega t - \varphi)
$$

These equations have two entirely separate parts (Figure 4.11):

- $P_{mean} = UI \cos \varphi$, a 'continuous', nonfluctuating value, known as the *mean power*; this is what is called the *active* power, or power *in watts*.

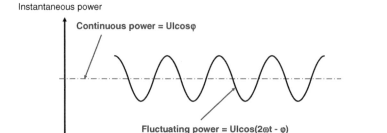

Figure 4.11   Diagram of the variations of instantaneous power as a function of time

- $UI\cos(2\omega t - \varphi)$, a power fluctuating (at a double frequency) as a function of time and having a mean of zero.

By definition, following the above conventions, let us also specify the following:

$$\text{Reactive power, } Q_{\text{reactive}} = UI\sin\varphi$$
$$\text{Apparent power, } S_{\text{apparent}} = UI$$

Now let us return to the initial equation for instantaneous power:

$$p(t) = u(t)i(t) = U\sqrt{2}\cos\omega t \, I\sqrt{2}\cos(\omega t - \varphi)$$

and write the values of $u(t)$ and $i(t)$ in their complex forms, in exponential notation. Thus:

$$\ddot{u} = U\sqrt{2}e^{j\omega t}$$

and

$$\ddot{i} = I\sqrt{2}e^{j(\omega t - \phi)}$$
$$\ddot{i} = I\sqrt{2}e^{j\omega t}e^{-j\phi}$$

and its conjugate:

$$\ddot{i}* = I\sqrt{2}e^{-j\omega t}e^{j\phi}$$

Now let us calculate the following two expressions:

$$\ddot{u}\ddot{i} = 2UIe^{2j\omega t}e^{-j\phi}$$

and

$$\ddot{u}\ddot{i}* = 2UIe^{j\phi}$$

the second of which can be written in conventional notation:

$$\ddot{u}\ddot{i}* = 2UI(\cos\varphi + j\sin\varphi)$$
$$\ddot{u}\ddot{i}* = 2UI\cos\varphi + j(2UI\sin\varphi) = \text{real part} + \text{imaginary part}$$

Now let us identify the two sets of equations, member by member:

$$\text{Real }(*) = 2UI\cos\varphi = 2P_{\text{mean}}$$
$$\text{Imag }(*) = 2UI\sin\varphi = 2Q_{\text{reactive}}$$

from which we can derive

$$P_{\text{mean}} = \tfrac{1}{2}\text{Re}(\ddot{u}\ddot{i})$$
$$P_{\text{mean}} = \tfrac{1}{2}(2UI\cos\varphi) = UI\cos\varphi$$

In the above equations, the values of U and I express the effective values of the alternating signal and the mean power expressed in watts (actually in mean watts).

If we wish to introduce into the equation for mean power, P_{mean}, the peak values/amplitudes U_0 and I_0 of the signals $u(t)$ and $i(t)$, each of whose values is of course $\sqrt{2}$ times greater than those of U and I, we can write the following, having taken care to introduce a correction factor of $^1/_2$:

$$P_{\text{mean}} = UI\cos\varphi = \frac{U\sqrt{2}I\sqrt{2}\cos\varphi}{2} = \frac{1}{2}U_0I_0\cos\varphi$$

depending on whether we are expressing

- U and I, using the effective values of the signal or
- U_0 and I_0, using the peak values of the signal.

Also, if $u(t)$ and $i(t)$ are in phase, φ is zero and $\cos\varphi$ is equal to 1. Then

$$P_{\text{mean}} = {}^1/_2 U_0 I_0$$

where P_{mean} is in watts and U_0 and I_0 are peak values.

In all these equations, we must be careful not to confuse or mix up the effective values U and I and the peak values U_0 and I_0 of the electrical signals, which are only separated by a small, but far from negligible, factor of $\sqrt{2}$ in each case (i.e. 141% of difference, or 3 dB!).

Finally, for information, let us calculate $\ddot{p}$:

$$\ddot{p} = \tfrac{1}{2}\ddot{u}\ddot{i}* = P_{\text{mean}} + jQ_{\text{reactive}}$$

and

$$S_{\text{apparent}} = UI$$

$$S_{\text{apparent}} = UI\sqrt{\cos^2\varphi + \sin^2\varphi}$$

$$S_{\text{apparent}} = \sqrt{U^2I^2\cos^2\varphi + U^2I^2\sin^2\varphi}$$

$$S_{\text{apparent}} = \sqrt{P_{\text{mean}}^2 + jQ_{\text{reactive}}^2}$$

$$S_{\text{apparent}} = |\ddot{p}| = \text{modulus of } \ddot{p}$$

4.10 Appendix 4: Brief Notes on Vectors

Scalar Product of Two Vectors

The scalar product ('dot product') of two vectors **A** and **B**, at an angle of θ to each other, is equal to the following scalar quantity (Figure 4.12):

$$\mathbf{A \bullet B} = |A|\,|B|\cos\theta$$

Clearly, when $\theta = 90°$, $\cos\theta = 0$, and therefore the scalar product of two orthogonal vectors is zero.

Vector Product of Two Vectors

The vector product ('cross product') of two vectors $\vec{A}$ and $\vec{B}$ at an angle of θ to each other is equal to a third vector $\vec{V}$, perpendicular to the plane formed by the two original vectors, according to the rule shown in Figure 4.13. Its modulus |V| is

$$|V| = |A|\,|B|\sin\theta$$

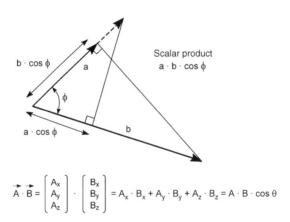

$$\vec{A} \cdot \vec{B} = \begin{bmatrix} A_x \\ A_y \\ A_z \end{bmatrix} \cdot \begin{bmatrix} B_x \\ B_y \\ B_z \end{bmatrix} = A_x \cdot B_x + A_y \cdot B_y + A_z \cdot B_z = A \cdot B \cdot \cos\theta$$

Figure 4.12   Scalar product of two vectors **A** and **B**

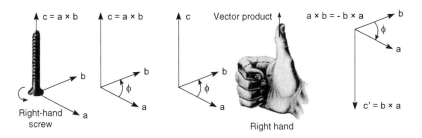

$$\vec{A} \times \vec{B} = \begin{bmatrix} A_x \\ A_y \\ A_z \end{bmatrix} \times \begin{bmatrix} B_x \\ B_y \\ B_z \end{bmatrix} = \begin{bmatrix} A_yB_z - A_zB_y \\ A_zB_x - A_xB_z \\ A_xB_y - A_yB_x \end{bmatrix}$$

Figure 4.13   Vector product of two vectors $\vec{A}$ and $\vec{B}$

Clearly, when $\theta = 0°$, $\sin \theta = 0$, and therefore the scalar product of two collinear vectors is zero.

Note

For the vector product, J. W. Gibbs used the multiplication sign ($\times$), which is currently used in the United States. The notation used more widely in Europe ($\wedge$) was developed by C. Burali-Forti. Gibbs also used bold dots ($\cdot$) for the scalar product (as above).

5

Wave Propagation in Free Space

5.1 Isotropic and Anisotropic Antennas

The radiating source (the base station) generally consists (Figure 5.1) of an amplifier that can deliver an electrical power (in W), called the conducted power, $P_{\text{cond bs}}$, to a load, which in the present case is the base station antenna. For the present, we assume that this antenna is lossless. Note that, for simple and well-known reasons of power adaptation, the output impedance of the amplifier and the load impedance (base station antenna) are often adapted (having identical and/or conjugate values).

5.1.1 What Is 'Isotropic', Anyway?

Let us start with a brief language lesson, on the definition of the adjective 'isotropic'. This word is made up of two very significant parts:

- *isos* = equal, identical, the same;
- *tropos* = way, method.

Therefore, by definition, an isotropic radiating source is a source that radiates and propagates waves equally, in the same way, and therefore uniformly in all directions. Now let us look at the specific case of the antenna. By definition, an antenna is isotropic if it transmits exactly the same power (flux density) in all directions ('omnidirectional uniform spherical radiation') (Figure 5.2(a)). In other words, this means that the radiated power flux density (*s*) is independent of the direction of radiation.

5.1.2 Anisotropic

Except in the case of an antenna with very small physical dimensions (such as a very small loop or hairpin antenna of the type used in mobile telephones or remote controllers for vehicles), the actual physical and mechanical properties of an antenna mean that it is rarely 'isotropic' in practice. It is therefore called 'anisotropic', where the 'an' is a negative prefix. In many cases

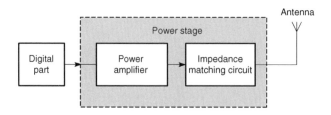

Figure 5.1   Structure of the power stage of a base station

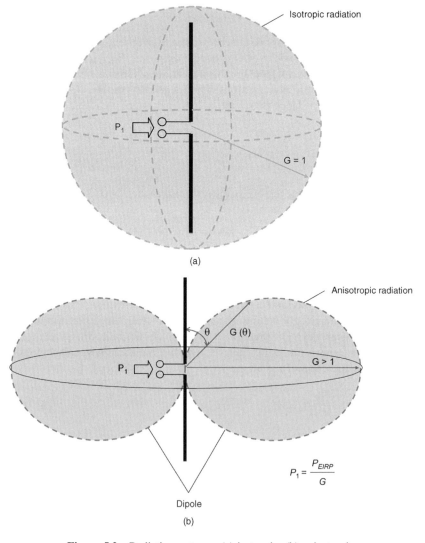

Figure 5.2   Radiation patterns: (a) isotropic; (b) anisotropic

the physical shape of the antenna causes it to radiate in privileged directions (known as lobes or beams, see Figure 5.2(b)), which contain all or some of the radiated power supplied by the amplifier's power stage. It therefore has a gain, or one of a number of gains, in one or more given directions, with respect to an antenna that is isotropic.

5.2 Antenna Gain

For a specified conducted electrical power, $P_{\text{cond bs}}$, supplied by the base station amplifier and for a specified observation point, we normally define the *antenna gain*, G, as the ratio between the power flux density produced by the actual antenna considered and/or used, Ant_2, in the direction in which this density is maximal, $s_{2\text{max}}$, and the power flux density, $s_{1\text{iso}}$, which would be produced by a hypothetical isotropic antenna, Ant_1, with unitary gain in the same conditions of power, $P_{\text{cond bs}}$.

The preceding paragraph can be summarized in two different ways, which differ in their presentation but are identical in their basic meaning:

- for a given electrical power $P_{\text{cond bs}}$ applied to the antennae:

$$\text{Antenna gain} = \frac{\text{power flux density radiated by Ant}_2 \text{ in its maximal direction } s_{2\text{max}}}{\text{power flux density radiated by the isotropic antenna } \text{Ant}_1 \text{ at the same distance } s_{1\text{iso}}}$$

- for a given effect product, at a given observation point:

$$\text{Antenna gain} = \frac{\text{power } P_{\text{cond bs}} \text{ that should be delivered by the amplifier with an isotropic antenn Ant}_1}{\text{power } P_{\text{cond bs}} \text{ of the transmitter with a directive antenna Ant}_2}$$

These ratios are usually expressed either in absolute values or in relative values (dB).

Note that this definition of the gain does not in any way imply that the antenna is 'active' or otherwise; it simply indicates a comparative aspect of the respective performances of the antennae.

5.2.1 Power, and All That

Since we are dealing with power, let us take a closer look at the concept. Of course, the official unit of power is the watt, but what kind of power are we talking about? Is it the r.m.s. power, the peak power (as in the previous chapter), the mean power, the electrical power available at the amplifier output, the radiated power, the power measured by a particular method (peak, quasi-peak, etc.), power X, or power Y? Moreover, to make matters clearer, we often use dB or dBm instead of watts! To sum up: there are many different forms of power. You should be very careful, then, when using the term 'power', because sometimes this can lead to serious misunderstandings, even among the leading experts in the field! (Yes, we know who they are!) In other words, things are not as simple as they appear.

5.2.2 Decibel (dB)

The *bel* is defined as the logarithm to the base 10 of the ratio between two powers. This ratio, expressed in decibels dB (i.e. in values ten times smaller), is stated in the following equation:

$$dB = 10 \log \frac{P_2}{P_1}$$

Table 5.1 shows, by way of example, some common values of the ratio P_2/P_1 in dB.

Table 5.1

P_2	1000 times	greater than P_1	30 dB
P_2	100		20 dB
P_2	10		10 dB
P_2	once	equal to P_1	0 dB
P_2	10 times	smaller than P_1	−10 dB
P_2	100		−20 dB
P_2	1000		−30 dB

5.2.3 dBm

For simple reasons of practicality and/or convenience, this power is often expressed in dBm – also known as 'dB milliwatt' – rather than dB. In this case, the reference value of P_1, at '0 dBm', is the level of 1 mW, rather than the enormous amount of one watt, which is rarely required in ordinary RFID applications:

$$dB = 10 \log \frac{P \text{ in mW}}{1 \text{ mW}}$$

Table 5.2 shows the conversion between mW and dBm that is most suitable for our RFID applications.

Table 5.2

mW	dBm	Some examples
4000 mW	+36 dBm	$2 \text{ W} = 2 \times 1 \text{ W} = 3 \text{ dB} + 30 \text{ dBm} = 33 \text{ dBm}$
1000 mW	+30 dBm	
500 mW	+27 dBm	
100 mW	+20 dBm	
10 mW	+10 dBm	
1 mW	0 dBm	$50 \text{ mW} = 100 \text{ mW}/2 = 20 \text{ dBm} - 3 \text{ dB} = 17 \text{ dBm}$
100 μW	−10 dBm	
10 μW	−20 dBm	
1 μW	−30 dBm	
100 nW	−40 dBm	
10 nW	−50 dBm	
1 nW	−60 dBm	
100 pW	−70 dBm	

5.2.4 Electrical Power, Conducted Power and Radiated Power

Here is a brief survey of the differences between the terms 'electrical power', 'conducted power' and 'radiated power', based on some examples.

A generator/amplifier delivers an electrical power to an ohmic resistance. This power is in watts and is dissipated as heat. (Some of you may of course say that it is also a radiated power, since the heat also creates infrared radiation.) For example, an amplifier delivers a power of 10 W to its ohmic load of 2 Ω. The power of 10 W is the conducted power P_{cond} that can be delivered by the amplifier. In our particular case of UHF and SHF RFID, the RF amplifier loads are, in physical terms, antennae that have a radiation resistance as explained in the previous chapter (for example, $\approx 75\,\Omega$ for a $\lambda/2$ antenna). Assume that we have an RF amplifier that can deliver 2 W to a load of 75 Ω. This means that this amplifier can deliver a conducted power in watts of 2 W to a purely ohmic resistance of 75 Ω, and that, if the ohmic resistance of 75 Ω is replaced with an antenna having a radiation resistance of 75 Ω, it will radiate an electromagnetic power. How? How much? We don't know. The reason is simple. We have not said what the antenna gain is. Is it an isotropic antenna? Or anisotropic?

Having introduced these concepts, I will now give a precise definition of the concept of radiated electromagnetic power.

5.2.5 Equivalent Isotropically Radiated Power (EIRP)

Let us return to the second statement of the antenna gain in the previous paragraph, for the same effect produced, at the same observation point:

$$\text{Antenna gain} = \frac{\text{power } P_{\text{cond bs}} \text{ that should be delivered by the amplifier with an isotropic antenn Ant}_1}{\text{power } P_{\text{cond bs}} \text{ of the transmitter with a directive antenna Ant}_2}$$

In other words, the product (electrical power $P_{\text{cond bs}}$ of the transmitter [base station] with its directional antenna Ant$_2$ × antenna gain) is equal to the electrical power $P_{\text{cond bs}}$ that should be delivered by the amplifier with an isotropic antenna Ant$_1$ having unitary gain.

We can conclude that 'the power $P_{\text{cond bs}}$ that should be delivered by the amplifier with an isotropic antenna Ant$_1$' represents the equivalent power that would be radiated by a source (amplifier + antenna) whose antenna is isotropic and has unitary gain.

Based on these assumptions and the hypotheses of the previous paragraphs, this power $P_{\text{cond bs}}$ is called the EIRP (equivalent isotropically radiated power).

Definitions
Isotropic Power, P_{iso}
What is known as isotropic power is the amount of power radiated by a source (referred to subsequently as 'isotropic'), consisting of a transmitter supplying a power $P_{\text{cond bs}}$ and having an antenna with a specified gain G whose radiation pattern is isotropic. Theoretically, the antenna gain is totally independent of its 'isotropic' radiation properties:

$$P_{\text{isotropic radiated}} = P_{\text{cond bs}} G$$

Equivalent Isotropic Radiated Power, P_{EIRP}
The fact that a radiating source is driven by an electrical power $P_{\text{cond bs}}$ and that the radiation pattern of the antenna may possibly not be isotropic or have a specific gain G, due to its

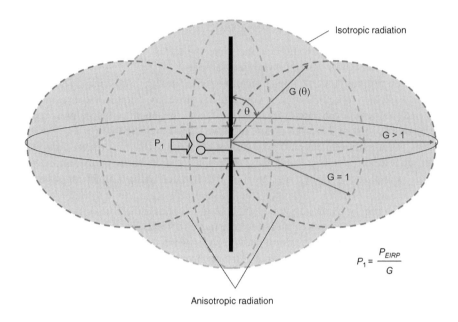

Figure 5.3   Equivalent isotropic radiated power, P_{EIRP}

characteristics and the anisotropy of its radiation, has led to the definition of equivalent isotropic radiated power (EIRP), such that this power P_{EIRP} produces the same effect, at the same observation point, as the power of the isotropic source described in the previous paragraph:

$$P_{\text{EIRP}} = P_{\text{cond bs}}G$$

It follows from the above that

$$P_{\text{iso}} \text{ at this point} = P_{\text{EIRP}} \text{ at this point}$$

With respect to a 'reference' antenna whose radiation pattern is isotropic and whose gain is unitary $(G = 1)$:

$$P_{\text{iso}} = P_{\text{cond bs}}$$

The power P_{EIRP} is equal to the product of the electrical power $P_{\text{cond bs}}$ supplied by the transmitter's amplifier to the antenna, multiplied by the gain of the antenna system in a given direction, relative to the power that an isotropic source would supply at the same observation point (Figure 5.3).

For a system relative to an isotropic source:

$$P_{\text{EIRP}} = P_{\text{cond bs}} \times \text{gain of the antenna}$$

Note

There are many versions of this term. Even the most authoritative reference sources (the IEEE dictionary, the ISO, etc.) offer several versions of the wording for EIRP. Here are the most common ones:

- equivalent isotropically radiated power;
- effective isotropically radiated power;
- equivalent isotropical radiated power;
- effective isotropical radiated power.

Examples

Assume that, for an RFID application, the local regulations stipulate a maximum power level of $P_{\text{EIRP max}} = +36\,\text{dBm}$ that is not to be exceeded. This level of $+36\,\text{dBm}$ (corresponding to an equivalent isotropic radiated power of 4 W) can be equated, for example, to:

- a conducted electrical power, $P_{\text{cond bs}} = 4\,\text{W}$, transmitted via an isotropic antenna with a gain of 1 (gain = 0 dB):

$$4000\,\text{mW}/1\,mW \rightarrow \ = 10\log 4000 = 10\log 1000 + 10\log 4 = 30 + 6 = 36\,\text{dBm}$$

- or a conducted electrical power $P_{\text{cond bs}} = 1\,\text{W}$, transmitted via an anisotropic antenna whose gain is 4 in the direction of measurement (gain = + 6 dB):

$$\text{`1 W'} + 6\,\text{dB} = \text{`1000 mW'} + 6\,\text{dB} = 30\,\text{dBm} + 6\,\text{dB} = 36\,\text{dBm}$$

An Example of 'Distorted' Information

In the chapter on current standards and regulations, I shall show that, in Europe (document ERC 70 03), for some SHF bands, the maximum authorized power P_{EIRP} (with an 'I' – not P_{ERP} without an 'I') at 2.45 GHz, without a licence, supplied continuously (at a mark–space ratio of 100%), is

$$P_{\text{EIRP max}} = 500\,\text{mW} = +27\,\text{dBm}$$

Sometimes, equipment manufacturers may announce 'miraculous' (and rather suspect) operating ranges that are far greater than those of their rivals. If we go through their specifications with a fine toothcomb, we will find that there has been a confusion of categories: the actual maximum conducted electrical power delivered by the amplifier does indeed conform to the authorized limit, but the manufacturers' measurements and demonstrations are based on an antenna with a nonunitary gain. For example:

- output conducted electrical power, $P_{\text{cond bs}}$: 27.00 dBm;
- base station antenna gain, G_{bs}: 2.14 dB, i.e. a dipole antenna;
- therefore, the total real P_{EIRP} is 29.14 dBm, a level that of course is not declared, because it exceeds the current regulations.

Notes

In this case, the true value of P_{EIRP} is 29.14 dBm, rather than the authorized 27 dBm. For information, 29.14 dBm = 820 mW, i.e. 500 mW × 1.64 = 64% more than the permitted level! Very helpful!

For anyone who still finds it difficult to deal with percentages and decibels, simply imagine these amounts translated into statements of earnings on a payslip. That will give you a much better idea of the differences involved!

5.3 Power Flux Density at One Point in Space

Let us consider a standard RFID situation where there is a source (base station) consisting of a transmitter delivering a conducted electrical power $P_{\text{cond bs}}$ (in W) – denoted P_{bs} below for the sake of simplicity – combined with an antenna having a gain $G_{\text{ant bs}}$, denoted G_{bs} for the same reason. At any point in space at a distance r, all of the power radiated by the base station is distributed over the surface of a radiation sphere with a surface area of $4\pi r^2$ (i.e. the Poynting vector flux).

Consequently, at a point in the space at a distance r from the source, the power flux density $s = |S|$ (expressed in W m^{-2}) available at any point in the 'free' space is as follows:

$$ s = |S| = \frac{P_{\text{bs}}G_{\text{bs}}}{4\pi r^2} = \frac{dP}{d\sigma} \text{ W m}^{-2} $$

where s is the power flux density, P_{bs} is the electrical power $P_{\text{cond bs}}$ supplied by the base station amplifier to its antenna, G_{bs} is the antenna gain $G_{\text{ant bs}}$ of the base station, r is the distance between the source and the measurement point, $4\pi r^2$ is equal to the surface area of the sphere of radius r and Σ is the surface element.

The product $(P_{\text{bs}}G_{\text{bs}})$ represents the EIRP, P_{EIRP} of the base station, this value being identical to the power that would be radiated by an equivalent isotropic source whose antenna gain G_{bs} is equal to 1:

$$ s = \frac{P_{\text{bs EIRP}}}{4\pi r^2} = \frac{P_{\text{bs iso}}}{4\pi r^2} \text{ W m}^{-2} $$

General Relationships Between the Electrical Field and the Radiated Power

I have shown that the Poynting vector S is equal to the vector product $E \times H$, and that, in RFID applications, in other words in far fields, these are orthogonal ($\sin 90° = 1$) and therefore the modulus $|S|$ of the vector S (i.e. its peak value) can be expressed as $|S| = |E|\,|H|$. I have also established that the relation between the electrical and magnetic fields across the impedance of a vacuum takes the form $E = Z_0 H$. From these two equations, we find that

$$ |S| = |E|\frac{|E|}{Z_0} $$

where $|S|$ and $|E|$ are still expressed as peak values, since the factor 1/2 is not present in this equation, and thus

$$ E_{\text{peak}} = \sqrt{s_{\text{peak}}Z_0} $$

I have also shown that the peak mean power flux density radiated by a source (which has been made isotropic) is as follows:

$$ s_{\text{peak}} = \frac{P_{\text{EIRP peak}}}{4\pi r^2} $$

Note, first of all, that there is no physical difference between the terms $P_{\text{mean tot peak}}$ and $P_{\text{EIRP rms}}$. As mentioned in the introduction to the concept of radiation resistance (see Chapter 4), these two terms refer to exactly the same phenomenon in physical terms. Furthermore, since I have pointed out that, in general, $P_{\text{EIRP rms}} = P_{\text{cond rms}}G_{\text{ant bs}}$, the above equation can be written as follows:

$$E_{\text{rms}} = \sqrt{\frac{P_{\text{EIRP rms}}Z_0}{4\pi r^2}}$$

Recalling that $Z_0 = 120\pi$, let us enter its value into the above equation. This gives

$$E_{\text{rms}} = \frac{\sqrt{30P_{\text{EIRP rms}}}}{r}$$

$$E_{\text{rms}} = 5.477\frac{\sqrt{P_{\text{EIRP rms}}}}{r} \text{ V m}^{-1} \text{ for an isotropic radiating source}$$

or, expressed as a peak value:

$$E_{\text{peak}} = \sqrt{2}E_{\text{rms}}$$

$$E_{\text{peak}} = 7.746\frac{\sqrt{P_{\text{EIRP rms}}}}{r} \text{ V m}^{-1}$$

To recap, in the previous chapter I showed that:

- for a Hertzian dipole,

$$E_{\text{peak Hertz}} = 9.487\frac{\sqrt{P_{\text{mean tot peak}}}}{r}$$

where E_{peak} is expressed in $V_{\text{peak}} \text{ m}^{-1}$ and $P_{\text{mean tot peak}}$ in $W_{\text{mean peak}}$;
- for a $\lambda/2$ dipole,

$$E_{\text{peak }\lambda/2} = 9.923\frac{\sqrt{P_{\text{mean tot peak}}}}{r}$$

where E_{peak} is expressed in $V_{\text{peak}} \text{ m}^{-1}$ and $P_{\text{mean tot peak}}$ in $W_{\text{mean peak}}$.

These values seem very strange at first sight. However, let us take a closer look. For an isotropic antenna, this corresponds to

$$E_{\text{peak}} = 7.746\frac{\sqrt{P_{\text{cond rms}}G_{\text{ant bs}}}}{r}$$

$$E_{\text{peak}} = 7.746\sqrt{G_{\text{ant bs}}}\frac{\sqrt{P_{\text{cond rms}}}}{r}$$

For a given conducted power, $P_{\text{cond eff}}$, applied to the antenna x used for the base station (regardless of its type, which may be isotropic, Hertzian dipole, $\lambda/2$ dipole, etc.) at a given observation point (at a given observation distance r), depending on the antenna used, we will find radiated power flux densities s_x (and thus received power levels P_x) for this antenna, which will be, by definition, the ratio of the gains of the antennae used, G_x. I have also shown on several occasions that, as a general rule, the electrical field E is proportional to the square root of the power. We can therefore write:

$$\frac{s_x}{s_1} = \frac{\text{gain}_x}{\text{gain}_1} = \frac{P_x}{P_1} = \left(\frac{E_x}{E_1}\right)^2$$

Assuming that, by definition, the antenna '1' is isotropic and therefore that its gain G_1 is equal to 1 and its electrical field is E_1, we can directly find the relative values of the electrical fields E_x radiated by the other antennae used:

$$\frac{\text{Gain}_x}{1} = \left(\frac{E_x}{E_1}\right)^2$$

$$E_x^2 = E_1^2 \, \text{gain}_x \rightarrow E_x = E_1 \sqrt{\text{gain}_x}$$

There is now no more need for diagrams. Table 5.3 summarizes all the values described in the preceding paragraphs.

Table 5.3

Type of antenna	Antenna gain	Coefficient k for peak values of E and P	Coefficient k for r.m.s. values of E and P
Isotropic antenna	1	$7.746\sqrt{1} = 7.746$	$7.746/\sqrt{2} = 5.47$
Hertzian dipole	1.5	$7.746\sqrt{1.5} = 9.487$	$9.487/\sqrt{2} = 6.71$
$\lambda/2$ dipole	1.64	$7.746\sqrt{1.64} = 9.923$	$9.923/\sqrt{2} = 7.017$

5.4 Effective Radiated Power P_{ERP}

By definition, the effective radiated power P_{ERP} of a radiating source, in a given direction, is equal to the net electrical conducted power supplied to the antenna, $P_{\text{cond bs}}$, multiplied by the relative gain that the antenna would have with respect to a $\lambda/2$ dipole for radiation in the same direction:

$$P_{\text{ERP}} = P_{\text{cond bs}} \frac{\text{gain of the antenna concerned}}{\text{gain of a } \lambda/2 \text{ dipole}}$$

Why such a strange definition?

The answer is simple. In many applications, mainly at UHF (including RFID), in order to simplify the mechanical production of the antennae, designers often use $\lambda/2$ antennae (see below), and therefore the use of ERP levels makes it easy to compare the results directly between systems.

Examples

Assume that the local regulations specify that we must not exceed a maximum power level of $P_{\text{ERP max}} = 2\,\text{W}$ in LBT conditions (see Chapter 16) (this has been the case in Europe for UHF RFID since 2005).

(a) If the antenna used is of the '$\lambda/2$ dipole' type, then, given that the gain of a $\lambda/2$ dipole antenna (see below in this chapter) is 1.64 (i.e. 2.14 dB), the transmitter amplifier can deliver an electrical power $P_{\text{cond max}}$ of 2 W, since

$$P_{\text{cond bs max}} \times \frac{1.64}{1.64} = 2\,\text{W} \times 1 = P_{\text{cond bs max}} = P_{\text{ERP max}}$$

(b) If the antenna used has a gain of 6 dB (i.e. an absolute gain of 4), we can write

$$P_{\text{cond bs max}} \times \frac{4}{1.64} = P_{\text{cond bs max}} \times 2.44 = 2\,\text{W}$$

and therefore

$$P_{\text{cond bs max}} = 0.82\,\text{W}$$

Conclusions from Examples (a) and (b)
With an antenna having a gain of 6 dB, if $P_{\text{ERP max}} = 2\,\text{W}$ max, then $P_{\text{cond bs max}} = 0.82\,\text{W}$. We can then calculate the $P_{\text{EIRP max}}$ associated with the $P_{\text{ERP max}}$:

$$P_{\text{EIRP max}} = 0.82\,\text{W} + 6\,\text{dB}$$
$$P_{\text{EIRP max}} = 0.82\,\text{W} \times 4 = 3.28\,\text{W}$$

Generalizing the Relation Between P_{EIRP} and P_{ERP}
As a general rule, regardless of the antenna used, if its gain is x dB (i.e. an absolute gain of y),

$$P_{\text{ERP}} = P_{\text{cond bs}} \times \frac{y}{1.84} \rightarrow P_{\text{cond bs}} = P_{\text{ERP}} \times \frac{1.64}{y}$$

giving us the value of P_{EIRP}:

$$P_{\text{EIRP}} = P_{\text{cond bs}} y \rightarrow P_{\text{EIRP}} = P_{\text{ERP}} \times \frac{1.64}{y} y$$

and therefore, finally:

in linear terms: $P_{\text{EIRP}} = 1.64\,P_{\text{ERP}}$

in dBm: $P_{\text{EIRP}}\,(\text{dBm}) = 2.14\,\text{dB} + P_{\text{ERP}}\,(\text{dBm})$

Note

Power levels are often expressed as P_{ERP} in UHF and as P_{EIRP} in SHF. The reason for this difference is that it is easy to produce a '$\lambda/2$ dipole' antenna for UHF, whereas for SHF (2.45 GHz) the antenna are often not pure dipoles and have a higher gain, in order to compensate for higher attenuation losses in the communication medium (mainly air in the case of RFID) (see Chapter 6).

Comparison of Performance Between Europe and the USA
There is often a war of numbers over the possible operating distances between Europe and the USA. This is totally pointless, for the following reasons:

- The comparison is often between 'chalk and cheese', since the maximum radiated power levels permitted by the regulations are not the same.
- Moreover, the power levels are often expressed in P_{ERP} on one side of the Atlantic and P_{EIRP} on the other!

I can assure you that this is no way to do business. To give a better idea of what is involved, let us look at a real-world example.

An example in UHF
Suppose that, at a given date, the current regulations permit:

- for the USA (902 to 928 MHz):

$$P_{EIRP\ USA} = 4\ W\ max = 36\ dBm$$

(Note that the maximum conditions of 1 W are conducted + 6 dB antenna gain.)
- and in Europe:

$$P_{ERP} = 2\ W\ max = 33\ dBm$$

i.e.

$$P_{EIRP\ Europe} = 1.64 \times 2 = 3.28\ W\ max = 35.14\ dBm$$

(*indoor* and *outdoor*, with a transmission mark–space ratio dc = 100%), for LBT operating conditions only, or, in another solution,

$$P_{ERP} = 0.5\ W\ max = 27\ dBm$$

$$P_{EIRP\ Europe} = 1.64 \times 0.5 = 0.820\ W\ max = 29.14\ dBm$$

(in *outdoor* conditions with dc = 10%).
   At a given application point, in *indoor* conditions (inside a building) and ignoring the different values of the mark–space ratio, and hence the possible problems of collision management, for the same received power flux density s, the operating ranges (according to the power flux density equation) will be proportional to the received electrical field E, and therefore to the ratio of the square root of the maximum possible EIRP levels, i.e. in the above cases:

- for $P_{ERP} = 2\ W$:

$$\frac{r_{Eur}}{r_{USA}} = \sqrt{\frac{P_{EIRP\ Europe}}{P_{EIRP\ USA}}} = \sqrt{\frac{3.28}{4}} \approx 0.9$$

- for $P_{ERP} = 0.5$ W:

$$\frac{r_{Eur}}{r_{USA}} = \sqrt{\frac{P_{EIRP\ Europe}}{P_{EIRP\ USA}}} = \sqrt{\frac{0.820}{4}} \approx 0.45$$

This simply means that, leaving aside all arguments, simple compliance with the regulations, everything else being equal, means that the same transponder can only operate in Europe at a maximum range of 90% of its possible operating range in the USA (e.g. 3.5 m in the USA and 3.15 m in Europe). These explanations will be pursued further in Chapter 6, to bring this ratio of 90% almost up to 100%. For now, you will have to be patient.

Readers are therefore advised to pay great attention to product specifications, the terms used and the regulations they have to comply with and the way that they are drawn up.

Examples of Calculation of the Antenna Gain

Antenna Gain of a Hertzian Dipole
As shown in Chapter 4, the effective mean power flux density for a Hertzian dipole with a length l that is small by comparison with the wavelength λ is as follows:

$$\langle s \rangle = \frac{15}{\pi} I_0^2 l^2 \frac{\sin^2\theta}{r^2}\ W_{peak\ mean}\ m^{-2}$$

and this function has a maximum $s_{mean\ radiated\ max}$, when $\theta = 1$ and therefore $\theta = \pi/2$, i.e. in the direction of the equatorial plane:

$$s_{mean\ max} = \langle s \rangle = \frac{15}{\pi}\left(\frac{I_0 l}{\lambda r}\right)^2$$

I have also shown, in the same chapter, that the total power radiated by a dipole is

$$P_{mean\ tot\ radiated} = 40\left(\frac{\pi I_0 l}{\lambda}\right)^2$$

However, by definition, for an isotropic antenna delivering the same total power, the isotropic power flux density s_{iso} would be

$$s_{iso} = \frac{P_{iso}}{4\pi r^2}$$

where P_{iso} denotes the total mean radiated isotropic power (in watts rms) and $4\pi r^2$ denotes the surface area (in m^2) of the sphere of radius r surrounding the isotropic antenna.

Dividing $P_{mean\ tot\ radiated}$ by $(4\pi r^2)$, we obtain

$$\langle s \rangle = 10\pi\left(\frac{I_0 l}{\lambda r}\right)^2\ W_{peak\ mean}\ m^{-2}$$

By establishing the ratio $s_{mean\ max}/s_{iso}$ (both expressed in $W_{rms}\ m^{-2}$), we have also calculated the relative gain provided by using the Hertzian dipole antenna as compared with a standard isotropic antenna with unitary gain – in the direction of the equatorial plane – when both antennae are driven in the same electrical conditions:

$$\text{Gain of the Hertzian dipole} = \frac{s_{\text{mean max}}}{s_{\text{iso}}} = 1.5$$

and therefore, to summarize, for an isotropic Hertzian dipole:

$$\text{Gain of the Hertzian dipole} = 1.5 \text{ or in dB} = 10 \log (1.5) = 1.76 \text{ dBi}$$

Antenna Gain of a λ/2 Dipole
In Chapter 4, I showed that, for a λ/2 dipole (i.e. with $n = 1$), the effective mean power flux density was given by

$$\langle s \rangle = \frac{15}{\pi} I_0^2 \frac{\cos^2[(\pi/2)\cos\theta]/\sin^2\theta}{r^2} \text{ W}_{\text{peak mean}} \text{ m}^{-2}$$

This function has a maximum value, $s_{\text{mean max}}$, for $\theta = \pi/2$, i.e. in the direction of the equatorial plane:

$$s_{\text{mean max}} = \langle s \rangle = \frac{15}{\pi} \left(\frac{I_0}{r} \right)^2$$

In the same chapter, I also showed that the effective total power radiated by a λ/2 dipole was

$$P_{\text{mean tot radiated } (\lambda/2 \text{ dipole})} = 36.564 I_0^2$$

However, by definition, for an isotropic antenna delivering the same mean total effective radiated power, the isotropic power flux density s_{iso} would be

$$s_{\text{iso}} = \frac{P_{\text{iso}}}{4\pi r^2}$$

where P_{iso} denotes the total mean radiated isotropic power (in watts rms) and $4\pi r^2$ denotes the surface area (in m²) of the sphere of radius r surrounding the isotropic antenna. Dividing the latter value by $4\pi r^2$, we obtain

$$s_{\text{iso}} = 2.911 \left(\frac{I_0}{r} \right)^2 \text{ W}_{\text{rms}} \text{ m}^{-2}$$

By establishing the ratio $s_{\text{average max}}/s_{\text{iso}}$, we can calculate the relative gain resulting – in the equatorial plane – from the use of a λ/2 dipole antenna, as compared with a standard isotropic antenna with unitary gain when both antennae are driven in the same electrical conditions:

$$\text{Gain of the } \lambda/2 \text{ iso dipole} = \frac{s_{\text{mean max}}}{s_{\text{iso}}}$$

$$\text{Gain of the } \lambda/2 \text{ iso dipole} = \frac{(15/\pi)(I_0/r)^2}{2.911(I_0/r)^2} = \frac{4.777}{2.911} = 1.641$$

For an isotropic $\lambda/2$ dipole:

$$\text{Gain of the } \lambda/2 \text{ dipole} = 1.64 \text{ or in dB} = 10\log(1.64) = 2.14\,\text{dBi}$$

We will make ample use of these values subsequently for evaluating antennas and the performance of transponders.

dB, dBi and dBd

If you have been very attentive, you will have noticed the appearance of some units in the form of dBi. Are there any other minor subdivisions of dB? There certainly are!

To make matters completely clear:

- To state the gain of a specific antenna with respect to an isotropic antenna, we express it in dBi, where 'i' stands for 'isotropic'.
- To state the gain of a specific antenna with respect to a $\lambda/2$ dipole antenna, we express it in dBd, where 'd' stands for 'dipole'.

Clearly, there is a relationship between these, which will be quite evident if you have read the sections above:

$$G(\text{dB}) = G(\text{dBi}) = G(\text{dBd}) + 2.14\,\text{dB}$$

Example

From the start of this chapter, where I have described an antenna, not quite correctly, as having a gain of $+6\,\text{dB}$, what I mean is that its gain is $+6\,\text{dBi}$ or $+3.86\,\text{dBd}$.

Directivity and Radiation Efficiency of an Antenna

As stated above, antennae have a gain, so we could generalize the equation as follows:

$$G(\theta, \phi) = \frac{(dP_{\text{radiated}}/d\Omega)}{(P_{\text{conducted}}/(4\pi))}$$

where Ω is the value of the solid angle at which the observation is made and $P_{\text{conducted}}$ is the total power supplied by the amplifier to the antenna terminals.

Note also that, throughout this chapter, we have assumed that the antenna was lossless, in other words that all the conducted power $P_{\text{conducted}}$ supplied by the amplifier is entirely radiated by the antenna. Unfortunately, this is never quite the case, since the antenna is never perfect, and has losses that can be expressed by losses in watts P_{loss} due to an equivalent resistance R_{loss} including all the losses (imperfect conductors, substrate, etc.). The radiated power, P_{radiated}, will therefore be given by

$$P_{\text{radiated}} = \eta_{\text{rad}} P_{\text{conducted}}$$

where η_{rad} is the radiation efficiency of the antenna.

Now, assuming that P_{radiated} is the total power radiated by the antenna, allowing for the preceding comments, we can also define what is known as the directivity of an antenna by establishing

$$D(\theta, \phi) = \frac{(dP_{\text{radiated}}/d\Omega)}{(P_{\text{radiated}}/(4\pi))}$$

Antenna manufacturers often state the maximum value of this parameter and supplement this with the global radiation pattern of the antenna (see Figure 5.4 for an example). By establishing the relation between the three equations above, we immediately find

$$G(\theta, \varphi) = \eta_{\text{rad}} D(\theta, \varphi)$$

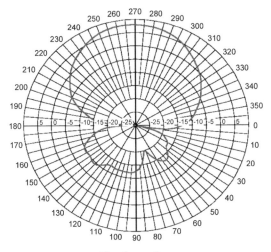

Elevation section

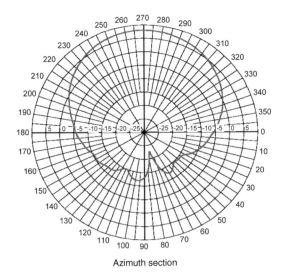

Azimuth section

Figure 5.4 Example of a radiation pattern (elevation/azimuth)

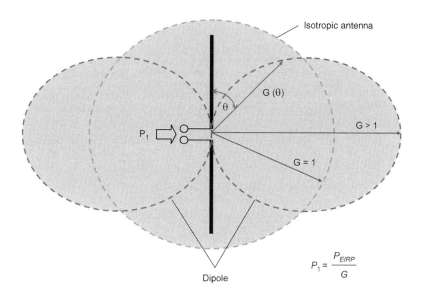

Figure 5.5   Relationship between the gain and the directivity of an antenna

or, alternatively,

$$G(\theta, \varphi) = \frac{P_{\text{radiated}}}{P_{\text{conducted}}} D(\theta, \varphi)$$

In the common case in which the antenna is assumed to have no losses (or very few), $P_{\text{loss}} = 0$ (none of the power supplied to the antenna unit has been lost in the resistance R_{loss}), the product $G_{\text{ant}}P_{\text{cond}} = P_{\text{EIRP}}$ is entirely radiated via the radiation resistance R_{ant} of the antenna (Figure 5.5). In this case, $\eta_{\text{rad}} = 1$, which clearly means that $G = D$.

Finally, it should be noted that, in many cases, the conducted power level is not what we would expect, since we must allow for the value of the reflection coefficient Γ due to imperfect matching between the amplifier and antenna impedances, which tends to reduce the actual value of the conducted power.

Beamwidth

The directivity D of an antenna is usually associated with a new parameter, describing the width of the radiated beam. This value (the *beamwidth*) is associated with the vertical and horizontal angles of the 'cone' of radiation of the radiated mean power density $\langle s_r \rangle$. To determine this value for a given type of antenna, we calculate the angle θ for which the value of $\langle s_r \rangle$ is divided by two (in other words, the angle at which the antenna gain is also reduced by 3 dB).

Examples

(a) For a Hertzian dipole, I have shown that

$$\langle s_{\text{peak}} \rangle = \langle |s_r| \rangle = \frac{15\pi}{\lambda^2} I_0^2 l^2 \frac{\sin^2\theta}{r^2}$$

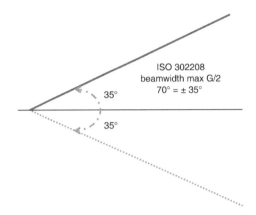

Figure 5.6 Beamwidth: the case of the ETSI 302 208 – LBT (listen before talk) – standard

In order for the value of $\langle\, s_r\,\rangle$ to be divided by two, in the vertical direction of radiation, all other things being equal, we must have $\sin^2\theta = \frac{1}{2}$, which is the case when $\theta = 45°$. The beamwidth of a Hertzian dipole antenna is therefore $2\,(90° - \theta) = 90°$.

(b) For a $\lambda/2$ dipole, I have shown that

$$\langle s_r \rangle = \frac{15}{\pi} I_0^2 \frac{(\cos^2[(\pi/2)\cos\theta])/\sin^2\theta}{r^2}$$

In order for the value of $\langle s_r \rangle$ to be divided by two, in the vertical direction of radiation, all other things being equal, we must have

$$\frac{\cos^2[(\pi/2)\cos\theta]}{\sin^2\theta} = \frac{1}{2}$$

which is the case when $\theta = 51°$. The beamwidth of a $\lambda/2$ dipole antenna is therefore $2(90° - 51°) = 78°$ (Figure 5.6). I will return to this subject in Chapter 16, when I discuss the ETSI 302 208 standard (2 W_{ERP} at UHF) in relation to the LBT principle.

(c) An amplifier can deliver a conducted power of 1 W to an 'antenna' load. If this amplifier is provided with an isotropic antenna with unitary gain, then by definition it will uniformly radiate the power delivered by the amplifier, namely 1 W_{EIRP}.

The same amplifier is connected to an antenna having a special shape. The power radiated along the principal axis of the antenna is then measured and, because of the special shape of the antenna, this power is found to be, for example, 4 W_{EIRP}. The gain of the antenna is then 4 W/1 W = 4 or, in dB, 10 log 4 = +6 dB; in other words, this antenna is rather directive and the beamwidth of the radiated beam is small.

This completes our survey of the pure propagation of RF waves in free space. Now we shall consider how the electrical power is recovered at the terminals of a receiving antenna.

6

Power Recovery at the Terminals of the Tag Antenna

This is one of the most important aspects of RFID applications. Simply because of the costs involved, tags for widespread use (for traceability applications, etc.) have to be remotely powered by the incident radiation from the base station.

The quality of the remote power supply and consequently the operating distance depend to a great extent on the amount of energy or power that the tag can recover from the base station in order to operate. Here we encounter one of the main problem areas of RFID at UHF and SHF.

6.1 Recovering the Transmitted Radiated Power (or Some of It)

As demonstrated previously, the (isotropic) antenna of the base station radiates energy, which travels towards the tag at the speed of light in all directions. At any instant, this energy is distributed uniformly over the whole surface of a sphere whose centre is the antenna of the base station and whose radius r is equal to the travelling time multiplied by the speed of light. Since the surface area of a sphere ($4\pi r^2$) is multiplied by four when its radius is multiplied by two, the energy density s_{EIRP} is divided by four whenever the radius is multiplied by two. This principle holds true throughout the universe.

Given the conducted electrical power that can be supplied by the base station amplifier, $P_{\mathrm{cond\ bs}}$, and the power, P_{EIRP} (in watts), radiated by the amplifier via the gain of its antenna, G_{bs}, we know the power flux density at a point in space, since by definition the isotropic power flux density s_{iso} for an isotropic antenna is

$$S_{\mathrm{EIRP}} = \frac{P_{\mathrm{EIRP}}}{4\pi r^2}\, W_{\mathrm{mean\ peak}}\, m^{-2}$$

We can now determine the power available for the tag P_t (actually in the load R_l that it represents) as a function of the frequency f, the distance r, the antenna gains $G_{\mathrm{ant\ bs}}$ and $G_{\mathrm{ant\ t}}$, as well as the respective effective areas of the transmission and reception antennas of the base stations and tags.

6.2 The Concept of Aperture or Surface

In order to be able to supply energy to its electronic circuits, the tag must collect the energy passing through a certain 'imaginary, immaterial' surface located around its antenna, which is called the *antenna aperture* or *effective area* σ_e of the tag's antenna.

Before going any further, let me introduce the concept of area (aperture), which we will need to use later on. For ease of understanding, let us assume for now that the receiving antenna of the tag is a horn (Figure 6.1) whose aperture is a 'mouth' with the area σ.

Let us expose this antenna to the radiated electromagnetic field created by a uniform plane wave. This is what happens at a long distance (in the far field, therefore) from a radiating transmitting source, where the radii of the spheres of propagation are assumed to have infinite values and we can therefore consider that plane waves are present. The power flux density (Poynting vector) of this wave is s (in W m^{-2}). By definition, all the power of the wave absorbed or extracted by the horn through the area σ of its mouth is therefore

$$P = \sigma s$$

To avoid confusion, it is useful to distinguish several different kinds of 'areas', i.e. apertures, depending on the phenomena we are studying:

- the effective area σ_e, as mentioned above;
- the dispersion (*scattering*) area σ_s, which is described below (in Chapter 8);
- the loss area σ_l, which is the area corresponding to the losses of the antenna;
- the collecting area σ_c, which is the sum of all the above areas;
- the physical area σ_{ph}, which is the mechanical area of the antenna.

6.2.1 Power Available to the Tag Load at a Point in Space

By definition, at a point in space, the power $P_{t\,rms}$ available to the (electrical) load placed across the terminals of the tag antenna is proportional to the modulus of the radiated effective mean power flux density s, in W$_{rms}$ m^{-2}, and to the maximum effective power capture area of the tag's receiving antenna, σ_e, in m^2.

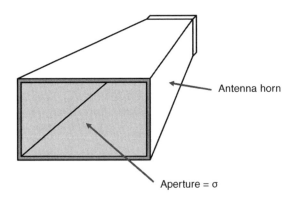

Figure 6.1   Schematic diagram of the aperture of the tag's receiving antenna

This 'effective area' (which in principle is entirely immaterial) therefore establishes the relationship between the power density at one point in space and the power that the tag's payload can absorb from the incident radiation. If a tag has an effective antenna area of σ_e, the total power that is received (or 'captured') and therefore available at the terminals of the load of the tag antenna P_t (in watts) will be as follows:

$$P_{t\,rms} = \sigma_e s \quad \text{in effective watts}$$

Now let us get down to the specifics by seeing how we can evaluate this area.

Note

Throughout the following sections, I will assume that the antennas of the base station and tag are orientated 'face to face', in other words that the peaks of the radiation patterns are aligned.

6.2.2 Equivalent Circuit of the Tag

Let us assume that:

- we are in the far field, and therefore in the presence of a plane wave, and that an electrical field with an effective strength of E_{rms} originating from the base station is present around the tag antenna;
- the dipole has an associated effective length l_{eff} (I will show you its value at the end of the chapter);
- when the antenna is in the no-load state, a potential difference $V_{equi\,rms}$ is created across its terminals ($V_{equi\,rms} = E_{rms} l_{eff}$);
- when a load Z_l is connected to the terminals of the antenna, a potential difference (PD) $V_{received\,rms}$ is developed across its terminals.

The tag assembly can be represented by the equivalent circuit shown in Figure 6.2. This consists of three parts:

- a Thévenin generator equivalent to the tag antenna assembly:
  - consisting of a voltage source delivering an equivalent voltage $V_{equi\,rms}$ in no-load conditions. Note that if the load is matched, $R_l = R_{ant}$ and therefore $P_{ant} = P_l$:

$$V_{equi\,rms} = 2V_{received\,rms} = 2\sqrt{P_t R_l}$$

  - having an internal impedance (impedance of the tag antenna) as follows:

$$Z_{ant\,t} = (R_{ant\,t} + R_{loss}) + jX_{ant\,t}$$

  where $R_{ant\,t}$ is the radiation resistance of the antenna of the transponder, R_{loss} is the ohmic resistance of losses and $X_{ant\,t}$ is the reactance of the transponder antenna;
  - this equivalent generator represents the total combination of $P_{conducted\,bs}$, $G_{ant\,bs}$, the attenuation due to the medium, the power flux density s in which the transmission takes place and finally $G_{ant\,t}$;

Forward link

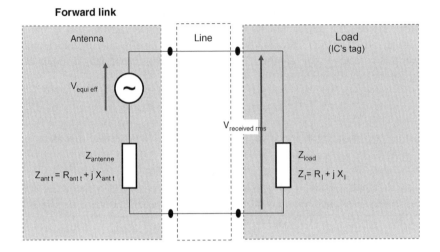

Figure 6.2 Equivalent circuit of the tag

- an external 'load' whose impedance $Z_1 = (R_1 + jX_1)$ is the impedance equivalent to all the circuits present across the terminals of the antenna (normally the entire integrated circuit forming the tag);
- finally, a 'wire' link (the printed circuit, soldered joints, etc.) between the antenna output and the physical input of the integrated circuit which forms, or can form, a distributed or lumped constant line at these frequencies (UHF, SHF). This is normally short, and therefore has no losses, or has losses that are considered negligible, but which, if present, must be taken into account for impedance matching and for quantifying the wave reflection coefficients if there is an impedance mismatch between the source and the load.

6.2.3 Power in the Load R_1

While the tag is illuminated by the incident electromagnetic radiation, the voltage $V_{received}$ creates a current I in the load ($Z_1 = R_1 + jX_1$):

$$I = \frac{V_{received}}{R_1 + jX_1}$$

This current therefore flows through the whole of the circuit equivalent to the combined load connected in series with the antenna ($Z_{ant\,t} = R_{ant\,t} + jX_{ant\,t}$). The voltage V_{equi} is therefore as follows:

$$V_{equi} = [(R_{ant\,t} + jX_{ant\,t}) + (R_1 + jX_1)]I$$

or, alternatively, if we enter the value of I in this equation:

$$V_{received} = \frac{R_1 + jX_1}{(R_{ant\,t} + jX_{ant\,t}) + (R_1 + jX_1)} V_{equi} = f(V_{equi}, Z_{ant}, Z_1)$$

Regardless of the impedances present, the complex form of the general equation for the radio frequency current I flowing in the circuit will be as follows (in the case of lossless short lines):

$$I = \frac{1}{(R_{\text{ant t}} + R_1) + j(X_{\text{ant t}} + X_1)} V_{\text{equi}}$$

We can use this equation to calculate the effective value I_{rms} of the current, by calculating the modulus (effective value) of the complex impedance of the circuit. This value is

$$I_{\text{rms}} = \frac{1}{\sqrt{(R_{\text{ant t}} + R_1)^2 + (X_{\text{ant t}} + X_1)^2}} V_{\text{equi rms}}$$

and, when raised to the square,

$$I_{\text{rms}}^2 = \frac{V_{\text{equi rms}}^2}{(R_{\text{ant t}} + R_1)^2 + (X_{\text{ant t}} + X_1)^2}$$

Regardless of the complex value of the load impedance $Z_1 = R_1 + jX_1$ and its modulus $|Z_1| = \sqrt{R_1^2 + X_1^2}$, we can calculate the real power (in watts) delivered by the generator (the antenna) to the load R_1 connected to its terminals. The general equation for this is

$$P_{1,\text{rms}} = R_1 I_{\text{rms}}^2$$

and therefore

$$P_{1\,\text{rms}} = R_1 \frac{V_{\text{equi rms}}^2}{(R_{\text{ant t}} + R_1)^2 + (X_{\text{ant t}} + X_1)^2} \quad \text{watts}$$

6.2.4 Effective Area σ_e of a Tag

The relationship between this power $P_{1\,\text{rms}}$ consumed in the load (and therefore available for an application) and the effective power flux density s of the incident radiation present at the point in space in question represents the effective area σ_e of the tag:

$$\sigma_e = \frac{P_{1\,\text{rms}}}{s}$$

and therefore, bringing this into the preceding equation,

$$\sigma_e = \frac{R_1 V_{\text{equi rms}}^2}{s[(R_{\text{ant t}} + R_1)^2 + (X_{\text{ant t}} + X_1)^2]}$$

This equation is generally applicable.

The Case of Conjugate Impedance Matching in the Source and Load

As usual, for any generator/load pair (see the inset below), the maximum power is supplied to the load when conjugate matching is provided between the source impedance $Z_{\text{ant t}}$ and load impedance Z_1, such that

$$R_1 = R_{\text{ant radiation}} + R_{\text{loss}}$$

and $X_1 = -X_{\text{ant}}$ and $R_{\text{loss}} = 0$, in other words the losses are negligible.

In this specific case, where $R_1 = R_{\text{ant t}}$ and $X_1 = -X_{\text{ant t}}$, we obtain

$$P_{\text{1 rms max}} = \frac{R_{\text{ant t}}}{(R_{\text{ant t}} + R_{\text{ant t}})^2} V_{\text{equi rms}}^2 = \frac{1}{4R_{\text{ant t}}} V_{\text{equi rms}}^2$$

By definition, when matching takes place, the ratio between the maximum power $P_{\text{1 rms max}}$ available in the load and the effective power flux density s of the incident radiation present at this point in space is called the 'effective area' of the tag antenna, $\sigma_{\text{e t}}$:

$$\sigma_{\text{e t}} = \frac{P_{\text{1 rms max}}}{s}$$

$$\sigma_{\text{e t}} = \frac{R_1 V_{\text{equi rms}}^2}{s(2R_1)^2}$$

$$\sigma_{\text{e t}} = \frac{V_{\text{equi rms}}^2}{s(4R_1)}$$

or

$$\sigma_{\text{e t}} = \frac{V_{\text{equi rms}}^2}{s(4R_{\text{ant t}})} \text{ m}^2$$

since this is a case of conjugate matching of the reactances X and resistances R. This area $\sigma_{\text{e t}}$ (also called the 'aperture') represents the 'effective' area of the space required to provide the maximum power of the incident radiation that can be supplied to the load (i.e. impedance matching) (Figure 6.3).

Note that $R_1 = R_{\text{ant}}$, and therefore

$$V_{\text{equi rms}} = 2V_{\text{received rms}}$$

$$V_{\text{equi rms}} = 2\sqrt{P_{\text{t rms}}R_1}$$

Let us now find the value of $\sigma_{\text{e t}}$.

Demonstration of the Maximum Power for $R_l = R_{\text{ant } t}$

The current flowing in the circuit shown in Figure 6.2 is

$$I = \frac{V_1}{R_1 + R_{\text{ant t}}}$$

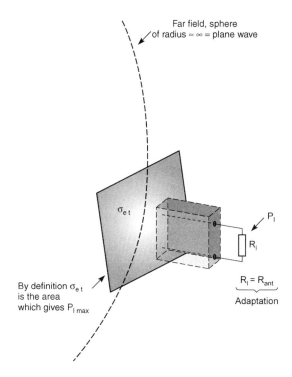

Figure 6.3 Effective power reception area of the tag

The PD at the terminals of the load resistance R_1 is therefore

$$V_1 = R_1 \frac{V_1}{R_1 + R_{\text{ant t}}}$$

and the power dissipated in it is expressed thus:

$$P_{R1} = V_1 I$$
$$P_{R1} = V_1^2 \frac{R_1}{\left(R_1 + R_{\text{ant t}}\right)^2} = f(R_1)$$

Before determining whether the power in the load P_{R1} passes through a peak, let us examine the derivative of this equation, in the form u/v. This is equal to $(u'v - v'u)/v^2$, i.e.

$$P'_{R1} = \frac{V_1^2 \left(R_{\text{ant t}} - R_1\right)}{\left(R_1 + R_{\text{ant t}}\right)^3}$$

and is cancelled when $R_1 = R_{\text{ant t}}$ and the function passes through a maximum for this value, which is as follows (Figure 6.4):

$$P_{R1 \text{ max}} = \frac{V_1^2}{4R_1}$$

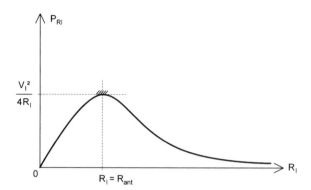

Figure 6.4   Variation of power transferred to the load as a function of the load resistance: $P_{Rl} = f(R_l)$

Equivalent/Effective Area σ_e of an Antenna

Let us take a closer look at the value of the effective area. We shall consider a link between two antennas, A and B, and these two possible cases of operation:

- Case 1: A sends and B receives.
- Case 2: B sends and A receives.

The antenna A is characterized by its gain G_A, its equivalent area σ_A and its radiation resistance R_A. The antenna B is characterized by its gain G_B, its equivalent area σ_B and its radiation resistance R_B.

The Reciprocity Principle

The reciprocity principle signifies that the problem is symmetrical, because a transmitting antenna is also a receiving antenna that has the same characteristics. Thus, in both cases we should find similar results for the ratio of the received voltage to the inducing current. Because of this reciprocity, we can write

$$\frac{e_1}{I_{01}} = \frac{e_2}{I_{02}}$$

where e_1 and e_2 are the peak voltages cases in 1 and 2 respectively, and I_{01} and I_{02} are the peak currents in 1 and 2 respectively.

Case 1

If $P_{E1\,rms}$ denotes the r.m.s. power transmitted at A and $P_{R1\,rms}$ is the r.m.s. power received at B, we can write

$$P_{E1\,rms} = \frac{1}{2}R_A I_{01}^2$$

where $P_{E1\,rms}$ is the conducted power of the transmitter. The factor $^1/_2$ is there because $I_{01} = $ peak amplitude and $P_{R1\,rms} = s_1\sigma_B$, i.e. in the case where the load B is matched to the radiation resistance R_B.

Given that

$$s_1 = \frac{P_{E1\ rms}G_A}{4\pi r^2}$$

it follows that

$$P_{R1\ rms} = \frac{P_{E1\ rms}G_A}{4\pi r^2}\sigma_B$$

Furthermore, since the load of the receiver is identical to/matched to the radiation resistance of the antenna, if we call the peak value at the antenna terminals e_1, we can also write

$$P_{R1\ rms} = \frac{[e_1/(V_2/2)]^2}{R_B} = \frac{e_1}{8R_B}$$

From this set of equations we can deduce the value of the ratio $(e_1/I_{01})^2$:

$$\left(\frac{e_1}{I_{01}}\right)^2 = \frac{8R_B P_{R1\ rms}}{(2P_{E1\ rms}/R_A)} = \frac{8R_B P_{R1\ rms}}{(2/R_A)((P_{R1\ rms} \times 4\pi r^2)/(G_A\sigma_B))}$$

and therefore

$$\left(\frac{e_1}{I_{01}}\right)^2 = \frac{G_A R_A R_B \sigma_B}{4\pi r^2}$$

Case 2
We can follow the same procedure in the reciprocal case, letting $P_{E2\ eff}$ be the effective transmitted power at B and $P_{R2\ eff}$ be the effective power received at A, so that

$$P_{E2\ rms} = \frac{1}{2}R_B I_{02}^2$$

where I_{02} is the peak amplitude and $P_{R2\ rms} = s_2\sigma_A$. Given that

$$s_2 = \frac{P_{E2\ rms}G_B}{4\pi r^2}$$

it follows that

$$P_{R2\ rms} = \frac{P_{E2\ rms}G_B}{4\pi r^2}\sigma_A$$

We can also write

$$P_{R2\ rms} = \frac{e_2^2}{8R_A}$$

i.e.

$$\left(\frac{e_2}{I_{02}}\right)^2 = \frac{G_B R_A R_B \sigma_A}{\pi r^2}$$

Given that

$$\frac{e_1}{I_{01}} = \frac{e_2}{I_{02}} \rightarrow \left(\frac{e_1}{I_{01}}\right)^2 = \left(\frac{e_2}{I_{02}}\right)^2$$

it follows that

$$G_A \sigma_B = G_B \sigma_A$$

or, alternatively,

$$\frac{G_A}{\sigma_A} = \frac{G_B}{\sigma_B}$$

Note
This equation is remarkable because it is independent of the intrinsic properties of the antennas. It is therefore valid in any circumstances, e.g. if A is a Hertzian dipole and B is an antenna of any type.

Making use of this important fact, let us now calculate the value of this ratio in the specific case of a Hertzian dipole, in order to find its general expression. Let us assume that we have a Hertzian dipole receiving antenna with a length of $l \ll \lambda$ positioned face to face with an identical Hertzian dipole transmitting antenna, and exposed to the electrical field E produced by the latter. Given that the rms mean power flux density s of a Hertzian dipole is

$$\langle s \rangle = \frac{1}{2} \times \frac{|E|^2}{120\pi}$$

where $\langle s \rangle$ is in watts r.m.s. and E is still expressed as a peak value.

If the length l (effective/equivalent length) of the Hertzian dipole is also assumed to be small with respect to the wavelength λ, so that the electrical field E resembles a uniform field around the dipole, the peak voltage ($V = V_{rms}\sqrt{2}$) collected at the terminals of the receiver antenna is $V = |E|l$. As usual, this dipole will deliver the maximum received power for a load (resistance) R_l when the value of the latter is equal to the equivalent (internal) impedance of the Hertzian dipole, R_{ant}. In this case, the maximum power $P_{1\ rms\ max}$ available in the load will be

$$P_{1\ rms\ max} = \frac{(V_{rms}/2)^2}{R_{ant}} = \frac{V_{rms}^2}{4R_{ant}}$$

$$P_{1\ rms\ max} = \frac{V^2}{8R_{ant}} = \frac{|E|^2 l^2}{8R_{ant}}$$

(The curious presence of '8' in the denominator, instead of the usual '4', is due to the fact that, in order to keep the equations uniform from the beginning of the previous chapter onwards, the symbol V denotes the peak voltage, not the r.m.s. voltage.)

In this optimal case of impedance matching between the source resistance (the radiation resistance of the antenna) and the load resistance, the maximum r.m.s. power $P_{1\ rms\ max}$

available to the load is equal to P_t, and thus

$$P_{t\,rms} = s\sigma_{e\,t} = \frac{1}{2} \times \frac{|E|^2}{120\pi}\sigma_{e\,t} = P_{l\,rms\,max} = \frac{|E|^2 l^2}{8R_{ant}}$$

I have also shown that the radiation resistance of a Hertzian dipole is

$$R_{ant\,dipole} = 80\left(\frac{\pi l}{\lambda}\right)^2$$

Finally, by introducing these two values into the last equation for $P_{t\,rms}$, we obtain

$$R_{ant\,dipole} = \frac{1}{2} \times \frac{|E|^2}{120\pi}\sigma_e = \frac{|E|^2 l^2}{8 \times 80(\pi l/\lambda)^2}$$

from which is derived the general expression of the equivalent or effective area of a Hertzian dipole antenna, $\sigma_{e\,ant\,doublet}$, in the case of maximum power transfer between the antenna and the load ($R_{ant} = R_l$):

$$\sigma_{e\,ant\,dipole} = \frac{3\lambda^2}{8\pi} = 0.119\lambda^2 \text{ for a Hertzian dipole}$$

The General Relationship Between the Equivalent Area and the Gain of an Antenna

To conclude, as mentioned above, since we know that the gain G of a Hertzian dipole is $1.5 = 3/2$, we can determine the ratio G_A/σ_A in this specific case, which is also, as I have said, the general case. We find that

$$\frac{G_{ant\,Hertzian\,dipole}}{\sigma_{e\,ant\,dipole}} = \frac{3/2}{3\lambda^2/(8\pi)} = \frac{4\pi}{\lambda^2} = \frac{G_A}{\sigma_A}$$

Since, as proved in the previous paragraphs, the value of G_A/σ_A is true regardless of the type of antenna, we can now generalize the relationship between the equivalent area σ_e of an antenna and its gain G:

$$\frac{G}{\sigma_e} = \frac{4\pi}{\lambda^2} \quad \text{or} \quad G = \frac{4\pi\sigma_e}{\lambda^2} \quad \text{or} \quad \sigma_e = \frac{\lambda^2 G}{4\pi}$$

> **Note**
> In reality, the gain of an antenna is slightly weaker than the theoretical value given by the above expression, for various reasons, including the efficiency of the antenna, the factor known as the 'illumination factor', representing the ratio between the flux actually emitted or captured by the real antenna and the theoretical flux, etc.

Equivalent Area of the Antenna of a Tag

In the case of our tag, in other words with maximum power transfer between the antenna and the load ($R_{ant\,t} = R_l$), where $G_{ant\,t}$ is the gain of the tag antenna, this proportionality is written thus:

$$\sigma_{e\,t} = \frac{\lambda^2}{4\pi} G_{ant\,t}\ m^2$$

Note

For the sake of completeness, since the gain of an antenna depends on its orientation and location in space, in the form $G_{ant\,t}\,(\theta,\varphi)$, the same will apply to $\sigma_{e\,t}$ in the form $\sigma_{e\,t}\,(\theta,\varphi)$.

Example

Table 6.1 shows an example of the effective areas of the antenna of an RFID tag, $\sigma_{e\,t}$, as a function of the frequencies and types of antenna used.

Table 6.1

f				866 MHz	2.45 GHz
λ				0.346 m	0.1224 m
Type of antenna	$G_{ant\,t}$		$\sigma_{e\,t}$	$\sigma_{e\,t}$	$\sigma_{e\,t}$
	Absolute value	dBi			
Isotropic	1	0 dB	0.079 λ^2	94.7 cm^2	11.9 cm^2
Hertzian dipole	1.5	1.76 dB	0.119 λ^2	142 cm^2	17.8 cm^2
$\lambda/2$ dipole	1.64	2.14 dB	0.13 λ^2	156 cm^2	19.5 cm^2
Specific antenna (example)	4	6 dB	0.318 λ^2	382 cm^2	17.6 cm^2

Notes

As the values in the table indicate, the effective area $\sigma_{e\,t}$ is quite independent of the actual physical area of the antenna $\sigma_{e\,ph}$, which is often negligible if the antenna is wire or a metal rod. Also, as indicated by the equation for $\sigma_{e\,t}$ and this table, the effective antenna area is inversely proportional to the square of the operating frequency ($f = 1/\lambda$). For information, if everything else is equal, then for a given operating distance the power transmitted by the base station at 2.45 GHz must be 7.4 times greater than the power transmitted at 900 MHz in order to compensate for the difference due to the operating frequency between 2.45 GHz and 900 MHz ((σ_{et} at 900 MHz)/(σ_{et} at 2.45 GHz) = $(2450/900)^2 = 7.4$!)

The ratio $\sigma_{e\,t}/\lambda^2$ is often used to facilitate comparison between systems. It relates the effective area to the wavelength used, in order to provide a comparative value that can be governed by standards. Also, if everything else is equal, then an increase in the gain of an antenna is accompanied by an increase in the area $\sigma_{e\,t}$ and in the directivity of the antenna.

Note

In the case of conjugate matching ($R_l = R_{ant\,t}$), since we know that

$$S_{rms} = \frac{P_{bs\,rms}\,G_{bs}}{4\pi r^2} \tag{6.1}$$

we can combine the three equations found previously:

$$P_{t\ rms} = \sigma_{e\ t}\ s_{rms}$$

$$\sigma_{e\ t} = \frac{\lambda^2}{4\pi}G_{ant\ t} \tag{6.2}$$

$$\sigma_{e\ t}\frac{V^2_{e\ qui\ rms}}{s(4R_1)} \quad or \quad = \frac{V^2_{equi\ rms}}{s(4R_{ant\ t})} \tag{6.3}$$

The first equation 6.1 and the second 6.0 give us

$$P_{t\ rms} = \frac{\lambda^2}{4\pi}G_{ant\ t}\frac{P_{bs\ rms}G_{bs}}{4\pi r^2}$$

and by introducing the third 6.3, we obtain

$$\sigma_{e\ t} = \frac{V^2_{equi\ rms}}{[P_{bs\ rms}G_{bs}/(4\pi r^2)](4R_{ant\ t})}$$

and if we identify this result with the second:

$$\sigma_{e\ t} = \frac{\lambda^2}{4\pi}G_{ant\ t} = \frac{\pi r^2 V^2_{equi\ rms}}{P_{bs\ rms}G_{bs}R_{ant\ t}}$$

and, therefore, if impedance matching is achieved, then $R_1 = R_{ant\ t}$.
Voltage at the terminals of the antenna in no-load conditions are

$$V_{equi\ rms} = \frac{\lambda}{2\pi r}\sqrt{P_{bs\ rms}G_{bs}G_{ant\ t}R_1}$$

$$V_{equi\ rms} = \frac{\lambda}{2\pi r}\sqrt{P_{EIRP\ rms}G_{ant\ t}R_1}$$

Example
In Europe, at a distance of 10 m and using the LBT (listen before talk) principle:

Frequency $(f = 867\ MHz)$: $\lambda = 0.34\ m$
Maximum permitted radiated power $P_{ERP} = 2\ W\ \rightarrow\ P_{EIRP} = 3.28\ W$
Tag antenna, $\lambda/2$ dipole: $G_{ant\ t} = 1.64$
Antenna matching: $R_{ant\ t} = R_1 = 73.128\ \Omega$
Voltage of antenna in no-load conditions: $V_{equi\ rms} = 137.4\ mV$ approx.

The Friis Equation
Still in the context of definitions of effective areas (conjugate impedance matching between antenna and load $\rightarrow P_1 = P_{ant\ t}$), in order to obtain the maximum power available at the terminals of the tag antenna, we now carry this value of $\sigma_{e\ t}$ back into the equation $P_t = \sigma_{e\ t}s$. Assuming for simplicity that $P_{ant\ t} = P_t$, we obtain

$$P_{t\ rms} = \sigma_{e\ t}s_{rms} = \frac{\lambda^2}{4\pi}G_{ant\ t}s_{rms} = P_{1\ rms} \quad watts$$

If we now replace s with its value, we obtain

$$P_{t\,rms} = \sigma_{e\,t}\,s_{rms}$$

$$P_{t\,rms} = \frac{\lambda^2}{4\pi}G_{ant\,t}\frac{P_{bs\,rms}G_{bs}}{4\pi r^2}$$

This now brings us – regardless of the type of antenna, as long as it is matched – to the famous equation that H.T. Friis devised and first published in May 1946:[1]

$$P_{t\,rms} = P_{bs\,cond\,rms}G_{bs}\left(\frac{\lambda}{4\pi r}\right)^2 G_{ant\,t} = P_{l\,rms} \quad \text{W r.m.s.}$$

For ease of understanding, in order to follow the physical sequence of events – from cause to effect – I have departed from the usual practice and placed the terms and factors of this equation in a precise order, so as to display:

- first of all, the product $(P_{bs\,cond\,rms}G_{bs})$, which is, of course, the $P_{bs\,EIRP\,rms}$ transmitted by the base station;
- then, in the brackets, the effect of the medium, referred to below as the attenuation coefficient of the medium in which the wave is propagated;
- and, finally, the effect of the gain of the (receiving) antenna of the tag.

This power represents the (maximum) power available in the load when $R_l = R_{ant\,t}$, thanks to the presence of the effective cross-section of the antenna $\sigma_{e\,t}$. This power may be totally absorbed by the load, or partially absorbed and/or reflected. This will be examined in Chapters 7 and 8 for cases in which the load is not matched, where the power mismatch factor q is introduced and the general equation becomes

$$P_{t\,rms} = qP_{bs\,EIRP\,rms}\left(\frac{\lambda}{4\pi r}\right)^2 G_{ant\,t} = P_{l\,rms} \quad \text{watts r.m.s.}$$

Explanation of the Friis Equation in Physical Terms

To make matters clear and avoid all confusion, the Friis equation, and particularly the way in which it is constructed, is such that it gives the value of the maximum power $P_{t\,rms}$ available to the load (R_l) connected to the output terminals of the tag's receiving antenna (since it includes the value $G_{ant\,t}$), for the (optimal) case in which the load (R_l) is equal to the radiation resistance ($R_{ant\,t}$) of the receiving antenna – simply because the calculation has been carried out in that way from the start. The principle of this calculation is such that any other interpretation of the formula is incorrect, especially if the load resistance of the tag is not equal to the radiation resistance of the receiving antenna, since in this case there would be a mismatch between the source and load, leading to the appearance of reflected waves, reflection coefficients, standing wave ratios and their consequences, as we shall see subsequently.

This representation or interpretation (Figure 6.5) signifies that, at the base station (if everything is correctly matched), the conducted power is converted into power that is radiated via the base station antenna. Then, at the point of reception, the power flux density received on

---

[1] H. T. Friis, 'A note on a simple transmission formula', *Proceedings of the IRE*, **41**, 1946.

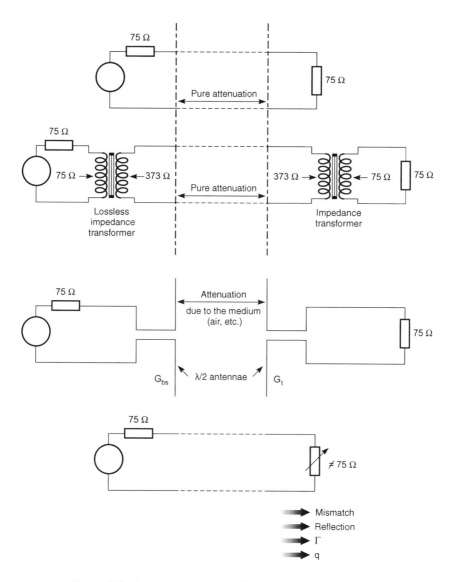

Figure 6.5   Interpretation of the Friis equation in physical terms

the effective area of the tag antenna is converted into maximum power available to a load matched to the radiation resistance of the tag's receiving antenna. This is exactly what the Friis equation says.

We can produce a highly simplified picture of the impedance matching described above, delivering the same useful power to the load, if we represent this whole system by an equivalent circuit including:

- an equivalent generator comprising:
  - a voltage source at $(2V_{1\ rms})$,

- an internal resistance of R_{ant},
  this equivalent generator representing the total combination of $P_{conducted\ bs}$, $G_{ant\ bs}$, the
  attenuation due to the medium in which transmission takes place, a received power flux
  density $\langle s \rangle$ and finally $G_{ant\ t}$;
- an external load comprising a load resistance $R_l = R_{ant}$.

In this matched case, therefore, although this is much simplified, all the power P_t (transferred
and received) will therefore be dissipated in its watt form in the resistance $R_l = R_{ant}$. In this case,
the PD $V_{l\ rms}$ present at the terminals of R_l (known as the 'received PD' present at the terminals
of the antenna) will be as follows:

$$P_{t\ rms} = P_{l\ rms} = \text{Friis equation} = \frac{V_{l\ rms}^2}{R_{l\ rms}}$$

Example
Consider, for example, a UHF RFID system operating in France or Europe in the
869.4–869.65 MHz band:

- the base station fitted with a type $\lambda/2$ antenna radiates, in accordance with the local regulations,
  a maximum power $P_{ERP\ max}$ equal to $P_{bs\ rms} = 500\,\text{mW}$, i.e. an EIRP of 820 mW EIRP;
- while at a distance of 2 m there is a tag, also fitted with a $\lambda/2$ antenna to whose terminals is
  connected a load resistance with a value matched to the radiation resistance of the tag
  antenna, i.e. 73 Ω.

Given:

$f = 869\,\text{MHz}$
$\lambda = 0.345\,\text{m}$
$P_{bs\ ERP\ rms} = 500\,\text{mW}$
$G_{bs} = 1.64 \rightarrow P_{bs\ EIRP} = 820\,\text{mW}$
$r = 2\,\text{m}$
$G_{ant\ t} = 1.64$
$R_l = 73\,\Omega$

let us calculate $P_{t\ rms}$:

$$P_{t\ rms} = P_{bs\ rms} G_{bs} \left(\frac{\lambda}{4\pi r} \right)^2 G_{ant\ t}$$

$$P_{t\ rms} = 0.5 \times 1.64 \left(\frac{0.345}{4 \times 3.14 \times 2} \right)^2 \times 1.64 = 255\,\mu\text{W r.m.s.}$$

We can calculate the following:

$$(V_{l\ rms})^2 = P_{t\ rms}\ R_l$$

$$(V_{l\ rms})^2 = 255 \times 10^{-6} \times 73 = 18\,615 \times 10^{-6}$$

Therefore

$$V_{1\,rms} = 136\,mV \rightarrow V_{equi} = 2\,V_{1\,rms} = 272\,mV$$

Note

If the input circuit of the tag is tuned with a quality factor $Q = 20$, the voltage present at the input of the tuned circuit will obviously be $20(136 \times 10^{-3}) = 2.720\,V$ r.m.s.

6.2.5 Attenuation Coefficient for Direct Line of Sight Transmission

The Friis equation introduced above indicates the relationship between the total power $P_{bs\,EIRP}$ radiated by the base station and the power $P_t = P_1$ available to the load connected to the terminals of the tag antenna, allowing for the respective gains of the two antennae when $R_1 = R_{ant\,t}$:

$$P_{t\,rms} = P_{bs\,EIRP\,rms}\left(\frac{\lambda}{4\pi r}\right)^2 G_{ant\,t} = P_{1\,rms} \quad \text{watts}$$

where the value

$$\text{Attenuation coefficient} = \frac{1}{[\lambda/(4\pi r)]^2} = \left(\frac{4\pi r}{\lambda}\right)^2$$

represents the absolute value of the 'attenuation coefficient' ('att') – in terms of power – due to the medium in which the electromagnetic wave is propagated between the base station and the tag. The Friis equation then takes the form

$$P_t = P_{bs\,EIRP}\frac{1}{att}G_{ant\,t}$$

Given that $\lambda = vT$ (v = velocity of propagation of the wave, approximately equal to the speed of light = c, and T = period of the wave), then if we replace T with its value, $T = 1/f$, the equation can also be written thus:

$$P_t = P_{bs}G_{bs}\left(\frac{v}{4\pi rf}\right)^2 G_{ant\,t}$$

or, alternatively,

$$\text{Attenuation coefficient} = att = \frac{1}{[v/(4\pi rf)]^2}$$

As indicated by this equation, the available power P_t decreases with the squares of distance and frequency. Everything else being equal, this attenuation coefficient is therefore much greater at 2.45 GHz than at 900 MHz. Moreover, the attenuation coefficient also depends on the medium passed through, because of the factor v, the real velocity of propagation of the waves in this medium.

The Attenuation Coefficient in dB

Now let us estimate the value of this attenuation, not in absolute terms, but in dB:

$$\text{att(in dB)} = 10 \log \frac{1}{[v/(4\pi r f)]^2} = -20 \log \frac{v}{4\pi r f}$$

where f is the frequency in Hz and r is the distance in metres. Assuming that the velocity of propagation of the waves, v, is substantially equal to the speed of light, i.e. approximately $3 \times 10^8 \, \text{m s}^{-1}$, we obtain

$\text{att (in dB)} = -[10 \times 2(+ \log v - \log 4\pi - \log r - \log f)]$

$\text{att (in dB)} = -[20(\log 3 \times 10^8) - 20 \log 12.56 - 20 \log r - 20 \log f]$

$\text{att (in dB)} = -(147.56 - 20 \log r - 20 \log f)$

$\text{att (in dB)} = -147.56 + 20 \log f + 20 \log r$, with f in hertz and r in metres

Because of the form of the attenuation equation, we often say that it decreases by '20 dB per decade' in frequency and distance.

Note

In fact, if we wish to be more precise, the attenuation for propagation in the line of sight is given by

$$\text{att (in dB)} = -147.56 + 20 \log f + 20 \log r + [E_s(p) + A_g]$$

where $E_s(p)$ is the correction factor due to the multiple paths of the transmitted wave and/or to certain wave focusing phenomena. For information:

$$E_s(p) = 2.6\left[1 - e\left(\frac{-r}{10}\right) \log \frac{p}{50}\right] = 0 \quad \text{for} \quad p = 50\%$$

A_g is the gaseous absorption. Generally, in free space, $E_s(p)$ and A_g are negligible and the general equation shown in the preceding section is often sufficient.

For strictly practical reasons, this equation is often written using different units, more appropriate for RFID applications and for the frequencies at which they operate. In this case, the general equation above becomes:

F = frequency expressed in GHz
r = distance expressed in km
$\text{att (in dB)} = -147.5 + 20 \log f + 20 \log r$
$\text{att (in dB)} = -147.5 + 20 \log (f \times 10^9)$
$\qquad + 20 \log (r \times 10^3)$
$\text{att (in dB)} = -147.5 + (20 \times 9) + 20 \log f$
$\qquad + (20 \times 3) + 20 \log r$
$\text{att (in dB)} = -147.5 + (180 + 60)$
$\qquad + 20 \log f + 20 \log r$

F = frequency expressed in GHz
r = distance expressed in m
$\text{att (in dB)} = -147.5 + 20 \log f + 20 \log r$
$\text{att (in dB)} = -147.5 + 20 \log (f \times 10^9)$
$\qquad + 20 \log r$
$\text{att (in dB)} = -147.5 + (20 \times 9)$
$\qquad + 20 \log f + 20 \log r$
$\text{att (in dB)} = -147.5 + 180 + 20 \log f$
$\qquad + 20 \log r$

Thus the same equation can be presented

either with f in GHz and r in km:
$\text{att (in dB)} = +92.5 + 20 \log f + 20 \log r$

or with f in GHz and r in m:
$\text{att (in dB)} = 32.5 + 20 \log f + 20 \log r$

Examples

To give you an idea of the real values involved, here are some examples of RFID applications at UHF and SHF (in air or a vacuum), using gigahertz and metres as the units.

At UHF
At 910 MHz (mean value of the 860–960 MHz band)
Using the approximate equation expressed in GHz and m:

$$att(dB) = 32.5 + 20 \log f + 20 \log r$$

$$att(dB) = 32.5 + 20 \log(0.91) + 20 \log r = 32.5 + 20(-0.041) + 20 \log r$$

$$att(dB) \text{ at } 910 \text{ MHz} = 31.68 + 20 \log r$$

with r in metres:

– at 1 m: att (dB) = 31.68 + 20 log(1) = 31.68 dB
– at 4 m: att (dB) = 31.68 + 20 log(4) = 43.72 dB

At 4 m = constant and in the limits of the 860–960 MHz frequency band
Using the approximate equation (in GHz and m):

$$att(dB) \text{ at } 4 \text{ m} = 32.5 + 20 \log f + 20 \log r$$

$$att(dB) \text{ at } 4 \text{ m} = 32.5 + 20 \log f + 20 \log(4) = 44.541 + 20 \log f$$

$$att(dB) \text{ at } 860 \text{ MHz} = 44.541 + 20 \log(0.86) = 43.23 \text{ dB}$$

$$att(dB) \text{ at } 960 \text{ MHz} = 44.541 + 20 \log(0.96) = 44\,186 \text{ dB}$$

Delta attenuation Δ in the 860–960 MHz frequency band
Assuming that all the parameters remain constant/perfect over the whole band in the above example (which is never the case, since the fixed length of the antenna is really $\lambda/2$ for a single frequency only!), at 4 m, between the centre of the band (considered as the reference) and the lower and upper ends, the results are, respectively,

$$att(dB) \text{ at } 910 \text{ MHz} = 0 \text{ dB, close to reference frequency in the USA}$$

$$att(dB) \text{ at } 860 \text{ MHz} = 43.72 - 43.231 = +0.482 \text{ dB},$$

with better performance at the lower end of the band (Europe)

$$att(dB) \text{ at } 960 \text{ MHz} = 43.72 - 44.186 = -0.466 \text{ dB},$$

with worse performance at the top end of the band (Japan)
i.e. the 'delta' between the top and bottom of the band is $\Delta = 0.955$ dB.

SHF, at 2.45 GHz
Similarly, using the approximate equation (in GHz and m):

$$\text{att(dB)} = 32.5 + 20 \log f + 20 \log r$$

$$\text{att(dB)} = 32.5 + 20 \log(2.45) + 20 \log r$$

$$\text{att(dB) at 2.45 GHz} = 40.3 + 20 \log r$$

with r in metres:

– at 1 m: att (dB) $= 40.3 + 20 \log(1) = 40.3$ dB
– at 10 m: att (dB) $= 40.3 + 20 \log(10) = 60.3$ dB

For information, as stated above, if everything else is equal (same observation distance, etc.), the attenuation is seven (7.4) times greater at 2.45 GHz than at 900 MHz (as indicated by the attenuation equation, in the square of the ratio of the frequencies): in other words, $(2450/900)^2 = 7.4$ or, in dB, $10 \log 7.4 = 8.7$ dB (!) $= 40.3 - 31.58 = 8.7$ dB.   Q.E.D.

Important note
Note also that, for simple reasons of internal (mechanical) molecular resonance, some chemical bodies are more effective than others at absorbing the energy carried by certain wavelengths. This is the case, for example, with the water molecule at the frequency of 2.45 GHz. If these molecules are present in the transmission medium, the resulting attenuation is considerable. For example, as explained in detail in Chapter 7, users of RFID are advised not to use the 2.45 GHz frequency for labelling packages that may remain on unloading platforms and be subject to damp, rain, ice or snow, or where labels may be deliberately wetted for fraudulent purposes, in order to make them 'disappear' during subsequent attempts to identify them by RFID.

6.3 Definition of the Main Parameters Required for an RFID Application

Before examining the detailed relationships between emitted power and minimum operating distances, etc., I will briefly define the universal properties of tags and their integrated circuits.

6.3.1 Sensitivity of the Transponder

Sometimes, the minimum rms power $P_{t\ min}$ that the effective area of the antenna $\sigma_{e\ t}$ must collect in order to enable the transponder to start operating correctly is called the 'transponder sensitivity'.

Figure 6.6 shows the arrangement by which the energy carried by the RF field and collected by the effective area of the antenna is generally converted into continuous power. As shown here, the antenna receives the electromagnetic energy and, in the case of a $\lambda/2$ dipole antenna, converts the impedance of the air (377 Ω) to 73 Ω, as seen from the ends of the antenna terminations. The two Schottky diodes, with low direct voltages, connected at the antenna terminals form a voltage doubler, such that the RF energy ($W = {}^1\!/_2 CV^2$) transmitted by the base station and received by the tag can be recovered in the form of continuous voltage (V_{dc}) at the terminals of the filter capacitor C.

Because of the low antenna impedance, the power required to operate the circuit depends on the voltage required by the integrated circuit and the impedance of the antenna.

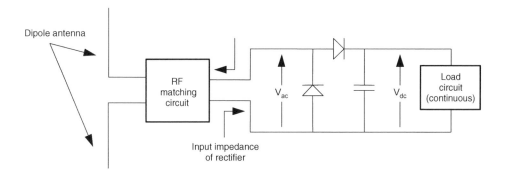

Figure 6.6   Equivalent block diagram of energy reception by a UHF–SHF tag

Example

For integrated circuits operating at x volts continuous (V_{dc}), using a voltage doubler circuit as mentioned above, where the direct voltage drops of the diodes are of the order of 2×0.3 V, it is sufficient to collect an effective voltage $V_{rms\ min}$ of $(x/2)/\sqrt{2}$ V effective (RF voltage), corresponding to a minimum r.m.s. power of $P_{t\ rms\ min}$ effective milliwatts at a tuned $\lambda/2$ antenna whose radiation resistance $R_{ant\ t}$ is 73 Ω.

Table 6.2 shows an example. Note that the voltages are shown in volts.

Table 6.2

V_{dc} required to supply the integrated circuit	V_{dc}	5	3	2	0.4
V direct of the two diodes of the voltage doubler	V	0.6	0.6	0.4	0.4
V_{peak} before rectification followed by filtering	V_{peak}	5.6	3.6	2.4	0.8
Requisite V_{peak} of incident voltage	V_{peak}	2.8	1.8	1.2	0.4
$V_{eff\ min} = (V_{peak}/2)/1.414$	$V_{rms\ min}$	1.980	1.273	0.848	0.28
$P_{t\ rms\ min} = \dfrac{V^2_{rms\ min}}{R_{l\ matched}}$	mW	53.7	22	10	1

To provide a specific example, when the impedance is matched, then during the unmodulated stage of operation the power threshold of the NXP Semiconductors U_code HSL integrated circuit made in CMOS 2 V_{dc} technology is 35 µW at 900 MHz and 120 µW at 2.45 GHz.

Now you know how to calculate, with a reasonable degree of accuracy, the power level likely to be received at a given distance. On to the next step.

6.3.2 *The Heart of the Matter*

Having brought you up to speed in that brief introduction, let us return to the real problems faced by RFID users on a daily basis. In fact, when matching takes place, since the minimum power $P_{min} = P_{t\ min}$ is defined as the minimum power required by the tag's integrated circuit to operate correctly, the problem is often one of determining:

- the maximum operating distance of a specific tag for a given $P_{bs\ EIRP}$;
- or the power $P_{bs\ EIRP\ min}$ to be delivered by the base station to achieve a desired operating distance;

- or the minimum electrical field strength, E_{min}, required to operate a given tag at a specified distance.

> **Some important notes**
> To be quite clear, the term 'correct operation of a tag' covers a variety of notions.
> The first of these may refer to a tag of the 'remotely powered' type where the power to be delivered to it, as described above, is the power that suffices to provide a correct local supply.
> The second may be due to the fact that a minimum power level must be supplied to the tag to make its input stage operate correctly (regardless of its power supply mode, which may be 'remote' or battery assisted).
> The third may relate to meeting both of the above conditions simultaneously.
> However, you should be careful not to confuse the notion of the distance for 'correct operation of the tag' (i.e. the tag is powered, can understand commands, respond to instructions, etc.) with the distance for 'correct operation of an RFID system', which covers the performance of the forward link (above) but also all the concepts relating to the return link (sensitivity of the base station input stage, etc.).

In these three cases, we must use the Friis formula in an 'inverse' form. Here it is again:

$$P_{t\ rms} = P_{bs\ EIRP\ rms} \left(\frac{\lambda}{4\pi r}\right)^2 G_{ant\ t} \text{ watts r.m.s.}$$

From this we shall extract many parameters used for defining and characterizing an RFID application in detail.

The Maximum Theoretical Distance for the Remote Power Supply and Operation of a Given Tag

This problem is certainly the best known and the one most frequently encountered by the user of UHF and SHF tags. It is raised in a context in which the parameters of the base station are specified ($P_{bs\ EIRP}$) and the power consumption required for the operation of the tag, $P_{t\ min}$, is known. In this context, we can directly extract several forms of the maximum theoretical value of r from the Friis equation:

$$r_{max} = \sqrt{\frac{P_{bs\ EIRP}G_{ant\ t}\lambda^2}{P_{t\ min}(4\pi)^2}}$$

$$r_{max} = \frac{\lambda}{4\pi r}\sqrt{\frac{P_{bs\ EIRP}}{P_{t\ min}}}G_{ant\ t}$$

$$r_{max} = \frac{\lambda}{4\pi r}\sqrt{\frac{P_{bs\ EIRP}}{P_{t\ min}}}G_{ant\ bs}G_{ant\ t}$$

Example 1

Figure 6.7 shows an example of the maximum theoretical distances, without any allowance for tolerances or dispersion, in the equatorial plane and on the principal axis, for a tag (NXP Semiconductors U_code HSL) requiring a power of $P_{t\ min\ nom} = 35\,\mu W$ at 900 MHz for UHF and $P_{t\ min\ nom} = 120\,\mu W$ at 2.45 GHZ for SHF, for the various levels of EIRP that can be used according to the current regulations in Europe and in the USA, where each of the base stations and tags uses $\lambda/2$ antennae tuned precisely to the incident frequency.

Range calculations for UHF & SHF systems (theorical, out of & including tolerances)

						Base station					Tag										OUTPUT	
remote power supply condition according to Friis equation	Frequency				Pbs cond	Pbs	Antenna gain Gbs			P ERP	P EIRP = Pbs Gbs	Antenna gain (dipole λ/2)		Min. Power for the chip		Antenna polarization		Antenna efficiency		Antenna matching		Operating range	
	f	λ	duty		max		Gbs	Gbs dBd	Gbs dBi	max	max	Gibl	equals	Pchip	equals	polar	equals	AntEff	equals	Match	equals	range	max
note	MHz	m	cycle			W				P ERP W	P EIRP W	(dipole λ/2)	dB	μW	dBm		dB	%	dB	%	dB	m	
UHF Europe in/outdoor a	869.4	0.345	10%			0,5	1,64	0	2,15	0,50	0,82	1,64	2,15	35	-14,56	100	0,00	100	0,00	100	0,00	5,38	
indoor only b	867,6	0,346	LBT			2	1,64	0	2,15	2,00	3,28	1,64	2,15	35	-14,56	100	0,00	100	0,00	100	0,00	10,79	
America in/outdoor c	915,0	0,328	100% FHSS			1	4		6,02	2,44	4,00	1,64	2,15	35	-14,56	100	0,00	100	0,00	100	0,00	11,30	
SHF Europe in/outdoor	2450,0	0,122	15%								0,5	1,64	2,15	120	-9,21	100	0,00	100	0,00	100	0,00	0,81	
indoor only	2450,0	0,122									4	1,64	2,15	120	-9,21	100	0,00	100	0,00	100	0,00	2,28	
America in/outdoor	2450,0	0,122									4	1	0,00	120	-9,21	100	0,00	100	0,00	100	0,00	1,78	
	2450,0	0,122									4	1,64	2,15	120	-9,21	100	0,00	100	0,00	100	0,00	2,28	

a in Europe, in this band, max. possible P ERP = 0.5 W - power described according to an antenna base 1/2 and must therefore be associated with Gbs of 1.64

b in Europe, in this band, max. possible P ERP = 2 W - power described according to an antenna base 1/2 and must therefore be associated with Gbs of 1.64 and must be associated with obligatory ETSI 302 208 standard for LBT (Listen Before Talk)

c the FCC stipulates Pmax EIRP = +36 dBm = 4W with P cond max = 1 W (30 dB) i.e. with a non-isotropic antenna with G = 4 (6 dB) …i.e. directive

Figure 6.7 Theoretical operating distances at UHF and SHF (remotely powered tags)

> **Note**
> For information, the power $P_{t\,min}$ depends on the technologies used by component manufacturers, the complexities or functionality of the integrated circuits of tags, etc. At UHF, this power is generally of the order of 10 to 150 μW, depending on the tag types and applications.

Example 2

Now consider the use of the same tag – same antenna and same integrated circuit using CMOS technology – and, assuming that the power consumption of the tag (the power P_t) increases (roughly proportionally) with its operating frequency in the band from 860 to 960 MHz, we can calculate the exact difference between the maximum operating distances in Europe and the USA by using the above equation again, with λ replaced by its value v/f.

Assuming that,

- in Europe:

$$f_{\text{central Europe}} = 867\,\text{MHz}, \; P_{\text{ERP}} = 2\,\text{W (in LBT mode)}, \; P_{\text{EIRP}} = 3.28\,\text{W}$$

- in the USA:

$$f_{\text{central USA}} = 915\,\text{MHz}, \; P_{\text{EIRP}} = 4\,\text{W}$$

Disregarding the variation of the tag antenna gain $G_{\text{ant t}}$ and its impedance mismatching due to this variation of gain, due to the fact that its length is never $\lambda/2$ at either of the frequencies, the theoretical value of the ratio between $r_{\text{max Europe}}/r_{\text{max USA}}$ can then be written thus:

$$\frac{r_{\text{max Europe}}}{r_{\text{max USA}}} = \frac{f_{\text{central USA}}}{f_{\text{central Europe}}} \sqrt{\frac{P_{\text{bs EIRP max Europe}}}{P_{\text{bs EIRP max USA}}}} \sqrt{\frac{P_{t\,915}}{P_{t\,867}}}$$

$$\frac{r_{\text{max Europe}}}{r_{\text{max USA}}} = \frac{915}{867} \sqrt{\frac{3.28}{4}} \sqrt{\frac{915}{867}}$$

and therefore

$$\frac{r_{\text{max Europe}}}{r_{\text{max USA}}} = [1.055 \times 0.9055] \times 1.027$$

$$\frac{r_{\text{max Europe}}}{r_{\text{max USA}}} = 0.955 \times 1.027$$

$$r_{\text{max Europe}} = 98\% \text{ of } r_{\text{max USA}}$$

From this result we can therefore conclude that we obtain practically the same maximum operating distance with 2 W ERP in Europe or with 4 W EIRP in the USA.

Minimum Power P_{bsEIRP} that the Base Station Must Supply for Correct Operation of the Tag at the Distance r

Conversely, in order to ensure that the least efficient tag operates correctly at a specified distance r, the base station must radiate a minimum EIRP ($P_{\text{bs EIRP min}}$) such that it can supply the tag with at least the minimum power $P_{t\,min}$ required for its correct operation, as indicated in the previous section. Returning to the inverse Friis equation, we immediately see that

$$P_{bs\,EIRP\,min} = \frac{P_{t\,rms}}{G_{ant\,t}} \left(\frac{\lambda}{4\pi r}\right)^2 \quad \text{watts r.m.s.}$$

Important

Since there is direct proportionality between P_t and $P_{bs\,EIRP}$, you must be aware of the differences between the published 'typical' and 'minimum and maximum' power levels guaranteed by manufacturers of integrated circuits, if you wish to avoid unpleasant surprises in the form of defective operation in RFID applications, and especially if you wish to avoid exceeding the maximum P_{ERP} and/or P_{EIRP} permitted by local regulations.

The Minimum Voltage $V_{received\,min}$ Required for Correct Operation of a Tag

What this means is that the voltage applied to the connecting terminals or pads of the tag's integrated circuit must exceed the minimum threshold voltage $V_{ic\,rms\,min}$ (often denoted $V_{ic\,threshold}$) specified by the manufacturer of the device.

Example

Let us take the example of the NXP Semiconductors U_code HSL circuit. The data sheet for this product shows that it requires a minimum threshold operating power $P_{ic} = 35\,\mu W$ at 867 MHz, and that at this frequency the complex value of its input impedance is $Z_{ic} = 35 - j720 = (R_{ic\,s} + jX_{ic\,s})$, in other words the equivalent of a circuit composed of a series resistance of 35 Ω in series with a capacitance such that $1/(C_{ic\,s}\omega) = 720\Omega$, i.e. a capacitance $C_{ic\,s}$ of 240 fF (Figure 6.8).

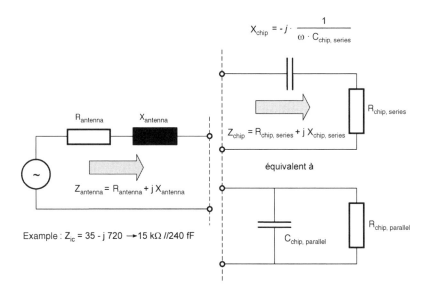

Figure 6.8   Equivalent diagram of the input stage of a UHF–SHF integrated circuit

On the other hand, the intrinsic value of the quality factor Q_{ic} of the integrated circuit is

$$Q_{ic} = \frac{X_{ic\,s}}{R_{ic\,s}}$$

$$Q_{ic} = \frac{1}{R_{ic\,s} C_{ic\,s} \omega}$$

$$Q_{ic} = \frac{720}{35} = 20.6$$

By definition, the power (in watts) of $35\,\mu W$ rms is entirely dissipated in the resistance of $35\,\Omega$. Given that $P = R_{ic\,s} I_{ic}^2$, let us calculate the equivalent current flowing in the resistance:

$$I_{ic\,eff} = \sqrt{\frac{P_{ic}}{R_{ic\,s}}}$$

$$I_{ic\,eff} = \sqrt{\frac{35 \times 10^{-6}}{35}} = 1\,mA\ rms$$

Since the current flows in the series circuit $R_{ic\,s}$, $C_{ic\,s}$, the potential difference $V_{ic\,eff}$ developed at the terminals of this system will be

$$V_{ic\,rms} = |Z_{ic}| I_{ic\,rms}$$

where, of course,

$$|Z_{ic}| = \sqrt{R_{ic\,s}^2 + X_{ic\,s}^2}$$
$$|Z_{ic}| = \sqrt{35^2 + 720^2} = 720.85\Omega$$

Note the strong predominance of the impedance of the capacitance:

$$V_{ic\,rms} = |Z_{ic}| I_{ic\,rms} = 720.85\ mV\ rms$$

or, more easily measured, $= 1.019\,V_{peak}$ or $2.038\,V$ peak-to-peak.

This is the actual voltage that must be present at the input of the integrated circuit to enable the circuit to operate correctly.

Note

If we had considered an equivalent parallel configuration of the input circuit of the integrated circuit, for the same power in watts dissipated in the circuit ($35\,\mu W$) this would have represented an equivalent parallel resistance of

$$P_{ic} = \frac{V_{ic}^2}{R_{ic\,p}}$$

Therefore

$$R_{\text{ic p}} = \frac{V_{\text{ic}}^2}{P}$$

$$R_{\text{ic p}} = \frac{(720.85 \times 10^{-3})^2}{35 \times 10^{-6}} = 14.85 \,\text{k}\Omega$$

and the capacitance $C_{\text{ic p}} \approx C_{\text{ic s}} = 240 \,\text{fF}$.

Important note
The above example is based on the NXP Semiconductors U_code HSL integrated circuit. And why not? However, you should note that there are other circuits available on the market, and in these the impedances $R + jX$ may be different and more or less suitable for other operating modes or applications of the tags.

See, for example, Table 6.3 (which is not exclusive and is only valid for a given date, etc., etc.), where the values are of course given for the same conditions of bare chips, for the same frequency of 867 MHz ($\omega = 2\pi f = 5.445 \times 10^9 \,\text{rd s}^{-1}$).

As shown in this table, the R, X and Q values of the various products differ widely, leading to different antenna designs (in terms of dimensions, shapes, technology, materials, etc.) to provide better adaptation of R_{ant} and conjugate reactance, in other words antenna inductance.

Now let us look at two examples, in the first of which the tag is not tuned, while in the second case it is tuned.

Tag not tuned
For the same current flowing in the 'antenna + input circuit of integrated circuit' system, this means that the electrical field develops a voltage $V_{\text{equi rms}}$ of ($V_{\text{ic rms}} + R_{\text{ant t}} I_{\text{c rms}}$) at the terminals of the antenna.

Tuned tag (they all are, more or less!)
Very frequently, the impedance $Z_{\text{ant t}}$ of the tag antenna is arranged to be of the form ($R_{\text{ant t}} + jX_{\text{ant t}}$) in order to make use – for the central operating frequency – of the well-known conjugate impedance matching in R and X. Consequently, the tag (the system formed by the antenna and the integrated circuit) becomes a tuned circuit and has its own quality factor (under load) $Q_{\text{tag}} = X/R$, where R is the global load resistance of the tuned circuit.

Table 6.3

	$R_{\text{ic s}}$ (Ω)	$X_{\text{ic s}}$ (Ω)	Q	$R_{\text{ic p}}$ (Ω)	$C_{\text{ic p}}$ (fF)
U_code HSL (ISO 18 000-6 B)					
Bare chip	35	$-j\,720$	20.6	14 850	255
In package (TSSOP 8)	12.7	$-j\,457$	36	16 460	401
U_code EPC C1 G2					
First generation bare chip	41	$-j\,865$	21.3	18 080	212
First generation in package (TSSOP 8)	22	$-j\,404$	19	7460	454
Second generation bare chip	17	$-j\,170$	10	1700	1100
ST µE - XRA000 EPC G1	7.4	$-j\,218$	29.45	6420	843

Evidently, the overvoltage due to this quality factor contributes to the actual development of a voltage Q_{tag} times greater at the input of the integrated circuit. The fact that the tag antenna and its load are matched means that $R_{ant\,t}$ and $R_{ic\,s}$ are equal to each other and that, in this case, $Q_{tag} = Q_{ic}/2$ (Figure 6.9).

The phase diagram in this figure clearly shows that:

- The two voltages present at the terminals of the antenna inductance and the integrated circuit capacitance are equal and opposite, and their sum is therefore zero, since $LC\omega^2 = 1$ ($C_{ic\,s} = 240\,fF$, meaning that $L_{ant\,s} = 130\,nH$ at $900\,MHz$).
- In this conjugate matching, the voltage $V_{equi\,rms}$ is (as usual) twice the voltage present at the terminals of R_{ic}; in other words, $(2R_{ic}I_{ic}) = 2 \times 35 \times 1 \times 10^{-3} = 70\,mV$ r.m.s.
- The voltage V_{ic} at the terminals of the integrated circuit is $(Q + 1)$ times the voltage across R_{ic}.

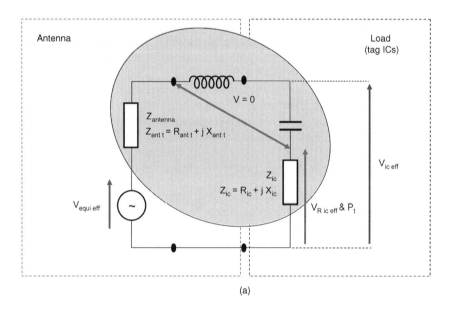

(a)

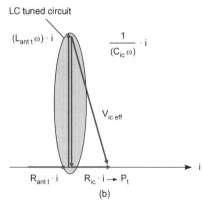

(b)

Figure 6.9 Equivalent circuit for the conjugate matching of the antenna and its load

- The LC tuned circuit (totally reactive) is entirely transparent with respect to the power transferred by the antenna to the load in watts R_{ic}, and therefore the Friis equation is directly applicable. In fact, the tuned circuit has the sole purpose of bringing the voltage at the terminals of the integrated circuit to the appropriate level to enable the electronics of the circuit to operate correctly. The technological compromise in the production of the tag is therefore a matter of designing a system such that the power consumption is as low as possible (e.g. 30 µW) and the capacitor is as small as possible in terms of area and technology, without reducing its value to such an extent that it would be impossible either to provide the voltage required for the correct supply of the integrated circuit or to facilitate fitting to the tag in the factory, in view of the positioning tolerances (1.1 pF, for example). Note also that the circuit chosen for our examples – the NXP Semiconductors U_code – contains a voltage doubler circuit, which prevents the use of an excessively high value of Q, which would make the bandwidth too selective for the planned RFID applications.
  To be continued...

Minimum Electrical Field Strength E_{min} Required for the Correct Operation of the Tag

This voltage will only be present if the strength of the radiated electrical field has reached the value of E_{min}. Thus what we need to do is to estimate the minimum electrical field strength E required for the tag. To do this, we start again with the equation for the power available to the load R_1 which by definition is as follows in conditions of matching:

$$P_{t\,rms} = \frac{V_{received\,rms}^2}{R_{ll}} = \sigma_{e\,t} s_{rms} \quad \text{W r.m.s.}$$

Given that the relation between the effective incident power flux density s and the electrical field E in the far field is

$$s_{rms} = \frac{1}{2}\frac{|E|^2}{Z_0} = \frac{E_{rms}^2}{Z_0} \quad \text{W m}^{-2}\,\text{r.m.s.}$$

and inserting the values of $\sigma_{e\,t}$ and s_{rms} into the equation for $P_{t\,rms}$, we obtain

$$P_{t\,rms} = \frac{\lambda^2}{4\pi}G_{ant\,t}\frac{E_{rms}^2}{Z_0} = \frac{V_{received\,rms}^2}{R_{ls}}$$

Using the above expression, and replacing Z_0 (impedance of the air) with its value (120π), we can find the minimum electrical field strength required to operate a tag whose minimum operating power is known, thus:

$$E_{min\,rms} = \sqrt{\frac{4\pi \times 120\pi \times P_{t\,rms}}{\lambda^2 G_{ant\,t}}} = \frac{4\pi}{\lambda_0}\sqrt{\frac{30 P_{t\,rms}}{G_{ant\,t}}}$$

Replacing the various parameters by the values in the above equations, we immediately find ($R_1 = R_{ant}$):

$$V_{received\,rms} = \frac{\lambda}{4\pi r}\sqrt{(P_{bs}G_{bs})G_{ant\,t}R_{ant}} \quad \text{volts}$$

which is the value we found before.
  Clearly, when $R_1 = R_{ant}$ the voltage at the terminals of the tag antenna is $V_{equi\,rms} = 2V_{received\,rms}$.

Example (continued)

At 900 MHz ($\lambda = 33.3$ cm), with a $\lambda/2$ dipole antenna having a gain of 1.64 and a tag with a minimum operating power of 35 µW, the tag must be immersed in the following minimum electrical field in order to operate correctly:

$$E_{\text{min rms}} = \frac{4\pi}{0.33} \sqrt{\frac{30 \times 35 \times 10^{-6}}{1.64}} = 0.963 \text{ V m}^{-1} \text{ r.m.s.}$$

If the current local regulations specify a power of $P_{\text{bs ERP max}} = 500$ mW, i.e. $P_{\text{bs EIRP max}} = 820$ mW, this minimum field is guaranteed at a maximum operating distance determined thus:

$$E_{\text{rms}} = \frac{\sqrt{30 P_{\text{bs EIRP rms}}}}{r} = \frac{5.478 \sqrt{P_{\text{bs EIRP rms}}}}{r}$$

Therefore

$$r = \frac{5.478 \sqrt{P_{\text{bs EIRP rms}}}}{E_{\text{rms}}}$$

$$r = \frac{5.478 \sqrt{0.820}}{0.963} = 5.15 \text{ m}$$

To be continued yet again ...

Effective or Equivalent Length of an Antenna

Using the two equations above, we can easily find the value $l_{\text{effective}}$ of the 'effective' length of the antenna, by a simple procedure. Given that, on the one hand,

$$s_{\text{rms}} = \frac{1}{2} \frac{|E|^2}{Z_0} = \frac{E_{\text{rms}}}{Z_0} \quad \text{W m}^{-2} \text{ r.m.s.}$$

and, on the other hand, $V_{\text{equi rms}} = l_{\text{effective}} E_{\text{rms}}$, we immediately obtain

$$l_{\text{effective}} = \frac{V_{\text{equi rms}}}{\sqrt{s_{\text{eff}} Z_0}}$$

Now we replace each term with its value:

$$V_{\text{equi rms}} = 2 V_{\text{receuved rms}} \quad \text{(when there is impedance matching, } R_1 = R_{\text{ant t}})$$

$$l_{\text{effective}} = \frac{2[\lambda/(4\pi r)] \sqrt{P_{\text{bs}} G_{\text{bs}} G_{\text{ant t}} R_{\text{ant t}}}}{\sqrt{[P_{\text{bs eirp rms}}/(4\pi r^2)] Z_0}}$$

and, noting that P_{bs}, G_{bs} and $P_{\text{bs EIRP rms}}$ all represent the same value, we can simplify the equation to

$$l_{\text{effective}} = \frac{\lambda \sqrt{G_{\text{ant t}} R_{\text{ant t}}}}{\pi \sqrt{120}}$$

The Case of a λ/2 Dipole Antenna

For a λ/2 dipole antenna, with a gain $G_{\text{ant t}} = 1.64$, when it is matched so that $R_1 = R_{\text{ant t}} = 73.128\,\Omega$, we obtain

$$l_{\text{effective}} = \frac{\lambda\sqrt{1.64 \times 73.128}}{\pi\sqrt{120}}$$

$$l_{\text{effective}} = \frac{\lambda\sqrt{120}}{\pi\sqrt{120}}$$

Thus

$$l_{\text{effective}} = \frac{\lambda}{\pi} = 0.318\lambda$$

in the case of a matched λ/2 dipole antenna

Example (continued and concluded)

For a given antenna whose length is known by definition, this minimum necessary field corresponds to a no-load PD ($V_{\text{equi rms}} = E_{\text{rms}}l_{\text{eff}}$) at its terminals whose value is divided by two when the load is matched, in order to be applied to the terminals of the integrated circuit. Also, if the tag is tuned, this PD is multiplied by the quality factor Q_{tag} ($= Q_{\text{ic}}/2$).

Take the example of a λ/2 antenna (i.e. $R_{\text{ant t}} = 73\,\Omega$) operating at 900 MHz. Its effective length is therefore

$$l_{\text{effective}} = \lambda/\pi$$

$$l_{\text{effective}} = 0.333/3.14 = 0.106\,\text{m}$$

$$V_{\text{equi rms}} = 0.963 \times 0.106 = 102\,\text{mV r.m.s.}$$

and with matching $R_1 = R_{\text{ant}} = 73\,\Omega$ (since the antenna is the λ/2 type)

$$V_{R1} = {}^1\!/_2 V_{\text{equi rms}}$$

$$V_{R1} = 51\,\text{mV} \ldots \text{across } 73\,\Omega$$

giving us a power level of *I*:

$$P_{R1} = \frac{V_{R1}^2}{R_1}$$

$$P_{R1} = \frac{2601 \times 10^{-6}}{73} = 35\,\mu\text{W}$$

Q.E.D.

Note

If you have followed everything carefully, you will have spotted a difference between the two calculated values of $V_{\text{equi rms}}$: the first is 70 mV r.m.s., but now we have a second value of 102 mV r.m.s.! Heavens above! Does that mean that the whole theory is wrong? It certainly deserves some explanation. Here it is:

- The power transmitted to the load is identical (35 μW) in both cases.
- The maximum operating distance is the same in both cases, according to the Friis equation and its context (the matching condition $R_{ant} = R_1$ is true in both cases).
- The local electrical field is therefore identical for both of the above examples.
- In the second case, we used a pure $\lambda/2$ antenna, and therefore $R_{ant\ t} = (73 + j0)\ \Omega$.
- In the first example, this was not the case, because $R_{ant\ t}$ had to be equal to 35 Ω to achieve matching. Therefore the antenna is not of the $\lambda/2$ type, and the equation for the effective length $l_{effective} = 0.318\lambda$ cannot be applied; in other words, in view of the resulting figures, the effective length of the antenna in the first case includes a series inductance, which appears to be as follows:

$$l_{effective} = k\lambda, \text{ where the value of k is to determined,}$$

$$V_{equi\ rms} = E_{rms}k\lambda$$

- Therefore

$$0.07 = 0.963 \times k \times 0.333$$

$$k = 0.218$$

In conclusion, this means that the provision of a tuned circuit for the conjugate matching of the tag did not adversely affect the operating distance, and enabled the electrical level at the input of the integrated circuit to be matched, by a combination of a reduction in the effective length of the antenna and simultaneous compensation and improvement by the quality factor Q of the tuned circuit.

End of example! At last!

Reducing the size of the Dipole Antenna

In UHF applications, the mechanical dimension (length) of a $\lambda/2$ dipole (15 to 20 cm) is often considered too large for use in a tag. Unfortunately, at a given frequency (wavelength), when the branches of the dipole are shortened so that the dipole no longer has a length of $\lambda/2$, the equation describing this shows that the efficiency of the energy transfer decreases very rapidly. The curve in Figure 6.10 shows the effect of such changes. For example, if the length is halved, the losses rise to 97%. This is something to ponder on.

It is worth knowing that some antenna technologies enable the length of the tag branches to be reduced, by increasing their widths. Each branch of the tag antenna then becomes a rectangular surface. Why not make use of this, especially as the total area occupied by the two surfaces just happens to be equal to that of an ISO card, and the losses by comparison with a $\lambda/2$ dipole are barely 10%. The implications are obvious. Now it is up to you to experiment with the best antenna design and field modelling software (such as the well-known HFSS tool developed by Ansoft).

A Brief Summary

Figure 6.11 is a table providing an example and summary of the main parameters discussed above. The two framed boxes show the maximum permitted values under US regulations. Note also the high electrical field strength ($> 10\ \mathrm{V\,m^{-1}}$).

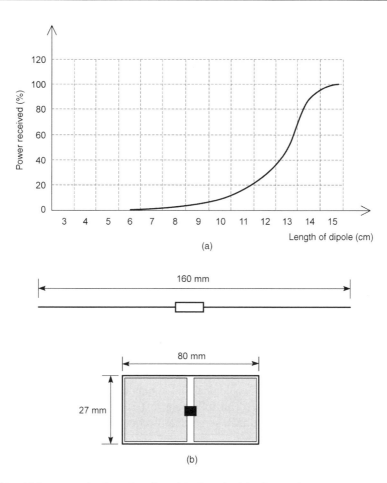

Figure 6.10 (a) Power received as a function of the length of the dipole. (b) An example 'substantially' equivalent to a dipole in smart card format

Desired distance	Frequency used	Wavelength	Characteristics of the tag used					Base station			Electrical field at the tag
			Gain of tag antenna. G_ant_t	Effective length of tag antenna	Minimum power required	Load resistance. R_l	Voltage received at the tag	P_eirp	P_eirp	P_erp	
m	MHz	cm		cm	µW eff DC	kohm	V_eff	W eff	dBm	dBm	V/m eff
1,00	433,00	69,28	1,64	22,03	35,00	14,85	1,70	39,00	45,91	43,76	34,20
1,00	866,00	34,64	1,64	11,02	35,00	14,85	1,70	155,99	51,93	49,78	68,41
1,00	915,00	32,79	1,64	10,43	35,00	14,85	1,70	174,14	52,41	50,26	72,28
1,00	2450,00	12,24	1,64	3,89	120,00	14,85	1,70	1248,52	60,96	58,82	193,53
1,50	433,00	69,28	1,64	22,03	35,00	14,85	1,70	87,75	49,43	47,28	34,20
1,50	866,00	34,64	1,64	11,02	35,00	14,85	1,70	350,98	55,45	53,30	68,41
1,50	915,00	32,79	1,64	10,43	35,00	14,85	1,70	391,82	55,93	53,78	72,28
1,50	2450,00	12,24	1,64	3,89	120,00	14,85	1,70	2809,18	64,49	62,34	193,53

Figure 6.11 Summary of the main parameters relating to tag performance

Minimum Assured Operating Distance of a Tag

This heading may look strange, but there may indeed be a minimum operating distance that must be guaranteed. This is because a component manufacturer always provides one or more maximum levels of current, voltage, power, temperature, etc., that must not be exceeded for fear of destroying or damaging the component. Depending on the application, this level, or these levels, may be reached when the tag is near, or very near, to the base station and the electromagnetic field is too strong.

RFID and Strong Fields

Throughout this book I have dealt with the efforts made to ensure that tags operate correctly in weak electromagnetic fields, so that they are remotely powered in the correct way and communication between base stations and transponders can be assured.

We now need to examine the other aspect of the application, namely the use of strong fields. This is because there are many applications in which the transponders move within the space containing the magnetic field, and may therefore be subjected to high magnetic field strengths when they physically approach the base station antenna.

Communication by Smoke Signals?

As mentioned previously, most base station manufacturers supply base stations that can operate over the longest possible distance without ever exceeding the maximum value specified by the standards and regulations, regardless of the sensitivity of the transponders that may be present in the electromagnetic field. Obviously, where strong fields are present, it is preferable for the tag not to use smoke signals to communicate with the base station, even if that would be an appealing picture, with all kinds of ecological and humanitarian implications for the preservation of Native American culture. The following text is intended to help you avoid any problems in this connection, which may sometimes become a 'burning' issue. (See also the discussion later in this chapter).

Note

No maximum values of the fields E and/or H are specified in the RFID standards (ISO 18000-4 or -6, EPC, or others). The only available specifications are found in local regulations for different countries, which may or may not be based on CEPT, ERC, ETSI, FCC, ARIB and other documents, and in 'Human Exposure' standards based on the conclusions of WHO (World Health Organization), which will be discussed in detail in Chapter 16.

The Function of the Controller and its Results

In order to be able to provide a wide choice of operating distance, nearly all the integrated circuits for tags available on the market have internal control systems of one kind or another, designed to keep, or tend to keep, the supply voltage at the input of the integrated circuit practically constant, based on the strength of the electromagnetic field in which the transponder is immersed.

There are different types of controller, including some of the parallel 'shunt' type (see Chapter 9). Let us look at the operating principle of this control system.

In the presence of a very weak electromagnetic field

When the distance is large (in other words, when the electromagnetic field is very weak), the voltage present at the terminals of the integrated circuit is too weak to reach the activation threshold of the controller, and thus follows in a linear way the present value of the electromagnetic field. In this case, the integrated circuit does not yet have a sufficient internal supply voltage to operate, and consumes no power (or very little); thus it has a high input impedance. The transponder system then shows a very high value of Q.

If the transponder is brought towards the base station again, the magnetic field strength increases and, before the controller activation threshold is reached, the transponder integrated circuit begins to be powered and therefore begins to operate.

In the presence of a weak electromagnetic field
Moving still closer to the base station, the voltage at the terminals of the tag's integrated circuit reaches the specified level for correct operation, a level at which the control circuit has not usually started to operate. This is the level (e.g. the level of 720 mV r.m.s. approximately that I have used throughout this book) that serves to define the minimum maximum operating distance. In this operating mode, the impedance of the transponder is known (in our case, $R_{ic\ s} = 35\,\Omega$; $C_{ic\ s} = 240\,\mathrm{fF}$ or $R_{ic\ p} = 14.8\,\mathrm{k\Omega}$; $C_{ic\ p} = 240\,\mathrm{fF}$), and therefore at this distance the value of Q_{ic} is known. In our example, the value of this factor is about 20, and the value of the factor for the tag, Q_{tag}, is about 10, following conjugate matching.

In the presence of medium and strong fields
Above this threshold value, the process becomes quite different. Regardless of the increasing strength of the electromagnetic field, the control circuit comes into action and tries (success-fully, if possible!) to keep the alternating voltage at the input terminals of the integrated circuit constant. To do this, the control circuit progressively decreases the shunt resistance connected in parallel with the input of the integrated circuit, thus progressively 'squashing' the incident signal from the base station. Clearly, the apparent input impedance of the integrated circuit will be modified; in this case it is decreased.

We can now see that, in the far field, the structure of this impedance is principally 'capacitive', $C_{ic\ p}$, since the value of $R_{ic\ p}$ is very high, and that it progressively becomes mainly 'resistive', since the apparent value of $R_{ic\ p}$ will decrease strongly. The value of the quality factor $Q_{tag\ under\ load}$ will therefore decrease considerably, and the initial value found with weak fields will be only a distant memory.

Back to the future
To make matters clear, let us return to our example of working with a tag having an NXP Semiconductors U_code circuit. The maximum specified value of the input current (in the input pins) of the U_code HSL integrated circuit is 30 mA r.m.s. (note that this is only 10 mA r.m.s. for U_code EPC C1 G2 circuits). Note, by the way, that nobody has mentioned whether the structure of this effective current is reactive or not. It is effective, which is all that matters. Whether it flows in a capacitance or a resistance is of little importance – but this is where all the difference lies!

In fact, in the presence of a weak electromagnetic field (at long distance), as mentioned above, owing to the very high value of the resistance $R_{ic\ p}$ (14.8 kΩ) with respect to the impedance of the capacitance $C_{ic\ p}$ (240 fF); $Z_{C\ ic\ p} = 1/(C_{ic\ p}\omega) = 737\,\Omega$ at 900 MHz, almost all the current flowing in the input pin of the integrated circuit passes into the integrated capacitance, and therefore only a very small part flows in the resistance of the integrated circuit. On the other hand, in strong fields, the current flowing in the input pin consists mainly of the current flowing in the small apparent resistance presented by the controller.

To quantify all the elements of the transponder in strong fields, we must start from the following two assumptions:

- The controller controls the system in a perfect way (this is not entirely true, but is a useful assumption for now). It therefore maintains a constant alternating voltage at the input of the integrated circuit V_{ic} (approximately 720 mV r.m.s., in our case).
- The manufacturer of the integrated circuit specifies the maximum effective value which the input current of the circuit must not exceed (in our case, 30 mA effective).

These assumptions immediately lead us to the following two conclusions:

- On the one hand, if, regardless of the strength of the electromagnetic field in which the transponder is immersed, the voltage $V_{ic} = 720$ mV r.m.s. is always present at the terminals of the integrated circuit, and if the input capacitance $C_{ic\,p}$ is unchanged (and there is no reason for it to change, as the applied voltage is constant), a current of $V_{ic}/(C_{ic\,p}\omega) = 720$ mV/ $737\,\Omega = 0.98$ mA will always flow in it.
- On the other hand, if we wish to avoid exceeding the maximum permitted input current (30 mA r.m.s.) in the integrated circuit, the rest of the current will go elsewhere, namely into the resistance R_{ic} formed by the parallel connection of the initial $R_{ic\,p}$ and the controller resistance.

This is self-evident. To be more precise, we could write Div $I = 0$, which may look better but means exactly the same thing.

Note that, in this extreme case, this remaining current (with allowance for the phase relations between the capacitive and resistive current), equal to approximately 29 mA, flows in the equivalent resistance provided by the control circuit.

Example of calculation of the equivalent resistance R_{ic} of the controller in the presence of strong electromagnetic fields
At the input of the integrated circuit, in terms of current equations, we can write

$$i_{R\,ic} = \frac{v_{ic}}{R_{ic}}$$

and

$$i_{C\,ic} = v_{ic}(jC_{ic}\omega)$$

At 900 MHz, with $C_{ic} = 240$ fF:

$$i_{C\,ic} = v_{ic} \times j(240 \times 10^{-15} \times 5652 \times 106) = v_{ic} \times j(1.356 \times 10^{-3})$$

Therefore

$$i_{ic} = i_{R\,ic} + i_{C\,ic}$$

$$i_{ic} = \left[\frac{1}{R_{ic}} + j(1.356 \times 10^{-3})\right] v_{ic}$$

or, alternatively, as an effective value

$$i_{ic\ eff} = \sqrt{\left(\frac{1}{R_{ic}}\right)^2 + (1.356 \times 10^{-3})^2}$$

Assuming, for example, that

- according to voltage regulator, V_{ic} remains substantially constant, $V_{ic} = 1$ V eff, and
- the maximum value of I_{ic} eff is known, I_{ic} eff $= 30$ mA eff,

we can find the value of R_{ic} from the above equation. Thus we obtain

$$\frac{I_{ic}^2}{V_{ic}} - (1.356 \times 10^{-3})^2 = \left(\frac{1}{R_{ic}}\right)^2$$

$$\frac{(30 \times 10^{-3})^2}{1^2} - (1.356 \times 10^{-3})^2 = \left(\frac{1}{R_{ic}}\right)^2$$

$$900 \times 10^{-6} - 1.8 \times 10^{-6} = \left(\frac{1}{R_{ic}}\right)^2$$

$$900 \times 10^{-6} = \left(\frac{1}{R_{ic}}\right)^2$$

Therefore

$$R_{ic} = \sqrt{\frac{1}{900 \times 10^{-6}}}$$

$R_{ic} = 33.3\ \Omega$ and thus $I_{R\ ic\ eff} = V_{ic}/R_{ic} = 1/33 = 30$ mA r.m.s.
We can again draw two main conclusions – and another one as well:

- This current, flowing in the equivalent (ohmic) resistance of the controller, will cause a heating of the integrated circuit P (in watts) of $V_{ic}I_{R\ ic} = 1$ V $\times$ 30 mA $= 30$ mW (compared with 35 μW approx. in a weak field).
- If the input voltage is controlled – and therefore constant – and the current flowing in the controller is known, we can determine the equivalent resistance that it represents, as follows: 1 V/30 mA $= 33.3\ \Omega = R_{shunt\ min}$, which almost entirely short-circuits $Z_{C\ ic\ p}$ (approx. 737 Ω at 900 MHz).
- Finally, given the ohmic value represented by the controller, we can calculate the new 'strong field' value of Q_p for the L, C, R system of the transponder, thus:

$$Q_p = \frac{R_p}{L_{ant\ t}\omega}$$

where, of course,

$$R_p = R_{L\,p}//R_{ic} = (14.8 \times 10^3)//24.8, \text{ i.e.approximately } 23\,\Omega$$

and therefore

$$Q_p = \frac{23}{130 \times 10^{-9} \times 2 \times 3.14 \times 900 \times 10^6}$$

$$Q_p = 0.03!!, \text{ instead of the initial level of } 10!$$

The quality factor is practically zero and therefore, in the presence of strong fields, the global transponder circuit is practically aperiodic. This is because, in strong fields, with a low value of Q_p due to the low value of R_p (because of the low value of R_{ic}), the input circuit is, for practical purposes, not tuned, but totally aperiodic. The electrical circuit consisting of the antenna and the integrated circuit then becomes a simple LR network.

Important conclusion

The discussion above clearly reveals the difficulties that have traditionally been encountered by system designers, and indicates to potential users all the complexities involved in the precise detailed specification of a system. It also clearly shows that the global electrical structure of the tag impedance varies markedly with the distance (between the maximum and minimum distance), changing from a quasi-pure LC tuned circuit to a quasi-pure aperiodic LR circuit, and that the global value of the quality factor Q of the tag undergoes a considerable change.

Maximum Current, Minimum Distance and Maximum Power and Temperature
After this voluminous introduction, let us move on at last to examine the value of the maximum current flowing in the input pins of the integrated circuit.

Maximum current
Let us see when (i.e. at what distance from the base station) we can expect to find this crucial value of 30 mA r.m.s. in the input circuit of the tag's integrated circuit. As a first approximation, assume that the elegant theory is applicable and can give us a reasonably accurate idea of the problem. Assuming that the antenna is a matched $\lambda/2$ dipole (which becomes less and less true because of the operation of the controller), let us calculate the theoretical distance at which the field E radiated by the base station antenna can supply the tag with a power $P_{t\,max}$ such that the maximum current appears in the tag. In this case, the Friis equation tells us that

$$P_{t\,max} = P_{bs\,EIRP\,rms}\left(\frac{\lambda}{4\pi r}\right)^2 G_{ant\,t} \quad \text{W r.m.s.}$$

and therefore

$$r^2 = \frac{P_{bs\,EIRP\,rms}}{P_{t\,max}}\left(\frac{\lambda}{4\pi}\right)^2 G_{ant\,t} \quad \text{m}$$

which, for a given wavelength λ and a given antenna, shows the relationship between $r_{min\,max}$ and $P_{bs\,EIRP\,max}$ corresponding to a specified local regulation for a given integrated circuit ($P_{t\,max}$).

Example
Still using the U_code circuit at 900 MHz, $\lambda = 0.333$ m, with an antenna gain $\lambda/2 = 1.64$:

$$P_{t\,max} = \text{approximately 1 V r.m.s.} \times 30\,\text{mA} = 30\text{mW r.m.s.}$$

We have kept the value of 1 V on the assumption that the controller is carrying out its task perfectly (although, for a variety of technical reasons, this is not entirely true).

Since the tag must be able to operate worldwide, let us look at the case of the USA, where the permitted power radiated by the base station (EIRP) according to the local regulatory bodies (FCC) is highest, at 4 W EIRP. We find that

$$r_{min}^2 = \frac{4 \times (0.333)^2 \times 1.64}{30 \times 10 \times (4\pi)^2} \quad \text{W r.m.s.}$$

$$r_{min}^2 = 0.154 \text{ and therefore } r\,min = 0.39\,\text{m}$$

As a first approximation, assuming that the tag operates in the way indicated above when the tag is located at 40 cm from the base station (6.3 times greater than $\lambda/2\pi = 5.3$ cm, and therefore already at the start of the 'far field' area), the base station can still supply 30 mW to the tag; in other words, it can provide the maximum current of 30 mA r.m.s. at the input of the integrated circuit.

Theoretically, a tag manufacturer should state that the tag can operate between $r_{min} = 40$ cm and $r_{max} = x$ m, depending on the current local regulations.

Maximum power and temperature of the tag and its integrated circuit
We have seen how the typical maximum power dissipated by certain integrated circuits of a tag could be, for example, approximately $P_{t\,max} = 30$ mW in the presence of strong electrical fields. If we imagine special cases where the controller is at its maximum control limit, with an accumulation of tolerances, etc., this power could become as much as

$$P = V = I_{ic\,rms}\text{max}$$

$$P = 3.5 \times 0.030 = \text{approx. } 100\,\text{mW}$$

At this power level, with a thermal resistance $R_{th\,j\text{-mb ic}}$ of the bare integrated circuit (the size of which is often less than 1 mm^2) of the order of 50 to 80 °C W^{-1} (see below), the 'thermal Ohm's law' predicts the following temperature rise ΔT between the junction 'j' of the chip and its mounting base ('mb'):

$$\Delta T = PR_{th\,j\text{-mb ic}} = P \times \Sigma \text{ of the } R_{th}$$

$$\Delta T = 100 \times 10^{-3} \times 80 = 8°\text{C - not a large amount!}$$

Working in an ambient temperature of 25 °C, many users are quick (too quick) to conclude that the maximum temperature of the integrated circuit will be 33 °C – but this is wrong, for the simple reason that the thermal resistance of the tag packaging is far from the ideal level of 0 °C W^{-1}.

Thus in real-life applications, using electronic labels affixed to packaging boxes, when the input current is about 20 mA r.m.s. (directly dependent on the maximum permitted

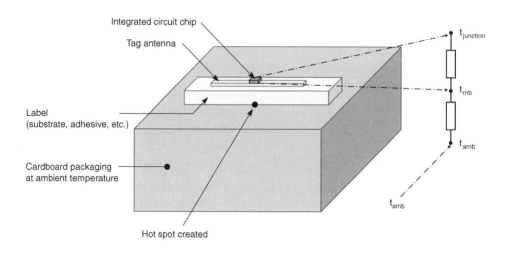

Figure 6.12   Thermal representation of the application of a tag to a package

electrical field), it is not unusual for heat probes with infrared detectors to measure maximum temperatures of the chip junction of the order of 55 °C or even 65 °C. This means that the temperature has risen by 30 °C (or even 40 °C) above the normal ambient temperature of 25 °C. To make this clearer, let us examine the sum of thermal resistances between the junction of the integrated circuit and the ambient temperature of the measurement location (Figure 6.12):

$$R_{\text{th j-amb}} = \left(R_{\text{th j-mb ic}} + R_{\text{th mb ic-package tag}}\right) + R_{\text{th package tag-amb}}$$

$$R_{\text{th j-amb}} = R_{\text{th tag}} + R_{\text{th package tag-amb}}$$

The tag manufacturer is solely responsible for determining the first bracketed term of the equation $R_{\text{th tag}} = R_{\text{th j-mb ic}} + R_{\text{th mb ic-package tag}}$, since this value depends on the choice of physical substrate (paper, polyester film, adhesive, etc.) on which is mounted the integrated circuit and antenna assembly that determines the value of $R_{\text{th mb ic-package tag}}$. If the tag substrate is paper, this value can be of the order of 80 to 100 °C W^{-1}, and the total thermal resistance of the tag will then be

$$R_{\text{th tag}} = R_{\text{th j-mb ic}} + R_{\text{th mb ic-package tag}}$$

$$R_{\text{th tag}} = 80 + 100 = 180 \,^{\circ}\text{C W}^{-1}$$

The end user of the tag is responsible for the second bracketed term of the equation, $R_{\text{th package tag-amb}}$. This is the thermal resistance of the package to which the label is affixed. If the package is made of cardboard, it will be a poor conductor of heat, and its thermal resistance may be as much as 200 °C to 300 °C W^{-1} (a very high level, because paper and cardboard are poor conductors of heat), giving a total of

$$R_{\text{th j-amb}} = \left(180 + 300\right) \,^{\circ}\text{C W}^{-1} = 480 \,^{\circ}\text{C W}^{-1}$$

The junction temperature of the integrated circuit will then be

$$T_{j\,max} = [25\,°C + (480 \times 100 \times 10^{-3})] = 73\,°C$$

which is a junction temperature that any commonly available integrated circuit can withstand easily, but, as indicated by the figure above, a hot spot will be created on the surface of the packaging box, immediately under the integrated circuit, at a temperature of $[25\,°C + (300 \times 100 \times 10^{-3})] = 55\,°C$. Sometimes this temperature ($55\,°C$) is enough to blacken some labels containing tags and/or give off wisps of smoke, depending on the type of packaging, if the packages are left too close to a base station antenna for too long.

The conclusion is simple: forewarned is forearmed!

Notes

Those who have read my previous books may recall that Murphy's Law number 243642b states that:

- a driver will always park his lift truck (with a pallet on the forks) as close as possible to a base station before going for his lunch break and
- whenever a conveyor breaks down for any length of time, there will always be a package and its electronic label immediately facing the antenna of the most powerful base station in the whole system.

7

Reality Check: How to Manage Everyday Problems

It is time to move on to the sober reality: after all, what we must always do is design systems that operate in specific circumstances!

Regarding propagation and free space, the general condition for maximizing the power transferred to the tag is, as we have seen, that propagation should take place in free space, that the axis of propagation of the base station and the axis of reception of the tag should be perfectly in line and that the impedances of the tag antenna and of the integrated circuit (with or without its casing) should be perfectly matched.

Sadly, the operating conditions are often far from perfect, leading to numerous problems arising from:

- wave absorption and reflection, due to the environmental conditions;
- losses due to problems of optical alignment between the base station antenna and the tag antenna (polarization);
- impedance mismatches:
  - between antennae and their driving circuits (at the base station),
  - between the antenna and the load (in reception at the tag) ('matching');
- all the associated losses measured in watts.

In the following part of this lengthy chapter, I will examine all these sources of mismatching . . . and the headaches they cause.

Note on the rest of this chapter

To avoid any confusion or any wrong impressions that you may gain from the following lines, you should know that the following paragraphs are not intended to be 'pro' or 'anti' UHF/SHF.

You may be surprised to discover the many problems of implementation that are often disregarded or only mentioned briefly in the specialist press, either because of a lack of knowledge of the problems of everyday reality or because of powerful lobbying by certain business organizations defending their own interests while demonstrating their chronic obscurantism concerning these technical problems.

RFID at Ultra and Super High Frequencies: Theory and Application   Dominique Paret
© 2009 John Wiley & Sons, Ltd

In my view, these problems are an integral part of the story of RFID applications at UHF/SHF (or even the whole story); furthermore, it is extremely difficult to alter the laws of physics and their consequences!

7.1 Effects of the Application Environment

Perhaps you thought that the problems had all been dealt with? No! In fact, when electromagnetic waves are propagated in a real medium, rather than an ideal one, they are generally reflected, refracted, diffracted, absorbed or damped by materials present in the environment of the RFID system.

The following paragraphs will detail the everyday problems encountered in the implementation of RFID at UHF and at 2.45 GHz.

7.1.1 Absorption

The phenomenon of wave absorption arises because the material is slightly conducting and therefore a current flows in it. The energy contained in the radiated field is not conserved, and some of it is lost in the absorbent material. Obviously, the presence of these materials between the base station antenna and the tag antenna alters and has a strong effect on the operating distance of the system. For more specific information, Table 7.1 shows the absorbent properties and the effects of these in some materials that are normally present in the industrial environments of RFID applications.

7.1.2 Reflection

Let us now consider the phenomenon of reflection of waves which, theoretically, can arise anywhere, but usually inside a building, in what is called an 'indoor' configuration. Theoretically, the field conserves the same energy when there is pure wave reflection (specular, diffuse, diffractive, etc.). In reality, there is always some absorption.

For a given frequency (f = constant) and in a specific environment, such as air (ε and μ given), and therefore for a known constant wavelength λ, because of the different lengths of the wave path, which depend on the observation point (the tag), signals that have undergone reflection create interference and constructive or destructive effects with the direct primary wave from the base station, according to the respective amplitudes and phases (Figure 7.1). You will note that the first part of the last sentence makes a lot of provisos about the assumed operating conditions. We will move away from this blueprint a little further on.

Constructive Interference

When the reflected wave is in phase or nearly in phase with the direct primary wave from the base station, the reflected signal is added to the direct signal. Figure 7.2(a) shows this situation.

This may mean that some very distant operating points can be found. In this case, we speak of 'super distances' and 'hot spots' (Figure 7.2(b)). These points, which are often isolated, sometimes lead to confusion concerning the stated operating distances, and they may occasionally be troublesome for the planned application, in spite of the benefits that they may offer. In one well-known example, components located at the back of a warehouse may be

Table 7.1

Liquids	
Pure water, H_2O (nonconducting)	No absorption, or very little
Water (conducting)	Strong absorption, often prevents communication
Oil (nonconducting)	No absorption, or very little
Tag wet or damp (because of rain or ice)	No problem; operation expected
Human body	
Human body in front of the tag	Strong absorption (the human body is approximately 80% water)
A hand placed on the tag	Strong absorption; tag does not operate
Tag or badge on the front of the body	Strong absorption; operates poorly or not at all
Tag a few cm (3 to 5) in front of the body	Absorption, but expected to operate
Metals	
Tag behind a sheet of metal	Does not operate. In fact, the metal does not absorb the incident wave, but reflects it, thus preventing it from passing; this is comparable with very strong absorption on the path from the base station to the tag
Tag stuck on metal	Not certain; depends on the metal
Tag separated slightly from the metal (by a few mm)	Operation is possible. The reflection of the wave striking the metal returns some of the field to the transponder, acting as a 'reflector'. This gives the transponder antenna a higher gain, and therefore greater directivity, which may be a good or a bad thing depending on the application. Also, if a sheet of metal is deliberately fixed a few millimetres behind the tag antenna, it also acts as an electrical screen against a field arriving from the rear. See also the section dealing with reflections
Rubber objects	
Adhesive tape	Low absorption
Adhesive bands specially designed for these applications	Absorption
Tag inserted into rubber (a tyre, for example)	Note that because of the permittivity of this type of object, the propagation velocity of the incident wave changes in this environment and therefore the wavelength (but not the frequency) also changes: see Chapter 4 again ($\varepsilon_0\mu_0c^2 = 1$). The mechanical dimensions of tags designed on the basis of $\lambda/2$ antenna technology therefore change to some extent (shorter for a tyre)

read constantly and automatically even though readings are only required in the middle distance.

Destructive Interference

When the reflected wave is in phase opposition or nearly in phase opposition with respect to the direct primary wave from the base station, the reflected signal is subtracted from the direct signal. Figure 7.3 shows this situation.

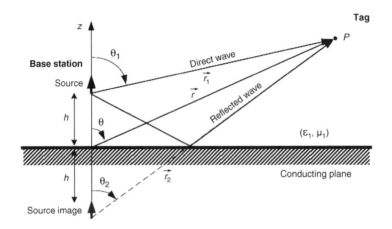

Figure 7.1   Phenomena of direct propagation and reflection of waves as a function of the environment

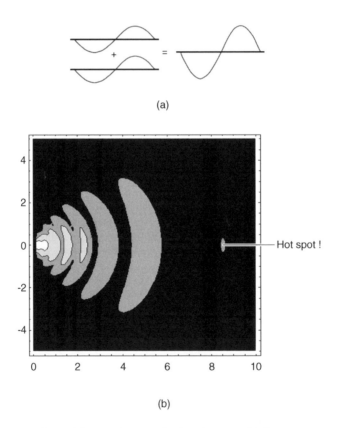

Figure 7.2   (a) Constructive interference. (b) Hot spot

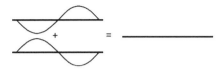

Figure 7.3 Destructive interference

This may mean that we find points where the tag can no longer be powered, and therefore cannot be read or written to. In this case we speak of 'black holes' (Figure 7.4). These points often occur repeatedly in the operating space of a system at multiples of the wavelength, and are therefore most troublesome, because, for example, if tags move rapidly through the field they are first powered, then unpowered, then powered again, and so on, which often gives rise to difficult problems of collision management if sufficient care is not taken. Some standards (ISO 18000-6) have now recognized these eventualities and have largely resolved the problems using special functional commands (see Chapter 15 on standards).

> **Note**
> This is a last note on the difference between this phase addition and subtraction. This only relates to a phase difference of 180°, which in terms of differences in the wave travel distance is equal to half the wavelength of the transmitted wave, in other words about $35/2 = 17.5$ cm at 900 MHz. The simple presence of a person who unexpectedly walks into the field may absorb some of the wave used for identification, such that identification is no longer possible in the presence of this person.

Some Examples
To give you more specific information, Table 7.2 shows the reflective properties of some materials normally present in industrial environments.

The Effects of Reflection from Metal in RFID Applications
As mentioned above, we can use the phenomenon of reflection to construct a reflector in order to increase the communication distance, by placing a metallic component just behind the tag. This increases the gain of the tag antenna, but also its directivity. Is this good or bad? It depends on the type of application planned.

Devices for guiding the propagated wave can be designed, using metal components that promote specular reflection. Similarly, a metal component can form a screen against the propagation of the wave because it is reflective. Thus we can have two views of the same problem, one positive and one negative:

- If the effect is desired:
  - it enables us to distinguish between different places where radio frequency identification is to be carried out;
  - it also enables us to conceal objects that are not to be identified, by wrapping them in conducting sheets.

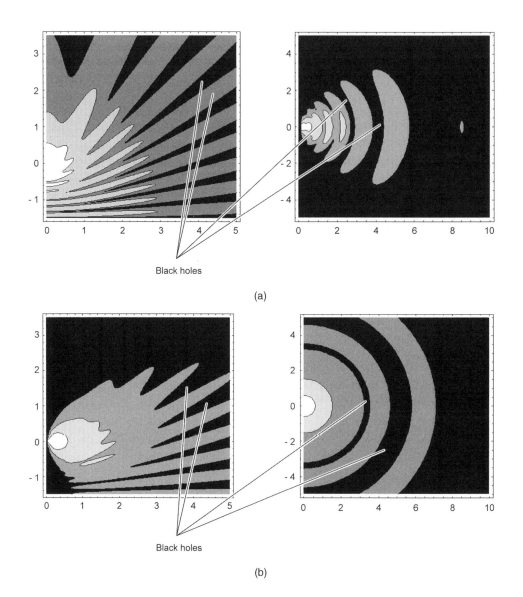

Figure 7.4   Examples of black holes

• If the effect is not desired:
  - it conceals objects from the system user, because they are wrapped in conducting sheets and therefore cannot be identified;
  - unless special measures are taken, it makes it virtually impossible to identify all the metal boxes placed inside, on or in a pallet.

In many applications, where the tag is concerned, it is best to form a small screen using a metal component that meets two requirements which are often complementary in applications:

Table 7.2

Metallic objects	
Metal	Highly reflective
Floor with metallic structure	Reflective
Metallic paint	Reflective
Water	
Water in the form of puddles or pools	Reflective
Other objects	
Paint	Depends on the material; often highly reflective
Wall	Reflective, depending on the material
Plastic sheet (plastic swing doors)	Reflective
Thin plastic film (packaging, etc.)	Reflective

on the one hand, the metal acts as a 'load bar' (reflector) while, on the other hand, it allows isolation with respect to other mediums (Figure 7.5). Table 7.3 is taken from ETSI documents and presents some examples giving the relationships between materials, losses and rnage/distances.

Some More Thoughts About Absorption and Detuning

To avoid confusion between these concepts, here are a few more details. We often hear comments such as: 'I put my tag behind my car windscreen and it really cut down the operating distance. I can't believe that glass 'absorbs' energy!'

This is both true and false. The explanation is that placing the tag behind the windscreen (lead glass, heat-absorbing glass or specially reinforced glass) has markedly altered the initial tuning of the tag (Figure 7.6, provided by RAFSEC, shows some effects of different materials on a tag initially tuned to 900 MHz. Note the detuning of about 200 MHz for 6 mm thick glass!).

Obviously, if the base station obstinately continues to transmit at 900 MHz, the apparent attenuation at this frequency will rise markedly, because of the detuning of the tag and also the number of dB per octave of the slope of the tuned circuit formed by the tag input circuit. However, if you had known at the start of your system planning that you would subsequently stick the tag behind the windscreen, you would have designed your tag to be tuned to 900 MHz with allowance for the effect of the glass, and then there would have been some attenuation due to the presence of the glass, but not so much!

Therefore you should not blame certain materials for faults that they do not have, or only have to a limited extent: you are responsible for the rest.

Important note
I suggest that you carefully read Section 7.11 at the end of this chapter concerning the real and/or apparent effect of the presence of liquid in RFID applications. We will return to all these matters in Chapter 20, which deals with drawing up data sheets and tag measurements.

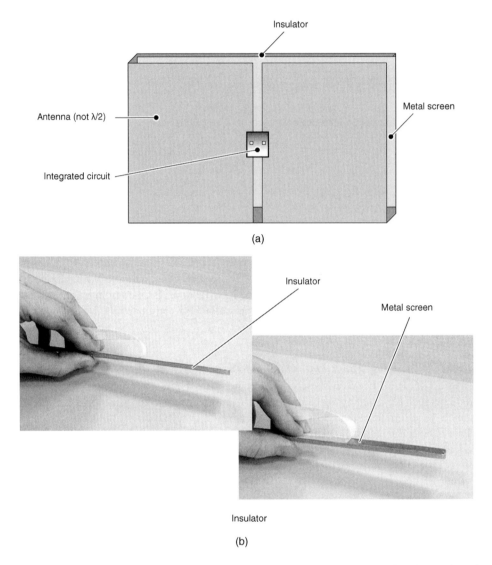

Figure 7.5 (a) Theoretical example. (b) Example of a metal screen used to isolate the tag from its environment

7.1.3 Refraction

The phenomenon of refraction of UHF and SHF electromagnetic waves (Figure 7.7) is of the same type as the familiar effect in optics. When the wave passes from one medium (air, for example) to another (water), the wave propagation velocity changes and we observe a change in direction of propagation, as in optics (with the refractive index), for the same physical reasons. Because of the topologies that are usually encountered in RFID, refraction does not cause too many problems in applications.

Table 7.3 In linear polarization (horizental or vertical)

Scenario	Reference distance (cm)	Range distance (cm)	$(R/R_{ref})^2$	Loss (dB)
Air	200	200	1.00	0.00
Tag on front of plastic case	200	180	1.23	0.92
Tag on front of plywood sheet	200	131	2.33	3.68
Tag on front of wood block 2.5 cm deep	200	120	2.78	4.44
Tag on front of paper 3 cm thick	200	108	3.43	5.35
Tag on front of empty plastic jug	200	149	1.80	2.56
Tag on rear of empty plastic jug	200	138	2.10	3.22
Tag on front of plastic jug filled with tap water	200	46	18.90	12.77
Tag on rear of plastic jug filled with tap water	200	31	41.62	16.19
Tag behind metal mesh 10×10 cm	200	28	51.02	17.08
Tag behind metal mesh 1×1 mm	200	10	400.00	26.02

From ETSI document. For the purpose of making these measurements the transmit level from the interrogator was set to a constant value.

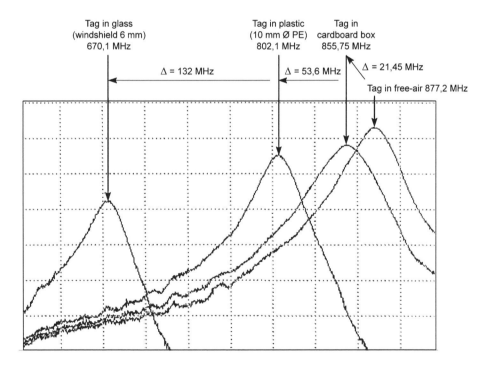

Figure 7.6 Example of the detuning of a tag due to a motor vehicle windscreen

Refraction

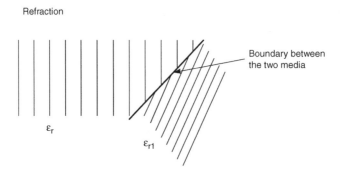

Boundary between
the two media

ε_r

ε_{r1}

Figure 7.7   The phenomenon of refraction of UHF and SHF electromagnetic waves

7.1.4 Diffraction

Similarly, the diffraction of electromagnetic waves due to passage through small apertures (Figure 7.8) does not give rise to major problems for RFID applications.

7.1.5 Indoor/Outdoor Models of Propagation

Moving back to technical matters, it is worth noting that different models of propagation are used, depending on whether we are planning to operate indoor environments (inside a room), urban environments (propagation through buildings in a city) or rural environments (in the open country). Our primary area of interest in RFID is the 'indoor environment', which we will now examine.

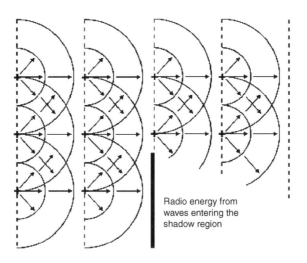

Radio energy from
waves entering the
shadow region

Figure 7.8   The phenomenon of diffraction of electromagnetic waves

Propagation Inside Buildings ('Indoor')

The outdoor propagation model is also called the 'free space' model, which has been fully described in Chapter 5, where I showed that the attenuation of the medium was conventionally equal to

$$\text{att(in dB)} = 10 \log \frac{1}{[v/(4\pi r f)]^2} = -10 \log \left(\frac{v}{4\pi r f}\right)^2$$

As for the indoor propagation model, this uses the above equation for attenuation in free space for distances r of less than 10 metres, in which the exponent of r is therefore a power of 2, but in the range of distances supported by the model (from 10 to 500 m) this value has to be corrected to about 3.5 (note that this value can vary from 1.6 to 4.5 depending on the type of building or materials). To illustrate this, consider an example of RFID at SHF at 2.45 GHz, which can easily be transposed to UHF. The equation of attenuation in free space as stated above is as follows (with f in GHz and r in m):

$$\text{att(dB)} = 32.5 + 20 \log f + 20 \log r$$

which, at 2.45 GHz and a distance of 10 m, gives the following attenuation (as shown above):

$$\text{att(dB)} = 32.5 + 20 \log(2.45) + 20 \log 10 = 60.3 \text{ dB}$$

to which we must add (in dB) the attenuation correction terms (which are actually multiplication factors in linear equations):

- due to a distance r in the range from 10 to 500 m:

$$10 \log \left(\frac{r}{10}\right)^{3.5}$$

- due to the absorption of the walls: M_{wall}.

Note that, for RFID at UHF and SHF, where the standard average thickness of walls or masonry is expected to be 22 cm, and depending on the materials used (plasterboard, glass wool insulation, standard or small bricks, concrete, reinforced concrete, etc.) with waves transmitted at about 10 to 20 dB, the final result is

$$\text{att(dB)} = 60.3 + 35 \log \frac{r}{10} + M_{\text{wall}}$$

Notes

This model is not applicable beyond 500 m because it is very unusual to find single buildings with distances of more than 500 m inside them! Sceptics should know that this indoor propagation model has been tested and proven by many companies and plenty of evidence in the form of measurements can be found in the RFID literature.

Final Comments (Before the Next Ones!)

The propagation of energy at UHF and SHF inside a building is radically different from what takes place outdoors, because propagation inside buildings is greatly influenced by many factors such as the shape, the construction materials, the type of building, the furniture and all the other elements inside the building.

Since the wavelengths of the radiation are small (e.g. 12 cm at 2.45 GHz and 35 cm at 900 MHz), the building will contain many objects whose dimensions are of the same order of magnitude as half a wavelength (i.e. 6 cm or 18 cm), or more, which can interact with the energy of the RF waves transmitted by the base station. Each of these objects is intrinsically a potential source of reflection, diffraction or dispersion of the radio frequency energy radiated by the base station.

In the case of base station antennae with low gain (a few dBi), such as those commonly used in RFID systems at UHF and SHF, their beam apertures are fairly wide (about 70 to 90°), and therefore they illuminate many reflective surfaces. Moreover, these surfaces are never arranged uniformly, but are orientated in many different directions and have all kinds of sizes and shapes. The result of all these reflectors is that the incident energy is reradiated in all directions. Thus the propagation of the energy radiated by the base station towards the tag does not take place in direct line of sight. Instead, the signal is reflected by the many surfaces present in the area illuminated by the antenna. Since the orientations of these reflective surfaces are uniformly distributed, and the surfaces are perfectly reflective, all the incident energy will be reradiated uniformly in all directions. This will have the immediate effect of entirely defocusing the initial radiation beam of the base station antenna and completely altering its real radiation pattern which, in this idealized scenario, will become totally isotropic. This may be good or bad – it all depends on the planned application. In reality, few objects are perfect reflectors at these frequencies, and many diffraction effects occur because of various obstructions on the path of the signal. Frequently, in fact, the 'effective gain' of the base station is somewhat less than 0 dBi. This is also in line with the experience of many retailers of RFID equipment!

I cannot claim to have reviewed all the problems that may arise in RFID applications, but we have completed our first approximate list of the main points concerning absorption and reflection. Having dealt with these problems, you will be able to progress at a faster pace.

Now let us review some of the ways in which these effects can be overcome.

7.1.6 Using the Spread Spectrum, or Frequency Agility, Technique

For many other reasons, explained fully in Chapter 13 which deals with the technology of spread spectrum, frequency agility, FHSS (frequency hopping spread spectrum) and DSSS (direct sequence spread spectrum), and the LBT (listen before talk) principle, we can overcome some of the above problems, especially those relating to communication 'black holes'.

In fact, owing to the regular frequency hopping caused by these techniques, the wavelengths of the propagated waves are constantly changing. Because of this, on the same propagation path as before, the final phase of the transmitted signal varies constantly as a function of the transmitted wavelength, and it will therefore be very likely that many of the previous local 'black holes' will be filled in at some of the transmitted frequencies. In fact, this will only be true of some of the frequencies in the band, and, since the transmitted data stream is carried on transmitted frequencies that vary randomly in time, there will be some bits that are perfectly received and assumed to be valid at any physical location of the tag, as well as losses of information (no bit received), although these will be less common than before. Compensation for residual communication errors must also be provided, e.g. by using CRC and automatic error correction systems.

Example

In the USA, according to the current FCC regulation, FHSS can be used to sweep the frequency band from 902 to 928 MHz, equivalent to wavelengths from 33.25 to 32.32 cm, i.e. a wavelength range of 0.93 cm.

If a 'perfect' black hole (total phase opposition) were present at a point in space at the minimum permitted frequency (902 MHz), then in order to have a signal completely in phase at the same observation point at the maximum permitted frequency, the signal would have to travel along an equal supplementary path $(n\lambda/2)$ at the maximum frequency, in other words $n \times (33.25/2)$ cm. At this frequency, the shortest of these supplementary paths is present when $n = 1$, i.e. 16.62 cm. When the maximum frequency of the permitted band is transmitted (928 MHz), this corresponds to a supplementary path of 16.16 cm. If one or more of the travel/ reflection paths is present in the application considered for 16.16 cm and integer multiples of this value (32.32 cm, etc.), everything will be perfect – otherwise, you will fall into the black hole and stay there!

Although the problem of black holes is considered to have been resolved by the use of solutions based on the FHSS technique, there is still a 'but', due to the fact that when these solutions are used the transmitted carrier has to be cut off for a few instants (obviously) when the transmitted frequency is changed. Unfortunately, this causes a momentary loss of power in some tags (particularly those in the vicinity of black holes) because of the cutting of the carrier; in some cases, these tags 'lose their memory' of what happened before, as a result of the power cut, and this can be very annoying, especially during the collision management procedure. Fortunately, the RFID standards experts in the ISO 18000-6 and -4 'air interface' group have allowed for this, by introducing some special commands into the communication protocols to overcome these physical deficiencies.

A final point: sadly, the use of FHSS is not authorized everywhere in the world. Although these methods are usable and used to a great extent in the USA where they are authorized, this is not the case in Europe, because the widths of the authorized RFID bands are insufficient, or practically so. Note, however, that there are now some limited prospects for the use of methods related to LBT (listen before talk – see Chapters 13 and 16) frequency agility techniques in Europe, according to ETSI 302 208. These methods can be used in a much smaller frequency band (2 MHz, divided into channels of 200 kHz) than the band authorized in the USA.

7.1.7 Multiple Antenna Systems

It is often necessary to use more than one base station, or a single base station with a number of antennas, in order to cover the whole of the desired space for the application without any black holes. It is generally preferable to use time division multiplexed systems and/or antennas whose operation is synchronized, monitored and controlled by a single base station provided for this purpose.

To avoid the loss of the internal states of the tags (due to interruptions in the radiated field in antenna switching sequences) between two phases of multiplexing, it is generally

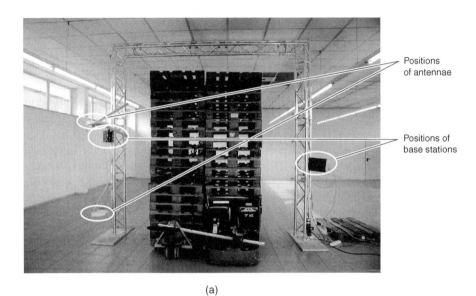

(a)

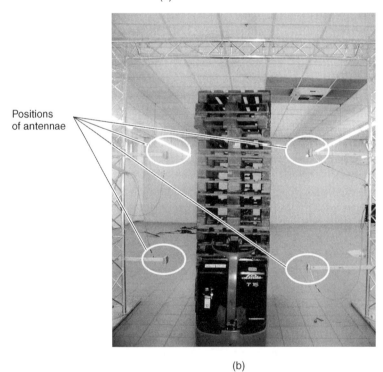

(b)

Figure 7.9   Examples of antenna positioning according to (a) FCC regulations (USA) and (b) ETSI/ERC 70-03 (Europe)

considered preferable for each of the multiplexing stages to be long enough for each of the antennas to process, in turn, all the tags present in the field, after which the host computer of the system concatenates all the results to manage missing and duplicated elements. The 'data exchange status bit' specified in ISO 18000-4 and -6 can be used as required for this purpose.

Positioning of Antennas

The mechanical positioning of the antennas must be such that the sum of the radiated fields covers the operational space or volume covered by each of the individual antennas. Increasing the amount of overlap improves the reliability of the radio frequency identification. The photographs in Figure 7.9 provide some specific examples of industrial implementations using two or four multiplexed UHF antennas.

If the objects to be identified move rapidly in the antenna field or fields, the multiplexing and switching of the antennas are often controlled by external movement detectors such as infrared or similar light barriers.

7.1.8 A Further Note on Differences between 13.56 MHz and UHF/SHF

I have often been asked about the spatial shaping of fields radiated between 13.56 MHz and UHF and the reproducibility of the spatial coverage of these. Because of the applicable physical laws, namely a strongly inductive link at LF and HF (125 kHz and 13.56 MHz), on the one hand (the Biot–Savart law), and radiation and propagation of electromagnetic waves at radio frequency, on the other hand (Maxwell's equations), the shaping of the magnetic fields used in applications at LF and HF is clearly more easily reproducible in terms of geometry than that of electromagnetic fields radiated at UHF/SHF, as the latter are much more dependent on numerous external parameters (reflection, absorption, etc.) that cannot be easily controlled in the applications (which may be subject to rain, or the necessary passage of people or objects through the space concerned, etc.). A comparative example, admittedly rather exaggerated but clearly demonstrating this difference, is given in Figure 7.10.

7.2 Tag Polarization Losses, $\theta_{\text{polarization}} = p$

The parameter representing losses due to a relative angular misalignment between the component of the electrical field E of the electromagnetic wave radiated by the transmitting antenna (the base station) and the principal axis of the tag's receiving antenna is called p or $\theta_{\text{polarization}}$. This factor p is introduced here to ensure that the user always remembers that, unfortunately, a tag is hardly ever positioned physically at the optimal angle with respect to the famous 'equatorial plane' and to the incident electrical field lines E; if it were, then all would be for the best in the best possible world of theoretical calculation! Clearly, the ideal value is $p = 1 = 100\,\%$, but in the least favourable case, its value may well drop to 0! I will not insult you by expecting you to design an application that operates close to the latter limit condition at all times. This factor is generally defined by the following mathematical expression:

$$p = \frac{\left|El_{\text{effective}}^{*}\right|^{2}}{\left|E\right|^{2}\left|l_{\text{effective}}^{*}\right|^{2}} \text{ as a \% of the maximum}$$

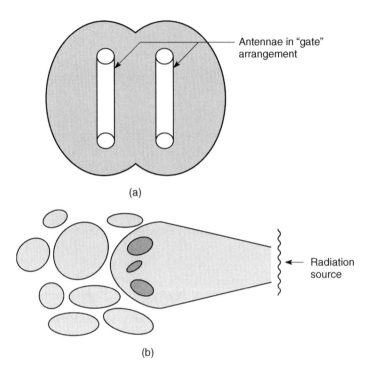

Figure 7.10   Examples of radiated fields (a) at LF and HF and (b) at UHF and 2.45 GHz

where $l_{\text{effective}}$ represents the effective length of the dipole and $l^{*}_{\text{effective}}$ is its conjugated value.

Since the product El represents a voltage, the determination of the ratio of the squares of this quantity requires that we estimate, at an instant t, for a given orientation of the tag antenna with respect to its optimal position, the power ratio between the power actually received and the maximum possible power. The power P_t actually received by the tag is therefore

$$P_t = pP_{t\,\text{max}}$$
$$p = \theta_{\text{polarization}} = (\text{Polar}_{\text{ant}\,t}\text{Polar}_{E\,\text{incident}}) = \text{scalar product of the two vectors:}$$
$$p = \theta_{\text{polarization}} = (|\text{Polar}_{E\,\text{incident}}||\text{Polar}_{\text{ant}\,t}|)\cos\theta\%$$
$$p = \theta_{\text{polarization}} = (|\text{Polar}_{E\,\text{incident}}|)(|\text{Polar}_{\text{ant}\,t}|\cos\theta)\%$$

where $\text{Polar}_{\text{ant}\,t}$ is the modulus of the orientation vector of the tag antenna, $\text{Polar}_{E\,\text{incident}}$ is the modulus of the polarization vector of the incident field E (linear H or V, circular right-hand, left-hand, elliptical, etc.) and θ is the angle between the two vectors described above; $\langle\langle\cdot\rangle\rangle$ represents the scalar product of these two vectors. As indicated in Figure 7.11, the term $(|\text{Polar}_{\text{ant}\,t}|\cos\theta)$ represents the projection of the vector collinear with the tag antenna on the axis collinear with the direction of the electrical field E radiated by the base station.

This is essentially a matter of precisely quantifying the well-known qualities represented by the three-dimensional radiation patterns (in reception) of the tag antennas as a function of the angles θ and φ used in polar coordinates relative to the incidence of the incident wave. In

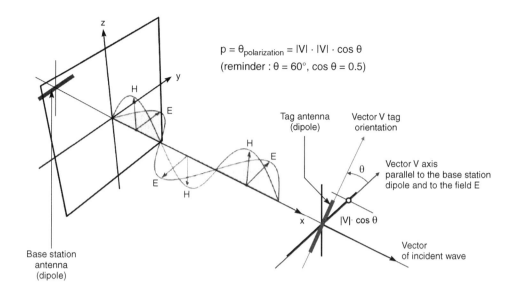

$$p = \theta_{polarization} = |V| \cdot |V| \cdot \cos \theta$$
$$(\text{reminder}: \theta = 60°, \cos \theta = 0.5)$$

Figure 7.11 Relative orientations of the base station and tag antennas in the far field

mathematical terms, this means generalizing the expression for the tag antenna gain in the form:

$$G_{ant\ t} = G(\varphi, \theta) = \ldots \text{ in our case, } (G_{ant\ t} \cos \theta)$$

Example

(a) If the angle of the axis of the tag's receiving dipole is orientated so as to receive the maximum electrical field E from the base station, the attenuation p due to 'polarization mismatching' is zero (0 dB). Figure 7.12 shows an example of an application of this kind in which the electrical field vector **E** of the plane wave propagated by the base station antenna (a Yagi antenna, with a high gain of about +6 to 15 dBi, and therefore directive) is horizontal (therefore in the far field of the base station) and perfectly aligned with the tag antenna.

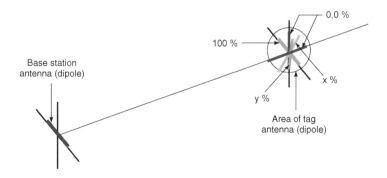

Figure 7.12 Examples of relative orientations of the base station and tag

Table 7.4   Angle expressed in degrees in linear polarization (horizontal or vertical)

Angle θ (deg)	$\cos\theta$	Attenuation (dB) $20\log(\cos\theta)$
Base station and tag antennas		
Aligned 0	1	0
15	0.966	0.3
30	0.866	1.25
45	0.707	3
60	0.5	6
75	0.259	11.74
Orthogonal 90	0	Infinite

(b) If the tag antenna remains in the equatorial plane, but has an orientation of $\theta = 60°$ with respect to the optimal position (i.e. an angle of 30° with respect to the incident wave), we have $\theta = 60°$ and therefore $\cos\theta = 0.5$; in other words, the attenuation will be $0.5 = 50\%$ ($-6\,\mathrm{dB}$), or alternatively, to generalize, the gain of the tag will appear to have decreased as follows:

$$G_{\mathrm{ant\ t}} = G(\varphi, \theta) = (G_{\mathrm{ant\ t}}\cos\theta)$$

$$G_{\mathrm{ant\ t}}(\mathrm{dB}) = G(\varphi, \theta) = G_{\mathrm{ant\ t}}(\mathrm{dB}) + [20\log(\cos\theta)]$$

and therefore, in our example,

$$G_{\mathrm{ant\ t}} = G(\varphi, \theta) = (G_{\mathrm{ant\ t}} \times 0.5)$$
$$G_{\mathrm{ant\ t}}(\mathrm{dB}) = G(\varphi, \theta) = G_{\mathrm{ant\ t}}(\mathrm{dB}) - 6\,\mathrm{dB}$$

(c) Finally, when the axis of the tag dipole is orthogonal to this optimal orientation, in other words, when the dipole axis is collinear with the direction of propagation of the wave or when it is positioned vertically with respect to the equatorial plane, the attenuation is infinite.

Table 7.4 summarizes the variations of the attenuation as a function of the relative orientations of the base station and tag.

(d) To improve matters in the last-mentioned cases, when permitted by local regulations, we can use circular or elliptical polarization waves, e.g. those constructed by using two dipoles mounted orthogonally with respect to each other and driven by voltages that may or may not be equal in value but in phase quadrature, thus creating rotating polarization (circular or elliptical) in the radiated wave (Figure 7.13).

Figure 7.14 shows the comparative performances of different wave polarizations. For example, if the two antennas (base station and tag) are linearly polarized and well aligned, the loss factor will be zero. However, if one is linear and the other is circular, this factor will be $-3\,\mathrm{dB}$.

Notes

(a) The properties described above are only true for the far field, when the propagated wave can be considered to be a true plane wave. At 900 MHz, for example, with a wavelength of about 35 cm, this is completely true beyond 1.5 m (about 4λ). In the near field, in other words near the antenna

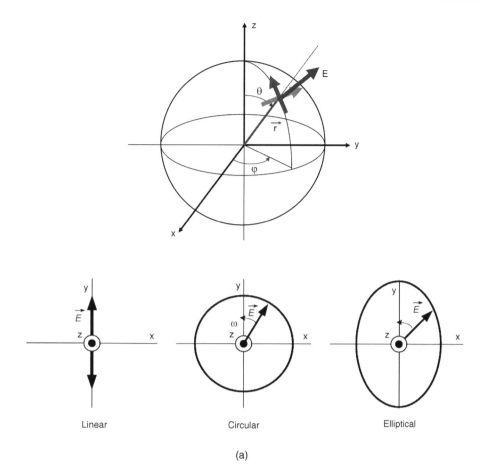

Figure 7.13   (a) Linear, circular and elliptical polarization. (b) Circular polarization

(from 0 to 1 m), as mentioned at the beginning of Chapter 4, Maxwell's equations indicate that the relative angles of the vectors **E** and **H** vary as a function of the distance, and, furthermore, the propagated wave is not a plane wave. The orientation of the tag therefore has less effect.

(b) If we now examine the problem in greater detail, at a location outside the equatorial plane the mean value of the power flux density $\langle s \rangle$ varies greatly (as a function of the shape of the radiation lobe, the angle of observation and consequently the associated directivity), and therefore the apparent gain decreases at a given point in the radiation space. This consequently leads to a loss of communication distance. Figure 7.15 shows the example of a pallet in which some of the well-arranged tags that it contains are located on the axis with a gain of G' such that $G' = G_{max}/2$. Also, the position of the tag is not optimal with respect to the electrical field vector **E** radiated at this point ($45°$ in the case shown in the diagram), causing a reduction in gain of $0.5 G_{max} \cos 45° = 0.5 G_{max} \times 0.707 = 0.35 G_{max} = 35\%$ of G_{max} – a decrease that is partially compensated for by the fact that the tag is located fairly close to the antenna, and is also on the boundary of the near and far fields, as mentioned at the beginning of the preceding section.

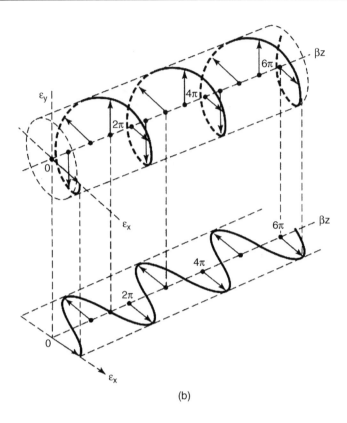

(b)

Figure 7.13 *(Continued)*

		Polarization of base station antenna		
		circular	vertical	horizontal
Position of tag	vertical	3 dB	0 dB	∞ dB
	horizontal	3 dB	∞ dB	0 dB
	inclined	3 dB	3 dB	3 dB
	parallel to beam	∞ dB	∞ dB	∞ dB

Figure 7.14 Performance of different polarizations

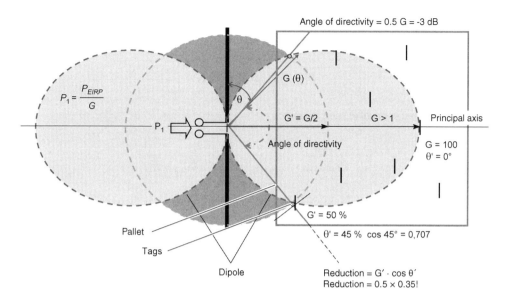

Figure 7.15   Detailed example of the angular positions of tags on a pallet

7.3 Antenna Load Mismatch Factor, $\theta_{\text{load matching}} = q$

The sole purpose of this section is to quantify the effect of the variation of nominal impedance mismatching (outside any tolerances) between the complex values of the output impedance of the tag antenna, on the one hand, and of the input impedance of the integrated circuit, on the other hand. For remotely powered tags, these values should theoretically be perfectly conjugated during the forward link. This is why the factor q appears. The effect of a deliberate impedance mismatch will be examined subsequently.

7.3.1 When the Source and Load Impedances Have any Values

As a general rule, if there is impedance mismatching (small or large, deliberate or otherwise) between the source and the load, the quantity q defines the power mismatch factor of the load (or $\theta_{\text{load matching}}$) as the ratio between the real power in watts dissipated in the load R_1 and the maximum possible power that would be available to it with optimal matching. This parameter therefore relates solely to power mismatching as seen from the 'load' end. We should be careful not to give it any other significance. By definition, it is as follows:

$$q = \frac{P_{1\,\text{rms}}}{P_{1\,\text{rms max}}} \quad \text{and therefore } P_{1\,\text{rms}} = qP_{1\,\text{rms max}}$$

where, of course,

$$P_{1\,\text{rms}} = R_1 I_{\text{rms}}^2 \text{ for any value } R_1 \text{ of the load}$$

$$P_{1\,\text{rms}} = \frac{R_1}{\left(R_{\text{ant t}} + R_1\right)^2 + \left(X_{\text{ant t}} + X_1\right)^2} V_{\text{equi rms}}^2 \text{ W r.m.s.}$$

Additionally, when there is conjugate matching between the source and load ($X_1 = -X_{\text{ant t}}$ and $R_1 = R_{\text{ant t}}$), the above equation is simplified, and we can find the value of $P_{1\text{ rms max}}$ thus:

$$P_{1\text{ rms max}} = \frac{V^2_{\text{equi rms}}}{4R_1} \text{ W}$$

The general expression of q then becomes

$$q = \frac{P_{1\text{ rms}}}{P_{1\text{ rms max}}} = \frac{4R_{\text{ant t}}R_1}{\left(R_{\text{ant t}} + R_1\right)^2 + \left(X_{\text{ant t}} + X_1\right)^2}$$

The factor q is very important, because it denotes the portion of the power consumed by the load as a function of its impedance mismatching with respect to the ideal load. This means that, when there is impedance mismatching between the source and load, only the power $P_{1\text{ rms}}$ (the real value of $qP_{1\text{ rms max}}$) is dissipated in the form of watts in the load, and because of the mismatching, as I will show in the fourth part of this chapter, the rest of this power, i.e. $(1 - q)P_{1\text{ rms max}}$, forms a part (but only a part) of the power reradiated towards the source (base station).

7.3.2 The Case in Which Source and Load Impedances are Tuned But Not Matched

In most RFID applications operating at UHF and SHF, the whole circuit formed by the antenna plus the tag integrated circuit is a tuned circuit for a given frequency ($X_{\text{ant t}} = -X_1$). In this particular case, the equation for q shown above is simplified to

$$q = \frac{P_{1\text{ rms}}}{P_{1\text{ rms max}}} = \frac{4R_{\text{ant t}}R_1}{\left(R_{\text{ant t}} + R_1\right)^2}$$

and therefore, assuming that

$$R_1 = aR_{\text{ant t}}, \text{ i.e. } a = R_1/R_{\text{ant t}}$$

$$q = \frac{P_{1\text{ rms}}}{P_{1\text{ rms max}}} = \frac{4a}{(1 + a)^2} = f(a) = f\left(\frac{R_1}{R_{\text{ant t}}}\right)$$

Some Physical Explanations About q

In order to quantify the load power mismatch by comparison with the theoretical results obtained from the Friis equation, we calculate the portion of the power $P_{1\text{ refl}}$ that is not absorbed by the load Z_1 and is therefore returned/reflected by the load towards the tag antenna. Obviously, the power actually transmitted to the load and dissipated in it, P_1, will be the difference between the maximum possible that it can receive, $P_{1\text{ max}}$ (equal to the received power P_t found from the Friis equation), and the power $P_{1\text{ refl}}$ reflected because of the impedance mismatching of the load:

$$P_1 = P_{1\text{ max}} - P_{1\text{ refl}}$$

(I will show subsequently that the power reflected at the load $P_{1\,\text{refl}}$ will form part of the power reradiated by the tag, and, as indicated below, this portion of the power will be used in the calculation of the radar cross-section (RCS) of the tag during this impedance mismatching.)

The ratio of the net effective power P_1 dissipated in the load at maximum incident power $P_{1\,\text{max}}$ is by definition equal to the 'antenna load mismatch factor', denoted by either q or $\theta_{\text{load matching}}$:

$$\theta_{\text{load matching}} = \frac{P_1}{P_{1\,\text{max}}} = 1 - \frac{P_{1\,\text{refl}}}{P_{1\,\text{max}}} = q$$

whose value varies in physical terms from 0 to 100 %. To quantify this value, we must calculate or measure the ratio $P_{1\,\text{refl}}/P_{1\,\text{max}}$, which is not as straightforward as it seems. But – for there is often a 'but' that sorts everything out – because of the general physical relationship between 'power' and 'voltage', $P = V^2/R$, we can logically write

$$\frac{P_{1\,\text{refl}}}{P_{1\,\text{max}}} = \frac{V_{1\,\text{refl}}^2}{V_{\text{inc}}^2} = \left(\frac{V_{1\,\text{refl}}}{V_{\text{inc}}}\right)^2$$

We must then quantify the phenomena of wave reflection, using the VSWR (voltage standing wave ratio), the reflection coefficient Γ, etc., due to the impedance mismatch between the source and load, and this is done in the same way as for distributed constant lines with negligible losses ('lossless').

7.3.3 The Relationship Between q and the Reflection Coefficient Γ

By definition, the ratio of complex voltage values between the reflected wave and the incident wave is called the reflection coefficient Γ (sometimes also denoted by ρ in the RF literature):

$$\Gamma = \frac{V_{1\,\text{refl}}}{V_{\text{inc}}} \text{ (in complex value, i.e. modulus and phase)}$$

where V_{inc} is the complex value of the voltage of the incident (received) wave in amplitude and phase, and $V_{1\,\text{ref}}$ is the complex value of the voltage of the reflected wave in amplitude and phase.

The value of $P_{1\,\text{refl}}/P_{\text{max}} = (V_{1\,\text{refl}}/V_{\text{inc}})^2$ is therefore equal to the square of the modulus of the complex value of Γ. Thus

$$\theta_{\text{load matching}} = \frac{P_1}{P_{1\,\text{max}}} = 1 - \frac{P_{1\,\text{refl}}}{P_{1\,\text{max}}} = 1 - \left(\frac{V_{1\,\text{refl}}}{V_{\text{inc}}}\right)^2 = 1 - \Gamma^2$$

or, to put it another way,

$$P_1 = (1 - |\Gamma|^2)P_{1\,\text{max}} = qP_{1\,\text{max}}$$

or alternatively

$$\theta_{\text{load matching}} = q = 1 - |\Gamma|^2 = \frac{4a}{(1+a)^2}$$

$$\theta_{\text{load matching}}(\text{in dB}) = 10\log q = 10\log(1 - |\Gamma|^2)$$

7.3.4 The Relations between q, Γ and a

Since the maximum power $P_{1\,max}$ that the load R_1 can receive in the matching condition is equal to P_t, let us calculate the value of the reflected portion of power $[(1 - q)P_{1\,max}]$ on this when there is impedance mismatching with respect to the optimal load and when R_1 absorbs $qP_{1\,max}$. To do this, we estimate the value of $(1 - q)$:

$$1 - q = 1 - \frac{4R_{ant\,t}R_1}{(R_{ant\,t} + R_1)^2 + (X_{ant\,t} + X_1)^2}$$

$$1 - q = 1 - \frac{(R_{ant\,t} - R_1)^2}{(R_{ant\,t} + R_1)^2 + (X_{ant\,t} + X_1)^2}$$

If the whole tag input circuit forms a tuned circuit, $X_{ant\,t} = -X_1$ (or alternatively $X_{ant\,t} = X_1 = 0$), and, assuming as before that $R_1 = aR_{ant\,t}$, i.e. $a = R_1/R_{ant\,t}$, the equation for x is simplified and becomes

$$1 - q = |\Gamma|^2 = \frac{(1 - a)^2}{(1 + a)^2} = \frac{(a - 1)^2}{(a + 1)^2}$$

In physical terms, this means that

$$\Gamma = \frac{a - 1}{a + 1} \quad \text{where } a = \frac{R_1}{R_{ant\,t}}$$

and that the value of Γ changes its sign according to whether a is larger or smaller than 1, in other words, according to whether the value of R_1 is greater or less than the value of $R_{ant\,t}$.

7.3.5 Summary

Table 7.5 gives a summary of the preceding paragraph.

Example

With a $\lambda/2$ dipole antenna connected to a load of 35 Ω, $R_1 = 35\,\Omega$:

$$\Gamma = (35 - 73)/(35 + 73) = (-38)/108 = -0.352$$

$$q = 1 - 0.124 = 0.876$$

Table 7.5

Matching conditions	R_1	$a = \dfrac{R_1}{R_{ant\,t}}$	$q = \dfrac{P_{1\,rms}}{P_{1\,rms\,max}} = \dfrac{4a}{(1 + a)^2}$	$|\Gamma|^2 = 1 - q$
Short-circuit	0	0	0	1
Matched	$R_1 = R_{ant\,t}$	1	1	0
Open	∞	∞	0	$(-1)^2 = 1$

7.4 Voltage Standing Wave Ratio (VSWR)

We shall now briefly consider the standing wave ratio (SWR) or voltage standing wave ratio (VSWR), often called S in the literature, which indicates the total impedance mismatching of a line as a function of its load. It quantifies the signal that is reflected due to the impedance mismatching, which itself is directly dependent on the amount of energy transmitted.

To determine the VSWR, we must calculate the ratio of the net power to the received power, which is equal to the ratio between the maximum and minimum voltage present on the transmission line:

$$\text{VSWR} = \frac{V_{\max}}{V_{\min}} = \frac{V_{\text{inc}} + V_{1\,\text{refl}}}{V_{\text{inc}} - V_{1\,\text{refl}}}$$

7.4.1 The Relationship between the Reflection Coefficient Γ and the Voltage Standing Wave Ratio (VSWR)

From the above equation we can easily deduce the modulus of the reflection coefficient $|\Gamma|$, because

$$\text{VSWR} = \frac{V_{\max}}{V_{\min}} = \frac{1 + (V_{1\,\text{refl}}/V_{\text{inc}})}{1 - (V_{1\,\text{refl}}/V_{\text{inc}})}$$

$$\text{VSWR} = \frac{V_{\max}}{V_{\min}} = \frac{1 + \Gamma}{1 - \Gamma}$$

and therefore

$$\Gamma = \frac{\text{VSWR} - 1}{\text{VSWR} + 1}$$

On the other hand, since

$$\Gamma = \frac{a - 1}{a + 1}$$

by identifying term by term, we conclude that

$$\text{VSWR} = \frac{V_{\max}}{V_{\min}} = a = \frac{R_1}{R_{\text{ant t}}}$$

$$\Gamma = \frac{(R_1/R_{\text{ant t}}) - 1}{(R_1/R_{\text{ant t}}) + 1}$$

$$\Gamma = \frac{R_1 - R_{\text{ant t}}}{R_1 + R_{\text{ant t}}}$$

or, to generalize,

$$\Gamma = \frac{Z_1 - Z^*_{\text{ant t}}}{Z_1 + Z_{\text{ant t}}} \text{ (in complex values, i.e. modulus and phase)}$$

where Z_1 is the complex impedance of the load, $Z_1 = R_1 + jX_1 =$ input impedance of the integrated circuit, $Z_{\text{ant t}}$ is the complex impedance of the tag antenna, $Z_{\text{ant t}} = R_{\text{ant t}} + jX_{\text{ant t}}$, and $Z^*_{\text{ant t}}$ is the conjugated impedance of the tag antenna, $Z^*_{\text{ant t}} = R_{\text{ant t}} - jX_{\text{ant t}}$.

The value of Γ becomes

$$\Gamma = \frac{(R_1 + jX_1) - (R_{\text{ant t}} - jX_{\text{ant t}})}{(R_1 + jX_1) + (R_{\text{ant t}} + jX_{\text{ant t}})}$$

and therefore

$$\Gamma = \frac{(R_1 - R_{\text{ant t}}) + j(X_{\text{ant t}} + X_1)}{(R_1 + R_{\text{ant t}}) + j(X_{\text{ant t}} + X_1)}$$

and its modulus is

$$|\Gamma| = \frac{\sqrt{(R_1 - R_{\text{ant t}})^2 + j(X_{\text{ant t}} + X_1)^2}}{\sqrt{(R_1 + R_{\text{ant t}})^2 + j(X_{\text{ant t}} + X_1)^2}}$$

Some Specific Cases

To illustrate this more clearly, here are some specific examples of values of Γ and of its modulus, denoted $|\Gamma|$ (be careful not to confuse this value with the symbolic 'absolute value'!). When a $\lambda/2$ antenna is used ($Z_{\text{ant t}} = R_{\text{ant t}} + X_{\text{ant t}}$ with $X_{\text{ant t}} = 0$), the equation for Γ becomes

$$\Gamma = \frac{(R_1 - R_{\text{ant t}}) + jX_1}{(R_1 + R_{\text{ant t}}) + jX_1}$$

Now let us look at some specific cases.

The Load Z_1 Is a Pure Resistance ($X_1 = 0$)

$$\Gamma = \frac{R_1 - R_{\text{ant t}}}{R_1 + R_{\text{ant t}}} = f(R_1)$$

- *When $R_1 = R_{\text{ant t}}$.* Note that this corresponds to the case where the complex impedances Z_1 and $Z_{\text{ant t}}$ are strictly conjugate. In this case, the reflection coefficient is zero and the maximum power permitted by the Friis equation will be transmitted to the load in the absence of any wave reflection between the source (tag antenna) and load. All the power is absorbed in R_1:

$Z_1 = Z_{\text{ant t}}$; line terminated on $Z_{\text{ant t}}$ (matching); $\Gamma = 0$, $|\Gamma| = 0$, no reflection

- *When R_1 has any value:*
  - Some of the incident power is absorbed in the resistance R_1 and some of the power is reflected, and the reflected wave has a smaller amplitude than the incident wave.
  - At the end of the line, the reflected wave is in phase or in opposition, depending on whether the value of R_1 is smaller or greater than $R_{ant\,t}$, since the sign of the numerator changes when $R_1 = R_{ant\,t}$.
- *When $R_1 = 0$,* the line is short-circuited and reflection occurs:

$$\Gamma = -1, \; |\Gamma| = 1: \text{ total subtractive reflection}$$

- *When $R_1 = infinite$,* the line is open and reflection occurs:

$$\Gamma = +1, \; |\Gamma| = 1: \text{ total additive reflection}$$

The load Z_l is a Pure Reactance ($Z_l = jX_l$)
In this case, the line is terminated on a pure reactance:

$$\Gamma = +j \text{ or } -j, \text{ according to whether the reactance is inductive or capacitive,}$$

$$|\Gamma| = 1: \text{ total reflection}$$

- No power in watts is absorbed by Z_l.
- The incident wave is totally reflected at the end of the line.
- The amplitude of the reflected wave is identical to that of the incident wave (total reflection).
- The phase differs according to the type of reactance:

$$Z_1 = \frac{-j}{C\omega}, \quad \Gamma = -j, \quad |\Gamma| = 1$$

$$Z_1 = \frac{j}{L\omega}, \quad \Gamma = +j, \quad |\Gamma| = 1$$

- The load Z_l has the general form $Z_l = R_1 + jX_1$.
- Some of the incident power is absorbed in the resistance R_1.
- The reflected wave can have any amplitude (provided that it is less than the incident wave).
- At the end of the line, the reflected wave has any phase, according to Γ.

Notes on RFID operating at UHF or SHF
Some information in advance – as I shall point out subsequently, if we wish to obtain the widest possible variation of reradiated power, and thus have the maximum variation of the radar cross-section of the tag, ΔRCS, it would be best to switch the tag antenna from a fully open position ($\Gamma = +1$) to the totally short-circuited position ($\Gamma = -1$), thus providing the largest possible global variation of Γ from $+1$ to -1, in other words a variation $\Delta\Gamma = [1 - (-1)] = 2$. However, this ideal plan is rarely implemented in current RFID practice. The reason for this is very simple: in order to produce inexpensive tags, we must be able to power the integrated circuit remotely on each tag. For

Table 7.6

	R_1	$a = \text{VSWR} = R_1/R_{\text{ant t}}$	Γ	$(1 - \Gamma^2)$	q	Power reduction $(1 - \Gamma^2) = q$	
						In %	In dB $10 \log q$
Short-circuit	0	0	-1	$1-1$	0	100	Infinite
	$1/3\ R_{\text{ant t}}$	0.33	-0.5	$1-0.25$	0.75	25	1.3
	$2/3\ R_{\text{ant t}}$	0.66	-0.2	$1-0.04$	0.96	24	0.2
Matched	$R_{\text{ant t}}$	1	0	$1-0$	1	0	0
	$3/2\ R_{\text{ant t}}$	1.5	0.2	$1-0.04$	0.96	24	0.2
	$2\ R_{\text{ant t}}$	2	0.33	$1-0.111$	0.889	11.1	0.5
	$3\ R_{\text{ant t}}$	3	0.5	$1-0.25$	0.75	25	1.3
	$5\ R_{\text{ant t}}$	5	0.66	$1-0.444$	0.556	44.4	2.6
	$10\ R_{\text{ant t}}$	10	0.818	$1-0.669$	0.331	66.9	4.8
Open	Infinite	Infinite	$+1$	$1-1$	0	100	Infinite

this purpose, in order to obtain the greatest possible operating distance, we must first establish the configuration corresponding to the maximum transfer of received energy, in other words provide perfect impedance matching. In this case, as I have shown, Γ will be equal to 0. If we then switch (as far as possible) to the antenna short-circuit position, so that $\Gamma = -1$, the maximum real $\Delta\Gamma$ will be $[0 - (-1)] = 1$, and the maximum power will be reradiated towards the antenna.

By way of a summary, Table 7.6 provides a numerical example of relations between a, VSWR, Γ and $q = (1 - \Gamma^?)$, the last of which describes the percentage mismatching of power transmitted to the load.

In RFID applications, it is commonly accepted that a power reduction of 4 % is 'tolerable' (since we know that it may be present); as indicated by the table, this corresponds to a mismatch in Γ from -0.2 to $+0.2$ or, in terms of VWSR, from 0.7 to 1.5.

Note
Where the base station is concerned, this mismatching is often due to the fact that the actual distance between the base station circuits and its antenna can vary, depending on the installation sites used. The lengths of the coaxial cables linking these components can therefore differ, which unfortunately causes a small change in the value of the termination impedance and consequently the VSWR. There are two possible ways of overcoming this:

- either by incorporating the base station circuits in the antenna casing and eliminating the connecting cables (or at least keeping their lengths known and constant)
- or by providing an electronic device in the base station to automatically provide perfect impedance matching between the output impedance of the amplifier and the antenna impedance, thus maintaining a VSWR of 1.

Example

I claimed at the outset that I would provide specific information, so here it is. Let us again take the example of the NXP Semiconductors U_code HSL integrated circuit. The complex input impedance of this bare integrated circuit (the chip without its casing), Z_{ic}, measured at 900 MHz using a network analyser (a highly expensive but essential piece of kit), has a typical value (i.e. without tolerances) of

$$Z_{ic} = R_{ic} + jX_{ic}$$

$$Z_{ic} = 35 - j720$$

giving an intrinsic quality factor $Q = X_{ic}/R_{ic} = 720/35 =$ approximately 20.

The symbolic depiction of this impedance represents an electrical network composed of a resistance ($35\,\Omega$) in series with a capacitance ($-j720$). In the present case, this series representation does not in any way correspond to the physical reality, which is a parallel configuration, composed of a resistance $R_{ic\ p}$ of approximately $15\,k\Omega$ connected in parallel ($R_{ic\ p} = Q^2 R_{ic\ s} = 20^2 \times 35$) with a capacitance of 240 fF (1 femtofarad $= 10^{-15}$ farad).

Let us return to our matching problem. First of all, if we use an easily constructed $\lambda/2$ dipole tag antenna whose impedance should be very close to $Z_{ant\ t} = 73 + j0$, it will be difficult to provide the matching directly. Assuming that we are successful in this, the reactive part '$-j720$' is intrinsically dependent on the exact operating frequency ($X_{ic} = -1/(c_{ic}\omega)$), and conjugate matching will therefore not be present, leaving aside any other tolerances, except at a specific operating frequency.

Unfortunately (or fortunately) for use, ISO 18000-6 states that, for a tag to conform to the standard, it must be able to operate correctly throughout the 860 to 960 MHz band,[1] in other words a fairly wide frequency band; it must therefore maintain the closest possible matching across the whole band. Unfortunately, as the following sections will make clear, the ideal often comes into painful conflict with the reality!

Example: Compliance with ISO 18000-6

ISO 18000-6 clearly stipulates that an RFID tag must operate correctly at UHF from 860 to 960 MHz in order to conform with the standard. Assuming that the tag is inexpensive (as we all hope!), this means that its technology must be as simple as possible. In this case, the equivalent input circuit of the integrated circuit will simply be a series network of resistance $R_{ic\ s}$ with a capacitance $C_{ic\ s}$, tuned to the median value of the band concerned, i.e. $(860 + 960)/2 = 910$ MHz.

To avoid excessively lengthy calculations, let us assume that the load resistance $R_{ic\ s}$ is always matched to the antenna radiation resistance $R_{ant\ t}$ over the whole frequency band in question, i.e. $R_{ic\ s} = R_{ant\ t}$. (In fact, this is not quite correct, as the value of $R_{ant\ t}$ for an antenna is a function of the wavelength λ, and obviously this varies as a function of the frequencies in the band.) Returning to the initial equation:

$$1 - |\Gamma|^2 = \frac{4R_1 R_{ant\ t}}{(R_1 + R_{ant\ t})^2 + (X_1 + X_{ant\ t})^2}$$

---

[1] Note that the requirement to operate correctly throughout the 860 to 960 MHz band is based on the fact that, in order to provide good functional interoperability, a label on a package must be readable by local readers worldwide, and these readers may operate under different local UHF regulations for frequencies ranging from 860 to 960 MHz.

and introducing $R_1 = R_{ic\ s} = R_{ant\ t}$, we find that

$$1 - |\Gamma|^2 = \frac{1}{1 + [(X_1 + X_{ant\ t})^2/(4R_{ant\ t}^2)]}$$

where $X_{ant\ t} = L_{ant\ t}\ \omega$:

$$X_1 = x_{ics} = \frac{-1}{C_{ic\ s}\omega}$$

Arranging matters so that at $f_0 = 910\,\text{MHz}$, $X_1 = -X_{ant}$, then

$$X_1 = x_{ics} = \frac{-1}{C_{ic\ s}\omega}$$

$$\omega_0^2 = \frac{-1}{L_{ant\ t}C_{ic\ s}}$$

Assuming now that $X = (X_{ant\ t} + X_{ic\ s})$, for a frequency ω_1 we can write

$$X = L_{ant\ t}\omega_1 - \frac{1}{C_{ic\ s}\omega_1} = L_{ant\ t}\left(\omega_1 - \frac{1}{L_{ant\ t}C_{ic\ s}\omega_1}\right) = L_{ant\ t}\left(\omega_1 - \frac{\omega_0^2}{\omega_1}\right)$$

$$X = L_{ant\ t}\frac{\omega_1^2 - \omega_0^2}{\omega_1} = L_{ant\ t}\frac{(\omega_1 - \omega_0)(\omega_1 + \omega_0)}{\omega_1}$$

Now, $(\omega_1 - \omega_0) = \Delta\omega$ and $(\omega_1 + \omega_0) = 2\omega_1$ if $\Delta\omega$ is small with respect to ω_0; thus $X = 2L_{ant\ t}\Delta\omega$, and therefore

$$1 - |\Gamma|^2 = \frac{1}{1 + (L_{ant\ t}^2\Delta\omega^2/R_{ant\ t}^2)}$$

By definition, the quality factor (in no-load conditions) of the tag antenna $Q_{ant\ t}$ is as follows:

$$Q_{ant\ t} = \frac{L_{ant\ t}\omega_0}{R_{ant\ t}}$$

and therefore

$$\frac{Q_{ant\ t}}{\omega_0} = \frac{L_{ant\ t}}{R_{ant\ t}}$$

If we transfer this value into the preceding equation, we obtain

$$q = 1 - |\Gamma|^2 = \frac{1}{1 + (Q_{ant\ t}^2\Delta\omega^2/\omega_0)}$$

or alternatively

$$q = 1 - |\Gamma|^2 = \frac{1}{1 + Q_{\text{ant t}}^2 (\Delta f / f_0)^2}$$

Application

Still considering the case in which we use the NXP Semiconductors U_code HSL circuit described above, at $f_0 = 910\,\text{MHz}$ (the central frequency of the 860–960 Hz band with Z_{ic} at $910\,\text{MHz} = 35 - \text{j}720$):

$$\omega_0 = 2 \times 3.14 \times 910 \times 10^6 = 5715 \times 10^6 \, \text{s}^{-1}$$

$$\omega_0^2 = 32.66 \times 10^{18}$$

$$C_{\text{ic s}} = 240\,\text{fF}$$

$$C_{\text{ic s}} = 7.8387 \times 10^6$$

If

$$L_{\text{ant}} C_{\text{ic s}} \omega_0^2 = 1$$

$$L_{\text{ant t}} = 0.1275\,\mu\text{H}$$

$$L_{\text{ant t}} \omega_0 = 729\,\Omega$$

$$R_{\text{ant t}} = 35\,\Omega$$

then

$$Q_{\text{ant t}} = 20.83$$

If the application has to conform to ISO 18000-6, and must therefore cover the 860 to 960 MHz band, without any indication of x dB possible losses at the ends of the band, then in this case

$$\Delta f = 910 - 860 = 50$$

$$\Delta f^2 / f_0^2 = 3 \times 10^{-3}$$

$$Q_{\text{ant t}}^2 \Delta f^2 / f_0^2 = 1.31$$

i.e. $q = 1/(1 + 1.31) = 0.43$, an enormous value, even in nominal conditions (not allowing for the tolerances of the components). In other words, at the limits of the band (860 and 960 MHz), the load mismatch in dB will be 10 log q, and therefore the load mismatch due to the frequency variation in the band will be 10 log $(0.43) = -3.68$ dB.

I will demonstrate subsequently that the communication distance is proportional to $\sqrt{q}$, and therefore

$$r \text{ at the limit of the band } = r \text{ at the central frequency} \times \sqrt{q}$$
$$= r \text{ at the central frequency} \times \sqrt{0.43}$$
$$= r \text{ at the central frequency} \times 0.656$$

giving a reduction in communication distance of $(1 - 0.656) = 34.4\,\%$ at the limits of the band (860 and 960 MHz) with respect to the central frequency of 910 MHz – and that is without considering the other factors. In short, if everything else is equal, a tag that is intended for worldwide use, and therefore centred on the central frequency, 910 MHz, of the UHF band (860–960), will work at 5 m in the USA (at 915 MHz) but only at about $5 \times 0.656 = 3.3$ m in France (868 MHz) and Japan (954 MHz).

Finally, given that $q = 1 - |\Gamma|^2$, we can find the associated value of Γ:

$$|\Gamma|^2 = 1 - q$$

$$|\Gamma|^2 = 1 - 0.43 = 0.57$$

$$|\Gamma| = \pm 0.755!$$

In other words, this tag will be nominally very poorly matched at the limits of the UHF band in France and Japan if no action is taken.

Note

It is always possible to correct or compensate the intrinsic selectivity of the tag over the width of the band, using the input matching circuit (inductance/capacitance), as is done with IF (intermediate frequency) amplifiers in television, but obviously this incurs costs, which are often not acceptable in mass-produced RFID systems.

The impedance of the load Z_l (the input circuit of the integrated circuit) often varies with frequency, and a component manufacturer will usually specify the values of R_{ic} and X_{ic} as a function of the frequency. This is sometimes enough to correct some or all of the tuning of the tag over the required range of frequencies. By way of example, let us consider the values for the integrated circuit (bare chip) of the U_code EPC C1 G2 circuit:

$$Z_l \text{ at } 867 \text{ MHz} = 41 - j865$$
$$910 \rightarrow 34.5 - j820$$
$$928 \rightarrow 36.5 - j794$$
$$2450 \rightarrow 11 - j295$$

The values of the imaginary part of the impedance Z_l indicated above, directly related to the frequency, indicate that the input capacitance of the integrated circuit changes very little with the frequency. On the other hand, we can see the marked variation of the real part, which means that the value of Q cannot be held constant.

If the value of Γ is deliberately changed from -0.75 to -1 in modulated operation, we must check that the variation in RCS (radar cross-section; see Chapters 8 and 9) agrees with the specifications of conformity standard ISO 18047-6 for ISO 18000-6.

For calibration of the type indicated above, we have just seen that a given tag will have widely differing nominal operating distances in the USA (where the central value is 915 MHz, very close to 910) and Europe and France, where it operates near the limit of the band (868 MHz). For the same EIRP, the operating distance is proportional to the square root of q, in other words in this case root $0.43 =$ about 65 % of the maximum distance at 910 MHz, i.e. in the USA.

We must remember that customers are not likely to specify 'Europe' tags, 'US' tags, 'US/Europe' tags, 'Japan' tags or even 'Worldwide' tags.

7.5 Losses Due to the Physical Design of the Antenna, $\theta_{antenna}$

I have already described several times how to determine the 'theoretical' effective area $\sigma_{e\,t}$ of a tag. As usual, however, the practice is rather different from the theory. This is because we have disregarded all the parameters that come into play in the complete estimation of the real value $\sigma_{e\,t}$, particularly the mechanical surface areas, the types of materials, etc., to be taken into account by this parameter, in addition to the ohmic losses of the tag antenna. These losses arise from dielectric factors, the Joule effect, corona effects, etc.

This parameter, which we call $\theta_{antenna}$, denotes the efficiency of the tag antenna ($P_{radiated}P_{inc}$) by showing the extent to which the theoretical equation $P_{t\,theoretical} = \sigma_{e\,t\,theoretical}\,s$ is incorrect in reality, because the tag really only receives

$$P_{t\,real}\sigma_{e\,t\,reals} = \theta_{antenna}\sigma_{e\,t\,theoretical}s$$

The value of $\theta_{antenna}$ is commonly around 0.7.

7.6 By Way of Conclusion

The power P_1 actually transferred to the load $R_1 = R_{ic\,s}$ (the integrated circuit) from the tag antenna must therefore allow for the specific matching efficiencies of the various parameters, which may be detuned and which are shown inside the parentheses below. The original Friis equation is modified to become

$$P_1 = P_{bs\,eirp}G_{ant\,t}\frac{\lambda^2}{(4\pi r)^2}\left(\theta_{load\,matching}\theta_{polarization}\theta_{antenna}\right)$$

All of these parameters will clearly have a direct effect on the maximum operating distance, according to the reciprocal formula presented previously:

$$r_{max} = \sqrt{\frac{P_{bs\,EIRP}G_{ant\,t}\lambda^2}{(4\pi)^2 P_1}\theta_{load\,matching}\theta_{polarization}\theta_{antenna}}$$

If you are not yet persuaded about the results of this, please see the specific examples in the following sections.

7.7 Real-World Examples of RFID at UHF and 2.45 GHz

Here, by way of illustration, are four examples (three at UHF and one at SHF) based on the use of an integrated circuit of the NXP Semiconductors U_code HSL class, whose required minimum reading power P_{ic} for the component is 35 μW at UHF and 120 μW at SHF (at 2.45 GHz).

The examples below are all based on these assumptions:

- tag antenna:
  - $\lambda/2$ dipole;
  - $G_{ant\ t} = 1.64$;
  - mismatching: $\theta_{load\ matching} = 0.8$ for operation in a planned frequency band;
  - $\theta_{antenna} = 0.7$ to allow for the losses of the nonideal antenna;
  - $\theta_{polarization} = 1.0$, assuming that the base station and tag antennae are perfectly aligned and located in the equatorial plane.

Notes

If the base station and tag antennas were not perfectly aligned, we should have to allow for:

- $\theta_{polarization}$, by multiplying the values below by the root of $\theta_{polarization}$ (e.g. $\theta = 45°$, $\cos \theta = 0.707$);
- $\sqrt{\theta_{polarization}} = 0.84$ (→ i.e. 16 % less distance!) and, purely for the sake of correction, we will temporarily forget to mention or allow for the temporal operating limits imposed regularly by the maximum duty cycles, which may or may not be permitted by the local regulatory bodies.

7.7.1 At UHF

- Example I, according to US regulations (FCC 47 part 15)

$$f = 915\,\text{MHz}$$

$$\lambda = 33\,\text{cm}$$

$P_{EIRP\ max} = 4\,\text{W}$ (1W conducted + 6 dB of base station antenna gain)

$$R_{max} = \sqrt{\frac{4 \times 1.64 \times 0.33^2}{(4\pi)^2 \times 35 \times 10^{-6}} \times 0.8 \times 1 \times 0.7} = 8.5\,\text{m}$$

- Example II, at UHF according to European regulations 2004 (CEPT − ERC 70 03)

$$f = 869\,\text{MHz}$$

$$\lambda = 35\,\text{cm}$$

$$P_{\text{ERP max}} = 500\,\text{mW} \rightarrow P_{\text{EIRP max}} = 820\,\text{mW}$$

$$R_{\text{max}} = \sqrt{\frac{0.82 \times 1.64 \times 0.35^2}{(4\pi)^2 \times 35 \times 10^{-6}}} \times 0.8 \times 1 \times 0.7 = 4.09\,\text{m}$$

- Example III, at UHF according to European regulations 2007 (ETSI 302 208 – LBT)

$$f = 869\,\text{MHz}$$

$$\lambda = 35\,\text{cm}$$

$$P_{\text{ERP max}} = 2\,\text{W} \rightarrow P_{\text{EIRP max}} = 3.28\,\text{W}$$

$$R_{\text{max}} = \sqrt{\frac{3.28 \times 1.64 \times 0.35^2}{(4\pi)^2 \times 35 \times 10^{-6}}} \times 0.8 \times 1 \times 0.7 = 8.18\,\text{m}$$

7.7.2 At 2.45 GHz

- Example I, according to US regulations (FCC 47 part 15)

$$f = 2.45\,\text{GHz}$$

$$\lambda = 12\,\text{cm}$$

$$P_{\text{conducted bs max}} = 1\,\text{W}, \; G_{\text{ant bs max}} = 4 \rightarrow P_{\text{EIRP max}} = 4\,\text{W}$$

$$R_{\text{max}} = \sqrt{\frac{4 \times 1.64 \times 0.12^2}{(4\pi)^2 \times 120 \times 10^{-6}}} \times 0.8 \times 1 \times 0.7 = 1.65\,\text{m}$$

Figure 7.16 shows many other examples (be sure to read the many very important notes printed under the table!).

Important

Some less scrupulous manufacturers may occasionally use base station antennas with a high gain (of 15, for example) so that they can claim remarkable performance levels, e.g. five- or six-element Yagi antennas used for television (Figure 7.17). This obviously enables the base station conducted power to be markedly reduced for the same authorized P_{EIRP}, while obscuring the fact that the radiation pattern becomes highly directive.

7.7.3 Initial Conclusions

Simply as a result of the introduction of the values of the three parameters developed in the preceding sections, namely q, p and 'antenna' (in the example, these have 'favourable' values of 0.8, 1 and 0.7 respectively, giving a product of $0.8 \times 1 \times 0.7 = 0.56$), Figure 7.16 clearly indicates that – even when the axes are optimally aligned in the equatorial plane – there is at

Range Calculation for UHF & SHF Systems (théorique, hors tolérances)

		INPUT		base station					tag							OUTPUT
remotely powered tag		Frequency		power - Pbs ERP		antenna gain Gbs		P EIRP = Pbs.Gbs	antenna gain tag		Min. Power for the chip		Antenna efficiency		Matching efficiency		Max. Operat range
		f MHz	lambda m	P W	Prdr W	Gbs	equals dB	P EIRP	Gtbl	equals dB	Pchip µW	equals dBm	AntEff %	equals dB	Match %	equals dB	range m
UHF europe in/outdoor a		900.0	0.333	0.5		1	0.00	0.50	1	0.00	60	-12.22	100	0.00	100	0.00	2.42
in/outdoor	b	900.0	0.333	0.5		1.64	2.15	0.82	1.64	2.15	60	-12.22	100	0.00	100	0.00	3.97
	c	868.5	0.345	0.5		1	0.00	0.50	1	0.00	60	-12.22	100	0.00	100	0.00	2.51
	d	868.5	0.345	0.5		1.64	2.15	0.82	1.64	2.15	60	-12.22	100	0.00	100	0.00	4.12
europe indoor only	e	872.0	0.344	2		1	0.00	2.00	1	0.00	60	-12.22	100	0.00	100	0.00	5.00
(future) indoor only	f	873.0	0.344	2		1.64	2.15	3.28	1.64	2.15	60	-12.22	100	0.00	100	0.00	8.19
america in/outdoor	g	915.0	0.328		4	1	0.00	4.00	1.64	2.15	60	-12.22	100	0.00	100	0.00	8.63
	h	915.0	0.328		4	4	6.02	4.00	1.64	2.15	60	-12.22	100	0.00	100	0.00	8.63
xxx	i	868.5	0.345		1	4	6.02	4.00	1.64	2.15	60	-12.22	100	0.00	100	0.00	9.09
xxx	j	868.5	0.345	2.44	2.44	1.64	2.15	4.00	1.64	2.15	60	-12.22	100	0.00	100	0.00	9.09
SHF europe in/outdoor		2450.0	0.122					0.5	1	0.00	150	-8.24	100	0.00	100	0.00	0.56
in/outdoor		2450.0	0.122					0.5	1.64	2.15	150	-8.24	100	0.00	100	0.00	0.72
europe indoor (15%)		2450.0	0.122					4	1.64	2.15	150	-8.24	100	0.00	100	0.00	2.04
america in/outdoor		2450.0	0.122					4	1	0.00	150	-8.24	100	0.00	100	0.00	1.59
		2450.0	0.122					4	1.64	2.15	150	-8.24	100	0.00	100	0.00	2.04
your own figures		860.0	0.349	0.5		1.64		0.82	1.64	2.15	60	-12.22	70	-1.55	80	-0.97	3.11

a   for information only, since an isotropic base station antenna with G = 1 does not exist and is meaningless where the definition of ERP is concerned
b   In Europe, P max possible = 0.5 W ERP. This power, described according to a lambda/2 base antenna, must therefore be associated with a Gant bs of 1.64
c   for information
d   the real value permitted in France
e
f   pending in Europe - for France, the ART [Telecommunications Regulation Authority] "has expressed reservations"
g   the FCC stipulates PmaxEIRP = + 36 dB, i.e. 4 W EIRP, meaning 4 W with an isotropic antenna having G = 1 ... which does not exist ... so purely theoretical
h   the FCC stipulates Pmax = + 36 dBm with Pmax = 1 W (30 dB) with an anisotropic antenna having G = 4 (6 dB) ... which does exist but is therefore highly directive
i   just for fun - the same calculation with the frequencies authorized in Europe
j   again for Pmax EIRP = + 36 dBm = 4 W but divided between P ERP = 2.44 W and a dipole antenna with G = 1.64, therefore less directive ... this is
    impossible because we are not allowed to exceed 1 W max conducted!

Figure 7.16   Examples of maximum operating distances as a function of the operating frequency and local regulations

least a (modest) discrepancy of -25% of the operating value between an ideal system and a real system (root of $0.56 = 0.75$, giving the value of -25%). To be very specific, let us look at the example of the system operating in Europe with $P_{EIRP} = 500\,mW$, $G_{ant\,bs} = 1.64$. In nominal conditions, we then move from 5.39 m to $(5.39 \times 0.75) = 4.04$ m, along the principal axis of the antenna only ($\theta_{polarization} = 1$), without any allowance for the tolerances of the various elements of the system.

7.8 Effects of the Mounting of the Integrated Circuit on the Tag Substrate

Perhaps you thought that the problems had all been dealt with at last? No!

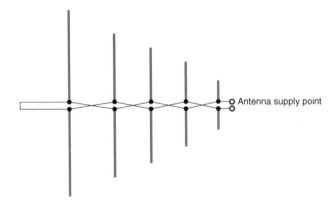

Antenna supply point

Figure 7.17   Example of a Yagi antenna for a base station

When the chip (integrated circuit) has been bought and delivered as a 'bare chip' or miniature casing, we must eventually connect it to the antenna that has been so carefully designed. It is at this point that new problems arise. I shall tell you all about these new adventures in Chapter 18!

7.9 By Way of Conclusion

Having outlined some solutions for overcoming several exemplary problems, I recommend that you always draw up the most comprehensive map that you can of the site where a UHF and SHF RFID system is to be installed, in order to avoid any surprises, whether good or bad. Allowance should also be made for the tolerances and dispersion (in time, moisture, etc.) of all the components involved (Figure 7.18):

- the real value $P_{bs\ EIRP}$ of the base station resulting from the possible mismatch between the amplifier output impedance and the base station antenna;
- the base station antenna gain $G_{ant\ t}$;
- the tag antenna gain $G_{ant\ t}$;
- the tag parameters;
- the guaranteed maximum consumption of the tag with respect to the nominal value;
- the tolerances of the complex input impedance values Z_{ic} of the integrated circuit;
- the mounting of the integrated circuit on the tag (precision, parallelism, inclination, etc.; see Chapter 18);
- the real orientation of the tag in the application;
- the absorption of the materials present in the environment;
- the reflection of the waves in the real application environment (gaps in reception, etc.) (the total order of magnitude of the effect of all these elements being 60% of the usable distance).

In such a system, at a given date, for a specific tag chip technology (power consumption), in Europe:

- Under the 2 W ERP regulations (CEPT/ETSI 302 208, LBT modulation), it is reasonable to specify an operating distance of about 4 m for real daily use, which is obviously less enticing for advertising purposes than the '10.72 m' (but approaching the US value of 11.30) indicated at the outset! At least this will have the merit of providing correct operation at all times, not just occasionally.
- Under the 500 mW ERP regulations (with a duty cycle, dc = 10 %), it is reasonable to specify an operating distance of about 2 m for real daily use, which is obviously less enticing for advertising purposes than the '5.39 m' (or 11.30 in US terms) indicated at the outset! At least this will have the merit of providing correct operation at all times, not just occasionally.

7.10 Example at UHF and SHF

To complete our discussion of this subject in relation to a real-life situation, Table 7.7 provides an example of commercial specifications for a particular type of remotely-powered transponder – in this case the NXP U_code_HFS integrated circuit that we have considered in all the examples in this chapter – showing the performance in terms of

Range calculations for UHF & SH-F systems (theorical, out of & including tolerances)

		INPUT		Base station						Tag									OUTPUT
Remote powering condition according to the Friis equation		Frequency		duty cycle	Pbs cond max	Antenna gain Gbs			P ERP max	P EIRP = Pbs Gbs max	Antenna gain (dipole λ/2)		Min. Power for the chip		Antenna polarization		Antenna efficiency loss		Antenna matching		Operating range max
	note	f MHz	λ m		Pbs W	Gbs	Gbs dBd	Gbs dBi	P ERP W	P EIRP W	G|bl dB	equals dB	Pchip µW	equals dBm	polar	equals dB	AntEff %	equals dB	Match %	equals dB	range m
UHF Europe in/outdoor	a	869.4	0.345	10%	0.5	1.64	0	2.15	0.50	0.82	1.64	2.15	35	-14.56	100	0.00	100	0.00	100	0.00	5.38
indoor only	b	867.6	0.346	LBT	2	1.64	0	2.15	2.00	3.28	1.64	2.15	35	-14.56	100	0.00	100	0.00	100	0.00	10.79
America in/outdoor	c	915.0	0.328	100% FHSS	1	4		6.02	2.44	4.00	1.64	2.15	35	-14.56	100	0.00	100	0.00	100	0.00	11.30
SHF Europe in/outdoor		2450.0	0.122	15%						0.5	1.64	2.15	120	-9.21	100	0.00	100	0.00	100	0.00	0.81
indoor only		2450.0	0.122							4	1.64	2.15	120	-9.21	100	0.00	100	0.00	100	0.00	2.28
America in/outdoor		2450.0	0.122							4	1	0.00	120	-9.21	100	0.00	100	0.00	100	0.00	1.78
		2450.0	0.122							4	1.64	2.15	120	-9.21	100	0.00	100	0.00	100	0.00	2.28
Example UHF applications with your own figures		869.4	0.345		0.5	1.64				0.82	1.5	1.76	60	-12.22	50	-3.01	70	-1.55	80	-0.97	2.08
		867.6	0.346		2	1.64				3.28	1.5	1.76	60	-12.22	50	-3.01	70	-1.55	80	-0.97	4.17
		915.0	0.328		1	4				4.00	1.5	1.76	60	-12.22	50	-3.01	70	-1.55	80	-0.97	4.37

a   in Europe, in this band, P ERP max possible = 0.5 W - power described for a l/2 antenna base, and must therefore be associated with Gbs of 1.64

b   in Europe, in this band, P ERP max possible = 2 W - power described for a l/2 antenna base, and must therefore be associated with Gbs of 1.64, and must be associated with the mandatory ESTI 302 208 standard for LBT (Listen Before Talk)

c   the FCC stipulates Pmax EIRP = + 36 dBm = 4 W with P_cond_max = 1 W (3c dB). i.e. with an anisotropic antenna having G = 4 (6 dB) ... which is therefore directive

Figure 7.18   Example of maximum operating distances with a U_{code} HSL circuit at UHF and SHF, allowing for tolerances and dispersion

Table 7.7

Frequency band (MHz)	Region	Note	Power	Duty cycle	Maximum reading distance	
					Maximum theoretical with simple antenna (8)	Maximum theoretical in difficult environments (8)
860–960 band						
869.4–869.65	Europe	(1)	0.5 W ERP	10%	4.0 m	2.0 m
865.5–867.6	Europe	(2)	2 W ERP	LBT	8.0 m	4.0 m
902–928	USA	(3)	4 W EIRP	100%	8.4 m	4.0 m
860–960	Others	(4)			0–3.5 m	0–4.0 m
2450 band						
2400–2483.5	Europe	(5)	0.5 W EIRP indoor and outdoor		0.6 m	0.5 m
2400–2483.5	Europe	(5)	4 W EIRP indoor only	15%	1.8 m	1.5 m
2400–2483.5	USA	(6)	4 W EIRP		1.8 m	1.5 m
2400–2483.5	Others	(7)				0 to 2 m

Notes:
(1) CEPT/ETSI regulations for NS SRDs: CEPT ERC 70 03, Annex1 – ETSI 300220-1.
(2) Regulations for RFID SRDs for Europe according to CEPT/ETSI 302 208.
(3) FCC 47 regulations, Part 15, Section 247.
(4) In many countries, regulations similar to FCC or CEPT/ETSI are applicable. In Japan, for example, the authorized band is currently 952 to 954 MHz.
(5) Current CEPT/ETSI regulations: ERC 70 03, Annex 11 – ETSI 300440-1.
(6) FCC 47 regulations, Part 15, Section 247.
(7) In many countries, regulations similar to FCC or CEPT/ETSI are applicable.
(8) The distances shown in this table are typical values for conventional tags and labels. Special antenna designs (high gain, directive antennas, etc.) may enable higher values to be achieved.

operating distance that we can expect from a given component, according to the part of the world where it is used and the current local regulations on power levels (ERP and/or EIRP).

Important

Evidently, the distances shown in the table allow for the fact that the intrinsic consumption of the integrated circuit is not the same at 900 MHz and 2.45 GHz. (For anyone who might doubt this, the power consumption of a CMOS integrated circuit increases with the frequency.) The figures in the last column of the table are typical operating values actually (physically) measured in 'difficult' environments, and not speculative levels from commercial texts or simulations 'on paper', as is too often the case.

Generally, the power to be supplied to the tag is greater for the stage of writing/etching the E2PROM memory of the tag, because of the occasional operation of additional special electronic circuits (charge pump, etc.), than for the reading stage. Here is one example: at a given date (or for a given technology, etc.), if we assume that the power required for writing, P_{write}, is two times the reading power P_{read}, then, everything else being equal, and given that the operating distance of a tag is inversely proportional to the square root of the power which it requires, the writing distance would be $1/(\sqrt{2})$ times $= 70\%$ shorter than the reading distance.

7.11 Appendix: Fact and Fantasy About UHF Tags and Water

How many times have I been subjected to questions, or ridiculous assertions, about tags operating at UHF in environments containing water or other liquids! Examples include: 'They say that water creates a lot of attenuation'; 'It won't work'; 'I've read that such-and-such a company had a competing tag that even works under water, but it can't be true' – I will omit the less sensible claims! A complete mishmash, in other words, and not the best way of promoting understanding of this market.

In view of this, the following section, while not attempting to be comprehensive or making any particular claims, will attempt to clarify your ideas about the many possible ways of determining and stating exactly what you want and when you want it, marketing permitting, regardless of the thorny details of this tricky subject. So brace yourself for more technicalities!

Note
To obtain the maximum benefit from the following explanations, you should carefully consult Figures 7.19 and 7.20 while reading the text.

7.11.1 The 'It Doesn't Work in Water!' Version

If, for example, we use a UHF RFID system whose base station transmits at 900 MHz and a (remotely powered) tag tuned by its designer to operate in air, also at 900 MHz, then, as I have shown in the preceding chapters, when the base station transmits a power of 4 W EIRP the tag will operate in air at, say, a distance of about 8 m in 'free space' (Figure 7.19, curve 1).

Now let us place this tag in water, or immediately behind a nonnegligible thickness of liquid. The electrical properties of this substance will detune the tuning frequency and also damp its resonance, thus reducing its quality factor. Following the change of environment from air to water, curve 2 of Figure 7.19 shows an example of a new position of the RF tuning of the tag in which the tag is detuned to 750 MHz, instead of the initial 900 MHz, and the peak of the new response curve is 40 dB below its initial level. And why not? It is simply a matter of physics.

Clearly, whether the propagation medium is air or water, the base station continues to transmit at 900 MHz. Therefore, as shown by curve 3 of the same figure, at this frequency, which is far from the new tuning of the tag, we again lose many dB, so that between the initial situation and the present case, there is a total discrepancy of -70 to -80 dB, making the operating distance virtually zero.

In physical, experimental and quantitative terms, the conclusion is plain: no communication – it doesn't work! The immediate, positive and definitive conclusion of some marketing departments is that a UHF tag does not work in the presence of water.

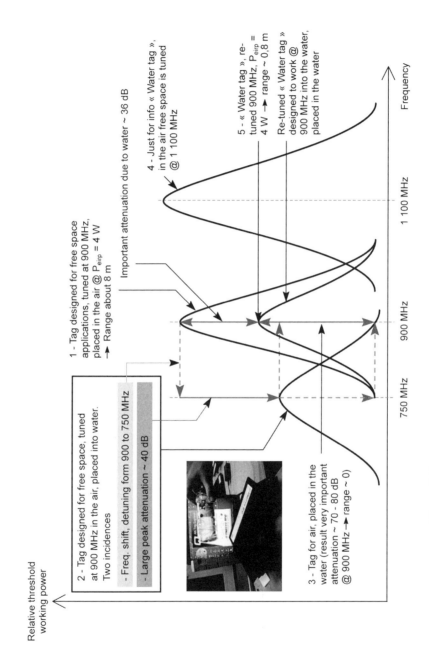

Figure 7.19 Effect and incidence of water on the tuning frequency of a tag

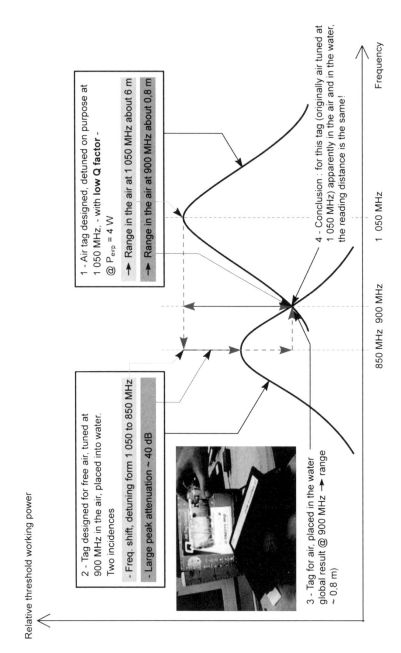

Figure 7.20 Minimizing and compensating the effects of water on the tuning of a tag

7.11.2 The 'But if it Works, Not So Well, but At Least it Works...' Version

Admittedly, if the user is clever or well-informed, then if he knows that water may be present in his application – which may cause frequent detuning of the tags – he can ask his tag supplier to deliver tags that are tuned to a higher frequency for operation in air, such as 1100 MHz (curve 4), so that the peak of the curve after detuning by water exactly matches the 900 MHz one (curve 5), and everything will be fine.

In this case, with the tag in the water, the peak of the new curve is locked to the 900 MHz carrier and the loss is only about 35 to 40 dB. This is still considerable, but at least it allows operation over distances of about 80 cm (in the same conditions as before, with the base station transmitting 4 W EIRP). Although the distance is obviously reduced from 8 m to 80 cm, this is often enough for many applications.

The positive and definitive commercial conclusion of some other marketing departments is that a UHF tag works less well in the presence of water, but at least it works!

7.11.3 The 'Oh No, it Works Just As Well!' Version

Of course, there will always be some awkward customers who expect a tag to have the same performance and functionality in air as in water, so that the effects of the medium are, or appear to be, totally transparent for the application and the user. However, we can cope with this as well (see Figure 7.20).

Given that we wish to achieve the same performance in air and water, with the base station still operating at 900 MHz, we must use a tag that has initially been tuned to 1050 MHz, for example, for use in the air (curve 1). If the base station transmitted at 1050 MHz, the operating distance of this tag would be about 6 to 7 m – but this cannot be the case. At 900 MHz, as shown by the response curve, we lose a large number of dB, reducing the operating distance so that the tag can only operate, for example, at 1 m in air.

Now pick up the tag and dip it in water. We have the same causes and the same effects as before: in other words, the tag is detuned (e.g. from 1050 to 850 MHz, with a loss of 40 dB at the peak of the curve: see curve 2). As before, the base station continues to transmit at 900 MHz, so now, in the presence of water, we have an operating distance of 1 m. The result is that the operating distances are identical in air and water.

This is the final, positive conclusion of certain other marketing departments: water has no effect on UHF tags!

I hope that these few lines will have helped you to understand certain differences in what we might call 'sensitivity', concealed somewhere between the physical phenomena and the claims of marketing departments.

8

Reflection and/or Reradiation of Waves and RFID Applications

In the preceding chapters, we have limited ourselves to a study of the effective area $\sigma_{e\ t}$, using this to find the power absorbed by the load from the incident wave, and we have demonstrated all the effects on applications while continuing to look through the 'wrong end of the telescope'. Unfortunately for you, this is not the end of the process. Don't despair, things can only get better!

As I have shown in the previous section, a propagated wave often meets obstacles on its path, giving rise to more complex problems. This is why I am now offering you this description of the specific problems due to the 'reflection' or 'reradiation' of the signal striking the tag, and their implications. We will therefore take a close look at what affects the reradiated power, namely the scattering aperture area, or radar area, or radar cross-section (RCS). We will therefore start by considering these matters:

- the value of the power reradiated by the tag, P_s;
- the general definition of the radar cross-section of the tag, $\sigma_{e\ s} = \text{RCS}$;[1]
- the value of this parameter during optimal power transfer $\sigma_{e\ match} = \sigma_{e\ structural}$ (the case of impedance matching);
- the value of $\sigma_{e\ s}$ in the case of impedance mismatching, $\sigma_{e\ mismatch}$.

In the next chapter we will go on to consider the principle and methods of back scattering and will conclude with the 'merit factor' that characterizes a tag, $\Delta\sigma_{e\ s} = \sigma_{e\ modulated} - \sigma_{e\ nonmodulated}$, which is one of the main parameters defining the ability of an RFID tag to communicate as a function of the distance.

8.1 The Physical Phenomenon of Wave Scattering

Let us start by defining our subject. When an electromagnetic wave encounters irregularities in the communication medium (a wall, the ground, an antenna wire, etc.), its propagation may be

---

[1] In the following chapters I will retain the notation $\sigma_{e\ s}$ for the radar cross-section of the tag, rather than RCS, in order to keep the notation uniform with that used for other areas.

altered in a random way. The general term for this phenomenon is 'scattering'. To be precise, in the field of radio wave propagation, the well-known reference authority IEEE defines scattering as 'a process by which a wave is scattered in different directions due to interaction with non-uniformities in the medium'.

In concrete physical terms, when an electromagnetic wave encounters an object, it sets up oscillating charges and currents in the surface of the object, thus creating local magnetic fields. The estimation of these fields is complicated, requiring mathematical tools that largely lie outside the scope of this book (numerical and analytical evaluation, approximation of tangential fields, etc.). In general, the spatial distribution of the scattered energy depends on the size, shape and composition of the object, on the wave and its incidence on the object, just as it does in optics when the same scattering phenomenon occurs.

In macroscopic terms, for the sake of simplicity, we can say that there are three main cases to be considered, according to the ratio between the size of the object and the wavelength of the incident wave.

8.1.1 The Size of the Object Encountered Is Very Small with Respect to λ

If the object is very small with respect to the transmitted wavelength, then we are in the presence of what is known as operation in the Rayleigh range. In this case, the effective radar cross-section of the object is very small, and the wave passes through the object (or seems to do so), because there are very few changes in the phase of the signal striking the surface of the object. Since the object essentially sees a quasi-static field, a magnetic moment of the dipole is created, causing the appearance of a scattered field.

8.1.2 The Size of the Object Encountered Is of the Same Order as λ

If the size of the object is of the order of magnitude of the incident wavelength (i.e. about one to ten times its value), there will be what is known as resonance range operation. In this case, the electromagnetic energy tends to remain attached to the surface of the object, thus creating surface waves, including propagated waves, creeping waves and transverse waves.

8.1.3 The Size of the Object Encountered Is Very Large with Respect to λ

If the object is very large with respect to the transmitted wavelength, then we are in the presence of what is known as operation in the optical range. In this case, as in optics, part of the wave is absorbed and part of it is reflected and diffracted in all directions as a function of the angle of incidence and the composition of the obstacle. The same concepts of specular, diffuse, multiple and refractive reflection, etc., are applicable here.

Note on UHF and SHF RFID applications

In RFID at UHF and SHF, the dimensions of the tag antennas are of the same order of magnitude as the wavelengths used, and so the main applications of these tags are in resonance range operating mode, where surface waves are preponderant. Different cases may arise in this area. The object may be a resonator (such as an antenna), but it may equally well absorb the wave completely. I shall show in this chapter how this depends, in particular, on the amount of load applied to the antenna.

8.2 Scattering Modes

Examining the problem at the microscopic level, we normally define three scattering modes:

- The *monostatic* or *back scattering mode*, in which the directions of the relevant incident and scattered waves coincide, but are opposite to each other. An object having monostatic scattering properties can be represented in a simple way by considering that it has an 'equivalent' scattering area, called the radar cross-section (RCS).
- The *forward (direct) scattering mode*, in which the two directions (incident and scattered) are identical.
- The *bistatic mode*, in which the directions of the incident and scattered waves are totally different.

Note that, in all of these modes, the resulting scatter contributes, or may contribute, to interference with other tags located in front of, behind or in the immediate environment of the tag concerned.

Note on RFID applications

For mechanical and financial reasons, in 99% of RFID applications the transmitter of the incident wave and the receiver of the scattered wave are physically located in the same position in a base station. Using these specific properties, RFID systems of the back scatter type provide communication by controlling (modulating) the response of the tag by changing the scattering properties of the tag. The scattering of the signal can also be used to find the geographical location of an object in space (as in radar).

To summarize, most RFID applications at UHF and SHF operate in the resonance range and in the monostatic (i.e. 'back scattering') mode.

A final point: the term 'scattering' is very broad-ranging. I will often use the terms 'reradiated', 'reflected', etc., which are roughly similar in a first approximation. This is because the incident wave induces electrical charges, and therefore magnetic fields, in the object. This leads to the flow of electrical currents in the object, producing power that the object can (re-)radiate, thus appearing to reflect part of the incident wave.

So much for the vocabulary. Now for the hard work!

8.3 Power Scattered/Reradiated/Reflected by the Tag, P_s

I will now show you in a few lines how the tag reradiates energy when it is illuminated by the incident wave from the base station. To quantify this radiation, we must first estimate the current and the conducted power supplied to the tag antenna.

8.3.1 Current Flowing in the Tag Antenna

In the previous chapter, I showed that the illumination of the tag by the incident electromagnetic wave from the base station enabled the greatest possible amount of power to be transferred via the effective area $\sigma_{e\,t}$ of the tag antenna to a matched (conjugated) load connected to the terminals of the antenna. I also provided a simplified equivalent model circuit of the system, shown in Figure 8.1.

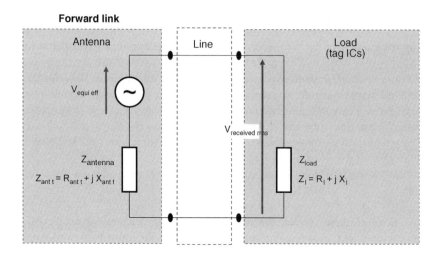

Figure 8.1 Equivalent circuit of the tag

The general equation for the radio frequency current flowing in the whole of this equivalent circuit (in the case of short lossless lines) is

$$I = \frac{V_{\text{equi}}}{(R_1 + jX_1) + (R_{\text{ant t}} + R_{\text{loss}} + jX_{\text{ant t}})}$$

where its modulus is

$$|I_{\text{rms}}| = \frac{V_{\text{equi rms}}}{\sqrt{(R_{\text{ant t}} + R_{\text{loss}} + R_1)^2 + (X_{\text{ant t}} + X_1)^2}}$$

and the phase φ is such that its tangent is as follows:

$$\tan \phi = -\frac{X_{\text{ant t}} + X_1}{R_{\text{ant t}} + R_{\text{loss}} + R_1}$$

For a given tag antenna – $R_{\text{ant t}}$, R_{loss} and $X_{\text{ant t}}$ – this value is a function of R_1 and X_1:

$$|I_{\text{rms}}| = f(R_1, X_1)$$

Let us now consider the power that appears in the impedance of the tag antenna, $Z_{\text{ant t}}$.

8.3.2 Conducted Power in the Antenna Impedance, Z_{ant}

Clearly, this current I_{rms} flows through the whole of the equivalent circuit, thus also creating, in the impedance of the tag's receiving antenna, $P_{Z\text{ant t}}$, a conducted power in watts in the whole of the real part of the antenna impedance $Z_{\text{ant t}}$, in other words in the sum of the radiation resistance $R_{\text{ant t}}$ and the loss resistance R_{loss}. Thus

$$P_{\text{Zant t}} = P_{\text{loss}} + P_{\text{Rant t}}$$

P_{loss}

The power in watts P_{loss} associated with R_{loss} ($P_{\text{loss}} = R_{\text{loss}}\, I^2_{\text{rms}}$) is quite simply dissipated as heat. In most RFID applications at UHF and SHF, these losses are small and often considered to be negligible or zero. We will return to the question of this power later in the book.

$P_{Rant\ t}$

As for the power in watts $P_{\text{Rant t}}$ dissipated in the (immaterial) resistance $R_{\text{ant t}}$, this does indeed exist. Its value is evidently

$$P_{\text{Rant t}} = R_{\text{ant t}}\, I^2_{\text{rms}}$$

If we square the value of I_{rms} and introduce this value into the equation for $P_{\text{Zant t}}$, regardless of the complex value of the load impedance Z_1, the general equation for this conducted power becomes

$$P_{\text{Rant t}} = P_{\text{ant rms}} = \frac{R_{\text{ant t}}}{\left(R_{\text{ant t}} + R_1\right)^2 + \left(X_{\text{ant t}} + X_1\right)^2}\, V^2_{\text{equi rms}} \text{ in W rms}$$

This conducted power in watts is dissipated in the tag antenna in the form of electromagnetic radiation. In other words, this power is reradiated or, if you like, dispersed (scattered). Thus the tag antenna also acts as a transmitter during the reception phase (forward link). To find the level of this reradiated power, we must start from the 'reciprocity principle', which states that an antenna receiving an electromagnetic wave is simultaneously a transmitting antenna. We must now consider the tag, not as a receiver but as a kind of miniature base station (i.e. a transmitter), in which a current I (the same as in the preceding section) flows through the antenna load with the resistance $R_{\text{ant t}}$.

8.3.3 The Power P_s (Re-)radiated by the Tag Antenna

Given that the gain of the antenna $G_{\text{ant t}}$ is still the same (it has not changed over the time interval), we can determine the effective equivalent reradiated isotropic power $P_{s\ \text{EIRP}}$ (in W) that will be dispersed or reradiated by the tag.

General Equation for the Power (EIRP) (Re-)radiated by the Tag

The general equation for this is therefore

$$P_{s\ \text{EIRP}} = P_{\text{Rant t}} G_{\text{ant t}}$$

$$P_{s\ \text{EIRP}} = R_{\text{ant t}}\, I^2_{\text{rms}} G_{\text{ant t}}$$

$$P_{s\ \text{EIRP}} = \frac{R_{\text{ant t}}}{\left(R_{\text{ant t}} + R_1\right)^2 + \left(X_{\text{ant t}} + X_1\right)^2}\, V^2_{\text{equi rms}} G_{\text{ant t}} \text{ W r.m.s.}$$

In the absence of loss in R_{loss}, this radiated or dispersed power $P_{s\ \text{EIRP}}$ (also known as the scattered power) is similar to that dissipated in the internal resistance of a generator supplying a load of equal value to its internal resistance.

Regardless of the origin or cause of the illumination of the tag by an electromagnetic wave, the general equation for $P_{s\,\text{EIRP}}$ shows that, independently of whether the impedances of the tag antenna and of the load are strictly matched ($R_{\text{ant t}}$ may be equal to or different from R_l), in order to absorb all the power retrieved by the antenna as a result of its effective area $\sigma_{e\,t}$, the tag will 'reflect', 'disperse' or 'reradiate' a power $P_{s\,\text{EIRP}}$ (in W) which will be proportional to the incident power flux density s arriving from the base station.

Because of this power $P_{s\,\text{EIRP}}$ reradiated by the tag, we can give the tag a value associated with the effective reradiation area, or scattering area, or scattering cross-section $\sigma_{e\,s}$, which I will define and calculate subsequently. First of all, here is a brief physical explanation of the phenomenon.

8.3.4 $P_{structural}$ and $P_{antenna\ mode}$

Not knowing the actual load impedance R_l connected to the tag antenna (it may vary from zero to infinite), we will carry out the mental operation of dividing the whole reradiated power P_s in two. For this purpose, we define one part, representing what the share of reradiated power would be if the impedances of the tag antenna and the tag load were matched, and call this the structural power ($P_{\text{structural}}$), and another, corrective part (to operate positively or negatively, as yet we do not know which!), in order to correct the first term according to the actual measurement of the real load; this second part is called the antenna mode power ($P_{\text{antenna mode}}$). In algebraic terms, therefore,

$$P_s = P_{\text{structural}} + P_{\text{antenna mode}}$$

$P_{structural}$

The power $P_{\text{structural}}$ is related to the operating mode of the tag antenna called the 'structural mode'. This means the operation of the whole tag when its antenna is connected to a load whose value is the conjugated value of its own impedance $Z_{\text{ant t}}$. In this case, $Z_l = Z_{\text{ant t}}^*$. Now, $Z_{\text{ant t}}$, and therefore Z_{ant}^*, is made up of $R_{\text{ant t}}$ and $X_{\text{ant t}}$, whose electrical values are simply functions of the mechanical and physical properties (material, shapes, dimensions) of the tag antenna; in short, they are functions of its geometrical and mechanical properties, etc. – hence the name.

For a given antenna, this term will be fixed and constant, and will represent the 'static' reference part of the global equation for P_s. Therefore its permanent contribution will never be taken into account when we are only considering dynamic variations of power reradiated during a possible phase of dynamic modulation of the load of the tag antenna.

As regards UHF and SHF RFID applications of the remotely powered type, I have made it clear already that the condition $Z_l = Z_{\text{ant}}^*$ is (practically always) fulfilled in the forward link, from the base station to the tag, if only as a way of attempting to recover the maximum power for the tag, and therefore that this term 'structural mode' is always present and corresponds to the 'matched' and/or 'nonmodulation' phase of the tag's load impedance. Looking back to the previous chapter, we therefore find that its value is

$$P_{\text{structural}} = \frac{1}{4R_{\text{ant t}}} V_{\text{equi rms}}^2 G_{\text{ant t}} \text{ and therefore } = P_{s\,\text{matched}}$$

$P_{antenna\ mode}$

The second term of the equation, called $P_{\text{antenna mode}}$, requires a little more explanation. It represents the difference (the mysterious corrective term of the preceding paragraphs) between the total power reradiated by the tag P_s and the structural power $P_{\text{structural}}$. This term, therefore, quantifies the variation (delta) of power reradiated between the real value, due to the real antenna load (a function of Z_l) (hence the name of 'antenna mode'), and the power reradiated when the impedances of the tag taken as the reference element are matched. Also, if the value of the structural power in the matched condition is known (see the preceding section), we can easily calculate the value of $P_{\text{antenna mode}}$ from the difference with respect to P_s.

In principle, there is no way of knowing the sign of $P_{\text{antenna mode}}$ (positive or negative) on the basis of the load Z_l applied to the antenna terminals, if nothing is specified in advance. It will have to remain a mystery for the time being!

Calculating $P_{antenna\ mode}$

Let us return to the general equation for P_s (for short; actually, we are talking about $P_{s\ \text{EIRP}}$):

$$P_s = \frac{R_{\text{ant t}}}{\left(R_{\text{ant t}} + R_1\right)^2 + \left(X_{\text{ant t}} + X_1\right)^2} V_{\text{equi rms}}^2 G_{\text{ant t}}$$

With tags tuned and matched (conjugate matching)

If the tag is tuned, $X_1 = -X_{\text{ant t}}$, and the equation for the reradiated power becomes

$$P_s = \frac{R_{\text{ant t}}}{\left(R_{\text{ant t}} + R_1\right)^2} V_{\text{equi rms}}^2 G_{\text{ant t}}$$

Furthermore, when $R_1 = R_{\text{ant t}}$ (conjugate matching) (the case where the reradiated power is called $P_{\text{structural}}$):

$$P_{s\ \text{matched}} = P_{\text{structural}} = \frac{1}{4R_{\text{ant t}}} V_{\text{equi rms}}^2 G_{\text{ant t}}$$

Now let us determine the power $P_{\text{antenna mode}}$ due to the possible mismatching of the load with respect to its optimal value ($R_1 = R_{\text{ant t}}$):

$$P_{\text{antenna mode}} = P_s - P_{\text{structural}}$$

$$P_{\text{antenna mode}} = \left[\frac{R_{\text{ant t}}}{\left(R_{\text{ant t}} + R_1\right)^2} - \frac{1}{4R_{\text{ant t}}}\right] V_{\text{equi rms}}^2 G_{\text{ant t}}$$

Let us develop the contents of the brackets, emphasizing the term $[V_{\text{equi rms}}^2 / (4R_{\text{ant t}})] G_{\text{ant t}}$ representing the power $P_{\text{structural}}$ reradiated by the antenna in conjugate matching conditions. Thus we obtain

$$P_{\text{antenna mode}} = \frac{4R_{\text{ant t}}^2 - \left(R_{\text{ant t}}^2 + R_1^2 + 2R_{\text{ant t}} R_1\right)}{\left(R_{\text{ant t}} + R_1\right)^2} P_{\text{structural}}$$

Table 8.1

R_1	a	Γ	$(1-\Gamma)^2$	$P_{\text{antenna mode}}$
0	0	-1	4	$3\,P_{\text{structural}}$
$R_{\text{ant t}}$	1	0	1	0
∞	∞	$+1$	0	$-1\,P_{\text{structural}}$

or alternatively, using the reduced variable $a = R_1/R_{\text{ant t}}$, the above equation becomes

$$P_{\text{antenna mode}} = \left[\frac{4}{(1+4a)^2}\right] P_{\text{structural}}$$

In previous sections I have specified that

$$\Gamma = \frac{a-1}{a+1}$$

from which equation we can easily derive the value of a:

$$a = \frac{1+\Gamma}{1-\Gamma}$$

Introducing the value of a into the equation for $P_{\text{antenna mode}}$, we obtain

$$P_{\text{antenna mode}} = [(1-\Gamma)^2 - 1]P_{\text{structural}}$$

$$P_{\text{antenna mode}} = \Gamma(\Gamma-2)P_{\text{structural}}$$

Table 8.1 shows some examples.

Important

Let me make it very clear that you should avoid any mistaken readings, and NEVER CONFUSE $(1-\Gamma)^2$ with $(1-\Gamma^2)$, an error that sometimes creeps into certain technical books or documents!

Note

Returning to one of the possible forms of the equation, $P_{\text{antenna mode}} = (\Gamma^2 - 2\Gamma)P_{\text{structural}}$, and calculating the differential of this function, we find

$$\frac{dP_{\text{antenna mode}}}{dP_{\text{structural}}} = 2(\Gamma-1)$$

This indicates the rate of change of P_{antenna} with respect to $P_{\text{structural}}$ according to the local value of Γ (Table 8.2).

Table 8.2

Load	Γ	$dP_{\text{antenna mode}}/dP_{\text{structural}}$
Antenna short-circuited	-1	-4
Matched load	0	-2
Antenna open-circuited	1	0

Note

When the value of R_1 is close to $R_{\text{ant t}}$ (*but only in this case*), in other words when R_1 varies very slightly around the value of $R_{\text{ant t}}$, and therefore a is close to 1 and Γ is close to zero, making the value of Γ^2 negligible compared with 2Γ, we obtain

$$P_{\text{antenna mode}} = -2\Gamma P_{\text{structural}}$$

In all other cases, the general equation is needed. A word to the wise. . . .

Calculating P_s

Now let us calculate the global value of P_s. To do this, we return to the general equation:

$$P_s = P_{\text{structural}} + P_{\text{antenna mode}}$$

and replace $P_{\text{antenna mode}}$ with its value:

$$P_s = P_{\text{structural}} + [(1-\Gamma)^2 - 1]P_{\text{structural}}$$

We obtain

$$P_s = (1-\Gamma)^2 P_{\text{structural}}$$

or alternatively

$$P_s = \frac{4}{(1+a)^2} P_{\text{structural}}$$

In the case of a tuned tag where $X_{\text{ant t}} = -X_1$, which is nominally true of at least 99%, if not 100%, of remotely powered tags in RFID applications, this equation gives the relationship between total power P_s reradiated by the tag and the reradiated structural power due to the specific type of antenna used. When we have determined the value of $P_{\text{structural}}$ for an antenna, we will have everything we need to modulate the value of R_1, and therefore of a to modulate the reradiated power, in other words to know the radar cross-section (RCS) of the tag.

For now, we can derive two obvious, but important, relations from the last equation:

(a) The ratio (immediate):

$$\frac{P_s}{P_{\text{structural}}} = \frac{4}{(1+a)^2} = (1-\Gamma)^2$$

(b) The difference $P_s - P_{\text{structural}}$, or ΔP, which is actually the same as $P_{\text{antenna mode}}$:

$$\Delta P = P_s - P_{\text{structural}} = P_{\text{antenna mode}}$$

representing the dynamic part (the variation of reradiated power) due to the load impedance mismatch, regardless of whether this is deliberate or involuntary. We will see that this difference in reradiated power reflects a variation in the tag's radar cross-section, ΔRCS, which will be calculated in the next chapter. In the particular case (often found in remotely powered RFID applications at UHF and SHF) where, in the voluntary load matching phase, R_1 changes abruptly from conjugate matching, $R_1 = R_{\text{ant t}}$, to $R_1 = 0$ (load short-circuited), this simultaneously gives rise to two conditions, $a = 0$ and $\Gamma = -1$, and therefore

$$\Delta P = P_{\text{antenna mode}} = [2 + (-1)^2] P_{\text{structural}} = 3 P_{\text{structural}}$$

Summary

$$P_{\text{structural}} = \frac{1}{4} \frac{V_{\text{equi rms}}^2 G_{\text{ant t}}}{R_{\text{ant t}}}$$

$$P_s = P_{\text{structural}} + P_{\text{antenna mode}}$$

$P_s = $ fixed part $+$ part that varies as a function of the actual load R_1

$$P_s = P_{\text{structural}} + [(1 - \Gamma)^2 - 1] P_{\text{structural}}$$

$$P_s = P_{\text{structural}} = (1 - \Gamma)^2 P_{\text{structural}}$$

Table 8.3 shows the values of P_s and $P_{\text{antenna mode}}$ according to R_1, a and Γ.

Table 8.3

Load	R_1	a	Γ	P_s	$P_{\text{antenna mode}} = P_s - P_{\text{structural}}$
Short circuit	0	0	-1	$4 P_{\text{structural}}$	$= 4 - 1 = 3\, P_{\text{structural}}$
Matched	R_{ant}	1	-0	$P_{\text{structural}}$	$= 1 - 1 = 0 \times P_{\text{structural}}$
Open circuit	∞	∞	$+1$	$0 \times P_{\text{structural}}$	$= 0 - 1 = -1 \times P_{\text{structural}}$

Back to the Roots
Now let us consider the deeper meaning of the power $P_s = (1 - \Gamma)^2 P_{\text{structural}}$, using Figure 8.2 (a), (b) and (c).

Load tuned ($X_{\text{ant t}} = -X_l$) and matched ($R_{\text{ant t}} = R_l$)
Figure 8.2(a) shows the whole tag in the matched state ($R_{\text{ant t}} = R_1$). In this case:

- Power dissipated in $R_1 = P_{\text{structural}} = P_{\text{matched}}$
- Power dissipated in $R_{\text{ant t}}$ (and therefore reradiated) $= P_{s \text{ structural}} = P_{\text{structural}}$.

The reradiated power $P_{s \text{ structural}}$ is known and is equal to the power P_t (Friis equation in the matching phase) multiplied by the gain $G_{\text{ant t}}$, as demonstrated above.

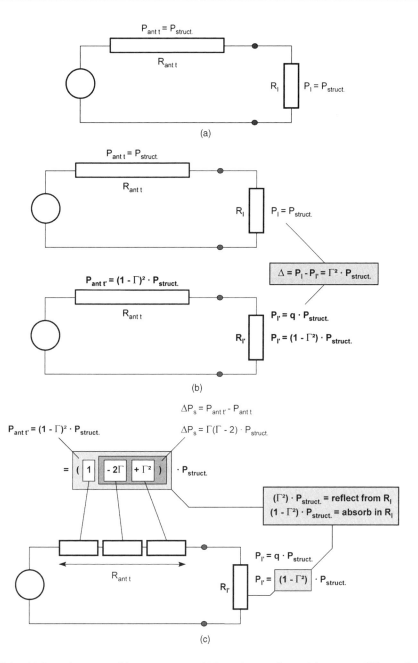

Figure 8.2 (a) Impedance matching: $R_{ant\,t} = R_l$. (b) Impedance mismatching: $R_{ant\,t}$ different from R_l. (c) Schematic representation of impedance mismatching: $R_{ant\,t}$ different from R_l

Load tuned ($X_{ant\ t} = -X_l$) and unmatched ($R_{ant\ t}$ different from R_l)

Figure 8.2(b) shows the whole tag when there is any mismatching ($R_{ant\ t}$ different from R_l). As a result of the impedance mismatching between the tag's antenna and the integrated circuit, part ($P_{1\ refl}$) of the maximum possible conducted power $P_t = P_{1\ max} = P_{structural}$ transmitted to the load R_1 will be reflected towards the power source.

As mentioned in a previous chapter, we obtain in this way:

- the power dissipated in $R_1 = qP_{structural} = (1 - \Gamma^2)P_{structural}$
- and, as mentioned above, the power dissipated in R_{ant} (i.e. the reradiated power), which is equal to $(1 - \Gamma)^2 P_{s\ structural}$.

The difference in useful power in the load R_1 between the 'matched' and 'unmatched' phases is, evidently,

$$P_{structural} - (1 - \Gamma^2)P_{structural} = \Gamma^2 P_{structural}$$

Figure 8.2(c) is essentially the same as Figure 8.2(b), but I have divided the reradiated power $P_s = (1 - \Gamma)^2 P_{s\ structural}$ into its different active components, to make it easier to understand the nature of the different effects. For this purpose, the resistance of the tag antenna $R_{ant\ t}$ has been divided into three resistances in series, R_1, R_2 and R_3, such that their sum is equal to $R_{ant\ t}$:

$$R_1 + R_2 + R_3 = R_{ant}$$

The values of R_1, R_2 and R_3 are chosen in such a way that:

- the power P_1 in $R_1 = 1P_{structural}$;
- the power P_2 in $R_2 = (-2\Gamma)P_{structural}$;
- the power P_3 in $R_3 = (\Gamma^2)P_{structural}$;
- the power P in $R_{ant\ t} = P_1 + P_2 + P_3 = (1 - 2\Gamma + \Gamma^2)P_{structural} = (1 - \Gamma)^2 P_{structural}$.

The load resistance R_1 remains unchanged: the power P_1 in R_1 is equal to

$$qP_{structural} = (1 - \Gamma^2)P_{structural}$$

The reradiated power P_3 represents the portion $(1 - q)P_{structural} = \Gamma^2 P_{structural}$ of power reflected because of the load mismatching:

- the power P_3 in $R_3 = \Gamma^2 P_{structural}$;
- the power P_1 in $R_1 = qP_{structural} = (1 - \Gamma^2) P_{structural}$. Thus

$$P_3 + P_1 = P_{structural} = P_{matched}$$

Where the antenna is concerned, the reradiated part of the power, P_1, equal to $P_{structural}$, is identical to that which was already present in the matched state. It is only the portions of the power called P_2 and P_3 that make a difference in the mismatched state, where P_3 represents the part reflected by the load and P_2 is due to the reflection of the load variation on the whole antenna circuit assembly.

The following sections will explain what happens in the case of mismatching.

$$x = P_{ant\ rms}/P_{ant\ rms\ max}$$

Now let us calculate the conducted power that can be reradiated by the tag antenna. The general equation for this is

$$P_{ant\ rms} = R_{ant\ t} I_{rms}^2$$

$$P_{ant\ rms} = \frac{R_{ant\ t}}{(R_{ant\ t} + R_1)^2 + (X_{ant\ t} + X_1)^2} V_{equi\ rms}^2 \ \text{W r.m.s.}$$

and the value of this reaches a maximum when $X_1 = -X_{1\ ant}$ and $R_1 = R_{ant\ t}$ (matching):

$$P_{ant\ rms\ max} = \frac{R_{ant\ t}}{(R_{ant\ t} + R_1)^2} V_{equi\ rms}^2$$

and therefore

$$P_{ant\ rms\ max} = \frac{V_{equi\ rms}^2}{4R_{ant\ t}}$$

Let x stand for the ratio $P_{ant\ rms}/P_{ant\ rms\ max}$ and let us calculate its value. We find

$$x = \frac{P_{ant\ rms}}{P_{ant\ rms\ max}} = \frac{4R_{ant\ t}R_{ant\ t}}{(R_{ant\ t} + R_1)^2 + (X_{ant\ t} + X_1)^2} = f(R_1, X_1) \text{ at } R_{ant\ t} \text{ and } X_{ant\ t} = \text{constant}$$

Therefore

$$P_{ant\ rms} = x P_{ant\ rms\ max}$$

This factor x is very important, because it denotes the relative portion of conducted power that is reradiated by the tag antenna as a function of the impedance mismatch, with respect to the ideally matched load.

The case of incompletely conjugate matching

If the whole input circuit of the tag (antenna + integrated circuit) forms a tuned circuit, $X_{ant\ t} = -X_1$ (or $X_{ant\ t} = X_1 = 0$), the general equation for x shown in the preceding section is simplified to

$$x = \frac{P_{ant\ rms}}{P_{ant\ rms\ max}} = \frac{4R_{ant\ t}R_{ant\ t}}{(R_{ant\ t} + R_1)^2}$$

Allowing for the fact that $a = R_1/R_{ant\ t}$, we obtain

$$x = \frac{P_{ant\ rms}}{P_{ant\ rms\ max}} = \frac{4}{(1+a)^2} = f(a) = f\left(\frac{R_1}{R_{ant\ t}}\right)$$

When a (i.e. R_1) varies from 0 to infinity, Table 8.4 shows the variations of x as a function of $a = R_1/R_{ant\ t}$.

Table 8.4

Load	R_1	a	x
Short circuit	0	0	4
Matched	R_{ant}	1	1
Open circuit	∞	∞	0

When q, the load mismatching value,

$$q = \frac{P_{1\,\text{rms}}}{P_{1\,\text{match max}}} = \frac{4}{(1+a)^2} = f(a) = f\left(\frac{R_1}{R_{\text{ant t}}}\right)$$

is compared with the value of x,

$$x = \frac{P_{\text{ant rms}}}{P_{\text{ant matched}}} = \frac{4}{(1+a)^2}$$

we derive, on the one hand,

$$q = ax = 1 - |\Gamma|^2$$

and, on the other hand,

$$x = \frac{q}{a} = \frac{1 - |\Gamma|^2}{a} = \frac{4}{(1+a)^2}$$

Note

If Γ is -1, i.e. if there is total reflection on a short circuit ($R_1 = 0$), the numerical value of the equation is of the type $(1 - |\Gamma|^2) = 0$ divided by $a = R_1/R_{\text{ant t}} = 0$, i.e. of the type 0/0. This indeterminacy is eliminated by using the 'true' general formula above, which gives the answer 4.

8.4 Radar Cross-Section (RCS) of the Tag, $\sigma_{e\,s}$

After these lengthy calculations relating to the power P_s reradiated by the tag, we can now go on to determine the radar cross-section of the tag.

8.4.1 Definition of the Radar Cross-Section, $\sigma_{e\,s}$ or RCS, of the Tag

The relationship between the power P_s reradiated or scattered by a tag and the modulus of the mean effective power flux density of the incident wave s received at the tag is represented in concrete terms by the definition of the associated equivalent radar cross-section $\sigma_{e\,s}$, thus:

$$\sigma_{e\,s} = \frac{P_s}{s} \rightarrow P_s = \sigma_{e\,s}s$$

(The index 's' of P_s stands for 'scattered'.)

The proportionality factor of this equation $\sigma_{e\,s}$ – the radar cross-section (RCS) of the tag, also known as the 'scattering aperture' (and therefore expressed in m²) – represents the equivalent area from which some of the energy (via the power flux density) collected by the object is reradiated towards the source in the case of a monostatic system. Its value is a measure of the object's capacity to reflect/reradiate, or not, the incident wave arriving from the source. Leaving aside the inessentials, the scattered power $P_{s\,\text{EIRP}}$ is therefore equivalent to what would be produced by an isotropic point source placed in the centre of the tag. Replacing s by this value,

$$\sigma_{e\,s} = \frac{P_{s\,EIRP}}{P_{EIRP\,bs}/(4\pi r^2)} = 4\pi r^2 \frac{P_{s\,EIRP}}{P_{EIRP\,bs}}$$

As I have already demonstrated, the radiated (and reradiated) power levels are directly related to the mean values of the moduli of the respective Poynting vectors in the form $s = \langle|\mathbf{S}|\rangle = |\mathbf{E} \times \mathbf{H}|$ (an equation in which the moduli of $\mathbf{E}$ and $\mathbf{H}$ are expressed as peak values). Theoretically, at any point in space, the value of $\sigma_{e\,s}$ is expressed by a vector value composed of amplitude and phase:

$$\sigma_{e\,s} = \lim_{r \to \infty} 4\pi r^2 \frac{E_s \times H_s}{E_i \times H_i}$$

In the case of RFID operation that we are concerned with, in the far field ($r \gg \lambda/2$), as I have stated, the vectors $\mathbf{E}$ and $\mathbf{H}$ are orthogonal to each other and the modulus of $\mathbf{H}$ is related to that of $\mathbf{E}$ by the relation $H = E/Z_0$. The vector product is therefore reduced to a simple scalar product, and the mathematical expression of the radar cross-section $\sigma_{e\,s}$ then becomes

$$\sigma_{e\,s} = \lim_{r \to \infty} 4\pi r^2 \frac{|E_s|^2}{|E_{EIRP\,bs}|^2}$$

Note

This equation assumes that the axes of the base station and of the reradiation/scattering from the tag are optimally aligned – in terms of the relative angle between the antennas of the base station and tag – but that the observation of the scattered power P_s will depend on the relative positions of the source of the illumination and the point from which the scattered power is observed. We will return to this matter subsequently in the discussion of monostatic and bistatic co-location or non-co-location of the transmitting and receiving antennas of the base station.

The value of $\sigma_{e\,s}$ depends on many parameters of the tag. Some of the most important ones are:

• its shape;
• its dimensions;
• the material from which it is made;
• the structure of the material forming its surface (no, the 'stealth' tag has not been invented as yet!);
• the wavelength of the transmitted wave;
• the polarization of the wave;
• the load impedance of the antenna that it represents.

The value of incident s has already been described in detail in the previous chapter; now we simply have to estimate the value of the power P_s in order to determine $\sigma_{e\,s}$.

8.4.2 Calculating the Radar Cross-Section of the Tag, $\sigma_{e\,s}$

For this purpose, we return to the general equation for P_s, in which we will assume that the losses due to the presence of R_{loss} are negligible, in order to avoid making the equations too cumbersome. We know that

$$P_s = \frac{R_{ant\,t}}{(R_{ant\,t} + R_1)^2 + (X_{ant\,t} + X_1)^2} V^2_{equi\,rms} G_{ant\,t}\ \text{W r.m.s.}$$

Dividing P_s by s, we find the value of $\sigma_{e\,s}$ directly:

$$\sigma_{e\,s} = \frac{P_s}{s} = \frac{P_{structural} + P_{antenna\,mode}}{s}$$

Clearly, this equation also includes, for the same reasons, two parts that are similar to those introduced previously when the equations for P_s were established:

$$\sigma_{e\,s} = \frac{P_s}{s} = \sigma_{e\,s\,structural} + \sigma_{e\,s\,antenna\,mode}$$

where $\sigma_{e\,s}$ = fixed part + part that is variable as a function of the actual load R_1:

$$\sigma_{e\,s} = \sigma_{e\,s\,structural} + [(1-\Gamma)^2 - 1]\sigma_{e\,s\,structural}$$

$$\sigma_{e\,s} = (1-\Gamma)^2 \sigma_{e\,s\,structural}$$

Figure 8.3 shows the variations of $\sigma_{e\,s}/\sigma_{e\,s\,structural} = f(\Gamma)$.

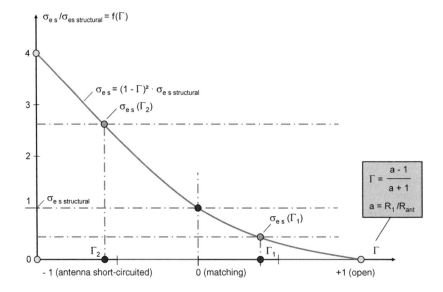

Figure 8.3   Graph of the relative variation of $\sigma_{e\,s} = f(\Gamma)$

Now we just need to calculate $\sigma_{e\,s\,structural}$ to find the variations of $\sigma_{e\,s}$ as a function of the load impedance Z_l, as we did for the reradiated power, as everything we have discussed in relation to the power aspect is also applicable to $\sigma_{e\,s}$.

Calculating $\sigma_{e\,s\,structural}$

To do this, we replace P_s with its value and find $\sigma_{e\,s}$. We obtain

$$\sigma_{e\,s} = \frac{1}{s} \frac{R_{ant\,t}\,V_{equi\,rms}^2}{(R_{ant\,t}+R_l)^2 + (X_{ant\,t}+X_l)^2}\,G_{ant\,t}$$

Using the Friis equation for the power supplied to the tag P_t, I have also shown previously that

$$P_{t\,rms} = \sigma_{e\,t}\,S_{rms}$$

$$P_{t\,rms} = \frac{\lambda^2}{4\pi}\,G_{ant\,t}\,S_{rms}\ \text{W r.m.s.}$$

and from this equation we can derive

$$S_{rms} = \frac{4\pi P_{t\,rms}}{\lambda^2 G_{ant\,t}}\ \text{W m}^{-2}$$

If we transfer the values of $P_{s\,rms}$ and s_{rms} into the above equation for $\sigma_{e\,s}$, we obtain

$$\sigma_{e\,s} = \frac{\lambda^2 G_{ant\,t}^2}{4\pi} \frac{R_{ant\,t}}{(R_{ant\,t}+R_l)^2 + (X_{ant\,t}+X_l)^2} \frac{V_{equi\,rms}^2}{P_{t\,rms}}\ \text{m}^2$$

This equation provides the general value of the radar cross-section $\sigma_{e\,s}$, also called the back modulation of the tag, regardless of the impedances of the antenna and the tag load.

Now let us examine the special case of this equation when $X_{ant\,t} = -X_l$ and $R_{ant\,t} = R_l$, corresponding to what is known as 'global conjugate matching' and to the 'structural mode' aspect of the argument. Consequently the form of the general equation for $\sigma_{e\,s}$ is simplified and becomes the value of $\sigma_{e\,s}$ in the conjugate load matching condition, called $\sigma_{e\,s\,structural}$:

$$\sigma_{e\,s\,structural} = \frac{\lambda^2 G_{ant\,t}^2}{16\pi R_{ant\,t}} \frac{V_{equi\,rms}^2}{P_{t\,rms}}\ \text{m}^2$$

When matching is present, the power P_t of the tag given by the Friis equation is equal to the power dissipated in the load and to that conducted in the tag antenna ($P_{R\,ant\,t} = P_l = P_t$), and therefore

$$P_{t\,eff} = \frac{V_{equi\,rms}^2}{4R_{ant\,t}}$$

If we replace $P_{t\ rms}$ with its value, the equation becomes

$$\sigma_{e\ s\ structural} = \frac{\lambda^2 G_{ant\ t}^2}{4\pi}\ m^2\ \text{(when the tag is impedance matched)}$$

Examples
The case of an isotropic antenna with unit gain
If the tag antenna is isotropic and has unit $G_{ant\ t}$, $\sigma_{e\ s\ structural} = 0.08\lambda^2 = f(\lambda)$.

The case of a λ/2 dipole antenna
If the tag antenna is of the $\lambda/2$ dipole type, and therefore $G_{ant\ t} = 1.64$, and if the load is matched, $R_l = R_{ant\ t} = 73.6\ \Omega$, then $\sigma_{e\ s\ structural} = 0.214\lambda^2 = f(\lambda)$.

Note
We could have found this value by a different method. For a $\lambda/2$ dipole, I have shown that

$$E_{rms} = \frac{7.017\sqrt{P_{EIRP}}}{r} = \frac{\sqrt{49 P_{EIRP}}}{r}$$

$$E_{rms}^2 = \frac{49 P_{EIRP}}{r^2}$$

Also, for a $\lambda/2$ dipole antenna with an effective length of λ/π:

$$V_{equi}^2 = \left(E_{rms}\frac{\lambda}{\pi}\right)^2 = \frac{49 P_{EIRP}}{r^2}\frac{\lambda^2}{\pi^2}$$

On the other hand, the received power flux density in the far field is

$$S_{rms} = \frac{P_{bs} G_{bs}}{4\pi r^2} = \frac{P_{EIRP}}{4\pi r^2}$$

Dividing V_{equi}^2 by s, we obtain

$$\frac{V_{equi}^2}{s} = \frac{49 \times 4\lambda^2}{\pi}$$

Also, taking into account the general equation for $\sigma_{e\ t\ \lambda/2}$:

$$\sigma_{e\ t\ \lambda/2} = \frac{R_t V_{equi}^2}{s[(R_t + R_{ant\ t})^2 + (X_t + X_{ant\ t})^2]}$$

$$\sigma_{e\ t\ \lambda/2} = \frac{R_t \times 49 \times 4\lambda^2}{\pi[(R_t + R_{ant\ t})^2 + (X_t + X_{ant\ t})^2]}$$

In the special (but frequently encountered) case of conjugate matching for a $\lambda/2$ dipole ($X_t = -X_{ant}$, $R_{ant} = R_t = 73.128\ \Omega$),

---

$$\sigma_{e\,t\,\lambda/2} = \frac{1}{4R_{ant\,t}}\frac{V^2_{equi}}{s} = \frac{49 \times 4\lambda^2}{\pi R_{ant\,t}} = 0.213\lambda^2$$

Table 8.5 shows an example of values of $\sigma_{e\,s\,structural}$, which of course has nothing to do with the actual physical surface area of the tag antenna, which is often negligible if it is the end of a wire or a simple metal rod.

The Relationship between the Effective Reception Area $\sigma_{e\,t}$ and the Reradiation Area $\sigma_{e\,s\,structural}$

Given that, if the tag is tuned to the carrier frequency in the same way, the effective reception cross-section $\sigma_{e\,t}$ of the tag is

$$\sigma_{e\,s\,structural} = \frac{\lambda^2}{4\pi}G_{ant\,t}$$

We can immediately deduce that the relation between the effective reception area and the reradiation area is

$$\sigma_{e\,s\,structural} = \sigma_{e\,t}\,G_{ant\,t}\,m^2 \text{ (when the tag is matched and tuned)}$$

Example of a λ/2 dipole antenna
In the case of a tag antenna of the $\lambda/2$ dipole type, with $G_{ant\,t} = 1.64$:

$$\sigma_{e\,s\,structural} = 1.64\,\sigma_{e\,t}\,m^2$$

Reflected/Reradiated Power, $P_{s\,structural}$

First of all, let us recall that, by definition, the general expression for reflected (scattered) or reradiated power is $P_s = \sigma_{e\,s}\,s$. If conjugate impedance matching is present (as in conventional applications), we can replace the effective reradiation area $\sigma_{e\,s\,structural}$ that we have just found and the incident power flux density s with their respective values, to give

$$P_{s\,structural} = \frac{\lambda^2 G^2_{ant\,t}}{4\pi}\frac{4\pi P_t}{\lambda^2 G_{ant\,t}} = P_t G_{ant\,t}$$

Recalling that (according to the Friis equation)

$$P_t = P_{bs}G_{bs}\left(\frac{\lambda}{4\pi r}\right)^2 G_{ant\,t} \text{ watts}$$

Table 8.5

f	λ	$\sigma_{e\,s\,structural}$ of an $\lambda/2$ dipole
866 MHz	0.346 m	256 cm^2
2.45 GHz	0.1224 m	32 cm^2

after simplification we obtain

$$P_{s\,structural} = P_{bs}G_{bs}\left(\frac{\lambda}{4\pi r}\right)^2 G_{ant\,t}^2 \text{ watts}$$

which is another way of expressing $P_{s\,structural}$, which is (fortunately) equal to

$$P_{s\,structural} = P_t G_{ant\,t}$$

as demonstrated above.

We have finally resolved a critical point in the physical interpretation of the story. When there is conjugate impedance matching between the antenna radiation resistance and the load:

- On one hand, some of the power, P_t (in watts), is dissipated in the load of the tag, and its value is given by the Friis equation.
- On the other hand, because of the flow of current due to the reception of the incident wave, exactly the same amount of 'conducted' power in watts is reradiated via the tag antenna in the form of $P_{s\,structural}$ ($P_{s\,structural} = P_t G_{ant\,t}$). This may seem strange. Even when maximum power is transmitted, an identical amount of (conducted) power is reradiated via the tag antenna.

This means that, in fact, when there is conjugate matching, the tag as a whole captures twice P_t and only half of this (one times P_t) can be used in the load, while the other half (one times P_t) represents the reradiated conducted power. Therefore, when measurements are made in the immediate environment of the tag, it is difficult to separate or isolate the part of the power radiated by the base station from the part of the power reradiated by the tag. This equality of the 'scattering area' (disregarding the gain) and the 'effective area' simply means that the tag antenna has at least the equivalent of two imaginary areas (apertures), of which one is called $\sigma_{e\,t}$, and is considered to be the area recovering the available power for the load, and the other is called $\sigma_{e\,s}$, and is responsible for the reradiated power.

Are there any more? Read on patiently and you will find out!

Note

While on this subject, we should remember that the effective reception cross-section of the tag $\sigma_{e\,t}$ was defined solely in the context of obtaining the power for use in the load, and therefore it cannot take the reradiated power into account (see the previous chapter).

If the conjugate matching condition is not met, as mentioned previously, some of the maximum possible power $P_{max} = P_t$ transferable to the load Z_l will not be transferred to it. In this case, the useful part of the power that is actually absorbed by the load is called qP_{max}, where q is the load mismatch factor as introduced previously:

- The missing amount will clearly be $(1 - q)P_{max}$ (remember that $(1 - q)P_{max} = \Gamma^2 P_{max}$).
- This amount $(1 - q)P_{max}$ will be reflected towards the power source.
- In other words, the latter amount $(1 - q)P_{max}$ will form part of the reradiated power (being more or less, according to the phase of the wave reflected on to the load, of the power already

reradiated in the case of matching), and will therefore apparently increase or reduce the reradiation area (RCS).

The Return Power Flux Density, $s_{back\ structural}$

By definition, when there is impedance matching between the source (tag antenna) and the load R_1, the power flux density of the reradiated return signal, which is available by reflection of the incident (forward) wave on the tag $s_{back\ structural}$, will be

$$s_{back\ structural} = \frac{P_{s\ structural}}{4\pi r^2} \text{ W m}^{-2}$$

However, in impedance matching conditions, if we replace $P_{s\ structural}$ with its value, we obtain

$$s_{back\ structural} = P_{bs}G_{bs}\frac{\lambda^2}{(4\pi)^3 r^4}G^2_{ant\ t} \text{ W m}^{-2}$$

The presence of the power 4 in the denominator of this equation thus indicates the huge effect of the distance r between the base station and tag[2] on the value of the power flux density of the reradiated back signal.

This equation is known as the 'radar equation'. This is because it defines the return power flux density due to the material, shape, wavelength, distance, transmitted power, etc., of the base station and tag system.

The Return Power, $P_{back\ structural}$, Recovered by the Base Station Receiver

Now that we know the value of the power flux density $s_{back\ structural}$ reradiated by reflection by the tag, we can calculate the power of the reradiated return signal $P_{back\ structural}$ that the receiving antenna of the base station with a gain of G'_{bs} will recover when the impedances of the tag are matched (and/or tuned). Let $\sigma_{e\ bs}$ stand for the equivalent/effective area of the base station. We find

$$\sigma_{e\ bs} = \frac{\lambda^2 G'_{bs}}{4\pi}$$

Be careful! I have deliberately written G'_{bs} because the receiving antenna of the base station may be co-located with the transmitting antenna but may have a different gain from the transmitting antenna of the base station, which has a gain of G_{bs}. However, in many cases $G_{bs} = G'_{bs}$!

Therefore

$$P_{back\ structural} = s_{back\ structural}\ \sigma_{e\ bs}$$

$$P_{back\ structural} = \left[P_{bs\ EIRP}G'_{bs}\frac{\lambda^2}{(4\pi)^3 r^4}G^2_{ant\ t} \right] \text{W}$$

---

[2] In this equation, it is assumed that the transmitting and receiving antennas of the base station are geographically co-located (monostatic antenna); otherwise, in the case of bistatic antennas, we would have to use the product $r_1{}^2 r_2{}^2$ (where r_1 is the forward distance and r_2 is the return distance) instead of r^4.

Rearranging the terms in a more convenient way, we obtain

$$P_{\text{back structural}} = P_{\text{bs EIRP}} G'_{\text{bs}} G^2_{\text{ant t}} \left(\frac{\lambda}{4\pi r} \right)^4 \text{ W}$$

Recalling that $[\lambda/(4\pi r)]^2$ represents the attenuation of the communication medium, we find

$$P_{\text{back structural}} = P_{\text{bs EIRP}} G'_{\text{bs}} G^2_{\text{ant t}} \left(\text{attenuation}^2 \right) \text{ W}$$

If we remember that we have already shown that

$$\sigma_{\text{e s structural}} = \frac{\lambda^2 G^2_{\text{ant t}}}{4\pi} \text{ m}^2$$

this expression is often also presented in the form

$$P_{\text{back structural}} = P_{\text{bs EIRP}} G'_{\text{bs}} \sigma_{\text{e s structural}} \frac{\lambda^2}{64\pi^3 r^4} \text{ W}$$

This equation, in its various forms, is also often known as the radar equation. This return power $P_{\text{back structural}}$ is the power due to the power $P_{\text{structural}}$ and, for a given base station/tag system, its value is constant/permanent.

Given that

$$\sigma_{\text{e s structural}} = \frac{\lambda^2 G^2_{\text{ant t}}}{4\pi} \text{ m}^2$$

we can also write this equation in two other forms:

$$P_{\text{back structural}} = s_{\text{incident}} \sigma_{\text{e s structural}} \left(\text{attenuation} \right) G'_{\text{bs}}$$

$$P_{\text{back structural}} = P_{\text{s structural}} \left(\text{attenuation} \right) G'_{\text{bs}}$$

Some Concluding Remarks

On the basis of my long experience in vocational training, I shall now suggest a final rewrite of this equation in a logical sequence corresponding to the physics of the forward and return transmission of the wave:

$$P_{\text{back structural}} = \left[P_{\text{bs EIRP}} \left(\frac{\lambda}{4\pi r} \right)^2 G_{\text{ant t}} \right] G_{\text{ant t}} \left(\frac{\lambda}{4\pi r} \right)^2 G'_{\text{bs}}$$

In physical terms, we have observed the following phenomena, starting from the beginning of this chapter:

- The base station supplies a radiated power $P_{\text{bs EIRP}}$.
- The forward signal undergoes a first 'attenuation' due to the medium.
- The gain of the tag antenna $G_{\text{ant t}}$ 'amplifies' the incident (forward) signal. All of this (the contents of the first set of brackets) is given by the Friis equation.
- The same tag antenna with the gain of $G_{\text{ant t}}$ contributes to the reradiation of the tag by scattering the incident signal.

- The return signal is then 'attenuated' a second time (by the same amount as in its forward travel).
- Finally, the radiated power of the return signal, received by the base station, is 'amplified' by the gain of its receiving antenna G'_{bs} (Friis again!).

This can all be expressed in dB in the following algebraic sum (but note that the attenuations have negative values):

$$P_{\text{back structural}} = P_{bs\,EIRP} + \text{attenuation} + G_{ant\,t} + G_{ant\,t} + \text{attenuation} + G'_{bs}$$

$$P_{\text{back structural}}\,(\text{in dB}) = P_{bs\,EIRP} + 2G_{ant\,t} + 2\,\text{attenuation} + G'_{bs}\,dB$$

which is simply the logarithmic version of our last equation:

$$P_{\text{back structural}} = P_{bs\,EIRP}G^2_{ant\,t}\,(\text{attenuation}^2)\,G'_{bs}$$

Example: UHF RFID in Europe
Be careful! In the example below, I have deliberately used two different antenna gains for the base station (bistatic antenna): the gain for the transmitting antenna is $+2.14\,dB$ and the gain for the receiving antenna is $+6\,dB$ (see the detailed example in Chapter 10):

$$f = 869.5\,\text{MHz}$$

$$r = 4\,m$$

$$\text{att at 869.5 MHz at 4 m(dB)} = 32.5 + 20\log(0.8695) + 20\log4$$

$$= 32.5 + [20(-0.051)] + [20(0.60206)]$$

$$= 43.52\,dB$$

$$P_{bs\,ERP\,max} = 27\,dBm\,(\text{i.e. } P_{ERP\,max} = 500\,mW\,ERP)$$

$$G_{ant\,bs\,at\,transmission} = +2.14\,dB\,(\text{i.e. a }\lambda/2\,\text{dipole antenna with } G_{ant\,bs} = 1.64)$$

$$P_{bs\,EIRP\,max} = +29.14\,dBm\,(\text{i.e. } 27 + 2.14\,dBm = 820\,mW\,EIRP)$$

$$G_{ant\,t} = +2.14\,dB\,(\lambda/2\,\text{dipole antenna with gain } G_{ant\,t} = 1.64)$$

$$G'_{ant\,bs}\,\text{at reception} = +6\,dB\,(\text{i.e. a different receiving antenna, with } G_{ant\,bs} = 4)$$

$$P_{\text{back structural}} = +29.14 + 6 + (2.14 + 2.14) + 2(-43.52)$$

$$P_{\text{back structural}} = -47.62\,dBm,\text{ of the order of a few tens of nW}$$

Budget for the Forward and Return Link
The last equation for $P_{\text{back structural}}$ given above can be used to quantify the global 'forward and return' energy budget of the link between the base station and the tag, which is therefore equal to

$$\frac{\text{Power received}}{\text{Power transmitted}} = \text{link budget} = \frac{P_{\text{back structural}}}{P_{bs\,EIRP}}G'_{bs}G^2_{ant\,t} \times \text{attenuation}^2$$

It is also very helpful to use this ratio when it is expressed in dB rather than in absolute values, thus:

$$\text{Link budget} = +G'_{\text{bs dB}} + 2G_{\text{ant t dB}} + 2\,\text{attenuation}_{\text{dB}}$$

Example: RFID at 2.45 GHz

Let us consider the example of a system operating at 10 m and at the frequency of 2.45 GHz. In order to compensate for the greater attenuation of the medium (air) due to the higher frequency, base stations are often fitted with directive transmitting and receiving antennae with gains G'_{bs} of the order of 4 or even 6 dB; thus they are directive. We have already seen that the attenuation of the 2.45 GHz at 10 m in air is 60.3 dB. Also, to enable the tags to be produced more cheaply, use is commonly made of $\lambda/2$ antennas with absolute gains $G_{\text{ant t}}$ of 1.64 or 2.14 dB. The global energy budget for the link is then

$$\text{Link budget} = +6 + (2 \times 2.14) + [2(-60.3)]$$

$$\text{Link budget} = -120.6 + 10.28 = -110.32 \text{ dB}$$

Assuming that this system is operating in the USA and the maximum permitted power $P_{\text{bs EIRP}}$ is $+36$ dBm (4 W), the received power $P_{\text{back structural}}$ at the base station will be

$$+36\,\text{dBm} - 110.32\,\text{dB} = -74.32\,\text{dBm, i.e. a few tens of pW}$$

Figure 8.4 shows the amplitude of the return signal in different situations.

As the last two examples have shown, the levels of power commonly received at the inputs of base stations are very low. To enable the receiving part of the base station to use these signals, the return signal from the tag must be accurately detected. This means that:

- The signal level must be at least higher than the level of the surrounding noise, which is referred to as the 'noise floor'.
- The ratio $P_{\text{back}}/P_{\text{bs EIRP}}$ must not be too low to allow the extraction and detection of the received signal.
- Also, the sensitivity of the base station receivers must be adequate.

distance	frequency	wavelength	P1	G_ant_bs	G_ant_bs	P_bs_eirp	G_ant_t	P_back	P_back/P1
		d'onde	conducted						
m	MHz	cm	W	-	dB	W	-	nW	dB
2	2400	12,5	1	4	6	4	1.6		-73
2	915	32,8	1	4	6	4	1.6		-56
2	866	34,6	1	4	6	4	1.6		-55
2	433	69,3	1	4	6	4	1.6		-43
4	2400	12,5	1	4	6	4	1.6		-85
4	915	32,8	1	4	6	4	1.6		-68
4	866	34,6	1	4	6	4	1.6		-67
4	433	69,3	1	4	6	4	1.6		-55

Figure 8.4 Amplitude of the return signal

From Figure 8.4 we can conclude that, in RFID at UHF and SHF, the power conditions for the return signal are present overall (the levels are higher than the conventional threshold of sensitivity of the input stages of the base stations) for relatively long distances, and therefore that, for operation in UHF back scattering mode, the main factors limiting the use of tags are found in the value of the maximum power $P_{bs\ EIRP}$, which can be transmitted for the purpose of remote power supply, the control of parasitic reflections, etc., which are parameters that have already been described fully.

If the signal level is below the noise level, reception will only be possible with the aid of an FHSS or DSSS device, whose operating principles are explained in Chapter 16. It is also clear that the use of battery-assisted tags enables the limitations of remote power supply to be overcome and allows reading to take place at longer distances, but this again raises the problem of the sensitivity of the input stage and the demodulator.

Summary

To sum up this long discussion, a system operating in the back scattering mode is primarily subject to two physical requirements:

- the success of the energy transfer for remotely powering the tag, based on the Friis equation;
- the ability of the base station to detect the small fraction of power that is returned to it, based on the radar equation.

We have now dealt with the 'structural' aspect. Let us go on to consider the 'antenna mode' part, due to the variation of the tag antenna load, and find the different values of $P_{back\ antenna\ mode}$ for different values of R_l.

8.4.3 Moving Away from Conjugate Matching

At last, we have finished with the subject of conjugate matching, the 'structural mode' and all its ramifications. Now let us see what happens in the 'antenna mode'.

So that we can use inexpensive tags in conventional RFID applications, based on the general equation for $\sigma_{e\ s}$, we shall now consider some individual cases of the values taken by $\sigma_{e\ s}$ as a function of the relation between the impedances of the source (the tag antenna) and the load (the integrated circuit of the tag, for example). We will therefore examine the standard cases of loads either in short circuit or in open circuit, where there is no conjugate matching between the source and load. After drawing some practical conclusions from these variations, we will see how to make the best use of these properties in RFID applications at UHF and SHF. For this purpose, we simply have to return to the general equation for $\sigma_{e\ s}$ and examine its variations as a function of R_l (or $a = R_l/R_{ant\ t}$). So here we go:

$$\sigma_{e\ s} = \frac{P_s}{s} = \sigma_{e\ s\ structural} + \sigma_{e\ s\ antenna\ mode}$$

where $\sigma_{e\ s}$ = fixed part + part that is variable as a function of the actual load R_l:

$$\sigma_{e\ s} = \sigma_{e\ s\ structural} + [(1-\Gamma)^2 - 1]\sigma_{e\ s\ structural}$$
$$\sigma_{e\ s} = (1-\Gamma)^2 \sigma_{e\ s\ structural}$$

So much for that part of the process. Given that

$$\sigma_{e\,s\,structural} = \frac{\lambda^2 G_{ant\,t}^2}{4\pi}$$

$$\sigma_{e\,s} = (1-\Gamma)^2 \frac{\lambda^2 G_{ant\,t}^2}{4\pi}$$

or alternatively, as a function of a, in other words as a function of the ratio $R_l/R_{ant\,t}$:

$$\sigma_{e\,s} = \frac{4}{(1+a)^2} \sigma_{e\,s\,structural}$$

or alternatively

$$\sigma_{e\,s} = \frac{G_{ant\,t}^2}{\pi(1+a)^2}\lambda^2 = f(a) = f\left(\frac{R_l}{R_{ant\,t}}\right)$$

Note
If $\lambda/2$ dipole antennas are used, with a constant gain of 1.64 independently of the wavelength used (because the antenna length is equal to $\lambda/2$!), we find

$$\sigma_{e\,s} = \frac{0.856}{(1+a)^2}\lambda^2$$

Tag Antenna Short-Circuited, $R_l = 0$ = Load Short-Circuited
Now let us short-circuit (abbreviation 'sc') the terminals of the tag antenna, in parallel with the tuned circuit: evidently, this means that $V_{rec} = 0$ and the load mismatch factor $q = 1 = 100\%$. In this case, $a = R_l/R_{ant\,t} = 0$ and $\Gamma = -1$, leading directly to

$$\sigma_{e\,s\,sc} = \sigma_{e\,s\,structural} + ([4-1]\sigma_{e\,s\,structural})$$

$$\sigma_{e\,s\,sc} = 4\sigma_{e\,s\,structural}$$

meaning that, as predicted, the power reradiated by the tag $P_{s\,sc}$ for the same received power flux density s will be 4 times greater than when the load R_l of the tag was matched:

$$P_{s\,sc} = 4P_{s\,structural}$$

Example of a $\lambda/2$ dipole antenna
In the case of a tag whose antenna is a $\lambda/2$ dipole, with $G_{ant\,t} = 1.64$ we have seen that, when the impedances are matched,

$$\sigma_{e\,s\,matched} = \sigma_{e\,s\,structural} = 0.214\lambda^2$$

and therefore that

$$\sigma_{\text{e s sc}} = 4\sigma_{\text{e s structural}}$$

$$\sigma_{\text{e s sc}} = 0.856\lambda^2$$

If $f = 866\,\text{MHz}$ and $\lambda = 0.346\,\text{m}$, $\sigma_{\text{e s sc}} = 1024\,\text{cm}^2$ (the area of a $32 \times 32\,\text{cm}$ square!)
If $f = 2.45\,\text{GHz}$ and $\lambda = 0.1224\,\text{m}$, $\sigma_{\text{e sc}} = 128\,\text{cm}^2$

which of course has nothing to do with the actual physical surface area of the tag antenna, which is often negligible if it is the end of a wire or a simple metal rod.

Note
Since every integrated circuit on the market has an input capacitance, the impedance of the tag antenna must include an inductive part in order to allow the tag to be tuned. Let us take one example.

As mentioned several times, the input impedance of the NXP Semiconductors U_code HSL integrated circuit is stated as $Z = 35 - \text{j}720$, which, translated into a parallel equivalent electrical circuit, gives a capacitance of about 240 fF connected in parallel with a resistance of about 15 kΩ. If we decide to tune the input circuit to 900 MHz, then

$$LC\omega^2 = 1$$
$$L = \frac{1}{C\omega^2}$$

$$L = \frac{1}{[240 \times 10^{-15} \times (2 \times 3.14 \times 900 \times 10^6)]^2} = 0.115\,\mu\text{H}$$

giving an impedance of

$$L\omega = 0.115 \times 10^{-6} \times (2 \times 3.14 \times 900 \times 10^6) = 450\,\Omega$$

which is very large with respect to 73 Ω!

Note also that, if the OOK modulation is carried out by the integrated circuit of the tag, in other words if the antenna load is short-circuited in line with the data traffic, the effective length of the antenna will decrease, and this will also slightly reduce the maximum value of $\sigma_{\text{e s antenna mode sc}}$.

Important notes
In the conditions of reradiation mentioned above (structural mode and antenna mode sc), the 'receiving' antenna of the tag acts as a very good power 'scatterer', and, if it is placed near another antenna, it may absorb or reradiate enough power to significantly alter the theoretical radiation pattern of the antenna of the transmitter considered in isolation. In these conditions, oddly enough, the receiving antenna can be seen either as a parasitic or perturbing element or as an advantage. This requires an explanation, since we must consider all the details for RFID applications.

The Favourable Effects
Let us start with the favourable effects. Depending on the phase of the current flowing in the parasitic element, this phenomenon of reradiation can be beneficial, by becoming a directing

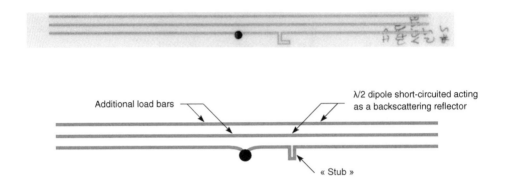

Figure 8.5   Example of tag using field strengthening bars

element offering greater directivity (as in the Yagi antennas used in television) or a reflecting/ radiating element enabling more energy to be recovered in another element. The latter phenomenon can be used to good effect by positioning 'bars' (actually short-circuited antennas, which therefore have a high level of reradiation) to increase the power received in the working element forming the tag's electronic circuit (see the photograph of an example of a design proposed by Intermec/NXP SC in Figure 8.5).

Less Favourable Effects
Now let us consider the less favourable effects. The fact that the antenna reradiates means that the field that it produces interferes with the incident field, and the resultant can create shadow areas in which other tags may cease to be powered or may lose their power supply and where others will be powered at a higher level than intended, causing them to reradiate more strongly and thus interfere with others even more.

This last comment is not to be taken lightly; because of the long communication distances mentioned in the preceding chapter, many users hope to operate with very many transponders (100, 200, 500, etc.) present simultaneously. Up to now they have been justified, since the collision management systems specified by the regulations (see Chapter 15) have been developed to overcome these problems. However, now let us consider some actual examples. Two cases are highly representative of the sober everyday reality of RFID.

Consider the case in which we want to read the contents of a pallet on which is placed a 'primary package' (a container) measuring 1 m^3 (= 1000 litres) containing 500 objects. To simplify matters, we shall assume that all the objects, made on a production line, are of the same type and are properly arranged (this is known as 'uniform' packaging). Each object therefore has an individual capacity of about 2 litres, and for the sake of simplicity we shall assume that they are cubes with edge measurements of 12.5 cm. If the boxes are properly arranged in the container, they will be placed at about 13 to 15 cm from each other, with all the labels placed in the same position on the objects, so that they are also spaced apart by the same amount. If the system operates at about 900 MHz ($\lambda/2$ = approximately 16 cm), then, owing to the reradiation that occurs between the labels, many of the shadow phenomena mentioned above are likely to occur, causing the labels to be masked during the read/write phases. And we have not even begun to consider the case where all the objects are packed in the loose state with many labels pressed against each other!

Another example concerns small objects arranged on a conveyor belt in small groups of five or six, but spaced apart, moving past a fixed base station. As I will show in Chapter 15, the technical specifications of the standards (ISO 18000-6-x) state that collisions between tags can be managed rapidly (with permanent inputs/outputs in the radiated field) up to about 100 to 500 tags per second. Let us be demanding and assume that we have 500 per second, and we will attempt to be realistic by assuming that the objects move in groups of five on a conveyor. This means that about 100 groups of objects pass the base station every second. To avoid the problems of proximity described in the first example, we must separate the groups from each other, by about 20 cm perhaps, to avoid any shadow areas caused by undesired reradiation, etc. A final remark: if we want to move 100 groups of objects with a spacing of 20 cm (making a total distance of 20 m) in one second, the conveyor belt will have to run at $20\,\mathrm{s}^{-1}$, i.e. $72\,\mathrm{km}\,\mathrm{h}^{-1}$! We are moving into the realms of flying carpets!

Information or Propaganda?

I have no arguments for or against UHF or SHF systems – certainly not – but I would ask potential users to be aware of the latent problems and others that may arise from their proposed applications, and to be sure that they can distinguish between 'paper' simulations of the possible number of collisions that can be managed per second and the sober reality. In this way they will be able to separate the information from the propaganda. More specifically, some companies or organizations have gained a lot of publicity by announcing collision management rates of up to 1000 or 1600 tags per second.

Why not? On paper, it is certainly possible, if we disregard the physical phenomena. What this means is that the timings described in the communication protocol for tag handling are executed in less than 1/1000 of a second (about a millisecond), but it does not in any way mean that if we place 1000 objects on a pallet (static, not moving) measuring $1\,\mathrm{m}^3$ we can read a thousand of them in one second. This is simply unrealistic because of the physical factors of proximity and interaction between tags, as described and explained in the preceding sections. Physical reality means that only between 100 and 200 tags can be recognized in such a space, without considering the time; alternatively, we can recognize 1000 tags in a second, but only if they are all separated from each other, meaning that they could never fit in a space measuring $1\,\mathrm{m}^3$.

Having embarked on this argument, let us follow it to its conclusion. In order to recognize all the objects while avoiding the problems of proximity, the objects must be spaced at least 30 centimetres apart, so we would have to envisage a pallet measuring $3 \times 3 \times 3\,\mathrm{m} = 9\,\mathrm{m}^3$. Not a common size! Moreover, in order to read a pallet like this, we would require reliable base station reading distances of the order of 5 m (because of the maximum aperture angle of the radiated beam required by ETSI 302 208), and this would entail EIRP levels that would often exceed the local regulations. We would also need to know how to manage weak collisions (reading collisions occurring between tags located very near the base station and those located far away), which is not a simple matter; otherwise we would have poor reading efficiency.

On the other hand, this value of 1000 tags per second may lead us to envisage a number of objects, spaced apart, moving rapidly past the base station. To picture this, consider the example in which, to achieve the maximum rate claimed, we place five objects, spaced apart, passing perpendicularly every 5 ms in front of the conventional working area of a base station with a distance of a few metres, assuming that the beam aperture is 1 m at a distance of 1.5 m. This means that the objects (all of them) travel 1 m in 5 ms, since we have enough time to manage five collisions in 5 ms. Translate this into km h^{-1}: 1 m in 5 ms is 200 m in 1 s and 3600 × 200 m in 1 h, i.e. 720 km h^{-1}! When do we hit the sound barrier? Please tell me if you know!

We can conclude that there are no miracle solutions, regardless of the systems and frequencies used.

Tag Antenna Open-Circuited, $R_l = \infty$ or $I = 0$

This is another special case of the 'antenna mode'. Theoretically, if no load is connected to the antenna terminals, we create an open circuit in parallel with the tuned circuit, which unexpectedly gives the tuned circuit a noninfinite quality factor (because it is limited by the antenna resistance) and a load power mismatch factor $q = 1 = 100\,\%$. It should be noted that, in physical terms, this case does not occur, or only occurs rarely, in real RFID applications, because there is always a minimum load (a high impedance of the integrated circuit). In this case, $a = R_l/R_{\mathrm{ant\ t}} = \infty$ and $\Gamma = +1$, leading directly to

$$\sigma_{\mathrm{e\ s\ \infty}} = \sigma_{\mathrm{e\ s\ structural}} + (0-1)\sigma_{\mathrm{e\ s\ structural}}$$

$$\sigma_{\mathrm{e\ s\ \infty}} = 0$$

which, as expected, produces no reradiated power! This is quite evident, because no current is flowing in the electrical circuit incorporating the load and the tag antenna:

$$P_{\mathrm{s\ \infty}} = 0$$

Conclusion

Table 8.6 summarizes this whole section by listing the values of $\sigma_{\mathrm{e\ t}}$ and $\sigma_{\mathrm{e\ s}}$ as a function of the ratio $a = R_l/R_{\mathrm{ant\ t}}$:

$$\sigma_{\mathrm{e\ s\ structural}} = \frac{\lambda^2 G_{\mathrm{ant\ t}}^2}{4\pi}$$

$$\sigma_{\mathrm{e\ s}} = \sigma_{\mathrm{e\ s\ structural}} + \sigma_{\mathrm{e\ s\ antenna\ mode}}$$

Table 8.6 Values of $\sigma_{\mathrm{e\ s\ x}}$ and $\sigma_{\mathrm{e\ s\ antenna\ mode}}$ as a function of R_l, a and Γ

Load	R_l	a	Γ	$\sigma_{\mathrm{e\ s\ x}}$	$\sigma_{\mathrm{e\ s\ antenna\ mode}} = \sigma_{\mathrm{e\ s\ x}} - \sigma_{\mathrm{e\ s\ structural}}$
Shortcircuit	0	0	-1	$4 \times \sigma_{\mathrm{e\ s\ structural}}$	$= 4 - 1 = 3 \times \sigma_{\mathrm{e\ s\ structural}}$
Matched	R_{ant}	1	-0	$\sigma_{\mathrm{e\ s\ structural}}$	$= 1 - 1 = 0 \times \sigma_{\mathrm{e\ s\ structural}}$
Open circuit	∞	∞	$+1$	$0 \times \sigma_{\mathrm{e\ s\ structural}}$	$= 0 - 1 = -1 \times \sigma_{\mathrm{e\ s\ structural}}$

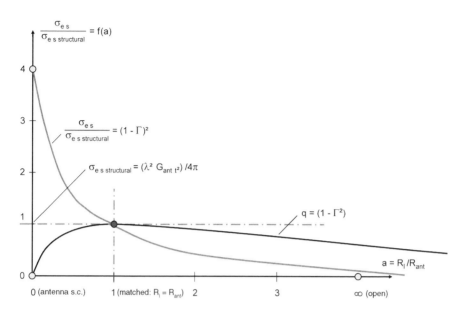

Figure 8.6 Graph of the variation of $\sigma_{e\,t}$ and $\sigma_{e\,s}$ as a function of the ratio $a = R_l/R_{ant\,t}$

where $\sigma_{e\,s}$ = fixed part + part that is variable (positive or negative) as a function of the actual load R_l:

$$\sigma_{e\,s} = \sigma_{e\,s\,structural} + [(1-\varGamma)^2 - 1]\sigma_{e\,s\,structural}$$

$$\sigma_{e\,s} = \sigma_{e\,s\,structural} = (1-\varGamma)^2 \sigma_{e\,s\,structural}$$

Figure 8.6 shows the variations of $\sigma_{e\,t}$ and $\sigma_{e\,s}$ as a function of the ratio $a = R_l/R_{ant\,t}$.

8.4.4 Variations of $\sigma_{e\,s\,antenna\,mode} = f(\varGamma)$

We can find $\sigma_{e\,s\,antenna\,mode}$ in the same way as $P_{s\,antenna\,mode}$:

$$\sigma_{e\,s\,antenna\,mode} = [(1-\varGamma)^2 - 1]\sigma_{e\,s\,structural} = \varGamma(\varGamma - 2)\sigma_{e\,s\,structural}$$

The variations of the 'dynamic' part representing the part due to the impedance mismatching of the antenna, $\sigma_{e\,s\,antenna\,mode} = f(\varGamma)$, with $-1 < \varGamma < +1$, will show a parabolic trend with a number of singular points already mentioned elsewhere. Table 8.7 shows the values of P_s and $P_{antenna\,mode}$ as a function of R_l and $\varGamma$.

Table 8.7 Values of P_s and $P_{antenna\,mode}$ as a function of R_l and $\varGamma$

	R_l	$\varGamma$	$(1-\varGamma)^2$	$\Delta\sigma_{es}$
Load short-circuited	0	-1	4	$3 \times \sigma_{e\,structural}$
Impedance matching	R_{ant}	0	1	$0 \times \sigma_{e\,structural}$
No load	∞	$+1$	0	$-1 \times \sigma_{e\,structural}$

8.5 Appendix

Let us assume that the input circuit of the tag consists of an LC tuned circuit and that the resistances $R_{ant\ t}$ and R_1 are strictly equal: $R_{ant\ t} = R_1$ (matched) $= 73\,\Omega$. Now let us examine the specific case of a UHF label that travels around the world, operating between 902 and 928 MHz in the USA one day, then stopping over in Europe and operating at about 868 MHz, subsequently passing through Japan (952–954 MHz) and finishing its world tour in China (operating between 952 and 954 MHz). Theoretically, the whole transponder will never be correctly 'conjugate' matched.

In mechanical and physical terms, the tag will always have an antenna of finite size, and if it is of the $\lambda/2$ type, it will only be $\lambda/2$ for a single frequency out of this whole band; thus, for a constant load R_1 of $73\,\Omega$, the antenna radiation resistance $R_{ant\ t}$ will not be strictly equal to $73\,\Omega$ over the whole band, and the equality $R_{ant\ t} = R_1$ will again be true for one specific frequency only. There will always be a system-wide mismatch between the source resistance $R_{ant\ t}$ and the load resistance R_1.

The same applies to the tuning of the LC tuned circuit. This is because X_1 will be different from $X_{ant\ t}$, since the equality $L\omega = 1/(C\omega)$, i.e. $LC\omega^2 = 1$, is true only for a given frequency, and not over the whole band from 860 to 960 MHz! We must therefore always allow for the famous mismatching factor, q.

Example

Let us assume that the tag, fitted with a $\lambda/2$ antenna, has to cover the whole band from 860 to 960 MHz, in other words a band of 100 MHz centred on 910 MHz (i.e. $2\Delta f/f_0 = 100/910 = 11\%$). Given that the bandwidth of a tuned circuit at $-3\,dB$ is equal to $2\Delta f = f_0/Q$, in order to meet the requirements of the application, the maximum value of the quality factor of the tuned circuit of the tag under load must be

$$Q_{tag} = 910/100 = 9.1$$

Now let us calculate the value of the inductance when the circuit is tuned to the central frequency and under load (matched):

$$Q_{tag} = \frac{L_{ant\ t}\,\omega}{R_{tot\ t}}$$

where $R_{tot\ t} = R_{ant\ t} + R_1 = 75 + 75 = 150\,\Omega$, and therefore L is

$$L_{ant\ t} = \frac{Q_{tag}R_{tot\ t}}{\omega}$$

$$L_{ant\ t} = \frac{9.1 \times 150}{2 \times 3.14 \times 910 \times 10^6} = 240\,nH$$

Now let us determine the capacitance C required for tuning to the centre of the band (910 MHz):

$$1 = L_{ant\ t}C_1\omega^2$$

and therefore

$$C_1 = \frac{1}{L_{\text{ant t}}\,\omega^2}$$

$$C_1 = \frac{1}{240 \times 10^{-9} \times (2 \times 3.14 \times 9 \times 10^8)^2} = 130\,\text{fF}$$

where $R_1'^2 = Q_{\text{tag}}^2 R_1$, incorporating the apparent resistance:

$$R_1'^2 = 9.1^2 \times 75 = 6210$$

9

The Back Scattering Technique and Its Application

As I have shown in the previous chapter, the power P_s reflected or reradiated by the tag (dependent on the value of the power flux density s) can be received and detected by the receiving antenna of the base station (which is often the same as the transmitting antenna), and can thus act as a signal informing the base station of whether or not an object or tag is present in the electromagnetic field. While the tag is illuminated, and regardless of whether it is remotely powered or locally battery assisted, provided that it has been designed to respond accurately via a specific modulation, we will have (re-)invented a communication device called 'back scattering modulation', which will be described more fully later on.

It is therefore useful to analyse the way in which the RCS is varied or modulated and to define the extent of its variation–the quantity $\Delta\sigma_{e\ s}$–as a function of a possible coding and a specific modulation, which will enable us to determine its ability to be understood correctly by the base station. We will then go on to determine what its merit factor is, or ought to be.

9.1 The Principle of Communication by Back Scattering Between the Base Station and the Tag

As a general rule, allowing for exceptions (and there are some), the communication model followed by standard RFID systems used at UHF and SHF is based on the RTF (reader talk first) principle, using the half duplex mode (an alternating link between the base station and tag). The stages of transmission are summarized in Figure 9.1.

9.1.1 The Forward Link: Communication from the Base Station to the Tag

During the first phase, known as the forward link, of the half duplex, the base station transmits the carrier frequency to power the tag remotely. At the same time, during this operating phase, the carrier is modulated (in the ASK, or amplitude shift keying, mode, for example) for the transmission of the command and interrogation codes to the tag.

RFID at Ultra and Super High Frequencies: Theory and Application   Dominique Paret
© 2009 John Wiley & Sons, Ltd

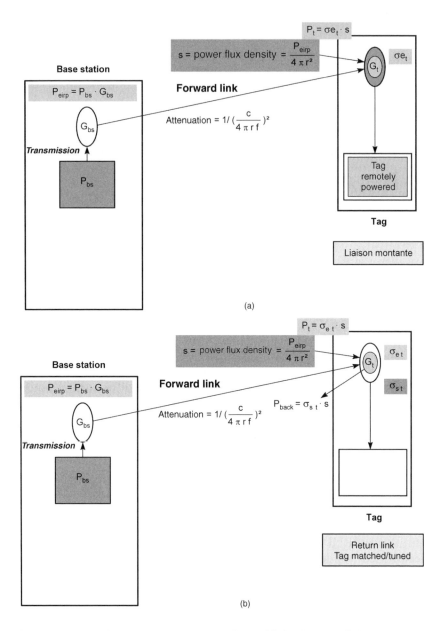

Figure 9.1 Principle of back scattering: (a) forward link; (b) return link, tag matched/tuned; (c) 'modulation' return link; (d) return link, tag mismatched; (e) return link

Note that, during this phase, the tag illuminated by the incident electromagnetic wave may either absorb the power that it receives or reradiate some of it, depending on the state of its antenna/load impedance matching. Generally, for remotely powered tags, the tag is made to absorb the maximum possible power (i.e. with no standing waves) during this phase, in order to provide the best possible remote power supply and thus achieve the highest possible operating

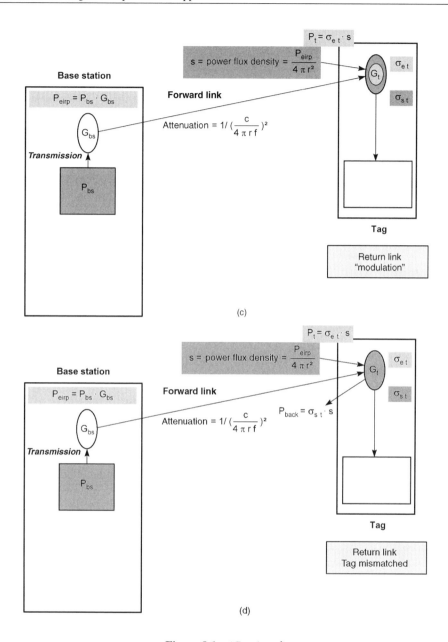

Figure 9.1 *(Continued)*

distance. However, it may reradiate during this phase, as explained in the previous chapter, according to its 'structural' aspect.

To sum up, during the forward link we have:

- conjugate impedance matching (in remotely powered tags);
- maximum power transmitted to the load (remote power supply);
- reradiation of the 'structural' type.

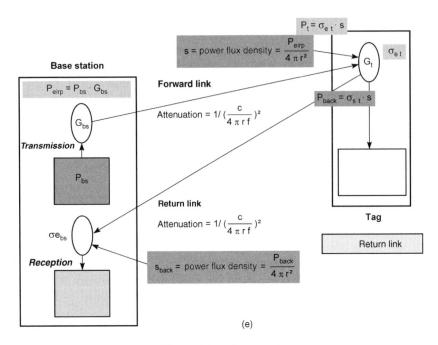

(e)

Figure 9.1 *(Continued)*

9.1.2 The Return Link: Communication from the Tag to the Base Station

During the second phase of the half duplex, called the 'return link', with communication from the tag, the base station initially supplies or maintains the sustained pure (unmodulated) carrier frequency to provide a physical support for the tag's response following the preceding interrogation commands. During this return link phase, two operating subphases may be present, depending on the binary information to be transmitted by the tag to the base station:

(a) Either the transmission of no useful information or the transmission of a logical '1'. Note that these are often identical in physical terms, since they correspond to the same phenomenon as that described in the previous section concerning the forward link.

(b) Elternatively, the transmission of a logical '0'. In this case, the tag's electronic circuit modulates the value of the load impedance $Z_l = R_l + X_l$ of the tag antenna at the rate of an OOK modulation corresponding to the logical data to be transmitted. Thus, at the tag, there will be an impedance mismatch between the source (the tag antenna) and its load, leading to the appearance of standing waves, and therefore to a new effective radar cross-section, and a variation of the RCS area, which will immediately modify the amount of power reradiated in a different way from that described in the previous section.

> **Notes**
> The receiving part of the base station examines the content of the return path (the return link) during subphases (a) and (b) only. To make this phenomenon easier to understand, the explanations above relate to the case of simple bit coding (NRZ) of the return link. This bit coding is often different (Manchester, BPSK, FM0, etc.). I will discuss these matters in detail in Chapter 15, which is concerned with standards.

To sum up, during the return link we have:

- deliberate mismatching of the antenna load impedance;
- a load power mismatch factor q;
- a change in the radar cross-section;
- the reradiation of a different power level by the tag, signifying the presence of a bit of opposite value.

The communication concept of the return link as outlined above is the main basis of UHF and SHF RFID systems operating in a mode of detecting the value of the reradiated/scattered return wave, known as the back scattering modulation mode.

... *Continuation*

Let us now examine the details of the phenomena produced by the tag in the return link, during the return wave modulation phase, e.g. the transmission of a logical '0' from the tag, starting from the optimal matching conditions mentioned above. Let us see what happens when we change the load impedance to values outside the optimal matching conditions mentioned above.

As indicated previously, the deliberate modulation of the load impedance Z_1 leads to an impedance mismatch between the source and the charge, on the one hand, and the presence of a wave reradiation phenomenon, on the other hand. Therefore we can also examine this problem of impedance modification in terms of a 'distributed constant line' and quantify this mismatch by using the reflection factor to calculate it.

9.2 The Merit Factor of a Tag, $\Delta\sigma_{e\,s}$ or ΔRCS

As shown above, in RFID applications, which operate on the back scattering principle and which are therefore based on the principle of modulating the wave reradiated by the tag, the effective radar cross-section $\sigma_{e\,s}$ of the tag varies when the impedance of the tag antenna is deliberately modified by changing the value of its resistive and/or capacitive part.

9.2.1 Definition of the Variation of the Radar Cross-Section RCS, $\sigma_{e\,s}$ or ΔRCS

The resulting variation of $\sigma_{e\,s}$ or RCS leads to the appearance of a new parameter, called $\Delta\sigma_{e\,s}$ or ΔRCS due to the modulation of the antenna impedance, which represents the difference between the two corresponding 'unmodulated' and 'modulated' values of $\sigma_{e\,s}$:

$$\Delta\sigma_{e\,s} \text{ or } \Delta\text{RCS} = \sigma_{e\,s\,\text{modulated}} - \sigma_{e\,s\,\text{modulated}}$$

Notes

1. In the technical literature, $\Delta\sigma_{e\,s}$ or ΔRCS are also called the 'merit figures' or 'merit factors' of an RFID tag.
2. Theoretically, the merit factor $\Delta\sigma_{e\,s}$ should be defined not as a scalar magnitude but as a vector value (i.e. of the complex variable) and therefore includes amplitude and direction (or modulus and phase), allowing for all the complex impedances involved.

This is because, depending on the different modulation states (particularly in the case of BPSK or QPSK), the values in watts of the reradiated scattered power may be equal while the phases of the scattered waves are different, and the difference of the scalar quantities $\Delta\sigma_{e\,s}$ between two states of modulation may be zero. To avoid this, the demodulator in the receiving part of the base station must allow for this possibility and must therefore be capable of demodulating either the amplitude or the phase of the received signal equally well.

Let us now take an overview of its variations as a function of the different parameters.

9.2.2 Estimation of $\Delta\sigma_{e\,s}$ as a Function of $\Delta\Gamma$

In an RFID application, we do not know in advance what the initial 'unmodulated' position of the tag corresponds to in physical terms (is it matched, nearly matched or completely unmatched?) or what the corresponding value of $\Gamma_1 = \Gamma_{\text{non modulated}}$ will be. To make matters clear, the 'unmodulated' physical state depends on the principles adopted for the design of the application. For example, the system designer may decide that the tag should be of the battery-assisted type (having an incorporated battery, but still of the passive type) because the application requires operation at a very long distance, in which case the tag antenna does not necessarily have to be matched in advance in the 'unmodulated' position to recover the maximum energy, since a battery is provided in the tag.

By contrast with many currently available books on this subject, I will cover every possible kind of application by stating, without any prior assumptions, that

$$\Gamma_1 = \Gamma_{\text{non modulated}} \text{ (initial value in the unmodulated position)}$$

$$\Gamma_2 = \Gamma_{\text{modulated}} \text{ (the value when the load impedance is switched, i.e. in the modulated position)}$$

$$\Delta\Gamma = \Gamma_2 - \Gamma_1 = \Gamma_{\text{modulated}} - \Gamma_{\text{nonmodulated}}$$

Using the general equation, we can write

$$\sigma_{e\,s} = \sigma_{e\,s\,\text{structural}} + \sigma_{e\,s\,\text{antenna mode}}$$
$$\sigma_{e\,s} = \text{fixed part} + \text{variable part (positive or negative) as a function of the load } R_l$$
$$\sigma_{e\,s} = \sigma_{e\,s\,\text{structural}} + [(1-\Gamma)^2 - 1]\sigma_{e\,s\,\text{structural}}$$
$$\sigma_{e\,s} = (1-\Gamma)^2 \sigma_{e\,s\,\text{structural}}$$

Because the structure of the equation $\sigma_{e\,s} = \sigma_{e\,s\,\text{structural}} + \sigma_{e\,s\,\text{antenna mode}}$ has a fixed part and a variable part, the difference between the 'modulated' and 'unmodulated' states is simply the algebraic difference between the two values of the variable part of the equation $\sigma_{e\,s\,\text{antenna mode}}$, namely $\sigma_{e\,s\,\text{antenna mode mod}}$ and $\sigma_{e\,s\,\text{antenna mode nonmod}}$ (Figure 9.2):

$$\Delta\sigma_{e\,s} = \Delta\text{RCS of the tag} = \sigma_{e\,s\,\text{antenna mode mod}} - \sigma_{e\,s\,\text{antenna mode nonmod}}$$

Now let us calculate the corresponding values of $\sigma_{e\,s1}$ and $\sigma_{e\,s2}$. We obtain

- in the 'unmodulated' phase: $\sigma_{e\,s1} = \sigma_{e\,s\,\text{structural}} + [(1-\Gamma_1)^2 - 1]\sigma_{e\,s\,\text{structural}}$;
- in the 'modulated' phase: $\sigma_{e\,s2} = \sigma_{e\,s\,\text{structural}} + [(1-\Gamma_2)^2 - 1]\sigma_{e\,s\,\text{structural}}$.

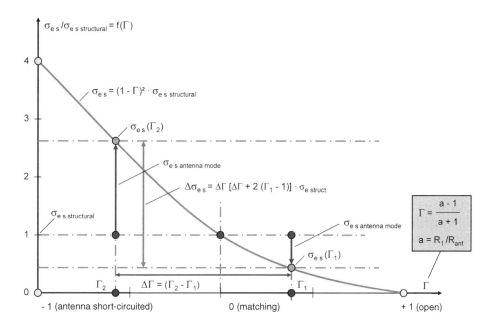

Figure 9.2   Calculation and variations of $\Delta\sigma_{e\ s} = \Delta RCS$

After reduction, the merit factor becomes

$$\Delta\sigma_{e\ s} = \sigma_{e\ s2} - \sigma_{e\ s1}$$

$$\Delta\sigma_{e\ s} = \Delta\Gamma[-2 + (\Gamma_2 + \Gamma_1)]\sigma_{e\ s\ structural}$$

Finally, replacing Γ_2 in this equation by its value $(\Delta\Gamma + \Gamma_1)$, we obtain

$$\Delta\sigma_{e\ s} = \Delta\Gamma[\Delta\Gamma + 2(\Gamma_1 - 1)]\sigma_{e\ s\ structural}$$

$$\Delta\sigma_{e\ s} = \Delta\Gamma[\Delta\Gamma + 2(\Gamma_1 - 1)]\frac{\lambda^2 G_{ant\ t}^2}{4\pi} = f(\Delta\Gamma, \Gamma_1) \qquad m^2$$

It is important to note that the function $\Delta\sigma_{e\ s} = \Delta RCS$ is simultaneously dependent on two elements: on the one hand the variable $\Delta\Gamma$ and on the other hand a parameter representing the initial value of Γ_1.

9.2.3   The variation $\Delta\sigma_{e\ s} = f(\Delta\Gamma, \Gamma_1)$

Let us return to the general equation for $\Delta\sigma_{e\ s}$ found above:

$$\Delta\sigma_{e\ s} = \Delta\Gamma[\Delta\Gamma + 2(\Gamma_1 - 1)]\sigma_{e\ s\ structural}$$

First of all, we should note that the value of $\Delta\sigma_{e\ s}$ is a function of two intercorrelated variables, and for the sake of simplicity we shall examine the case where $\Delta\sigma_{e\ s} = f(\Delta\Gamma)$ with $\Gamma_1 = $ constant.

Let us start with an observation. Theoretically, the maximum variation of $\Delta\Gamma$ is limited to the range of 0 to 2 inclusive, because the value of Γ can vary from -1 to $+1$. In practice, in the case

Table 9.1

Γ_1	Γ_2	$\Delta\Gamma = \Gamma_2 - \Gamma_1$	$\Delta\sigma_{e\ s} = \Delta\Gamma\ [\Delta\Gamma + 2(\Gamma_1 - 1)]\sigma_{e\ structural}$
$+1$	0	-1	$= -1[-1 + 2 \times (1-1)] = -1 \times \sigma_{e\ structural}$
$+1$	-1	-2	$= -2[-2 + 2 \times (1-1)] = +4 \times \sigma_{e\ structural}$
$\mathbf{0}$	$\mathbf{-1}$	$\mathbf{-1}$	$= -1[-1 + 2 \times (0-1)] = +3 \times \sigma_{e\ structural}$
$\mathbf{0}$	$\mathbf{+1}$	$\mathbf{+1}$	$= +1[1 + 2 \times (0-1)] = -1 \times \sigma_{e\ structural}$, as expected!
-1	0	$+1$	$= +1\ [+1 + 2 \times (-1-1)] = -3 \times \sigma_{e\ structural}$
-1	$+1$	$+2$	$= 2\ [2 + 2 \times (-1-1)] = -4 \times \sigma_{e\ structural}$

of numerous RFID applications of the remotely powered type, this range of variation will be smaller (from 0 to -1) because of the need to recover some of the incident energy to supply the remotely powered tags (see the list in Table 9.1).

Now let us take a closer look at the two very common cases shown in bold type in Table 9.1, which represent the great majority of conventional RFID applications.

The Usual Case of 'Remotely Powered' Applications, $\Gamma_1 = 0$ and $\Gamma_2 = x$

By far the greatest number of RFID applications operate in the 'remotely powered' mode and therefore start from the initial unmodulated position with conjugate matching, i.e. $R_1 = R_{ant\ t}$ and $a = 1$, in other words $\Gamma_1 = 0$, and then switches to a value of R_l different from that of R_{antt}, giving rise to a new value, Γ_2. In this case, but in this case only, $\Delta\Gamma = \Gamma_2$, and the above equation is simplified and reduced to the form:

$$\Delta\sigma_{e\ s} = \Gamma_2[-2 + \Gamma_2]\sigma_{e\ structural}$$
$$\Delta\sigma_{e\ s} = \Gamma_2(\Gamma_2 + 2)\sigma_{e\ structural}$$
$$\Delta\sigma_{e\ s} = \left(\Gamma_{mod}^2 - 2\Gamma_{mod}^2\right)\sigma_{e\ structural}, \text{ as mentioned above}$$

Table 9.2 shows examples of values for $\Delta\sigma_{e\ s}$.

Let us look at two subcases that frequently arise when $\Gamma_1 = 0$.

Subcase 1

R_1 is often switched all the way from the conjugate matching condition, $R_1 = R_{ant\ t}$, $a = 1$, to $R_1 = 0$ (load short-circuited) in UHF RFID, in order to maximize the variation of the value of $\Delta\sigma_{e\ s}$. This simultaneously gives rise to two conditions, $a = 0$ and $\Gamma = -1$, and therefore

$$\Delta\sigma_{e\ s} = \sigma_{e\ s} - \sigma_{e\ structural} = (4-1)\sigma_{e\ structural} = 3\sigma_{e\ structural}$$

Subcase 2

In RFID, it is sometimes desirable to reach a compromise between the variation of the radar cross-section and the power consumption of the tag, in order to optimize the operating distance

Table 9.2

R_1	a	Γ_2	$\Gamma_2 (\Gamma_2 - 2)$	$\Delta\sigma_{e\ s}$
0	0	-1	-3	$3 \times \sigma_{e\ s\ structural}$
R_{ant}	1	-0	-0	0
∞	∞	$+1$	-1	$-1 \times \sigma_{e\ s\ structural}$

of the system, and therefore R_1 is made to change only very slightly about the matching value $R_{\text{ant t}}$. In this case only, this means that $a = R_1/R_{\text{ant t}}$, i.e. close to or substantially equal to 1, and that Γ_2 is (very) close to 0 (matching). In this case, Γ_2^2 will be small with respect to $-2\Gamma_2$. Consequently, in this particular case only,

$$\Delta\sigma_{\text{e s}} = \sigma_{\text{e s antenna mode}} \cong -2\Gamma_2\,\sigma_{\text{e structural}}$$

which fits very well on the curve (the slope of the curve where $a = 1$, i.e. $\Gamma = 0$).

The Case of 'Battery Assisted' Applications, $\Gamma_1 = +1$ and $\Gamma_2 = -1$

If there is no need to provide a remote power supply to the tag, because the applications concerned use battery-assisted tags, then our aim will clearly be to benefit from the wider variation of $\Delta\sigma_{\text{e s}}$, by switching the load condition from fully 'open' to completely 'short circuited'. If the initial value of Γ in the unmodulated condition, Γ_1, is equal to 1 (i.e. there is an open load)–and only in this case–the above equation becomes

$$\Delta\text{RCS} = \frac{\lambda^2}{4\pi} G_{\text{tag}}^2 \Delta\Gamma^2$$

and clearly we can write

$$\Delta\text{RCS} = \frac{\lambda^2}{4\pi} G_{\text{tag}}^2 (\Gamma_2 - \Gamma_1)^2$$

an equation that appears in many books and documents in this field, but unfortunately with no indication of the limits of its validity (battery-assisted tags only). Therefore beware of unforeseen results! Table 9.3 shows the values for $\Gamma_1 = 1$.

An Example in RFID (Remotely Powered Tag)

In general and specific terms, in order to modulate the area $\sigma_{\text{e s}}$ of the tag, we reduce the value of R_1 by what is known as 'load modulation', or else we modify the value of the tuning capacitance upwards or downwards, using a variable capacitance diode. Very often it is impossible not to modify both of these simultaneously, as the resistive part of the load $R_1 = R_{\text{ant t}}$ takes a value of $(R_{\text{ant t}} - dR)$ and its reactive part $X_1 = -X_{\text{ant t}}$ takes a value of $(-X_{\text{ant t}} + dX)$. Let us return to the original general equation for the current I flowing through the equivalent circuit:

$$I = \frac{1}{(R_{\text{ant t}} + R_1) + j(X_{\text{ant t}} + X_1)} V_{\text{equi}}$$

and multiply above and below by the conjugated quantity of the denominator:

$$I = \frac{(R_{\text{ant t}} + R_1) - j(X_{\text{ant t}} + X_1)}{(R_{\text{ant t}} + R_1)^2 + (X_{\text{ant t}} + X_1)^2} V_{\text{equi}}$$

Table 9.3

Γ_1	Γ_2	$\Delta\Gamma$	$\Delta\sigma_{\text{e sx}} = \Delta\Gamma\,[\Delta\Gamma + 2(\Gamma_1 - 1)]\sigma_{\text{e structural}}$
$+1$	-0	-1	$= -1\,[-1 + 2 \times (1 - 1)] = +1 \times \sigma_{\text{e s structural}}$
$+1$	-1	-2	$= -2\,[-2 + 2 \times (1 - 1)] = +4 \times \sigma_{\text{e s structural}}$

It now becomes

$$I'' = \frac{[R_{\text{ant t}} + (R_{\text{ant t}} - dR)] - j[X_{\text{ant t}} + (-X_{\text{ant t}} + dX)]}{[R_{\text{ant t}} + (R_{\text{ant t}} - dR)]^2 + [X_{\text{ant t}} + (-X_{\text{ant t}} + dX)]^2} V_{\text{equi}}$$

$$I'' = \frac{(2R_{\text{ant t}} - dR) - jdX}{[4R_{\text{ant t}}^2 + (dR)^2 - (4R_{\text{ant t}} dR)] + (dX)^2} V_{\text{equi}}$$

Hypothetical Cases of Tag Operation

Let us assume that the total load impedance is only slightly different from that of the matching condition, in other words that dR is small with respect to $R_{\text{ant t}}$ and/or that dX is small with respect to X_{antt}, and therefore that the terms $(dR)^2$, $(dX)^2$ and $(4R_{\text{ant t}} dR)$ of the equation above are negligible with respect to the others.

Thus we obtain

$$I'' = \frac{(2R_{\text{ant t}} - dR) - jdX}{4R_{\text{ant t}}^2} V_{\text{equi}}$$

and therefore, finally,

$$I'' = \frac{V_{\text{equi}}}{4R_{\text{ant t}}} \left[\left(2 - \frac{dR}{R_{\text{ant t}}}\right) - j\frac{dX}{R_{\text{ant t}}} \right]$$

Now let us calculate the new effective radar cross-section $\sigma''_{\text{c s}}$ during this phase of impedance modulation. We can do this by using the same type of calculation as before. We know that, by definition, the new power reradiated by the tag P''_s is equal to

$$P''_s = (R_{\text{ant t}} I''^2) G_{\text{ant t}}$$

Since we know the new complex value of I'':

$$I'' = \frac{(2R_{\text{ant t}} - dR) - jdX}{4R_{\text{ant t}}^2} V_{\text{equi}}$$

we can calculate its effective value (in other words, the value of its modulus):

$$|I''| = \frac{\sqrt{(2R_{\text{ant t}} - dR)^2 + (dX)^2}}{4R_{\text{ant t}}^2} V_{\text{equi rms}}$$

and then square it:

$$|I''|^2 = \frac{(2R_{\text{ant t}} - dR)^2 + (dX)^2}{16R_{\text{ant t}}^4} V_{\text{equi rms}}^2$$

In this case (transponder impedance modulation), the reradiated power P''_s will therefore be

$$P''^2_s = R_{\text{ant t}} \frac{(2R_{\text{ant t}} - dR)^2 + (dX)^2}{16R_{\text{ant t}}^4} V_{\text{equi rms}}^2 G_{\text{ant t}}$$

We also know that the total structural power P_t received by the tag from the base station is

$$P_t = \frac{\lambda^2}{4\pi} G_{\text{ant } t} s \qquad \text{watts}$$

and therefore

$$s = \frac{P_t \times 4\pi}{\lambda^2 G_{\text{ant } t}} \qquad \text{Wm}^{-2}$$

and that the reradiated power P_s'' will now be

$$P_s'' = \sigma_{e\,s}'' s$$

and therefore

$$\sigma_{e\,s}'' = \frac{P_s''}{s}$$

Now we can transfer the value of s into the equation for $\sigma_{e\,s}$ and then replace P_s'' with its value, which gives us

$$\sigma_{e\,s}'' = \lambda^2 G_{\text{ant } t}^2 \frac{(2R_{\text{ant } t} - dR)^2 + (dX)^2}{4\pi \times 16R_{\text{ant } t}^3} \frac{V_{\text{equi rms}}^2}{P_t}$$

We have seen that, when the impedances of the tag antenna and the load are matched, the power P_t is equal to the power dissipated in the load, i.e.

$$P_t = \frac{V_{\text{equi rms}}^2}{4R_{\text{ant } t}}$$

If we transfer this value into the preceding equation, we obtain

$$\sigma_{e\,s}'' = \lambda^2 G_{\text{ant } t}^2 \frac{(2R_{\text{ant } t} - dR)^2 + (dX)^2}{4\pi \times 4R_{\text{ant } t}^2}$$

If we expand the numerator, provided that dR is assumed to be small, in other words that dR^2 and dX^2 are negligible because they are of the second order with respect to the other terms of the equation, we find

$$\sigma_{e\,s}'' = \lambda^2 G_{\text{ant } t}^2 \frac{R_{\text{ant } t}^2 - R_{\text{ant } t} dR}{4\pi \times 4R_{\text{ant } t}^2}$$

$$\sigma_{e\,s} = \frac{\lambda^2 G_{\text{ant } t}^2}{4\pi} \left(1 - \frac{dR}{R_{\text{ant } t}}\right) \qquad \text{m}^2, \text{ with tag not matched}$$

$$\sigma_{e\,s} = \sigma_{e\,s\,\text{structural}} \left(1 - \frac{dR}{R_{\text{ant } t}}\right) \qquad \text{m}^2, \text{ with tag impedance not matched and not tuned}$$

In conclusion, given the structure of the resulting equation, we can identify $(1 - dR/R_{\text{ant } t})$ with $(1 - \Gamma)^2 = (1 - 2\Gamma + \Gamma^2) \sim (1 - 2\Gamma)$, because $dR \ll R_{\text{ant } t}$ and therefore the value of Γ^2 is much smaller than that of Γ. Therefore

$$\Gamma = \frac{dR}{2R_{ant\ t}}$$

We could have expected this, because

$$\Gamma = \frac{a-1}{a+1} \qquad \text{where } a = \frac{R_1}{R_{ant\ t}}$$

and if R_1 is close to $R_{ant\ t}$ (and therefore dR is small), a is close to 1:

- Since a is close to 1, the numerator is $(R_{ant\ t} + dR) - R_{ant\ t} = dR$.
- Since a is close to 1, the denominator is $(R_{ant\ t} + dR) + R_{ant\ t} = 2\,R_{ant\ t}$, because dR is small, and therefore

$$\Gamma = \frac{dR}{2R_{ant\ t}}$$

A few notes

I may have spent a long time in setting out and presenting all the foregoing equations, only achieving useful approximations at the end of a series of proofs, but this has not been done in a spirit of excessive masochism. I have acted in this way purely because I wanted to allow readers and users to provide their own simplifications in the course of their calculations, in line with the specific circumstances of their RFID applications.

I should also point out that this equation does not depend on dX, and therefore that, in this case, the variation of the radar cross-section is essentially due to the load variation dR, and the phase does not change to any great extent.

Another Note

I could have followed a direct argument based on Γ:

$$\Gamma = \frac{Z_1 - Z_{ant\ t}}{Z_1 + Z_{ant\ t}}$$

$$\Gamma = \frac{(R_1 + jX_1) - (R_{ant\ t} - jX_{ant\ t})}{(R_1 + jX_1) + (R_{ant\ t} + jX_{ant\ t})}$$

$$\Gamma = \frac{(R_1 - R_{ant\ t}) + j(X_1 + X_{ant\ t})}{(R_1 + R_{ant\ t}) + j(X_1 + X_{ant\ t})}$$

If we now multiply the top and bottom of this expression by the conjugate of the denominator, we obtain

$$\Gamma = \frac{[(R_1 - R_{ant\ t}) + j(X_1 + X_{ant\ t})][(R_1 + R_{ant\ t}) + j(X_1 + X_{ant\ t})]}{(R_1 + R_{ant\ t})^2 + j(X_1 + X_{ant\ t})^2}$$

If there is optimal matching, in other words if the output impedance of the antenna and the input impedance of the integrated circuit are conjugate, i.e. $R_1 = R_{ant\ t}$ and $X_1 = -X_{ant\ t}$, then

$\Gamma = 0$ and the maximum available power will be transferred to the load. Before going more deeply into the theory, let us return to the specifics of UHF and SHF RFID applications. The tag impedance will be matched and then slightly mismatched; in other words, the resistive part of the load $R_1 = R_{ant\ t}$ will take the value of $(R_{ant\ t} - dR)$ and its reactive part $X_1 = -X_{ant\ t}$ will move to a value of $(-X_{ant\ t} + dX)$.

Replacing R_1 and X_1 with their new values in the equation for Γ, with no simplification, we obtain

$$\Gamma = \frac{(-2R_{ant\ t}dR + dR^2 + dX^2) + j(2R_{ant\ t}dX)}{4R_{ant\ t}^2 - 4R_{ant\ t}dR + dR^2 + dX^2}$$

Assuming that the variations of resistance dR and reactance dX are small (a small detuning of the tag), then in a first approximation we can disregard the terms dR^2, dX^2 and the product $(dR\ dX)$, which are all of the second order. After simplification, we have

$$\Gamma = \frac{-dR + jdX}{2(R_{ant\ t} - dR)}$$

and since $dR \ll R_{ant\ t}$:

$$\Gamma = \frac{-dR + jdX}{2R_{ant\ t}}$$

Given that, when the tag is only very slightly mismatched,

$$\Delta\sigma_{e\ s} \approx (-2\Gamma)\sigma_{e\ s\ structural}$$

$$\Delta\sigma_{e\ s} = \sigma_{e\ s\ structural}\frac{dR - jdX}{R_{ant\ t}}$$

The value in brackets is the tag modulation merit factor, and shows the real and imaginary parts that may be included in the value of $\Delta\sigma_{e\ s}$.

Important Note about UHF RFID Systems Using Phase Modulation
It should be noted that, based on the assumptions stated at the start of these explanations, if the value of dR is very small or even zero, and if only the value of dX is significantly modified (e.g. by modifying or modulating only the internal capacitance of the integrated circuit of the tag or, in other words, by keeping Z_{ant} equal to Z_1 in the forward link from the base station to the tag), the value of $\Delta\sigma_{e\ s}$ is a purely imaginary quantity. Essentially, this means that there is no change in the reradiated power in watts (back scattering) between the forward and return link phases of communication and that only the phase of the signal reradiated by the tag is modified by the (reactive) impedance modulation of the integrated circuit. In this case, the base station receiver has to carry out a phase demodulation of the back scattering signal, instead of performing amplitude variation demodulation of the ASK type on the received power as before. For your information, 99 % of commercially available base stations, which for many other reasons have I and Q demodulators (see Chapter 19), always carry out simultaneous amplitude and phase modulation.

Matching Factor

To conclude this example, let us now calculate the value of the matching factor (actually the mismatching factor) of the tag:

$$\theta_{\text{matching}} = 1 - |\Gamma|^2 = q$$

To do this, we start by calculating the modulus of this expression:

$$|\Gamma| = \frac{\sqrt{(\mathrm{d}R)^2 + (\mathrm{d}X)^2}}{2(R_{\text{ant t}} - \mathrm{d}R)}$$

and then square it:

$$|\Gamma|^2 = \frac{(\mathrm{d}R)^2 + (\mathrm{d}X)^2}{4(R_{\text{ant t}} - \mathrm{d}R)^2}$$

$$1 - |\Gamma|^2 = 1 - \frac{(\mathrm{d}R)^2 + (\mathrm{d}X)^2}{4(R_{\text{ant t}} - \mathrm{d}R)^2} = q$$

Note

Theoretically, the input impedance R_1 of the integrated circuit is not equal to $73\,\Omega$. Consequently, there is always an impedance matching circuit (a transformer or an LC circuit), which adjusts the impedance represented by the integrated circuit (about $35\,\mu\text{W}$ at $2\,\text{V}$ with $P = U^2/R_1 \rightarrow R_{1\,\text{ic}} = 80\,\text{k}\Omega$) to $73\,\Omega$.

9.2.4 Variations of $\Delta\sigma_{e\,s} = f(a)$

For reasons of simplicity, we often prefer to consider the variations of $\Delta\sigma_{e\,s}$ not as a function of Γ but rather as a function of $a = R_1/R_{\text{ant}}$. We can do this by returning to the previous general equation, $\Delta\sigma_{e\,s} = f(\Gamma)$, and replacing Γ with its value as a function of a. Given that

$$\sigma_{e\,s} = \frac{4}{(a+1)^2} \sigma_{e\,s\,\text{structural}}$$

we can now calculate $\Delta\sigma_{e\,s} = (\sigma_{e\,s} - \sigma_{e\,s\,\text{structural}})$, for the case of optimal matching:

$$\Delta\sigma_{e\,s} = \sigma_{e\,s} - \sigma_{e\,s\,\text{structural}} = \left[\frac{4}{(a+1)^2} - 1\right] \sigma_{e\,s\,\text{structural}}$$

Now we are 'theoretically' ready to apply the back scattering principle to RFID technology, but first of all we had better confront the cold hard facts of life, if only for a few moments.

9.2.5 After the Theory: The Cold Hard Facts of RFID at UHF and SHF!

We have now concluded our study of the theory of $\sigma_{e\,s}/\text{RCS}$ and other forms of $\Delta\sigma_{e\,s}$ and ΔRCS, which is absolutely necessary, but not sufficient. Everything I have said about the variations of

$\sigma_{e\ s}$/RCS relates to the best possible world, in which the tag can effortlessly handle very small fields E and H (when the tag is a long way from the base station) and very large fields E and H (when the tag is located very near the base station).

In fact, if the tag is going to be able to operate correctly at both long and short distance, we must include a new element in the integrated circuit, namely a 'shunt regulator', to ensure that excessive voltages do not appear across its terminals in the presence of strong fields (I have already mentioned this extra element in the first part of the book when discussing dissipation).

Everything I have discussed over the many pages above relates to the phase of operation when this regulator is inactive–which is important, since it defines the maximum operating distance of the system. As the tag approaches the base station, the shunt regulator increasingly comes into action in parallel at the input of the integrated circuit, thus mismatching the load from the radiation resistance of the antenna R_{ant}, causing a displacement of the reference point (where there is no modulation), which therefore approaches the vertical axis of the curve, making the value of ΔRCS smaller than before. This makes it harder to guarantee the variation ΔRCS for operation in proximity.

It is worth examining these matters in detail for remotely powered tags (used in most applications, therefore mainly with $\Gamma_1 = 0$), as this will help us understand the phenomena occurring in RFID at UHF and SHF. In order to do this, we need to examine the true equivalent electrical circuit of the tag, while also considering the effects of its distance from the base station (see Figure 9.3(a)).

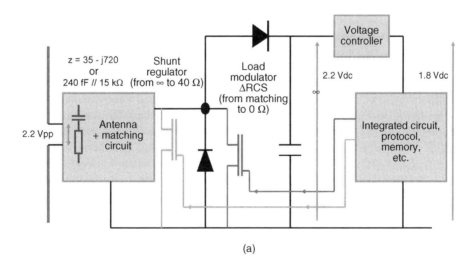

(a)

Figure 9.3 (a) UHF tag: block diagram (whole circuit). (b) Tag in the threshold field. (c) Graph showing the variation of ΔRCS in the case of very weak fields. (d) Optimizing the variation of the load resistance for the highest value of ΔRCS. (e) Tag in the medium field: the shunt regulator starts to conduct normally. (f) Variations of the position of Γ_1 as a function of the field strength. (g) Tag in a strong field: the shunt regulator is fully conducting. (h) The limiting case of the position of Γ_1 in a strong field

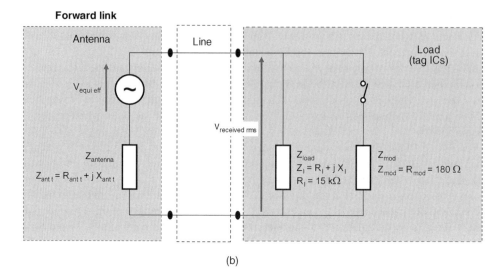

(b)

Figure 9.3   *(Continued)*

Tag Located Very Far from the Base Station

In this case, the tag does not receive (does not capture) enough energy to be remotely powered, so nothing happens–and the problem is solved!

Tag Exactly on Its Operating Threshold (Maximum Distance)

The tag begins to receive exactly the correct amount of power. The shunt regulator does not come into action (R_{shunt} is infinite), meaning that the tag benefits from all the possible incident energy supplied by the wave from the base station (Figure 9.3(b)).

The 'load modulation transistor' comes into action as required when data have to be sent from the tag to the base station. It then switches, thus changing the load impedance of the tag antenna and moving, in the case of a remotely powered tag, from the 'matched' value   ($a_{non\ mod} = R_l/R_{ant\ t} = 1$) (in order to obtain the maximum power) to the value $a_{mod} = (R_{match}$ in parallel with $R_{mod})/R_{ant\ t}$, enabling the initial back scattering area $\sigma_{structural}$ to be modified to a higher value ($\sigma_{e\ s}$/RCS) in order to reradiate more of the incident wave. The graphs in Figure 9.3(c) summarize the variation of ΔRCS in the case of very weak fields.

The resistance R_{mod} must be chosen in such a way that it allows the base station to understand and interpret the data (in the form of power variations) transmitted by the tag, and therefore the value of ΔRCS shown in the diagram must be above a minimum level. Consequently, there is a maximum ohmic value of $R_{mod\ max}$, and therefore of a_{mod} and Γ_{mod}. An example of a calculation is given below for UHF applications according to ISO 18000-6 (UHF) and 18000-4 (2.45 GHz), in other words with ΔRCS_{min} of 50 cm^2 (see the example below).

Of course, we could consider making life simpler by having R_{mod} equal to 0 Ω, thus immediately providing (Figure 9.3(d)) the greatest possible variation of the area ΔRCS, provided that this has no effect on the energy consumed by the tag – but that is another story!

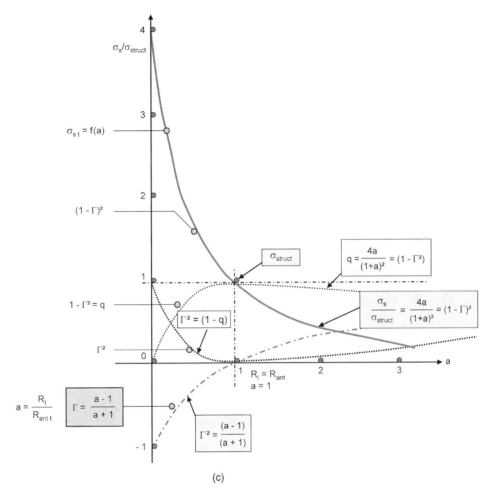

(c)

Figure 9.3 *(Continued)*

Example: $\Delta\sigma_{e\,s}$ and Conformity with ISO 18000-6 in the Threshold Field

To ensure high compatibility of tags and base stations, the 'Tag Parameter: 7d' of ISO 18000-6 states that 'the tag ΔRCS (Varying Radar Cross Sectional area) affects system performance. A typical value is greater than $0.005\,\text{m}^2 = 50\,\text{cm}^2$.' This means that, regardless of the frequency within the 860 to 960 MHz band (the mandatory range for conformity with ISO 18000-6), the minimum typical value of $\Delta\sigma_{e\,s}$ is $50\,\text{cm}^2$. Additionally, given that the value of $\sigma_{e\,s\,\text{structural}}$ is $0.214\,\lambda^2$ for a $\lambda/2$ dipole tag antenna with a gain of 1.64, and that

$$\Delta\sigma_{e\,s} = \left[\frac{4}{(a+1)^2} - 1\right]\sigma_{e\,s\,\text{structural}}$$

we can estimate the minimum value of $a = R_l/R_{\text{antt}}$ to satisfy the equation $\Delta\sigma_{e\,s} = 50\,\text{cm}^2$.

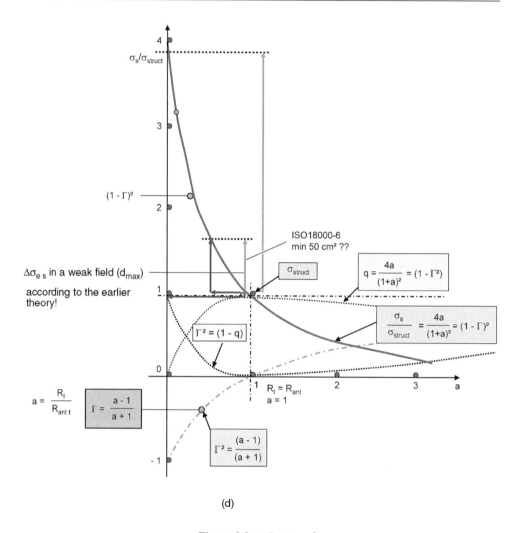

(d)

Figure 9.3   *(Continued)*

(a) If the frequency $= 960\,\text{MHz}$, $\lambda = 0.3125\,\text{m}$ and $\sigma_{e\ s\ structural} = 209\,\text{cm}^2$:

$$50 = \left[\frac{4}{(a+1)^2} - 1 \right] \times 209$$

and therefore

$$a = 0.797$$

(b) If the frequency $= 860\,\text{MHz}$, $\lambda = 349\,\text{m}$ and $\sigma_{e\ s\ structural} = 260\,\text{cm}^2$:

$$50 = \left[\frac{4}{(a+1)^2} - 1 \right] \times 260$$

and therefore

$$a = 0.83$$

where $R_{\text{ant t}} = 73.128\,\Omega$ and $R_1 = aR_{\text{ant t}}$, we can find the maximum value of the minimum value of R_1.

In case (a): $R_{1\,\text{min}} = 58.28\,\Omega$ and therefore $\Gamma_{\text{min}} = -0.113$.
In case (b): $R_{1\,\text{min}} = 60.7\,\Omega$ and therefore $\Gamma_{\text{min}} = -0.093$.

Note

With these very low values of Γ_{min} (about -0.1), we are in the area where Γ can be expected to approach zero. Therefore

$$\Delta\sigma_{\text{e s}} = -2\Gamma\sigma_{\text{e s structural}}$$

and therefore

$$\Delta\sigma_{\text{e s}} = (-2) \times (-0.1) \times 250 = +50\,\text{cm}^2 \qquad \text{Q.E.D.}$$

Tag Entering Its Normal Operating Range

The tag is now correctly supplied with power, even above its strict minimum level. The shunt regulator comes into action (or comes back into action) to limit the voltage at the input of the integrated circuit (Figure 9.3(e)).

For this purpose, the value of R_{shunt} decreases (by a few kΩ) and is connected directly in parallel with the load resistance of the antenna, thus causing a structural mismatch of the antenna, even if there is no modulation by the modulation transistor, and a decrease of the nominal value (without modulation) of a, which thus becomes $a' = (R_1$ in parallel

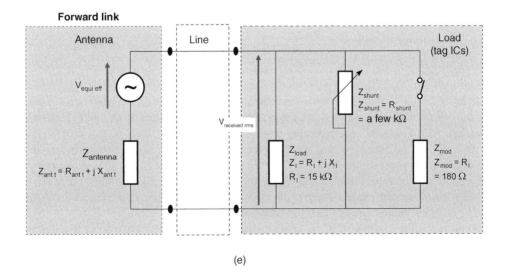

(e)

Figure 9.3 *(Continued)*

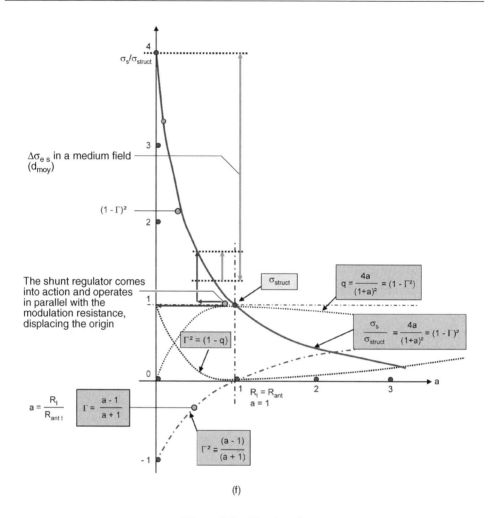

Figure 9.3   *(Continued)*

with $R_{\text{shunt}}/R_{\text{ant t}}$. This is equivalent to the displacement of the initial operating point of $a = 1$, $\Gamma_1 = 0$ (the point corresponding to the operating threshold) towards $a' < 1$ and Γ_1 other than zero, and with a slightly negative value (Figure 9.3(f)).

According to the data to be transmitted to the base station, the modulation transistor of the tag comes into operation and switches the above value (a') (which is no longer the matched value, so the maximum power is no longer received, although this does not upset the tag because it is closer to the base station) to the value R_{mod}, thus modifying the back scattering area (RCS) in order to reradiate more of the incident wave.

As shown in the diagram, this modulation changes the value of a, but also causes a further variation of ΔRCS, which is smaller than the previous variation. Furthermore, the value of ΔRCS decreases as the tag approaches the base station. The question that then arises is: Will we fall below the minimum ΔRCS required by the standard? To answer this question, we must consider the most unfavourable limiting case, which is present when. . ..

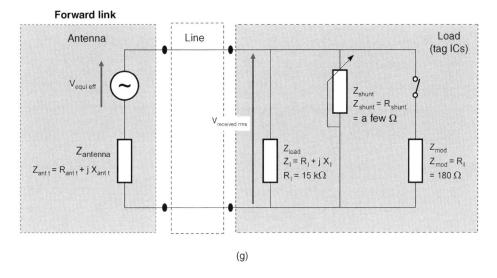

Forward link

(g)

Figure 9.3 *(Continued)*

The Tag is Located Extremely Close to the Base Station

The tag is correctly powered, even well above its absolute minimum level (Figure 9.3(g)). The shunt regulator is in full operation, thus limiting the voltage at the input of the integrated circuit. For this purpose, the value of R_{shunt} becomes very low (at a few ohms to a few tens of ohms) and is connected directly in parallel with the load resistance of the antenna, thus causing a very large structural mismatch of the antenna, even if there is no modulation by the modulation transistor, and therefore a very large decrease of the nominal value (without modulation) of a which thus becomes $a'' = (R_1$ in parallel with $R_{shunt})/R_{ant\,t}$. This is equivalent to shifting the initial operating point (at the operating threshold) of $a = 1$ towards a'', which is located very close to the vertical axis in Figure 9.3(h), and bringing Γ_1 close to the value of -1.

According to the data to be transmitted to the base station, the modulation transistor continues to attempt to operate, switching the value a'' to the value $(R_{adapt}//R_{shunt}//R_{mod} \approx R_{shunt}//R_{mod})$ to modify the back scattering area (RCS) as much as possible in order to reradiate some of the incident wave.

As shown in the diagram, this modulation changes the overall value of a and Γ, but also causes a further variation of ΔRCS, which is even smaller than the previous variation. Being very close to the vertical axis, the value of ΔRCS is even smaller, but must still be greater than the minimum value of ΔRCS required by the standard. This is what limits both the minimum distance and the value of R_{shunt}, and therefore the maximum strength of the field E that the tag can accept. Also, for simple reasons of power dissipation, chip manufacturers specify a maximum input current for their integrated circuits, e.g. 10 or 30 mA eff (refer back to the first part of this book if necessary for more information).

Example of a Real Case and Conformity with ISO 18000-6 in Strong Fields

As we have just seen, as the tag approaches the base station the shunt regulator comes increasingly into operation, and the value of Γ_1 (in the phase without modulation of the antenna

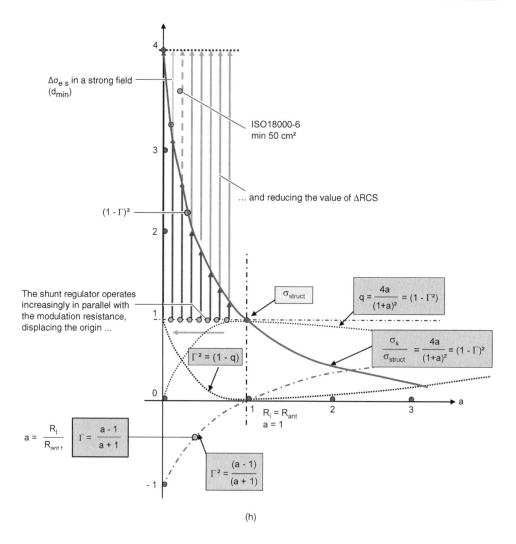

(h)

Figure 9.3   *(Continued)*

load) tends towards −1. If we wish to comply with ISO 18000-6, even in strong fields, the value of $\Delta\sigma_{e\ s\ typical}$ must be at least equal to $50\ cm^2$. Using a $\lambda/2$ dipole in which the value of $\sigma_{e\ structural}$ is approximately $200\ cm^2$ at 900 MHz, and short-circuiting the antenna entirely in the modulation phase, this means that the aim is to obtain $\Gamma_2 = -1$ during the modulation. We can use the equations shown above to determine the critical value of Γ_1 corresponding to this condition:

$$\Delta\sigma_{e\ s} = \sigma_{e\ s2} - \sigma_{e\ s1}$$
$$\Delta\sigma_{e\ s} = \Delta\Gamma[-2 + (\Gamma_2 + \Gamma_1)]\sigma_{e\ structural}$$
$$\Delta\sigma = [-2(\Gamma_2 - \Gamma_1) + (\Gamma_2 - \Gamma_1)(\Gamma_2 + \Gamma_1)]\sigma_{e\ structural}$$

Transferring the values:

$$50 = [-2(-1-\Gamma_1) + (-1-\Gamma_1)(-1+\Gamma_1)] \times 200$$

we obtain a second–degree equation in Γ_1, $[-\Gamma_1^2 + 2\Gamma_1 + 2.75]$, whose roots are

$$\Gamma'_1 = \frac{-2+3.873}{-2} = -0.89$$

$$\Gamma''_1 = \frac{-2-3.873}{-2} = +2.93 \qquad \text{a physically impossible value}$$

i.e. $\Gamma_1 = -0.89$ (tag unmodulated, in a strong field, with shunt fully operational).
We can now determine the value of a as follows:

$$a = \frac{1+\Gamma}{1-\Gamma}$$

$$a = \frac{1-0.89}{1+0.89}$$

$$a = 0.058 = R_1/R_{\text{ant t}}$$

which shows that, in order not to exceed the minimum value required by ISO 18000-6, the minimum value of the shunt resistance must not fall below $R_{\text{shunt min}} = 0.058 \times 73 = 4.24\,\Omega$ (with $R_{\text{ant t}}$ of 73 Ω on matching). The position of this point is shown in Figure 9.3(h) above.

By Way of Conclusion

As shown above, the value of $\Delta\sigma_{\text{e s}}$ decreases as we approach the base station. Precise measurements show that the measured variation of $\Delta\sigma_{\text{e s}}$ is substantially proportional to r^2. Additionally, the variation of the power reradiated by the tag, ΔP, is equal to $s\Delta\sigma_{\text{e s}}$. Now, the incident power flux density s is equal to $P_{\text{EIRP}}/(4\pi r)^2$, indicating that the ΔP reradiated by the tag is substantially constant regardless of the distance at which it operates, this property being mainly due to the presence and operation of the shunt regulator.

This concludes the major theoretical part of this section of the chapter, which is primarily concerned with reflection, absorption and the back scattering transmission principle and their effects. Unfortunately, as you will certainly have realized, the concepts of RCS and $\Delta\sigma_{\text{e s}}$ (or ΔRCS) are rather difficult to grasp, and the values involved are not very easy to measure (something that I have not discussed as yet). This all seems like a lot of work for nothing. It requires a certain amount of courage for the author to confess this so fully in his book!

Let me be quite honest with you. As I have shown above, when the tag circuit modulates the value of RCS, this is done by using a transistor operating in switch mode, in other words on an 'on/off' basis, and therefore using square wave signals (with or without subcarriers; see Chapter 16), which, when antenna mismatching takes place, produce a frequency spectrum including sidebands that are located on either side of the carrier frequency and which thus represent the modulating signal.

Most of the energy of the signal reradiated by the tag and representing the transmitted data is therefore in these sidebands, which is also the location of the power present in the return signal. If we keep strictly to the conceptual aspect of the RCS (measurement of the return power contained in the incident carrier only) we may find it very difficult to recover the signal! Also, we can use return modulation of the Manchester subcarrier coding (SCM) or BPSK type, as for RFID operating at 13.56 MHz (ISO 14443 and 15693), in an attempt to escape from the use of the unwieldy carrier signal for the amplification and demodulation of the very weak return signal. All this will be discussed in detail in Chapter 19, which is specifically concerned with the architecture of RFID base stations for use at UHF.

Example of a Method for Measuring ΔRCS

Returning to our topic, we can obtain the value of $\Delta\sigma_{e\,s}$ (often called ΔRCS in the literature) by, for example, using the measurement set-up shown in Figure 9.4, in which a base station supplies/transmits a constant isotropic power $P_{b\ \mathrm{EIRP}} = P_{\mathrm{cond}}G_{\mathrm{ant\ bs}}$. The power flux density radiated by the base station and present at the tag, at a distance r_1 from the base station, is, as expected,

$$a = \frac{P_{\mathrm{EIRP}}}{4\pi r^2}$$

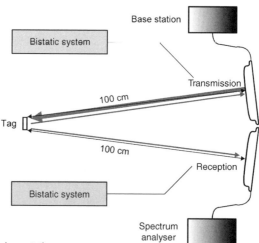

P_1	conducted power from the base station
G_s	gain of the transmitting antenna of the base station
R_1	distance between the tag and the transmitting antenna of the base station
R_2	distance between the tag and the measurement receiver antenna
G_R	gain of the receiving antenna of the measurement receiver
$P_{s,1}$	power measured in the first side band of the spectrum
λ	wavelength of the carrier wave

Figure 9.4 Method of measuring ΔRCS

Let $P_{s\ tag\ EIRP}$ denote the (difference in) global power EIRP reradiated by the tag when the return wave is modulated by modulation of the tag impedance using a square wave signal with a peak amplitude h. This is analysed in a conventional way into the Fourier series

$$f(x) = \frac{4h}{\pi}\left(\cos x - \frac{\cos 3x}{3} + \frac{\cos 5x}{5} - \cdots\right)$$

indicating that the amplitude of the first harmonic (the fundamental frequency) of the function that produces it is greater than $4/\pi$ ($=1.27$) at the initial value h of the square wave signal. Also, assuming that this square wave signal creates an amplitude modulation (AM) of the incident UHF/SHF carrier, the two reradiated sidebands created by this modulation support the signal.

Given that:

- U_{max} is the maximum amplitude in the phase of dynamic modulation of the tag;
- U_{min} is the minimum amplitude in the phase of dynamic modulation of the tag;
- U_b is the amplitude of the carrier present when there is no dynamic modulation of the tag;
- U_c is the sum of all the amplitudes of the signals making up the square wave;

we can write

$$(a)\ U_{max} = U_c + 2U_h$$

$$(b)\ U_{min} = U_c - 2U_h$$

and $U_{min} = 0$, when we perform $100\,\%$ amplitude modulation (ASK). If we now simplify matters by identifying the real square wave signal U_h with its first harmonic, and call the amplitude of its first harmonic (fundamental) $U_{s,1}$, Fourier analysis gives us

$$U_h = U_{s,1} \times \frac{\pi}{4}$$

and if we transfer this value into the above equations we obtain

$$U_{max} = U_c + 2U_{s,1} \times \frac{\pi}{4}$$

$$U_{min} = U_c - 2U_{s,1} \times \frac{\pi}{4}$$

Since U_{min} is equal to 0 in amplitude modulation (AM) of the $100\,\%$ ASK type, we can combine the last two equations to give

$$U_{max} = U_{s,1} \times \pi$$

This voltage-based description can be rephrased in terms of power (proportional to the square of the voltage). This entails a relationship such that the power reradiated by the tag (and therefore received by any receiver) P_{max} corresponding to U_{max} is equal to

$$P_{max} = P_{s,1} \times \pi^2$$

ΔRCS measurement spectrum

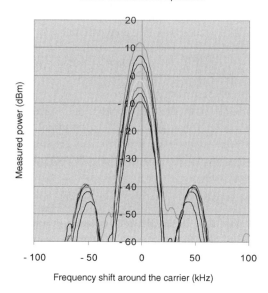

Figure 9.5   Example of a spectrum of the reradiated power

where $P_{s,1}$ is the power contained in the first sideband reradiated by the tag, which is about 10 times greater than that expected from a purely static (or slow) modulation of the tag impedance, provided that the return signal is only read by a detector whose analysis aperture has a narrow bandwidth (a few kHz) to ensure that only the sideband(s) due to the modulation signal are observed. Figure 9.5 provides an idea of the spectrum reradiated by the tag as a result of modulation by a conventional square wave signal.

If we only consider the reradiated power $P_{s,1}$ contained in the first harmonic of the spectrum, this will act as an equivalent isotropic transmission source for the back scattering signal, $P_{s,1\ \text{tag EIRP}}$, part of which, P_{recept}, is recovered (in the first harmonic of the spectrum) at the receiver with a gain of $G_{\text{antrecept}}$ located at a distance r_2, according to the Friis equation:

$$P_{\text{recept}} = P_{s1\ \text{tag EIRP}}\left(\frac{\lambda}{4\pi r_2}\right)^2 G_{\text{ant recept}}$$

and therefore

$$P_{s1\ \text{tag EIRP}} = P_{\text{recept}}\left(\frac{4\pi r_2}{\lambda}\right)^2 \frac{1}{G_{\text{ant recept}}}$$

By definition, the variation of the RCS of the tag, ΔRCS, representing the ratio between retransmitted power and the incident power flux density will be

$$\Delta\text{RCS} = \frac{P_{s\ \text{modul}} - P_{s\ \text{nonmodul}}}{s} = \frac{P_{\text{max}}}{s} = \frac{P_{s1\ \text{tag EIRP}} \times \pi^2}{s}$$

Combining the last of the above equations, we finally obtain

$$\Delta\sigma_{e\,s} = \Delta\text{RCS} = P_{\text{recept}} \times \pi^2 \frac{(4\pi)^3 r_2^2 r_1^2}{\lambda^2} \frac{1}{P_{\text{bs EIRP}} G_{\text{ant recept}}}$$

where P_{cond} is the conducted power of the base station and G_{antbs} is the gain of the base station transmitting antenna:

$$P_{\text{bs EIRP}} = P_{\text{cond}} G_{\text{ant bs}}$$

r_1 is the distance between the tag and the transmitting antenna of the base station, r_2 is the distance between the tag and the antenna of the measuring receiver, $G_{\text{ant recept}}$ is the gain of the receiving antenna of the measuring receiver, P_{recept} is the power received and measured in the first sideband of the spectrum and λ is the wavelength of the carrier wave. By applying this last formula to the measured value P_{recept}, we can find the value of $\Delta\sigma_{e\,s}$ that can be used in practice.

Note that all of these proposed measurement methods (Philips SC) have been accepted as the basis for the measurement of ΔRCS in ISO 18047-6 entitled 'Conformance Tests' for RFID at UHF.

For information, an example of values of $\Delta\sigma_{e\,s}$ (measured and then calculated) as a function of the power contained in the first harmonic of the switched signal is shown in Figure 9.6.

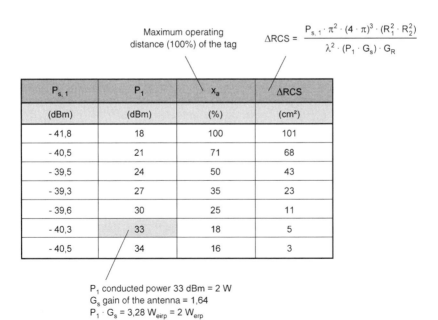

Maximum operating distance (100%) of the tag

$$\Delta\text{RCS} = \frac{P_{s,\,1} \cdot \pi^2 \cdot (4 \cdot \pi)^3 \cdot (R_1^2 \cdot R_2^2)}{\lambda^2 \cdot (P_1 \cdot G_s) \cdot G_R}$$

$P_{s,\,1}$	P_1	X_a	ΔRCS
(dBm)	(dBm)	(%)	(cm²)
- 41,8	18	100	101
- 40,5	21	71	68
- 39,5	24	50	43
- 39,3	27	35	23
- 39,6	30	25	11
- 40,3	33	18	5
- 40,5	34	16	3

P_1 conducted power 33 dBm = 2 W
G_s gain of the antenna = 1,64
$P_1 \cdot G_s$ = 3,28 W_{eirp} = 2 W_{erp}

Figure 9.6   Example of measured and calculated values of $\Delta\sigma_{e\,s}$

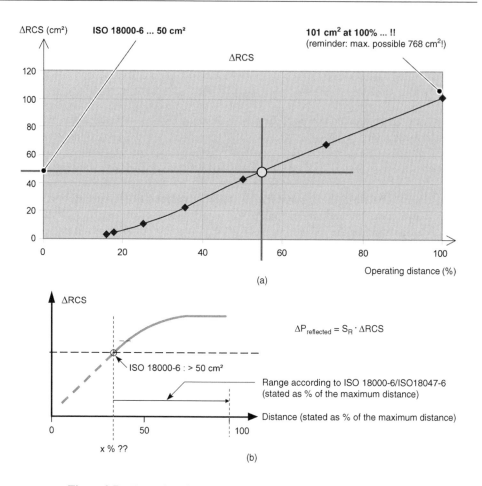

Figure 9.7   Examples of measured values of $\Delta\sigma_{e\ s}$ in commercial tags

As shown above, the value of $\Delta\sigma_{e\ s}$ becomes increasingly difficult to determine as we approach the base station. 'Up to what point?' you may well ask. Figure 9.7 shows a few 'bad examples'. Note two remarkable facts about this diagram:

- The horizontal axis is graduated according to the maximum distance for operation (100 %) of the tag.
- The diagram shows the value of $\Delta\sigma_{e\ s\ min}$, which must be conformed with according to ISO 18000-6.

To finish with this subject (without, hopefully, finishing you off), you should always remember that, as mentioned above, because of the necessary presence of a regulator in the tag to enable it to operate correctly in weak (far) fields and strong (very near) fields, the return modulation index will also depend on the distance, and therefore it is not as simple as it may appear to ensure a minimum value of ΔRCS! There is plenty of work to be done!

Developers of UHF and SHF base stations have therefore familiarized themselves with all these methods of signal amplification, selection and processing. If this is not your main area of expertise, I suggest that you consult the many specialist books about HF signals – or, if all else fails, e-mail the author!

The 'Radar' Equation

Now let us consider the second point, namely the radar equation (see the preceding chapter).

Power Reradiated by the Tag and Power Received by the Base Station
Remember that, as a general rule,

$$\Delta\sigma_{e\,s} = \Delta\Gamma\,[\Delta\Gamma + 2(\Gamma_1 - 1)]\sigma_{e\,s\,\text{structural}} = f(\Delta\Gamma \text{ and } \Gamma_1)$$

the value of which depends on both the variation of $\Delta\Gamma$ and the initial point of this variation Γ_1. Therefore, when the load impedance of the tag antenna circuit is modulated, the difference in power reradiated by the tag, ΔP_{back}, between the 'nonmodulation' and 'modulation' phases will be

$$\Delta P_{\text{back}} = P_{\text{back mod}} - P_{\text{back nonmod}}$$
$$\Delta P_{\text{back}} = \Delta\Gamma[\Delta\Gamma + 2(\Gamma_1 - 1)]P_{s\,\text{structural}}$$

and the difference in power received at the base station (allowing for the gain of its antenna) will be

$$\Delta P_{\text{back received}} = \Delta\Gamma[\Delta\Gamma + 2(\Gamma_1 - 1)]P_{s\,\text{structural}} \times \text{attenuation} \times G'_{bs}$$

$$\Delta P_{\text{back received}} = \Delta\Gamma[\Delta\Gamma + 2(\Gamma_1 - 1)]P_{bs\,\text{EIRP}} \frac{\lambda^2}{(4\pi r)^2} G^2_{\text{ant t}} \left(\frac{\lambda}{4\pi r}\right)^2 G'_{bs}$$

Given that $P_{bs\,\text{EIRP}} = P_{bs\,\text{cond}}\,G'_{bs}$:

$$\Delta P_{\text{back received}} = P_{bs\,\text{cond}}G^2_{bs}G^2_{\text{ant t}}\left(\frac{\lambda}{4\pi r}\right)^4 \Delta\Gamma[\Delta\Gamma + 2(\Gamma_1 - 1)]$$

This general equation enables us to design the receiving stages of the base station to make them capable of receiving and detecting the small fraction of power ΔP_{back}, which is reradiated by the tag via the variation of the area ΔRCS (see the example in the preceding chapter).

9.3 Appendix: Summary of the Principal Formulae of Chapters 7, 8 and 9

Antenna Gain
For a Hertzian dipole: gain $= 1.5$, or in dB $= 10\,\log(1.5) = 1.76$ dB.
For a Hertzian $\lambda/2$ dipole: gain $= 1.64$, or in dB $= 10\,\log(1.64) = 2.14$ dB.

Power

$$P_{\text{EIRP bs}} = P_{\text{cond bs}}G_{\text{ant bs}}$$
$$P_{\text{EIRP}} = 1.64\,P_{\text{ERP}}$$

Power Flux Density Produced by the Base Station

$$s = |S| = \frac{P_{out}G_{ant\ bs}}{4\pi r^2} = \frac{dP}{d\sigma} \qquad \mathrm{Wm}^{-2}$$

$$s = \frac{P_{bs}G_{ant\ bs}}{4\pi r^2} = \frac{P_{EIRP}}{4\pi r^2} \qquad \mathrm{Wm}^{-2}$$

Effective Area of the Tag

$$\sigma_{e\ t} = \frac{\lambda^2}{4\pi}G_{ant\ t} \qquad \mathrm{m}^2$$

Power Received by the Tag

$$P_t = \sigma_{e\ t}s \qquad \mathrm{W}$$

The Friis Equation

$$P_t = P_{bs}G_{bs}\left(\frac{\lambda}{4\pi r}\right)^2 G_{ant\ t} \qquad \mathrm{W}$$

$$P_t = P_{EIRP\ bs}\left(\frac{\lambda}{4\pi r}\right)^2 G_{ant\ t} \qquad \mathrm{W}$$

Attenuation Coefficient (in Air)

$$\frac{1}{[v/(4\pi r)]^2} = \mathrm{att} = \text{attenuation coefficient}$$

$$\mathrm{att}\ (\text{in dB}) = -147.56 + 20\log f + 20\log r \qquad \text{where } f \text{ is in Hz and } d \text{ is in m}$$

Power Reflected by the Tag

$$P_s = \sigma_{e\ s}s$$

Effective Area, or Radar Cross-Section

$$\sigma'_{e\ s} = \frac{\lambda^2 G_{ant\ t}^2}{4\pi} \qquad \mathrm{m}^2 \text{ with tag tuned}$$

$$\sigma'_{e\ s} = \frac{\sigma_{e\ t}}{2}G_{ant\ t} \qquad \mathrm{m}^2 \text{ with tag tuned}$$

Power Reflected/Scattered/Reradiated by the Transponder

$$P_s = \frac{P_{bs}G_{bs}}{2}\left(\frac{\lambda}{4\pi r}\right)^2 G_{ant\ t}^2 \qquad \mathrm{W}$$

Power Flux Density Reradiated by the Tag

$$s_{back} = \frac{P_{bs}G_{bs}}{2} \frac{\lambda^2}{(4\pi)^3 r^4} G^2_{ant\ t} \qquad W$$

Return Power Received by the Base Station (Three Ways of Writing the Same Equation)

$$P_{back} = P_{EIRP\ bs}\ G_{bs}G^2_{ant\ t}\left(\frac{\lambda}{4\pi r}\right)^4$$

$$P_{back} = P_{EIRP\ bs}G_{bs}\sigma_{e\ s}\frac{\lambda^2}{64\pi^3 r^4}$$

$$P_{back} = P_{EIRP\ bs}\left(\frac{\lambda}{4\pi r}\right)^2 G_{ant\ t}G_{ant\ t}\left(\frac{\lambda}{4\pi r}\right)^2 G_{bs}$$

Merit Factor of the Tag

$$\frac{\Delta\sigma_{e\ s}}{\sigma_{e\ s}} = \Delta\Gamma[\Delta\Gamma + 2(\Gamma_1 - 1)] \qquad \text{where } \Delta\Gamma = (\Gamma_2 - \Gamma_1)$$

10

RFID Case Studies Summarizing the Preceding Chapters

In this chapter, which concludes the second part of the book, I shall examine two typical examples of the estimation of link budgets that are frequently encountered in RFID applications operating at UHF and SHF.

10.1 Case 1: Application to a 'Remotely Powered Passive Tag'

These cases are characteristic of RFID applications for reading labels applied to packages and for the identification and monitoring of boxes, packaging cases or pallets.

Example 1a. In the first example, at 868 MHz, we shall calculate the global link budget of a system using a 'passive tag' (in fact, this will be done independently of its power supply device), operating at a distance of 4 m in a 'free field'.

Example 1b. In the second example, again relating to the use of a 'remotely powered passive tag', we shall calculate the power (EIRP) required to enable the base station to radiate over a given communication distance. To provide a little variation, in this example the frequency is 2.45 GHz and the operating distance is 1 m.

Of course, you are free to make whatever changes you like in the values shown in the final tables.

10.2 Case 2: Application to a 'Battery-Assisted Passive Tag'

The cases examined in the following text are characteristic of 'very long distance' RFID applications, which is why they feature battery-assisted systems, which still operate with a passive return link: in other words they use 'battery-assisted passive tags'. An example is RF identification based on the reading of tags placed on 20 foot steel shipping containers stored on quaysides, or access badges (incorporated in vehicles) for entering garages, etc.

In this case we shall calculate the transmission power that the base station requires in order to provide an operating distance of about 15 m. Again, you are free to make whatever changes you like in the values shown in the final table.

RFID at Ultra and Super High Frequencies: Theory and Application   Dominique Paret
© 2009 John Wiley & Sons, Ltd

10.3 Examples 1a and 1b: Application to a 'Remotely Powered Passive Tag'

This specific example is dedicated to all dreamers, visionaries, and other gullible and innocent people who often fall beneath the spell of enticing and fantastic technical and financial claims, which, regrettably, are often trotted out by the press and certain businesses to create astonishing headlines along the lines of 'Nobody will shave in the future!' Unfortunately, any true professional in this field will tell you that physical realities cannot be altered, and neither can the technology and the regulations!

10.3.1 Example 1a

To give you a picture of this 'harsh reality', here is a typical, professional and realistic example of the preliminary theoretical calculations[1] that need to be carried out for the estimation of the principal parameters for the operation of an RFID application at UHF at a distance of 4 m, such as the monitoring of labels on pallets or on packaging cases, conventionally used to resolve problems of traceability in supply chain management. This is a generalized example, and you are very welcome to create your own Excel table on your PC and modify the values of the UHF or SHF frequencies, ranges, etc.

This example, developed in accordance with one of the values shown in European Regulation ERC 70-03, Annex 11 (applicable throughout Europe) for SRDs (short range devices) also allows for conformity with ETSI 300 220-1. Again, you are free to include the values in ETSI 302-208 relating to RFID for Europe (subject to local practice and LBT) or those in FCC 47 Part 15 for the USA and Canada (see Chapter 16 on local regulations).

Our example provides valuable information on the maximum power consumption that the tag must not exceed for a feasible application in the remotely powered mode, together with the constraints relating to the forward and return link parameters (the forward + return link budget) for the tag operating 'passively' in the back scattering mode. This example will be supplemented in our study of the specific problems associated with the remote powering of the tag in Example 1b.

The Assumptions Underlying Example 1a

Table 10.1 shows the assumptions taken in Example 1a. For details of the everyday reality, including nominal operating frequencies, tolerances, reflection, multiple reflection, absorption due to the environment, etc., see Chapter 7. Roughly speaking, you should divide the operating distance by 2!

Results

The results are shown in Figure 10.1.

---

[1] It is worth knowing that, at UHF and SHF, leaving aside the application environment, the experimental results are not very different from the theoretical values of the signals and that these preliminary calculations generally give us a good idea of the reality, in what is known as 'free space'.

Table 10.1

Operating frequency	f	868 MHz	–
Wavelength	λ	34.6 cm	–
Proposed operating distance	r	4 m	–
Maximum conducted power supplied by the RF amplifier	$P_{cond\ max}$	205 mW	23.14 dBm
Gain of the base station antenna	G_{bs}	4	6 dB
$P_{ERP\ max}$ authorized (ERC 70-03)	P_{ERP}	500 mW	27 dBm
Gain of the tag antenna: $\lambda/2$ dipole	$G_{ant\ t}$	1.64	2.15 dB
Radiation resistance	$R_{ant\ t}$	73 Ω	

Summary of forward and return links and operation in back scattering mode in free space, not allowing for reflection, absorption and tolerances		Parameter symbol	Equation	Units	Absolute values	Relative Units values
Examples based on the NXP U_code HSL circuit						
System data						
Operating frequency		f		MHz	868,00	- -
Wavelength of the transmitted wave				cm	34,56	- -
Proposed operating distance		r		m	4,00	- -
Forward link	**Base station transmission ==> tag**					
Base station transmission						
P_ERP_max authorized in Europe	ERC 70 03 - SRD non specific	P_ERP		mW	2000,00	33,01 dBm
P_EIRP	max. equivalent to P_ERP	P_EIRP	1,64 x P_ERP	mW	3280,00	35,16 dBm
Gain of the base station antenna	slightly directive antenna	G_bs			4,00	6,02 dB
Maximum conducted power of the amplifier		P_cond	P_EIRP / G_bs	mW	820,00	29,14 dBm
Power (EIRP) radiated by the base station		P_bs_EIRP				
Standing wave ratio of the base station		VWSR				
Medium						
Medium					air	
Attenuation of air at @f and @r		Att			21145,45	43,25 dB
Tag						
Integrated circuit					U_code HSL	
Parameters of the input circuit resistance		R_I		ohm	35,00	
resistance		X_I		ohm	-720,00	
Quality factor of the integrated circuit only		Q_ic			20,57	
Minimum power consumption of the integrated circuit		P_t_min		µW	35,00	
Tag load quality factor (tag matched for best matching)		Q_tag			11,00	
Power flux density near the tag		s		mW/m²	16,31	
Tag antenna						
Gain $\lambda/2$ dipole		G_t			1,64	2,15 dB
Radiation resistance		R_ant_t		ohm	73,00	
Effective length						
Equivalent area of the tag		se_t		cm²	156,00	
Power received by the tag (Friis equation)		P_t		µW	63,70	-11,96 dBm
Voltage at the terminals of the unloaded antenna (no-load condition)		V_eff		V		
Voltage at the terminals of the integrated circuit (during matching)		V_reçue		V		
Direct voltage of rectifier diodes		V_direct		V		
Minimum electrical field required		E_min		V/m		
Max. possible operating distance						
Polarization max. angle				°	60,00	
Polarization				%	50,00	
Mismatch		q		%	no	
Losses				%		

(a)

Figure 10.1 Summary of the forward and return links and operation in the back scattering mode

Return link	tag ==> base station						
Tag	**Non-modulation phase**						
	Equivalent radar cross section (RCS) of tag (structural area)	se_structural		cm²	128,00		
	Load impedance Z_l = Z_ic						
	resistance	R_l		ohm	35,00		
	reactance	X_l		ohm	-720,00		
	Quality factor of IC load	Q			20,57		
	Voltage present at the input of the integrated circuit	Q x V_eff		V			
	γ						
	Power re-radiated by the tag — during pure carrier (tuned)	P_structural		µW	50,00	16,99	dBm
	Modulation phase						
	Load impedance during modulation	R_l mod		ohm	0, court circuit		
	Resistance mismatch ratio R_l/R_ant	a					
	Radar cross section (RCS) during modulation	s_s_mod		cm²	512,00		
	γ			W		########	dBm
	Merit factor of the modulation — delta RCS	delta s		cm²	384,00		
	Power re-radiated by the tag — during the modulation of the tag	P_s		mW			
	ΔP s			W		########	dBm
	Re-radiated power flux density	s_back		W/m²			
Medium							
	Attenuation of air at @t and @r	Att			21145,45	43,25	dB
Base station reception							
	Power flux density received at the base station	se_bs		W/m²			
	Base station antenna gain	G_bs'			4,00	6,02	dB
	Effective area of the receiving antenna of the base station.			cm²			
	Power received at the base station	P_back_bs		mW	0,00	-50,00	dBm
	unmodulated			nW	10,00	-50,05	dBm
	modulated				40,00		
Global							
	Global link balance	P_back/P_EIRP			83000000,00	79,19	dB

(b)

Figure 10.1 (*Continued*)

Conclusions of Example 1a

This example clearly shows that, in order to implement the planned application, with a free-field distance of 4 m:

- The tag must not consume more than 63 µW (for information, most commercially available UHF tags have a power consumption of 10 to 50 µW).
- The return level of the signal is −50 dBm, which is very easily handled and poses no major problem for the design of the base station circuits (see Chapter 19 which is entirely devoted to this subject).

10.3.2 Example 1b

Regardless of whether or not we know if the return signal can or will be detected by the base station circuits, let us calculate the power EIRP that the base station must develop in order to provide the correct remote power supply to the tag.

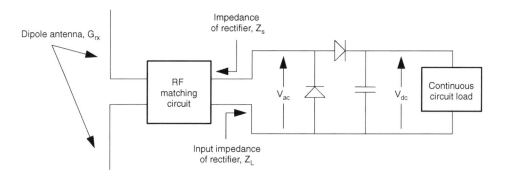

Figure 10.2   Equivalent circuit of the tag

Clearly, this problem can only be stated – and therefore resolved – if we know the supply voltage required for the correct operation of the integrated circuit to be implemented in the tag, as well as its power consumption (power and/or current).

Conversion of the RF Field to a Continuous Supply Voltage

Figure 10.2 shows one of the usual ways of constructing a system for recovering a continuous voltage based on the reception of an electromagnetic radio frequency field. This system, consisting of a capacitor and two diodes for rectifying the alternating voltage V_{ac} using a voltage doubler system, is used here to recover a filtered continuous voltage V_{dc} that is sufficient to supply the continuous load represented by the integrated circuit(s) of the tag. Naturally, if this is not the case, you should feel free to adapt the following text to your own circumstances, which should not cause you any great problems.

The antenna dipole, with a gain of $G_{ant\ t}$, collects the received RF power and delivers it through an impedance matching circuit (which is assumed to have no active loss, and therefore to be purely reactive) so as to maximize the supply of RF power to the rectifier circuit while optimizing the alternating value V_{ac} present at the terminals of the rectifier circuit.

As I have mentioned many times, this maximum power is produced when the source impedance Z_s of the matching circuit is equal to the complex conjugate value of the complex input impedance of the rectifier circuit Z_l, i.e. $Z_s = Z_l^*$, in other words at the moment when the global impedance is purely real and the maximum power in watts is recovered in the useful load.

Important notes
The calculations below are independent of the complex impedance of the tag antenna, since it is assumed that the impedance matching circuit matches the antenna impedance perfectly to the impedance represented by the voltage doubler rectifier circuit. It is also assumed that the tag (i.e. its integrated circuit) stops working when the continuous voltage applied to it falls below a threshold voltage V_{th}, i.e. when $V_{dc} < V_{th}$, corresponding to a minimum recovered power $P_{th\ supp}$.

Calculation of the Minimum Radiated Power EIRP Required from the Base Station

For these calculations, the following definitions are used:

- λ is the wavelength;
- r is the distance between the base station and the tag;

- $P_{\text{cond bs}}$ is the power of the base station transmitter;
- $G_{\text{ant bs}}$ is the gain of the base station antenna;
- P_t is the RF power received by the tag antenna (receiver);
- $G_{\text{ant t}}$ is gain of the tag antenna;
- Z_s is the complex impedance of the rectifier circuit;
- Z_l is the complex input impedance of the voltage doubler circuit;
- R_l is the real part of the input impedance of the voltage doubler circuit;
- $V_{\text{ac rms}}$ is the rms alternating voltage applied to the rectifier circuit;
- $V_{\text{ac peak}}$ is the peak alternating voltage applied to the rectifier circuit;
- V_{dc} is the continuous voltage obtained after rectification;
- V_{on} is the direct voltage of the rectifier diode;
- V_{th} is the continuous threshold voltage for the operation of the tag.

Assuming that the tag antenna is perfectly aligned with the base station antenna, the power received by the tag, P_t, is as follows (see the Friis equation in Chapter 6):

$$P_t = G_t G_{\text{bs}} P_{\text{bs}} \left(\frac{\lambda}{4\pi r} \right)^2 \tag{10.1}$$

As mentioned above, the maximum voltage developed across the terminals of the load Z_l (which forms the input impedance of the voltage doubler circuit) is produced when the matching circuit (purely reactive and therefore not dissipative) provides a source whose impedance is the 'conjugate' of the load circuit and whose available received power P_t is dissipated in the real part of the voltage doubler load. In this best case, the current in the load is

$$I_l = \sqrt{\frac{P_t}{R_l}}$$

$$P_t = R_l I_l^2$$

The amplitude of the r.m.s. voltage at the terminals of the load, $V_{\text{ac amp}}$, is equal to the product of the current and the impedance represented by the doubler circuit:

$$V_{\text{ac rms}} = |Z_l| \sqrt{\frac{P_t}{R_l}}$$

The peak amplitude $V_{\text{ac peak}}$ of the alternating voltage is related to the r.m.s. value by the well-known equation:

$$V_{\text{ac amp}} = \sqrt{2}\, V_{\text{ac rms}} = |Z_l| \sqrt{\frac{P_t}{R_l}} \tag{10.2}$$

and, allowing for the direct voltage drops of the diodes, the rectified continuous voltage at the output of the doubler will be

$$V_{\text{dc}} = 2(V_{\text{ac amp}} - V_{\text{on}})$$

This equation shows that the continuous voltage depends on the amplitude of the radio frequency signal (and therefore on the operating distance, via the value of P_t, see Equation 10.1) and on the direct voltage drops of the rectifier diodes.

To enable the tag to work properly, this continuous voltage must be equal to or greater than the voltage threshold V_{th} from which the tag begins to be correctly powered and therefore capable of operation:

$$V_{th} = 2(V_{ac\ amp} - V_{on}) \Rightarrow V_{ac\ amp} = \frac{V_{th}}{2} + V_{on}$$

Transferring the values obtained previously (Equations (10.1) and (10.2)) into the above equation, we find

$$\sqrt{2}|Z_1|\sqrt{\frac{G_t G_{bs} P_{bs}}{R_1}}\frac{\lambda}{4\pi r} = \frac{V_{th}}{2} + V_{on}$$

and consequently the minimum conducted power $P_{cond\ bs}$ that the base station must be able to supply is

$$P_{cond\ bs} = \frac{R_1}{2G_t G_{bs}}\left(\frac{V_{th}}{2} + V_{on}\right)^2 \left(\frac{4\pi r}{\lambda |Z_1|}\right)^2$$

or alternatively

$$P_{EIRP\ bs} = P_{cond\ bs}G_{bs} = \frac{R_1}{2G_t}\left(\frac{V_{th}}{2} + V_{on}\right)^2 \left(\frac{4\pi r}{\lambda |Z_1|}\right)^2 \tag{10.3}$$

The Assumptions Underlying Example 1b

Let us estimate the minimum power EIRP to be supplied by a base station operating at 2.45 GHz to remotely power a tag that is to operate at a distance of 1 m, where:

- $r = 1$ m
- $f = 2.45$ MHz
- $\lambda = 12.4$ cm at 2.45 GHz
- $G_{ant\ bs} = 4$ (directive antenna) or alternatively $= 6$ dB
- $G_{ant\ t} = 1.64$ (dipole antenna) $= 2.14$ dB
- $Z_s = (20 + j180)\ \Omega$ (at 2.45 GHz this gives an inductance L of 11.7 nH)
- $l_1 = (20 + j180)\ \Omega$ (at 2.45 GHz this gives a capacitance C of x fF)
- $R_1 = 20\ \Omega$
- $V_{on} = 0.49$ V
- $V_{th} = 2.2$ V continuous

Results

We obtain the following minimum required base station power $P_{EIRP\ bs}$:

$$P_{EIRP\ bs} = \frac{20}{2 \times 1.64}(1.1 + 0.49)^2 \frac{(4 \times 3.14 \times 1)^2}{\left[0.124 \times [20^2 + 180^2]^{1/2}\right]^2}$$

$$= 4.85\text{W EIRP} = 36.85\text{ dBm}$$

Note
In the chapter about standards and regulation, Chapter 15, I will show you that the example defined above may possibly not be viable (at the present date, without licence) simply because the maximum power EIRP authorized in the USA (with an indoor/outdoor duty cycle of 100%) is 4 W (with a maximum conducted power of 1 W) and is even less in Europe ($P_{EIRP} = 500$ mW with an indoor duty cycle of 15%). Moreover, this power level does not indicate that the return (back scattered) signal can be detected or demodulated by the base station. The only conclusion that we can draw is that the tag will at least be remotely powered. For your information, given that

$$P_t = G_t G_{bs} P_{bs} \left(\frac{\lambda}{4\pi r} \right)^2$$

the value of the power P_t received by the tag would be

$$P_t = 1.64 \times (4 \times 1.218) \left(\frac{0.124}{4 \times 3.14 \times 1} \right)^2$$

and therefore

$$P_t = 778\,\mu W$$

For information, this power represents a power consumption of about 350 µA with a 2.2 V supply.

At present (2009), tags available on the market consume about 80 to 150 µW at 2.45 GHz.

Other Views
To give you an idea of another possible approach, let us look at the same problem in terms of 'dBm'.

(a) We can express the power received at the tag in dBm:

$$P_t = 10 \times (\log 0.778) \text{ dBm}$$
$$P_t = 10 \times \log (7.78 \times 10^{-1}) = -10 + 10 \times (0.89)$$
$$P_t = -1.1 \text{ dBm}$$

or, alternatively, if we work directly in dBm:

$$P_t = G_t G_{bs} P_{bs} \left(\frac{\lambda}{4\pi r} \right)^2$$

and therefore, in the order of appearance of its factors,

$$P_t = (2.14) + (6 + 30.85) + 10\log(9.710^{-5})$$
$$P_t = 2.14 + (6 + 30.85) + (-50 + 9.87)$$
$$P_t = -1.14 \text{ dBm (rounded)}$$

(b) In Chapter 6, I showed that the attenuation of the signal sent by the base station was as follows, at 1 m and at 2.45 GHz:

$$\text{att at 1 m at 2.45 GHz} = 40.3 + 20\log 1 = 40.3 + 20 \times 0 = 40.3 \text{ dB}$$

Given that

$$P_{\text{bs EIRP}} = P_{\text{cond bs}}G_{\text{ant bs}} = 36.85 \text{ dBm}$$

at 1 m, the total available power at the tag antenna ($P_t - G_{\text{ant t}}$) will be

$$P_{\text{bs EIRP}} - \text{attenuation at 1 m} = 36.85 - 40.3 = -3.45 \text{ dBm}$$

Also, the equation (yes, the same one!)

$$P_t = G_t G_{\text{bs}} P_{\text{bs}} \left(\frac{\lambda}{4\pi r}\right)^2$$

indicates that the power we have just calculated, expressed in dBm, is equal to the difference ($P_t - G_{\text{ant t}}$), and therefore, expressed in dBm, it is

$$P_t - G_{\text{ant t}} = P_{\text{bs EIRP}} - \text{attenuation at 1 m}$$
$$-1.28 - 2.14 = 36.85 - 40.3$$
$$-3.42 = -3.45 \text{ dBm (rounded)}$$

The choice of calculation method is up to you.

10.4 Example 2: Application to a 'Battery-Assisted Passive Tag'

Now let us look at the second set of examples. Consider the case where the tag is locally supplied by batteries or accumulators. This is particularly common in applications requiring a 'very long' communication range (about ten to a hundred metres), such as electronic toll systems and the identification of very bulky objects (shipping containers, etc.).

Figure 10.3 shows the standard block diagram of an RFID system with battery-assisted passive tags, used conventionally at UHF and SHF. To allow for the use of simple and inexpensive tags (using only two components, namely the antenna and the diode included in the integrated circuit), the communication between the tag and the base station does not use a transceiver and therefore takes place in the 'passive' (back scattering) mode.

The diode carries out the functions of:

- detecting/demodulating the signal from the base station;

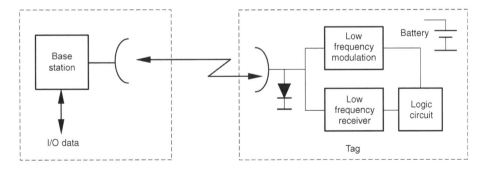

Figure 10.3 Block diagram of an RFID system with battery-assisted passive tags, used conventionally at UHF and SHF

- acting as an element for reradiating the incident signal towards the base station (backscatter tag).

10.4.1 Link Budget for a Battery-Assisted Passive Tag

Where a battery-assisted passive tag is used, the analysis of the link budget is simplified, because we no longer have to consider the remote powering of the tag, but only the fact that:

- On the one hand, the tag must be capable of interpreting the information received from the base station. For this purpose, the power of the signal received at the tag must be such that it exceeds the operating power threshold of the tag, $P_{t\,min}$. This threshold corresponds, on the one hand, to the smallest level that can be detected by the tag and, on the other hand, to a certain minimum power $P_{bs\,th}$ that the base station must transmit in order to ensure that the received power $P_{th}(r)$ is at least equal to $P_{t\,min}$ for operation at the desired distance r.

Warning! Important note

You should note that this power $P_{t\,min}$ is much lower than the power that the whole tag consumes in order to provide all its functions, because the power $P_{t\,min}$ only comprises the power required for the correct wake-up of the input stage, which notifies the tag of the presence of an incident signal from the base station (the order of magnitude is a few nW). In many cases, the detection of this incident signal also acts as the wake-up element and starts all the other functions of the tag, which draw power directly from the battery incorporated in the tag.

You should be careful, therefore, not to confuse $P_{t\,min}$ in the following sections with the value P_t examined in the previous examples, which indicated the global power to be supplied to the tag (order of magnitude of 30 to 80 μW) in a case where the tag did not have an on-board battery.

You should also be careful not to be misled by enticing offers from businesses that are expressly designed to create confusion between these two power levels, only drawing the customer's attention to the better one (i.e. the lower one). It may be very pleasant to dream, but reality can be a shock sometimes.

- On the other hand, the sensitivity of the base station demodulator must be sufficient to process the reradiated/reflected signal received from the tag. To do this, we require a detailed calculation of the whole conventional link budget for the path between the base station and the tag, and for the path of the portion of the wave reradiated/reflected by the tag towards the base station receiver, while naturally allowing for a safety margin.

Minimum Power to be Radiated by the Base Station $P_{bs\,th}$ in Order to Achieve the Tag Operation Threshold

We will start by examining the problem of determining the minimum power EIRP that the base station must deliver in order to exceed the tag's operating threshold, at the distance r at which the tag is required to operate (in other words, the distance at which the tag must be capable of understanding, or at least detecting, the commands from the base station, and activating itself to respond). To do this, before embarking on the calculations themselves, we need to know some specific electrical characteristics of the tag, particularly:

- $G_{ant\ t}$ = gain of the tag antenna
- tag_{sens} = lowest power, in dBm, that can be detected by the tag's detector (the diode)
- S/N_{tag} = signal-to-noise ratio (in dB) for a given bit error rate (BER), e.g. a BER of 10^{-6} (for a given protocol, CRC, etc.)

These parameters can be used to determine the minimum power level $P_{t\ min}$ required for the correct operation of the tag, which in dBm is as follows:

$$P_{tag\ min} = tag_{sens} + \frac{S}{N_{tag}}$$

Knowing this, we can find the minimum threshold of radiated power EIRP, $P_{ns\ th}$ (in dBm) which the base station must deliver in order to make the tag operate at a distance r:

$$P_{bs\ th} = P_{tag\ min} + att(r) - G_{ant\ t}$$

where $att(r)$, in dB, is the attenuation coefficient of the medium between the base station and tag as a function of r:

$$att(r) = 32.5 + 20 \log f + 20 \log r \qquad (\text{where } f \text{ is the GHz and } r \text{ is in m})$$

Transferring the values of the above equations into each other, we obtain the following formula, in dBm:

$$P_{bs\ th} = tag_{sens} + \frac{S}{N_{tag}} + (32.5 + 20 \log f + 20 \log r) - G_{ant\ t} = f(r)$$

These calculations can, of course, be carried out in the other direction.

The Assumptions Underlying Example 2
In line with the previous sections, and using the specific values of a particular tag, we can now calculate, for a given frequency, the power $P_{bs\ th}$ corresponding to the operating threshold of this tag as a function of the desired operating distance r (in m).
   Let us assume that:

- operating frequency = 2.54 GHz
- intrinsic sensitivity of the tag, $tag_{sens} = -50$ dBm (10 nW)
- tag minimum signal-to-noise ratio, $S/N = 17$ dB
- gain of the tag antenna, $G_{ant\ t} = 2$ dB

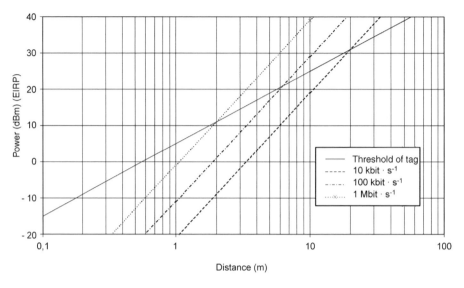

The solid curve indicates the threshold operating power of the tag
The curves in broken lines indicate the link budget of the second part

Figure 10.4   Graph of the variations of $P_{bs\ th} = f(r)$

Results

For an application operating at $f = 2.45$ GHz, the last equation (in dBm) becomes

$$P_{bs\ th} = tag_{sens} + \frac{S}{N_{tag}} + (40.2 + 20\log r) - G_{ant\ t}$$

The graph of $P_{bs\ th} = f(r)$ for this specific case is shown in full in Figure 10.4.

Note

In the example above, if the intrinsic sensitivity of the tag is $tag_{sens} = -50$ dBm, if $10\log P$ (in mW) $= -50$ dBm, then $P = 10$ nW.

For your information, if the impedance $Z = 50\,\Omega$, this power corresponds to the following PD:

$$P = \frac{U^2}{R} \text{ therefore } U = \sqrt{PR}$$

$$U = \sqrt{50 \times 10^{-8}} = 710\,\mu V$$

and the power

$$P_{tag\ min} = tag_{sens} + \frac{S}{N_{tag}} \text{ dBm}$$

$$P_{tag\ min} = -50\,dBm + 17\,dB = -33\,dBm = 10\log(0.5 \times 10^{-3})\,dBm$$

or, in watts,

$$P_{tag\ min} = 500\,nW$$

Global Budget for the Forward and Return Link
Given that, in this case, the tag can understand the incident messages of the forward link, let us examine the global budget of the forward and return link of the reflected wave.

Minimum Radiated Power $P_{rad}(r)$
In order to determine the minimum radiated power $P_{bs\ EIRP\ min}(r)$ (in dBm) that must be delivered by the base station for the desired operating distance r, we start with the following definitions:

- $P_{rx\ min}$ is the minimum power of the signal to be supplied to the receiver, including a margin to achieve a BER of 10^{-6}.
- $att(r)$ is the attenuation coefficient of the path (r) between the base station and tag.
- $G_{ant\ t}$ is the gain of the tag antenna.
- CL_{tag} is the conversion loss of the tag (see the note below).
- G_{rx} is the gain of the (receiving) antenna of the base station.

Note

The parameter CL_{tag} represents the 'conversion loss' of the tag, in other words the ratio (in dB) between the incident power received from the base station and the power reflected/reradiated by back scattering by the tag (of the order of magnitude of $-7\,dB$, i.e. a ratio of about 4.5) (if necessary, refer back to Section 5.4 of Chapter 5, which describes the relationship between the incident power and the variation of reradiated power due to the gain of the tag antenna, in other words the conversion loss of the tag).

The value in dBm of the minimum radiated power that the base station must be able to deliver is determined by the following formula (see the preceding chapter for a proof, if you are unconvinced):

$$P_{bs\ EIPR\ min}(r) = P_{rw\ min} + 2att(r) - 2G_{ant\ t} + CL_{tag} - G_{xx}$$

$$P_{bs\ EIRP\ min}(r) = f(r)$$

As indicated by the factor '2' multiplying the attenuation, this equation allows for the forward *and* return losses due to the back scattering, as described very fully above.

Minimum Value of the Signal Power $P_{rc\ min}$
To complete our calculation, we need to estimate the minimum power of the signal $P_{rx\ min}$ that must be applied to the input of the base station receiver to enable the system to operate correctly. It is expressed as follows:

$$P_{rx\ min} = (kT + BW_{dr}) + NF + \frac{S}{N} + LBM\ dBm$$

In this equation, the sum $kT + \mathrm{BW_{dr}}$ expressed in dBm represents in its 'disguised' logarithmic form the standard product $kT\,\mathrm{BW_{dr}}$ representing the 'noise power' (and therefore shown in W or dBm) present at the input of an amplifier. This can also be written thus:

$$10 \log(P \text{ input noise}) \text{ (in dBm)} = 10 \log (kT\,\mathrm{BW_{dr}}) \text{ (in dBm)} = 10 \log(kT) + 10 \log \mathrm{BW_{dr}}$$

where:

- $k =$ Boltzmann constant, $k = 1.38 \times 10^{-23}\,\mathrm{J\,K^{-1}}$
- $T =$ absolute temperature in degrees Kelvin at $T\,(^{\circ}\mathrm{C}) = 17\,^{\circ}\mathrm{C}$
- $T(\mathrm{K}) = 273 + 17 = 290\,\mathrm{K}$
- kT at $17\,^{\circ}\mathrm{C} = 4.002 \times 10^{-21}\,\mathrm{J}$, which, expressed in dB, gives $10 \log (kT)$

$$d = -210 + 10 \log (4)$$
$$= -204\,\mathrm{dB}, \text{ and, expressed in dBm J, gives} -204 + 30 = -174\,\mathrm{dBm}$$

- $\mathrm{BW_{dr}}$ represents the bandwidth of the receiver, in Hz. Its value is determined by the relation

$$\mathrm{BW_{dr}} = C\,\mathrm{DR}\,\mathrm{Hz}$$

- where C represents the spectral efficiency of the bit coding used, in Hz $\mathrm{bit^{-1}\,s^{-1}}$). This is the ratio of the bandwidth occupied (in Hz) to the communication bit rate (in bits $\mathrm{s^{-1}}$).
- DR $=$ bit rate of the transmitted data, in bits $\mathrm{s^{-1}}$, as follows:

$$\mathrm{BW_{dr}} - 10 \log (C\,\mathrm{DR})\mathrm{dBHz}$$

Note
The physical unit of the product $(kT\mathrm{BW})$ is the J Hz $=$ joule per second $=$ power in watts. The unit of the product (kT) is therefore the joule $(\mathrm{J} = \mathrm{W\,Hz^{-1}})$ which, expressed in dB, becomes dB $\mathrm{Hz^{-1}}$ or, alternatively, if the expression of the power is converted to mW, becomes dBm $\mathrm{Hz^{-1}}$. Therefore, at $17\,^{\circ}\mathrm{C}$, for $1\,\mathrm{Hz} \rightarrow kT = -174\,\mathrm{dBm\,Hz^{-1}}$.

- NF $=$ noise factor of the receiver stage, in dB
- $S/N =$ signal-to-noise ratio of the receiver in dB, for a given error rate, such as BER $= 10^{-6}$ in the example above
- LBM $=$ safety margin, or 'link budget margin' (tolerances, etc.) of the link budget, in dB

If we introduce all these values into the preceding equation, we obtain

$$P_{rx\,\mathrm{min}} = [kT + 10 \log (C\,\mathrm{DR})] + \mathrm{NF} + \frac{S}{N} + \mathrm{LBM}$$

We can finally determine the relationship between the power threshold $P_{\mathrm{bs\,EIRP\,min}}$ of the base station as a function of the distance and the bit rates used by replacing $P_{rx\,\mathrm{min}}$ with its value in

$$P_{\text{bs EIRP min}}(r) = P_{\text{rw min}} + 2\text{att}(r) - 2G_{\text{ant t}} + \text{CL}_{\text{tag}} - G_{\text{rx}}$$

giving the following extended equation:

$$P_{\text{bs EIRP min}}(r, \text{DR}) = [kT + 10\log(C\,\text{DR})] + \text{NF} + \frac{S}{N} + \text{LBM} + 2\text{att}(r) - 2G_{\text{ant}\,t} + \text{CL}_{\text{tag}} - G_{\text{rx}}$$

or, alternatively, in a shorter form:

$$P_{\text{bs EIRP min}} = f(r, \text{DR})$$

The Assumptions Underlying Example 2 (Continued)

At an operating frequency of 2.45 GHz, taking the following (realistic) values for the parameters stated above:

- kT (at $17\,^{\circ}\text{C}$) $= -174\,\text{dBm Hz}^{-1}$
- $G_{\text{ant t}}$ (gain of the tag antenna) $= 2\,\text{dB}$
- CL_{tag} (conversion losses of the tag) $= 7\,\text{dB}$
- $\text{att}(r)$ (attenuation in air at 2.45 GHz) $= 40.2 + 20\log r$
- DR (uncoded data rate) $= 10, 100, 1000\,\text{kbit s}^{-1}$
- C (coding efficiency, Manchester coding) $= 2$
- NF (noise factor of the receiver) $= 10$
- S/N (signal-to-noise ratio of the receiver [BER $= 10^{-6}$]) $= 17\,\text{dB}$
- $G_{\text{ant bs rx}}$ (gain of the antenna of the base station receiver) $= 6\,\text{dB}$
- LBM (link budget margin) $= 6\,\text{dB}$

Example

For a data rate of DR $=$ kbits s^{-1}, the last equation gives

$$P_{\text{bs EIRP min}}(\text{in dBm}) = \{-174 + 10\log(2 \times 100 \times 10^3) + 10 + 17 + 6 + [2(40.2 + 20\log r)]\}$$
$$-(2 \times 2) + 7 - 6$$

$$P_{\text{bs EIRP min}}(\text{in dBm}) = -174 + 3 + 5.0 + 10 + 17 + 6 + 80.4 + 40\log r - 4 + 7 - 6$$
$$= -10.6 + 40\log r$$

(Examples: r at 1 m, $P = -10.6\,\text{dBm}$; r at 10 m, $P = +29.4\,\text{dBm}$; see the curve shown as a chained line in Figure 10.4.)

Results (Continued)

The curves of the equation $P_{bs\ EIRP\ min} = f(r, DR)$, for data rates DR of 10 and 100 kbits s^{-1} and 1 Mbit s^{-1}, are shown in dotted lines in Figure 10.4 for distances from 10 cm to 100 m.

Conclusions of Example 2

With the parameter values chosen and used in the examples above (if you are unhappy with these values, please use your own – the whole purpose of showing you all these equations is to enable you to adapt them to your specific requirements), the curves in Figure 10.4 indicate the minimum power EIRP which the base station must radiate to ensure the correct operation of the system.

These curves are to be interpreted in two different ways:

- Firstly, the power EIRP depends solely on the threshold power of the tag for systems operating and distances of less than approximately 15 m, and also for RFID applications in which the bit rates are lower than 20 kbits s^{-1}.
- Secondly, above this distance, where higher rates are to be used, it is the global link budget that is the decisive factor for the (higher) minimum power EIRP required for the base station.

For example, a power EIRP of $+27$ dBm will give an operating distance of about 8 m for a passive (battery-assisted) tag at a bit rate of 100 kbits s^{-1}. Since the regulations (FCC, ERC 70-03, ETSI, etc.) clearly specify maximum values of EIRP or ERP, it is a simple matter to determine whether the passive tag has to be battery assisted or remotely powered in any planned application.

In conclusion, here is a final example relating to another application, namely the application governed by ISO 10374, which is used worldwide for identifying 20-foot containers, requiring the use of a power of 500 mW EIRP ($+27$ dBm), an operating distance of 13 m and a bit rate of about 65 kbits s^{-1}. These systems operate with a BER (bit error rate) which is better than 10^{-6}. The high level of data security is only achieved at the cost of multiple successive interrogations. However, these systems operate with a high antenna gain and in read-only mode.

It should also be noted that all the above calculations are rather 'optimistic' because they relate to applications using the model of propagation in free space, which is hardly ever the case. In practice, the operating distances will be slightly smaller, owing to the numerous reflections due to the environment.

10.4.2 General Conclusions from these Examples

We have now seen that the operation of a passive tag in the back scattering mode is subject to three conditions:

1. The condition that the threshold power is reached, allowing the tag to interpret commands from the base station.
2. The condition that power is transmitted to the tag, so that it is correctly remotely powered.

3. The condition that the link budget (forward and return) is satisfied according to the minimum detection threshold of the receiving part of the base station.

Provided that conditions 1 and 3 are met simultaneously, we can expect to achieve operation in the 'battery-assisted passive tag' mode. If we can do even better and meet condition 2 (which is dependent on local regulations), we can also hope to operate in the 'remotely powered passive tag' mode!

Now that you are fully up to date with the facts, please feel free to use them as you wish!

Part Three

Communication and Transmission, Baseband Signals, Carrier Modulation and Interleaving

Let me start by pointing out that Chapters 11 to 14 constitute the third part of this book, which can be seen as a single unit. It does not aim to provide a full course in telecommunications (the reader should consult the appropriate textbooks if necessary) but rather a summary of the essential factors to be considered when planning to develop RFID applications. On this point, I will just remind you that RFID systems operating in the UHF and SHF frequency bands use a 'radiation/propagation' mode, not an 'inductive/magnetic coupling' mode, as in the case of 125 kHz and 13.56 MHz, and therefore we must take the standard phenomena of wave propagation into account, by examining and allowing for problems of low S/N ratio, multiple paths, echo, reflection, absorption, noise, interference, jamming, etc.

The previous chapters have already given you many examples of possible RFID applications. Each of these is subject to strict constraints, which require the designer to make specific choices. The aim of these chapters is to make the reader aware of the main parameters and elements to be taken into account during the development of a project, so as to help him to make his choice among the maze of possible solutions, according to the problems arising and the nature of the planned applications.

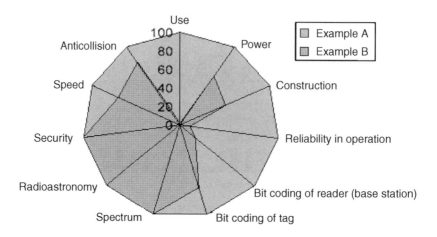

Figure Spider graph representing the performance of RFID systems

The diagram below shows an example of a special form of representation, called a 'spider graph', showing in graphic form the result of an optimal process of determining the global performance of a specific system.

Remaining within the field of RFID at UHF and SHF, the following chapters will examine the reasons for selecting and using particular solutions according to the planned applications, in which, despite what many people believe, the choice of carrier frequency is actually only a very small part of the solution to the problem.

As for the methods used in response to the problems, it will not be giving away too many secrets to tell you that, in spite of the vast range of proposed and feasible solutions, you will soon discover that the systems fall into just three principal categories: the first is concerned with industrial applications at low or medium bit rates, the second with short distance applications at high bit rates, and the third with long distance applications with their problems of electromagnetic pollution, low S/N ratio, collision and weak collision management, etc.

So the problem becomes more complicated. However, this is not the end of the matter. Now that we have considered what is 'strictly necessary' for an understanding of the phenomena of propagation, reflection and back scattering of signals at the carrier frequencies used in RFID at UHF and SHF, we must go on to examine the question of data, their coding, their structure and the spectra occupied by them in the baseband (in other words before carrier modulation), and then look at the types of modulation to be applied to the carrier frequencies in order to obtain the best performance in terms of the propagation, reception and demodulation of the signal. In the course of this investigation, I shall need to explain the operation of frequency hopping (FHSS) and spread spectrum (DSSS) techniques, since they are commonly used, at least in the USA, and certain frequency agility techniques (LBT), since these are authorized in most European countries for some of the UHF bands reserved for RFID applications.

To make these matters more approachable, I have divided this part into four chapters, which aim to explain and help you understand the reasons for the choices to be made in respect of:

- the digital aspects of the communication system, in terms of bit coding and baseband signals;
- the various types of carrier modulation;
- frequency hopping and agility and spread spectrum techniques;
- the consequences of their use and the question of interleaving.

As you will appreciate, these problems cover a very wide area.

11

Digital Aspect: Bit Coding and Baseband Signals

Many books have been devoted to the subject of the digital aspects of communication. I have no intention of commenting on these or summarizing their contents. I am simply concerned with the field of contactless RFID applications at UHF and SHF and their specific constraints, in other words the requirements for optimal energy transfer, speed of communication, security and reliability of transactions, and conformity with the current standards and regulations governing authorized signal levels and the nature of their associated spectra.

The purpose of the following sections is to analyse, illuminate and clarify these subjects. We will therefore be concerned with bit coding, carrier modulation and spectral content.

11.1 Bit Coding

I will start by considering the signals forming the digital content, at the bit level, of the message forming what is known as the 'baseband' transmission signal.

11.1.1 Some Terminology and Definitions

Bit

Regardless of its physical form (electrical, optical, pneumatic, etc.), the bit represents a binary logic value: all/nothing; open/closed; 1/0; yes/no; $-1/+1$; $+90°/-90°$, etc.

We can translate a data element into binary values. For example:

- The letters 'toto' translated into binary according to ASCII coding (7 bits + 1), where t = 01110100, o = 01101111, can also be written in hexadecimal notation as '7 4 − 6 F − 7 4 − 6F'.
- The number 253 converted to pure binary is 1111 1101 and can be converted to binary using 4-bit binary words as 0010 0101 0011.

RFID at Ultra and Super High Frequencies: Theory and Application   Dominique Paret
© 2009 John Wiley & Sons, Ltd

Bit Rate

The bit rate, also inaccurately referred to as 'speed', represents the number of logical bits per second transmitted between the transmitter and the receiver, in our case from the base station to the tag in the forward link and from the tag to the base station in the return link. The conventional units used are the bit per second, kbits s^{-1}, and other multiples of the same family.

Baud

The baud is the unit of the 'modulation rate' of a link. The modulation rate is the number of signal elements per second, in which all these elements are of equal length and each element represents one or more bits.

For many modems operating at or below 1200 bits s^{-1}, the modulation rate expressed in bauds is usually lower than the bit rate, because more than one bit is carried in each signal element.

Baud Rate

The baud rate is the unit representing the number of signal elements per second. If the signal element is a single bit, the representation of a binary state, the baud is equivalent to the bit rate.

11.1.2 Data

Data

The term 'data' denotes a parameter or a value, e.g. the 'toto', '235' or '1001101'.

Data Rate

The data rate is the rate (often also called the 'speed') at which data are transmitted between the transmitter and the receiver, in our case from the base station to the tag for the forward link and from the tag to the base station for the return link. Depending on whether the data consist of a bit, a byte, a symbol or other unit, in any format, the conventional units of measurement are bits per second, bytes per second, words per second or symbols per second. Let us note in passing that, according to ISO standards:

- 1 kB s^{-1} = 1000 bytes per second → k = factor multiplying by 1000;
- 1 KB s^{-1} = 1024 bytes per second → K = factor multiplying by 1024.

Author's note

Clearly, for most people, the difference of 2.4% is nothing remarkable – although if it was on a pay slip they would soon notice it! So you should take care to distinguish k and K. While on this subject, it is worth noting that, in the context of optimizing code for a microcontroller, it is always very useful to find this difference of 24 bytes between 1 kB and 1 KB when we are trying to 'shoe-horn' the last bytes into the memory space! Everyone working in this field will know exactly what I mean.

Now let us consider the different types of bit coding.

11.1.3 Coding and Physical Representation of the Bit

As in most digital communication devices, the bit coding principle is fundamental and is particularly important in contactless systems. The aim of the following sections is not to provide a full exposition of the comparative advantages or drawbacks of different types of bit coding, and indeed I must make it clear that I cannot lay claim to universal knowledge and will not attempt to examine the scientific intricacies of each coding that is discussed (leaving me open, no doubt, to the objections of purists). However, in the context of the RFID systems that concern us, we do need to examine the intrinsic performance of some types of coding more closely, in order to discover which are most suitable for the different operating phases (forward or return links) of the base station/tag system, as I will now explain.

In RFID, the choice of bit coding principle has a considerable effect on the quality of energy transfer and signal recovery (synchronization, etc.), and also on compliance with the RF radiation/pollution standards and local regulations, according to the data rate, the choice of carrier modulation principle, the collision management used and the resulting spectrum. This extensive field will be examined more closely in a later part of this chapter.

Let us start by seeing what the technical effects of our choice of a specific form of bit coding will be in the forward or return link, or in the collision management phase.

Expected Performance of Bit Coding in a Forward Link (From Base Station to Tag)

For data transmission from the base station to the tag, the expected performance of the bit coding of a forward link must be such that:

- After modulation of the carrier frequency, the radiated signal is present for as long as possible, thus providing the most efficient transfer and supply of energy (for remote powering); in other words, the remote powering and communication distances are as high as possible for a given amount of power radiated by the base station or, to put it another way, the transmitted power and radiation can be reduced to their minimum levels for a given distance.
- The signal radiated by the base station has the maximum number of possible transitions (from 1 to . . .), so that the tag can, if necessary, extract elements for synchronizing the electronics for digital decoding.
- A good or reasonable signal-to-noise ratio can be provided (this also relates to the types of modulation, which will be discussed subsequently).
- Long operating periods are provided and/or offered to enable the tag to handle the communication protocol and carry out its own task simultaneously and without difficulty.
- After modulation, the spectrum radiated by the carrier which is consecutive to, or associated with, this bit coding is the most suitable in terms of spectral efficiency and energy, and falls, as far as possible, within the ranges specified by local regulations.
- And so on.

Now let us consider the return link.

Expected Performance of Bit Coding in a Return Link (From Tag to Base Station)

In this phase of operation, the base station is ready to collect the signal from the tag and, if the tag is at a significant distance, the received signal is generally strongly affected by noise, making it harder to extract it from the noise and interpret it. In this case, the expected performance of the bit coding of a return link (from the tag to the base station) is such that we must consider bit codings such that:

- They include the greatest possible number of transitions during the bit time (as in a Manchester coded subcarrier, BPSK, and FM0, for example) to enable the base station to identify, extract and detect the signal easily even in the presence of noise.
- After the tag load modulation, the spectrum reradiated by the tag falls as far as possible within the ranges currently specified by local regulations (mainly when the tag is very close to the base station and the reradiated signals are strong);
- They minimize the overall power consumption of the tag during its response phase.
- It is easy to detect the presence of several tags simultaneously present in the electromagnetic field, in order to provide time-efficient collision management, and so on.

During the Phase of Testing for the Presence of Multiple Tags and Collision Management
When several tags are present simultaneously in the electromagnetic field mentioned above, it is also necessary to distinguish them, or at least to tell the base station (which is 'blind') that there are several of them. To do this, it is useful if the bit coding (in addition to the transitions mentioned above) can also indicate whether or not one or more collisions have occurred during this communication phase (which is also often affected by noise). This often makes it necessary to introduce a 'subcoding' into the bit, which, during the modulation of the transfer carrier, causes the appearance of a new signal, e.g. in the form of another new frequency or frequencies, generally known as subcarriers. Other techniques, such as the time slot technique, can also be used. I will describe this subject more fully in a later part of this chapter.

11.2 Different Types of Bit Coding for Use in RFID at UHF and SHF

The term 'bit coding' signifies the theoretical definition (on paper) of the form representing the signal (in the baseband) corresponding to the logical values of the bit. The definition of this coding is used subsequently to define its physical form (electrical, optical, pneumatic, smoke signals, etc., etc.). For the electronic engineer, this establishes the relationship between the theoretical values representing the bit coding and the corresponding physical electrical signals (electrical quantities, voltages, currents, electrical fields, magnetic fields, etc.).

The notes and comments in the previous sections should be supplemented with a remark about bit coding in general. It is widely known that the term 'bit' means binary information and 'binary' implies 'two'. However, in well-designed contactless applications (such as contactless smart cards, close relatives of the tags in RFID), 'binary' actually signifies 'four'! This is because we often wish to use bit coding with higher performance, such that we cannot only distinguish a '0' from a '1' but also detect a collision between a '0' and a '1', as well as another element indicating 'no information present'.

Having established the background, let us consider the main bit coding types proposed and/or used for contactless applications in RFID at UHF and SHF, with the specific details of their associated spectra in the baseband. At the end of this chapter I will briefly summarize the different performance levels of these types of bit coding and indicate their preferred applications.

> **Important**
> To avoid increasing the length of this book, the following pages will only describe the main types
> of bit coding used in RFID at UHF and SHF. There are many other forms of bit coding used in RFID,
> at the < 135 kHz and 13.56 MHz frequencies, for example. Readers interested in these particular
> codings should consult two of my other books (References 1 and 2 mentioned in the Preface).

11.2.1 For the Forward Link

The Bit Coding Family with '0' and '1' Bits of Equal Duration
NRZ (Nonreturn to Zero) and NRZI (Nonreturn to Zero Inverted)
In NRZ (nonreturn to zero), as the name indicates, the electrical quantity representing the
value of the bit remains constant throughout its duration. This bit coding is shown in graphic
form in Figure 11.1.

This coding, which is the simplest possible type, is notorious for its asynchronism, due to the
lack of transitions during the transmission of a sequence of bits of the same value, making it
theoretically necessary to transmit the continuous component and eventually retrieve it from
the radiated spectrum.

The use of this type of coding in contactless technology is often associated with certain
forms of RF modulation (FSK, 100% ASK or 10% ASK). This is because, in some cases, its
(statistical mean) energy efficiency is considered to be 'correct to fairly correct' for a forward
link (from the base station to the tag).

Examples
FSK $f_1 = $ '0'; $f_2 = $ '1'.
The carrier, of the same amplitude, is constantly present.
100% ASK on average
Statistically, one '1' bit is present for 50% of the time and therefore, on paper, it has mean
efficiency in terms of the remote power supply quality. Note that it is a fairly simple matter to
avoid this constraint by using various technological fixes.
10% ASK
Acceptable.

RZ (Return to Zero) and RZI (Return to Zero Inverted)
The principle of this coding:

- The value corresponding to the binary '0' remains constant throughout the bit time.
- The value corresponding to the binary '1' remains at level 1 for half the bit time and then
  returns to zero for the other half of the bit time (hence the name).

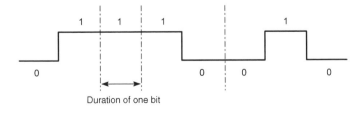

Duration of one bit

Figure 11.1   NRZ bit coding

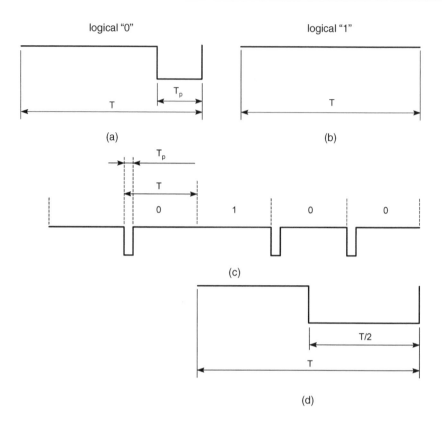

Figure 11.2 Pulses coded RZI bit coding: (a) logical '0'; (b) logical '1'; (c) example of binary sequences; (d) start bit

Without more ado, let us now consider a variant of this coding type that improves its performance in contactless applications.

Pulse Coded RZI

This type of coding belongs to the previous family (RZ), but the duration of the change of state during the binary '1' is shorter, which is why it is known as pulse coded RZI:

- The logical state '1' remains at rest.
- The logical state '0' includes a brief pulse at the start (or end) of the bit and then returns to the resting state.

Figure 11.2 shows the principle of this coding. I have deliberately talked about values that are the inverse of those in the description of RZ. This is because, as I will subsequently demonstrate, the level of the '1' bit generally corresponds to nonmodulation of the carrier and the pulse duration corresponds to the modulation of the carrier frequency.

In a transmission using this type of coding, it is necessary to provide a start bit (Figure 11.2(d)) to differentiate the start of the transmission from a logical '1' if the transmission begins with the binary value '1'. The start bit leaves the carrier unmodulated during its first half and modulates it during the other half of the bit time.

For RFID applications, this type of bit coding has the following properties:

- The communication rate is constant.
- It allows high rates to be achieved (a very large part of the bit time can be reserved for calculation).
- It is highly efficient for energy transfer (interruptions in the carrier are brief).
- It imparts considerable immunity to the movements of the transponders.
- It allows the use of high-quality factors for the base station antennas.
- It is easily decoded at the transponder (small chip area).
- It is easily decoded at the base station.
- It establishes a constant modulation time (except in the case of the start bit).
- The detection of the start bit is simple.

The Bit Coding Family with '0' and '1' Bits of Different Duration
Binary Pulse Length Modulation (BPLM), or Width or Repetition Coding
Binary pulse length modulation (BPLM) is based on a pulse of constant duration whose repetition cycle differs according to whether the signal represents a logical '0' or '1'. Generally, the duration of each bit is an integer multiple of a clock period.

Clearly, the instantaneous communication rate (in kbits s^{-1}) depends on the actual binary content present in the exchange. We can simply specify a 'worst case' rate (for the case of a message containing only the bits with the longest times, namely the '1' in our example) or state an average rate called 'equiprobable' (in other words, having as many logical '0' as '1' in the exchange, with the bit time defined as the mean bit time [('0' + '1')/2]).

Note that this bit coding, unlike those of the NRZ, RZ and other types, makes it easy to create a third state corresponding to the absence of a bit, thus facilitating the provision of a stop condition, which is different from '0' or '1'.

The members of this family of BPLM codings (PIE, inverse PIE, etc., as described below) have the following features:

- The bit time can be varied and its time value can be adjusted according to the spectra permitted by local regulations.
- The presence of the bits is easily detected because of the transitions due to the large number of pulses.
- Plenty of free time is available for remote powering and calculation in the time intervals when nothing is happening.

This type of coding is often used for the forward link between the base station and transponder in many commercial RFID systems, because it has a longer average charging time than that of other coding types, providing a high level of energy transfer to the tag and therefore, all other things being equal, a potentially longer communication distance.

Pulse Interval Encoding (PIE)
The bit coding type known as pulse interval encoding (PIE), an integral part of the BPLM family, is created using pulses of constant duration Pw, separated by variable time intervals, thus creating what are commonly called 'symbols'. These symbols are quantified with respect to integer multiples of a time interval called a 'Tari' (for historical reasons, related to ISO 18000-6 Part A). Figure 11.3(a) illustrates this bit coding.

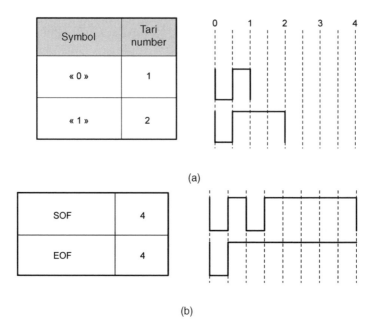

(a)

(b)

Figure 11.3 (a) Bit coding of the PIE type. (b) PIE symbols used for start of frame (SOF) and end of frame (EOF)

To start and terminate a communication frame, we must define new symbols other than the symbols '0' and '1' above, so that they can be identified easily using code violation detection systems. Figure 11.3(b) shows the PIE symbols used for start of frame (SOF) and end of frame (EOF). To determine the nature of the transmitted symbols, the tag measures the interpulse time between the high/low transitions shown in these diagrams.

This coding is often used in the forward link (from base station to tag), for the reasons already stated in the previous sections.

Inverse Pulse Interval Encoding (PIE)
This is as before, but the other way round (Figure 11.4).

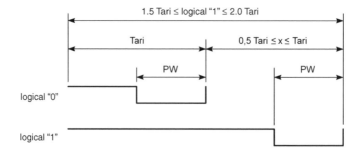

Figure 11.4 Bit coding of the inverse PIE type

Pulse Number Position Coding
The principle of this coding:

- A logical '1' is represented by one pulse.
- A logical '0' is represented by two pulses close to each other in time.
- There is a long time interval between data bits.

   Clearly, the time interval separating the bits must be longer that the duration of the logical '0', to ensure that these can be differentiated. While this may slow down the rate, it offers the possibility of modulating this time interval, if required, to adapt the data rate or the spectral form that is created.

Pulse Position Modulation (PPM) Coding
We should remember that many other types of coding are used for the forward link in RFID, for example:

- byte coding by fast mode position encoding;
- '1 out of 256' position encoding;
- '1 out of 4' position encoding;
- '1 out of 16' position encoding.

   For further details, you should consult one of my previous books (Reference 1 mentioned in the Preface).

11.2.2 For the Return Link

The Miller Family
Using many different technical devices, the form of initial bit coding, known as 'Miller encoding', has produced a large family of derivatives with numerous applications.

Miller Encoding
Let us start with the original version. The basic principle of Miller bit coding is shown in Figure 11.5(a). This bit coding, in which the '1' and '0' are of equal duration, is characterized by:

- a transition in the middle of the logical '1' bit;
- no transition in the middle of the logical '0' bit;
- a transition at the end of the bit for a '0' if the following bit is also a '0'.

   This bit coding has the advantage of having a baseband spectrum whose energy is limited to $1.5f_0$ (where f_0 is the frequency of the bit clock signal).

Important note
This spectrum, which is narrower than its predecessors, is more suitable for:

- transmissions in media where the bandwidth is limited, which is generally the case with RFID because of the narrow bandwidths of the authorized transmission channels;
- the electronics of the two circuits used in base stations and tags.

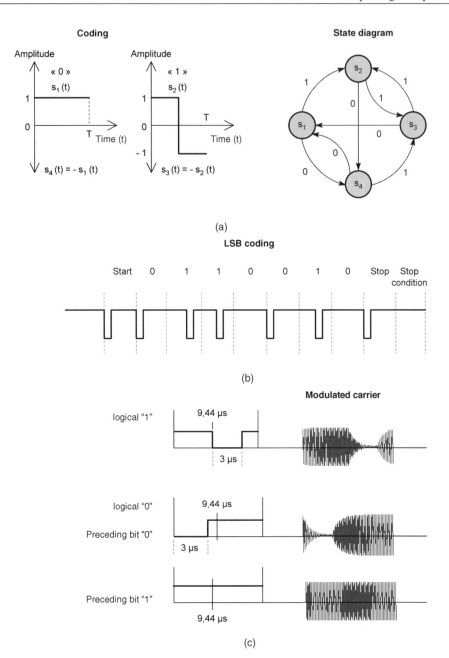

Figure 11.5 (a) Miller bit coding. (b) Modified Miller bit coding. (c) Examples of bit sequences

There is also a 'modified Miller' bit coding (Figure 11.5(b)), while Figure 11.5(c) shows some examples of bit sequences to illustrate the principle of this bit coding.

Miller Encoded Subcarrier
In order to create more transitions in the transmitted signal (especially when this is the reradiated signal from the tag which is very weak and often lost in noise), a supplementary signal is commonly introduced into the initial Miller bit coding described in the previous section. The frequency of this signal is a higher integer multiple M of the value of the bit. This creates a frequency called the subcarrier in the baseband signal spectrum, and the value of this, which depends on M, will be very useful afterwards for adapting the width of the radiated spectrum to different local regulations and also for extracting the received signal regardless

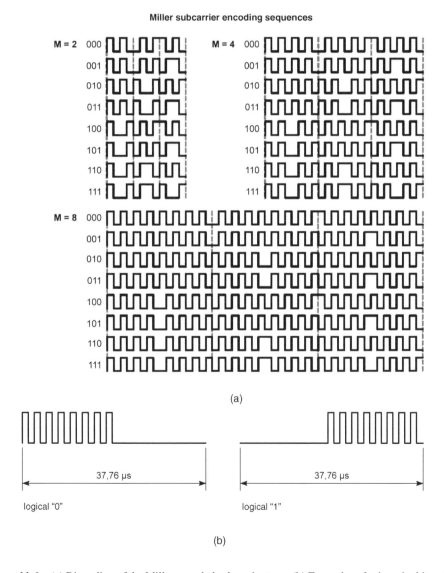

Figure 11.6   (a) Bit coding of the Miller encoded subcarrier type. (b) Examples of subcarrier bit coding

of whether noise is present. Figure 11.6 shows some examples of this type of bit coding where the values of M are 2, 4 and 8.

This bit coding, known as 'Miller encoded subcarrier', can be used in RFID in the link from the tag to the base station (the return link) to control the impedance modulation of the tag antenna in the back scattering phase, since this bit coding has many advantages:

- It is an excellent way of achieving a very good S/N ratio (and therefore a low BER, or bit error rate), because of the numerous transitions in the bit and the tag antenna impedance modulation which is equivalent to BPSK/DPSK modulation.
- It offers greater immunity to movements of the tags in the field.
- It facilitates collision detection at the bit level.
- The possible values of M can be used to achieve great flexibility in the choice of communication rates for the return link (from tag to base station), and thus to facilitate the adaptation/optimization of this link according to local regulations and/or in difficult environments (subject to noise, reflection, etc.). These matters will be discussed later on.
- The presence of the subcarrier facilitates the filtering and processing of the signal from the tag and its decoding (the base station hardware is independent of the desired data rate).
- It is easily generated in the transponder.
- Implementation of different data rates at the transponder is easy.
- The power consumption of the tag is practically independent of the subcarrier frequency (the value is low with respect to the value of the incident wave) and is not affected during the tag impedance modulation. For a given operating distance, this makes it possible to reduce greatly the value of the filter capacitor in the transponder and to reduce the area, and thus the cost, of the transponder itself.

The Large Family of Bi-Phase or Split-Phase Bit Codings
Bit codings of this generic family, called 'bi-phase', are among those in which the physical state of the bit can change at the start, at the end or in the middle of the bit period. There are three main forms:

- bi-phase level;
- bi-phase mark;
- bi-phase space.

These three forms of bit coding have similar spectral appearances, since they have no continuous component. Time information can also be extracted from the transmitted bi-phase codes, even in the presence of a sequence of bits with the same values, in order to assist with an understanding of the data.

Bi-phase bit coding requires more bandwidth than NRZ coding.

Bi-Phase Level, or Manchester, Encoding
The best known of these is the bi-phase level code, often called simply the bi-phase code or Manchester code. In this bit coding, the state of the signal during the first half of the bit indicates the value of the data (e.g. '0' if high, '1' if low). In the middle of the bit period, there is a transition that changes the level to the opposite electrical value and acts as a time signal.

To encode a bit by Manchester encoding, it is necessary to provide an exclusive OR (XOR) between a clock signal and the data signal:

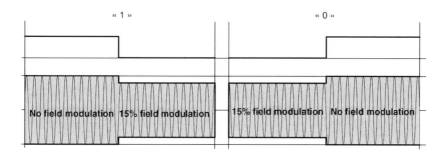

Figure 11.7   Manchester bit coding

- The '0' always has the same form with a negative transition in the middle of the bit.
- The '1' always has the same form with a positive transition in the middle of the bit.

Figure 11.7 illustrates this encoding.

Manchester encoding has the advantage of including a regular transition in the middle of each bit (regardless of its value), which provides synchronization at the receiver and makes its detection easier in noise. This last property makes it suitable for use in the return link, from the tag to the base station.

Differential Bi-Phase Coding (DBP) and Conditioned Di-Phase Procedure (CDP)
DBP, or CDP, coding is also well known. The procedure for this bit coding is as follows:

- A clock signal is used as the reference for data timing.
- Regardless of the value of the data bit to be encoded, a transition takes place at the start of each bit.
- A logical '0' always has a transition in the middle of the bit.
- A logical '1' never has a transition during the bit time.
- A transmission sequence always begins with a '0' data bit.

In this bit coding, transitions can always be provided during each '0' bit, by contrast with the transmission of a '1'. In other words, the logical '1' signal has the same timing as the bit clock and the timing of the logical '0' signal is double that of the bit.

This type of coding can be used to transmit from the transponder to the base station, since many transitions are present and the spectral energy content is very considerable.

Bi-Phase Space Coding, Also Called FM0
The bi-phase space bit coding technique is also known as FM0. This coding principle has the following properties (Figure 11.8(a)):

- A logical '0' has three transitions, including one at the start and one in the middle of the bit time.
- A logical '1' has one transition at the start of the bit time.

According to this description, transitions take place regularly at the starts of all the bits and produce different binary sequences, according to the preceding conditions, for any given data

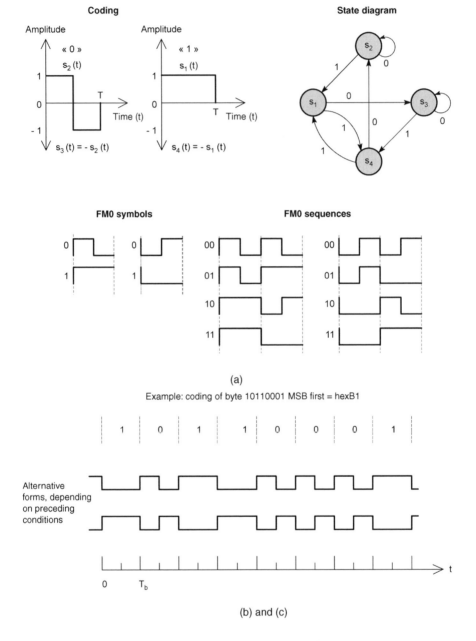

Figure 11.8   (a) FM0 bit coding. (b, c) Examples of coding sequences

value transmitted. Figures 11.8(b) and (c) show examples of coding sequences. Note that the diagrams show two possible forms of FM0 coding, because they naturally depend on the prior conditions.

This coding is very commonly used for the return link from the tag to the base station.

Binary Phase Shift Keying - BPSK-NRZ

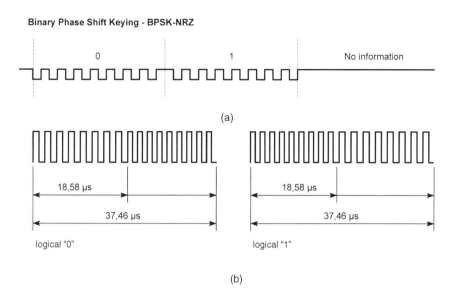

(a)

logical "0"   logical "1"

(b)

Figure 11.9   (a) Example of BPSK-NRZ bit coding. (b) Example of BPSK Manchester coded bit coding

Bi-Phase Mark
A logical '1' has a transition at the start, in the middle and at the end of the bit time. A logical '0' has a transition at the start and at the end of the bit time.

Binary Phase Shift Keying (BPSK)
In RFID applications, this type of coding is actually a bit subcoding. It uses an initial bit coding of the NRZ, Manchester or other type, using skip coding, according to the instantaneous value of the bit, the phase (0° and 180°) of a subcarrier frequency usually being generated by division of the carrier frequency or being an integer multiple of the bit frequency if its duration is not linked to the carrier frequency.
    To make matters clearer, let us consider two examples:

- The example of BPSK-NRZ coding, in which the phase changes with every change of the binary value of the bit (Figure 11.9a).
- The example of BPSK coding, using Manchester coding this time, in which a phase change takes place inside the bit itself (Figure 11.9b).

    This type of bit coding is used for the return link (from the tag to the base station), but, unlike the Manchester coded subcarrier coding, it makes no provision for easy detection of bit collisions at the base station, and it is therefore necessary to develop other strategies for collision detection (by verification of CRC values, for example) and management (by time slot methods, for example), if multiple tags are present within the electromagnetic field.

11.3 Summary of the Different Types of Bit Coding

Table 11.1 summarizes the discussion above by listing the main systems favoured by the ISO standards for the forward and return links used in RFID at UHF and SHF.

In the next chapter we will examine the choice of carrier modulation technique and its consequences.

Table 11.1

Frequency band	Standard	Link	
		Forward link bit coding	Return link bit coding
433 MHz	ISO 18000-7	Manchester	Manchester
860–960 MHz	ISO 18000-6 A	PIE	FM0
	ISO 18000-6 B	Manchester	FM0
	ISO 18000-6 C (EPC C1G2)	PIE (inverse)	FM0
			Miller SC
2450 MHz	ISO 18000-4 mode 1	Manchester	FM0
	ISO 18000-4 mode 2	Not applicable	Not applicable

12

Analogue Aspect: Carrier Modulation Methods

12.1 Type of Modulation

In this chapter, we shall see what types of carrier modulation can be used in RFID at UHF and SHF. However, first we need to take a look at the authorized, and feasible, frequencies and frequency bands.

12.1.1 Feasible and Authorized Frequencies and Channel Widths

There are few free UHF and SHF frequency bands around the world in which the use of proper RFID applications can be envisaged. It is frequently necessary to use those that are targeted at 'low-grade' applications, known as 'short range devices' (SRD) or 'nonspecific' (NS); only rarely can we use those that are purely dedicated to RFID. There are certain bands that are mainly used in contactless identification system. These are the UHF frequency of 433 MHz, and in some cases frequencies in the band from 860 to 960 MHz, and SHF frequencies at 2.45–5.8 and 24 GHz. For historical reasons, frequencies measured in GHz, in addition to UHF, are used more commonly in the USA than in Europe, because the authorized transmission levels are (generally) higher in America. There is a specific authorized bandwidth (Δf) around each of these nominal carrier frequencies, and the choice of bit coding type, data rate and modulation is crucial if the radiated signal spectrum is to fit in the authorized channel. These matters will be discussed in detail in Chapter 16.

12.1.2 Choice of Types of Carrier Modulation

A completely bare carrier is meaningless if it is not modulated, and modulation implies a spectrum associated with the modulating signal (in terms of shape and frequency) and with the type of modulation used. In our case, there are two elements – the base station and the tag – whose modes of operation are quite different.

RFID at Ultra and Super High Frequencies: Theory and Application   Dominique Paret
© 2009 John Wiley & Sons, Ltd

From the Base Station to the Tag: the Forward Link
The purpose of the base station is to transmit its signal and to receive information from the tag. In the transmission phase, in the forward link, the carrier is modulated and the radiated spectrum depends on the type of bit coding, the bit rate, the bandwidth of the modulating signal in the baseband, the type of modulation, the spread spectrum techniques used, the bandwidth of the antenna, the transmission level, etc., and everything must conform to the current standards and regulations.

From the Transponder to the Base Station: the Return Link
As mentioned above, during the phase of communication from the tag to the base station, antenna impedance modulation is carried out by the integrated circuit of the tag according to a specific bit coding, causing the wave to be reradiated (back scattering).

12.1.3 Mandatory Standards and Regulations

Whenever radio waves are transmitted, it is necessary to comply with the requirements of the law, regulations and standards relating to the disturbances that may be caused. These requirements relate to the authorized transmission levels for the frequency bands in which the waves are produced, and depend to a great extent on the country in which they are applicable. The same applies to standards and limits for the levels of radio waves in respect of the health and safety of individuals.

You should consult Chapters 15 and 16 for full details of this aspect of contactless identification, but in any case you should bear in mind that we shall be taking these levels into account throughout this chapter.

12.1.4 Effects of Types of Modulation, Bit Coding, Bit Rate, Standards, Protocols and Regulations

The specific constraints on contactless RFID applications have been stated at the beginning of this book, and these will guide us in establishing the possible principles to be adopted.

12.2 Types of Carrier Modulation for the Forward Link from the Base Station to the Tag

Within the field of RFID applications for the forward link (from the base station to the tag), we can use a very large number of types of carrier modulation to transmit a binary stream: e.g. amplitude shift keying (ASK) of the 100% or x % type, DSB, SSB, PR, etc., frequency shift keying (FSK), phase shift keying (PSK), binary phase shift keying (BPSK), quadratic amplitude modulation (QAM), quadratic phase shift keying (QPSK), etc.

Although some systems use FSK modulation, most RFID applications use ASK carrier modulation, and the spectrum of the modulating signals present in the baseband is completely or partially transposed by the RF modulation (Figure 12.1), a feature that we shall take into account in the rest of this description.

Let us start with amplitude modulation.

12.3 Amplitude Modulation

There are many variations of amplitude modulation (AM). If the modulating wave is of the analogue type, we use the term AM. However, if the modulating signal varies in discrete steps, in other words by successive shifts of amplitude, we speak of ASK (amplitude shift keying) modulation. There are many variants of ASK, classified according to the modulation index of the carrier (or the depth of modulation of the wave). If there are only two discrete levels, with no other kind of process, the modulation is generally termed OOK (on–off keying)/ASK, which is theoretically independent of the modulation index used. There are different forms of OOK: 100% OOK/ASK and *m* % OOK/ASK. 'OOK/ASK' is often used, inaccurately, to denote a 100% modulation index.

Let us take a closer look at these matters in the context of RFID applications.

12.3.1 ASK (Amplitude Shift Keying) Modulation

As mentioned above, ASK (amplitude shift keying) is a method of amplitude modulation of a carrier frequency using shifts of amplitude. The sidebands are well known and reflect the spectrum of the modulating signal produced in the baseband; therefore, if the signal has a narrow spectrum in the baseband, it will still be narrow when transposed to HF. On the other hand, if the index, or sometimes the depth or even the rate of modulation increases, the level of the sidebands (in dB) increases proportionately.

In contactless technology, we normally distinguish two types of ASK modulation, namely ASK 100%, on the one hand, and ASK *x* %, on the other (for information, ASK 30 to 100% is defined in ISO 18000-6 type A, ASK 18 to 100% in ISO 18000-6 type B and ASK 100 and 10% in ISO 18000-3 operating at 13.56 MHz), and a number of subvariants, DSB, SSB, PR, for very specific reasons, which are described below.

ASK 100% Modulation
After modulation, the form of the resulting signal follows the curve shown in Figure 12.2(a) (for values relating to Figure 12.2 see Table 12.1). Whenever the carrier is 100% modulated, there is an absence, or 'pause' of the carrier.

For many reasons (such as the carrier cut-off time, the rise time and the application time of this technique), this pause is rarely perfect. Furthermore, Figure 12.2(b) shows how the

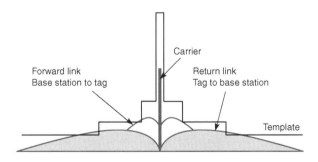

Figure 12.1   Spectra of the forward and return links

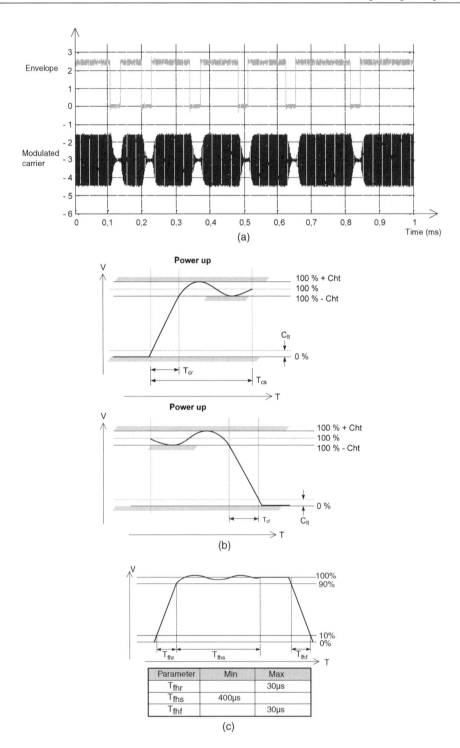

Figure 12.2 (a) Example of ASK 100% modulation (with 'modified Miller' bit coding). (b) The power up and power down phases of the carrier. (c) Power up and power down applications in FHSS frequency hopping systems

Table 12.1

	Parameter	Minimum	Maximum
Power up	T_{cs}		1500 μs
	T_{cr}	1	500 μs
	C_{ht}		10%
	C_{lt}		1%
Power down	T_{cf}	1	500 μs
	C_{ht}		±5%
	C_{lt}		1%

establishment time (power up) and cut-off/extinction time (power down) of the carrier is described in terms of time in ISO 18000-6, in the case of systems operating with frequency spread techniques using frequency hopping (FHSS, see below).

Figure 12.2(c) shows an example relating to the use of frequency hopping systems (FHSS, LBT, etc.).

ASK x % Modulation

Instead of modulating the carrier 100% ('on/off'), we can use a partial carrier modulation by a certain percentage. Figure 12.3(a) shows an example of an ASK x % modulated carrier.

There are many ways of describing this x % partial modulation, and much confusion is created as a result of incorrect terminology or ignorance. To ensure that we use the correct definitions, let us start with the terminology used by the ISO for RFID (in the ISO 18000-x standards).

The Modulation Index

If $V_{max} = a$, the voltage present before modulation of the carrier, and $V_{min} = b$, the voltage present during modulation (Figure 12.3(b)), the value of the modulation index m (in %) is given by the following equation:

$$m = \frac{V_{max} - V_{min}}{V_{max} + V_{min}} = \frac{a-b}{a+b} = \text{modulation index } (\%)$$

Modulation Rate

We can easily move from the modulation index m to the modulation rate, where the rate is V_{min}/V_{max} (%):

$$m = \frac{V_{max} - V_{min}}{V_{max} + V_{min}} = \frac{1 - (V_{min}/V_{max})}{1 + (V_{min}/V_{max})}$$

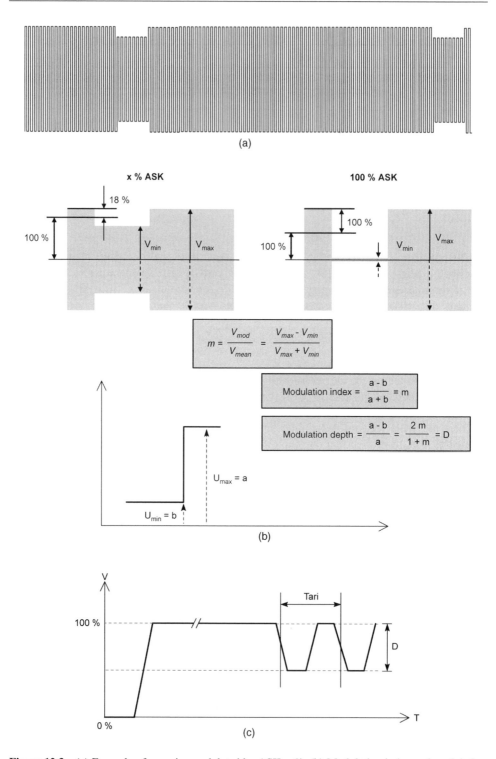

Figure 12.3 (a) Example of a carrier modulated by ASK x %. (b) Modulation index and modulation depth in ASK x %. (c) Example of modulation depth

and therefore

$$m\left(1 + \frac{V_{\min}}{V_{\max}}\right) = 1 - \frac{V_{\min}}{V_{\max}}$$

$$\text{Rate} = \frac{V_{\min}}{V_{\max}} = \frac{1-m}{1+m} = \frac{b}{a} = \text{modulation rate } (\%)$$

For example:

- If $m = 10\%$, $b/a = 81.8\%$ (not 90% as is widely believed!).
- If $m = 14\%$, $b/a = 75.4\%$.

In the terminology used for contactless technology and RFID, it is only the value m – the modulation index – that is used to specify ASK m % modulation, not the 'modulation rate' described above.

Modulation Depth
Sometimes we also define a modulation depth D, the value of which is given by

$$D = \frac{V_{\max} - V_{\min}}{V_{\max}} = 1 - \text{rate} = \text{modulation depth } (\%)$$

$$D = \frac{a-b}{a} = 1 - \frac{b}{a} \text{ expressed in } \%$$

Replacing b/a by its value, we have

$$D = 1 - \frac{1-m}{1+m}$$

and therefore, finally:

$$D = \frac{2m}{1+m} = \text{modulation depth } (\%)$$

or alternatively

$$D(m+1) = 2m$$

$$D = 2m - Dm = m(2-D)$$

$$m = \frac{D}{2-D}$$

Example
Unlike all the other standards, which specify modulation index values, ISO 18000-6 type A (for RFID at UHF from 860 to 960 MHz) provides a minimum modulation depth D of 27%, i.e. 0.27

(see Figure 12.3(c)), and therefore

$$m = \frac{0.27}{2-0.27} = 15.6\%$$

which is actually slightly lower than the level according to ISO 18000-6 type B, which directly states a minimum value of m of 18 %! For information, the same version of ISO 18000-6 type B can be viewed in terms of D to give the following value for $m_{min} = 18\% = 0.18$:

$$D = \frac{2 \times 0.18}{1+0.18} = \frac{0.36}{1.18} = 0.305; \text{ therefore } D = 30.5\%$$

Effects of the Chosen Modulation Index on the Radiated Spectrum

For ASK m % modulation, regardless of the carrier frequency, Figure 12.4(a) shows in a general way the variation of the reduction (in dB) of the level of the sidebands of the modulated signal as a function of the modulation index of the signal.

By way of example, all other things being equal, Figure 12.4(b) shows (in RFID, ISO 18000-3a for 13.56 MHz is identical for technical purposes to ISO 15693 for smart cards) the corollary of the above equation, in the form of the relationship between the read/write distance and the maximum modulation index that can be achieved without exceeding the maximum sideband level defined by the scale in the ETSI 300–330 pollution standard. In UHF, the curve would have a similar shape for the limits according to ETSI 300 220.

12.3.2 Derivatives of ASK

When the conventional amplitude modulation (AM) has been carried out by an analogue or digital signal, it is then possible to retain all or part of the resulting signal, by filtering or other methods, at the power stage of the base station.

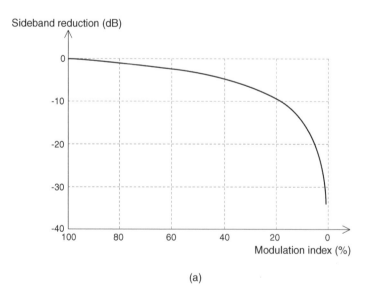

(a)

Figure 12.4 (a) Effect of the modulation index on the sideband level. (b) Optimal value of the modulation index for the best signal-to-noise ratio complying with the limits

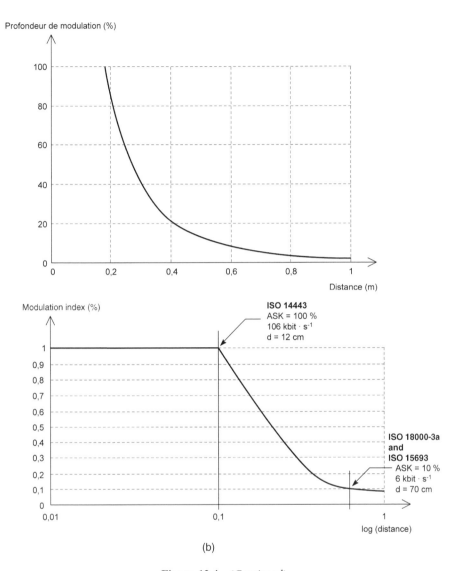

Figure 12.4 *(Continued)*

Double Sideband ASK (DSB-ASK)
This is the most common result of ASK modulation, in which the two sidebands (hence 'double sideband', or DSB) are retained following the modulation of the carrier by the modulating signal (Figure 12.5(a)).

Single Sideband ASK (SSB-ASK)
In some cases, SSB-ASK modulation is used in order to limit the width of the spectrum occupied by the radiation of the base station antenna after modulation. In this case, one of the initial two sidebands is suppressed or eliminated by suitable filtering in the final phase before transmission (Figure 12.5(b)).

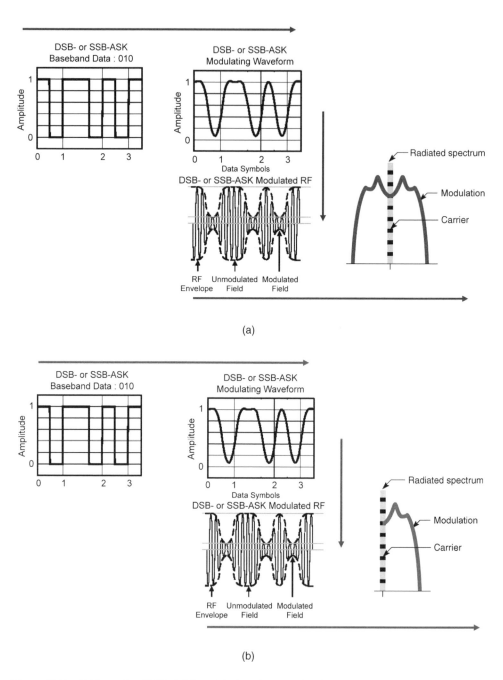

Figure 12.5 (a) Example of DSB-ASK modulation. (b) Example of SSB-ASK modulation. (c) Example of PR-ASK modulation/demodulation

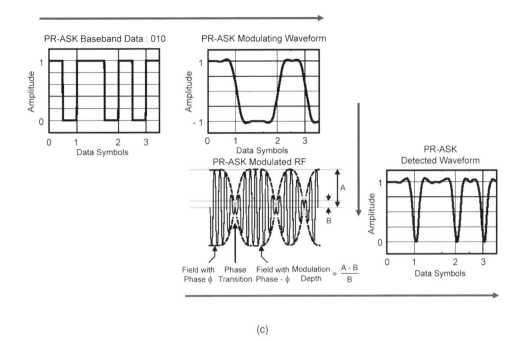

Figure 12.5 *(Continued)*

Phase Reversal (PR-ASK)

The form of modulation known as phase reversal (PR-ASK) is very like a convenient mixture of BPSK and conventional amplitude modulation. It is often said that a good picture is worth a thousand words (see Figure 12.5(c)). This diagram clearly shows the abrupt change of phase (total inversion) – hence the name PR – of the modulated wave at each transition of the binary value of the incident modulating signal and the simultaneous ASK modulation.

Note that the derivatives of conventional ASK modulation described above can be useful when adapting to the different channel widths specified in the local regulations of different countries (FCC, ETSI, ARIB, etc.).

12.4 Frequency Modulation and Phase Modulation

There are numerous methods of frequency modulation (FM, FSK, etc.) and phase modulation (PM, PS, BPSK, etc.) (Figure 12.6). These forms of carrier modulation are rarely used in RFID at UHF and SHF. Some systems use FSK carrier modulation, but the great majority use ASK *m* %.

This will surprise many readers, since the terms FSK and BPSK are frequently bandied about in the field of contactless technology. In these cases, however, what is being discussed is the method of modulation (of the subcarrier) within the bit coding used

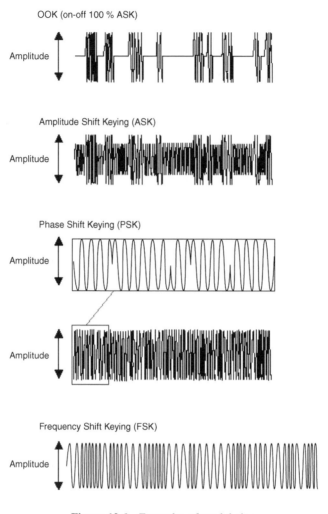

Figure 12.6   Examples of modulation

for the load modulation of the return link, not the modulation of the carrier transmitted by the base station, which is always modulated by ASK 100% or x %. Be careful to avoid confusion!

12.4.1 Frequency Modulation

(Analogue) modulation of the carrier by conventional FM is practically never used. However, its digital 'on–off' equivalent, known as OOK (on–off keying), is occasionally used. We are, of course, concerned with FSK (frequency shift keying) in this case.

Frequency Shift Keying (FSK)

FSK modulation is to FM as ASK-OOK is to AM. In other words, it is an on–off modulation of the frequency according to the two binary values of the bit.

12.4.2 Phase Modulation

Since the advent of digital television broadcast by cable and satellite, and mobile telephony, phase modulation (PM) and its variants, BPSK, QPSK, QAM, etc., have been very popular. Many variants of conventional phase modulation (PM) are now in use in numerous fields, including RFID. These variants include PSK, BPSK and DPSK.

Phase Shift Keying (PSK)
This is the simplest digital form of phase modulation, using on–off (OOK) modulation. The binary states 0 and 1 are represented by discrete changes in the phase of the carrier frequency according to a predetermined number of values.

Binary Phase Shift Keying (BPSK)
BPSK (binary phase shift keying) modulation uses a constellation diagram having only two points, e.g. $0°$ and $90°$ or $0°$ and $180°$, etc.

Differential Binary Phase Shift Keying (DBPSK)
The data are differentially encoded before the BPSK modulation is carried out.

QPSK xPSK and xQAM
QPSK, xPSK and xQAM are well-known types of carrier modulation that are widely used for transmitting digital mobile telephone and digital television (DVBT) signals via satellite, cable and terrestrial links. The constellations representing these modulations have numerous points, i.e. 2, 4, 32, 64, 128, 256, . . ., and the digital symbols carried are made up of 2, 4, 8, . . . bits according to the type of modulation.

Minimum Shift Keying (MSK)
MSK (minimum shift keying) is a special type of FSK phase modulation. In this modulation, the difference between the frequency of a logical '0' and that of a logical '1' (corresponding to the peak phase difference) is always equal to half the data rate, which is the minimum frequency offset for two FSK signals. For MSK modulation, therefore, the modulation index is 0.5. This type of modulation uses a continuous phase modulation scheme in which the frequency changes take place at the zero transition point of the carrier without phase reversal.

Gaussian Minimum Shift Keying (GMSK)
GMSK modulation is a derivative of the MSK modulation described above, in which the incident signal has previously been shaped by Gaussian filtering. This is because one of the main problems of MSK modulation in high-rate applications is that the signal produced by this modulation is not usually compact enough to fit in the available bandwidth. To make this modulation more efficient, the energy in the sidebands must be reduced by using low-pass filters. To do this, a Gaussian filter is used (with a response in the form of a bell-curve or Gaussian distribution) to obtain a cut-off frequency such that the response to pulses gives rise to very few excess oscillations (overshoots). Gaussian minimum shift keying (GMSK) is much more efficient than MSK in its use of bandwidth and transmitted power, because the transmitted harmonic content and bandwidth are much smaller.

For further information, you should consult the relevant literature[1] on signal processing.

12.5 Conclusion

We may conclude by saying that, in many cases, ASK modulation provides a satisfactory compromise between simplicity of the transponder detection circuits, a good S/N ratio, appropriate sideband values and levels according to local regulations such as FCC, ETSI and ARIB, and the maintenance of a high transmission rate.

---

[1] An example of a book dealing with these different types of modulation is *Électronique Appliquée aux Hautes Fréquences* by François de Dieuleveult (Dunod, Paris, 2007), which will give you all the theoretical and practical support you need to understand and apply these techniques correctly.

13

Spread Spectrum Techniques

13.1 Frequency Hopping and Agility Systems and Spread Spectrum Techniques

As I have just demonstrated, carrier modulation is an operation relating to the physical layer of the OSI model, which processes the working signal to be transmitted (a digital signal, in our case) so that it can be transmitted at radio frequency. Sometimes the methods described above cannot meet the performance requirements of applications, and we need to consider other complementary techniques. Because of this, some contactless devices in RFID systems operating at UHF (433 and 860–960 MHz) and SHF (2.45–5.8 and 24 GHz) use the techniques of frequency hopping (FH), frequency agility (listen before talk, or LBT), and the direct spread spectrum (based on a direct phase modulation sequence, called DSSS), which are generally described, often rather inaccurately, as 'spread spectrum' (SS) methods.

Note
You should note that these techniques are commonly used in RFID in the USA for tags operating at UHF and SHF, because the local regulations (FCC) offer much more scope for their use than the European regulations (CEPT, ERC, ETSI).

In the following sections I will provide a detailed explanation of these techniques, focusing on the appearances and shapes of the spectra radiated by the signal transmitted by the base station (the interrogator) and their effects. My description will cover two main areas:

- Firstly, the types and principles of frequency hopping or agility systems, which spread the radiated spectrum of carriers modulated by what is known as 'narrowband' modulation – named to point up the difference from the other type.
- Secondly, systems using direct spread spectrum techniques, which are given this name to distinguish them not from 'wideband' methods but from 'spread spectrum' methods.

More specifically, I will describe FHSS and LBT as examples of the first category and DSSS as an example of the second; all of these are widely used in RFID at UHF and SHF.

RFID at Ultra and Super High Frequencies: Theory and Application   Dominique Paret
© 2009 John Wiley & Sons, Ltd

Note that, in professional RFID circles, mention is also made of the potential use of UWB (ultra wideband) devices, which, in certain very specific conditions, can occupy a very wide frequency band, or even the whole authorized frequency band, and can contribute to the creation of new geolocation devices based on RFID.

13.2 Spread Spectrum Techniques (Spread Spectrum Modulation, SS)

For certain applications, the conventional modulation principles such as AM, FM, PM and their derivatives (ASK, ASK-OOK, QAM, FSK, BPSK, DPSK, QPSK, MSK, GMSK), as described in the previous section, are no longer adequate. This is because, in all these modulation modes, the baseband signal (representing the digital data to be transmitted) modulates a carrier having a given frequency, and the radiated spectrum is the direct result of this operation. The spectrum created by this modulation tends to be wide, but does not cover a very wide range (from hundreds of kilohertz to a megahertz). In this case, we speak of 'narrowband' operation.[1]

As in all systems based on the propagation of radiated waves, the waves are subject to problems of multiple paths or routes and the presence of noise, parasitic signals, interference, jamming, etc., and these problems increase at higher frequencies. This is the case with RFID at UHF and SHF.

Some complementary techniques, described below, have specific properties, which considerably improve the communication performance of a system in such ordinary operating conditions. They include spread spectrum techniques, which are widely used in radio frequency identification.

Spread Spectrum Techniques

As well as carrying out the pure narrowband amplitude modulation (AM), frequency modulation (FM), or phase modulation (PSK, QPSK, etc.) of the carrier frequency, as described above, it may be useful to 'spread' the spectrum of the radiated signal.

The aim of such techniques is to 'distribute', 'disperse' or simply 'spread' the spectral and energy content of the data as uniformly as possible in time, by carrying the signal over a considerably wider frequency range than that required for conventional narrowband communications (AM, FM, etc.), thus making it possible to retrieve these data in highly unfavourable conditions of interference and/or noise, as described below.

Properly speaking, spread spectrum techniques are not always carrier modulation techniques, but by using them we can take a stream of useful data (the data to be transmitted) in the form of an electrical signal with a known baseband spectrum and establish transmission techniques such that the final radiated signal spectrum is much wider than that of the signal of the initial data stream.

Thus they are forms of 'supplementary modulation' or 'supplementary coding' (simple, but inaccurate terms, that I will continue to use now and then, for want of any better words) – see the detailed explanations below – by which the mean transmitted spectral power density is 'dispersed' or 'spread' in a random or pseudorandom way (in other words it is either

---

[1] For a more detailed explanation, you should refer back to the previous section and also consult the specialist literature, such as the book by François de Dieuleveult, *Électronique Appliquée aux Hautes Fréquences*, 2007, available from Dunod, Paris.

uncorrelated, or only slightly correlated, with the data signal) over a frequency band whose width (bandwidth) is much greater than what is strictly required for the frequency spectrum of the useful data to be transmitted.

A Brief Historical (and Musical) Note

Surprising though it may seem (Figure 13.1), the best known spread spectrum technique (using frequency hopping, subsequently called FHSS, see below) was invented and patented during the Second World War by Hedy Lamarr, an Austrian-born American film actress and jazz and popular music singer, with the assistance of her pianist and composer, George Antheil.

This system was originally designed to prevent the jamming of guidance signals of submarine-launched torpedoes during the Second World War. The sequence used to define the spectrum spread was a sequence of notes based on a musical melody, virtually random in the case of the jazz improvisations of the famous pianist and singer, transcribed on to the player piano rolls used at that time.

Soon afterwards, the US Navy purchased all the patents and classified them as 'defense secrets', using these spread spectrum techniques up to the 1960s to keep their communications secret. These techniques only recently became public, and have played a major part in the expansion of Wireless LAN, GSM, IEEE 802.11 and such systems – and now RFID as well! It is surprising what music and musical improvisation can lead to. . .. (You may like to know that the author also wrote a treatise on musical harmony many years ago. You see how easy it is for people to be manipulated by the charms of music.)

The General Properties and Advantages of Spread Spectrum (SS)

My introduction to SS in the preceding sections may have left you with the impression that it does not make much of a contribution, or even that it has the major drawback of occupying a much greater bandwidth for a given useful signal than narrowband transmission. Considered in isolation, this is true, but if you take a wider view you will perceive all the qualities concealed within this 'defect'. In fact, these techniques offer many benefits. The fact that the spreading sequence is random (or pseudorandom, or weakly correlated) and unique for each user means that:

- Only a receiver or user that knows the spread spectrum sequence code used by the transmitter can select the desired transmission channel and decode the data correctly;

> **Note**
> As I will show later, the receiver must always know this frequency change sequence when the back scattering mode of RFID between the base station and transponder is used, since the return wave is a reflected part of the incident wave.

- Since each transmitter uses a unique spreading sequence, several transmitters can operate at the same time in a single frequency band; the probability of more than one transmitting at exactly the same frequency at the same instant is virtually zero.
- This means that, although the band occupied by a transmitter is much wider than in narrowband modulation, an infinite (or almost infinite) number of transmitters can operate simultaneously in the same bandwidth.
- This technique therefore provides inherent protection from jamming or hostile transmitters, regardless of whether they are narrowband or wideband (Figure 13.2(a), (b) and (c)), and is

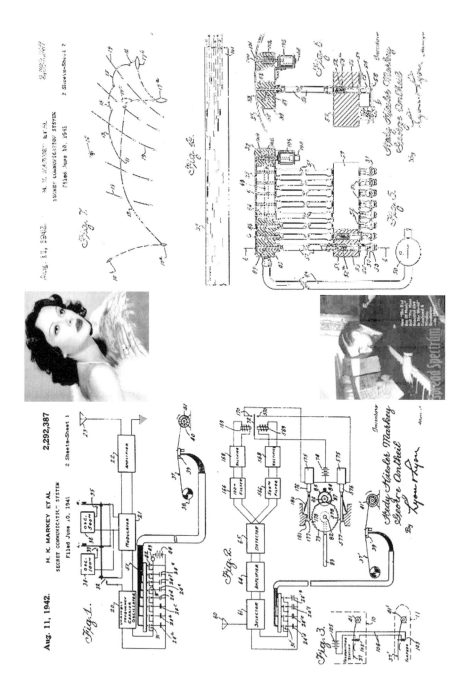

Figure 13.1 Hedy Lamarr and George Antheil and the historic original patents for FHSS

widely used in military applications because of its high level of transmission security. It has been increasingly used for several years in large-scale civilian applications such as GSM mobile telephony, Bluetooth, the IEEE 802.11× family, WiFi, etc.

• These techniques are used in code division multiple access (CDMA) systems in which all users simultaneously share the same bandwidth and can start their transmissions at any time without the need for specific medium access procedures.

• Because of the spread spectrum and the large number of frequencies contained in the radiated signal, these techniques provide multiple access paths (in terms of propagation)

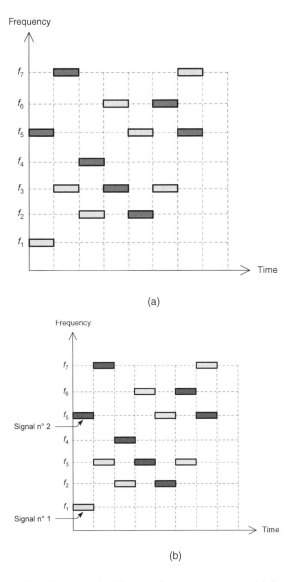

Figure 13.2 Examples of problems resolved by spread spectrum systems: (a) the principle of frequency hopping; (b) multiple transmitters in a single band; (c) jamming

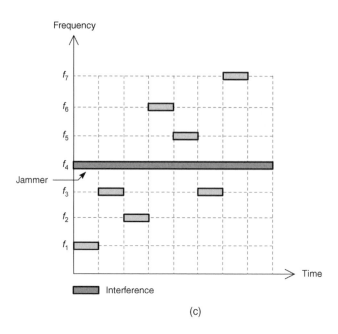

(c)

Figure 13.2 *(Continued)*

for communication, which, statistically speaking, increases the immunity of the signal-to-noise ratio, echoes and interference, thus ultimately improving the signal-to-noise ratio and BER.

- Since the signal is spread over a wide frequency band, the spectral power density is very small, so that other communications systems operating in the narrowband mode in the same frequency band are not affected by the simultaneous presence of this type of communication. However, the Gaussian noise is greater because of the width of the occupied band.
- To conclude this first list of the properties of SS systems, a wider transmission band means more possibilities for increasing the data rate.

Again, I suggest that any readers requiring full details of the theory of these techniques should consult the specialist literature.

Note that these techniques are widely used in the USA for RFID tags or industrial devices operating at UHF (902 to 928 MHz) or SHF (2.45 and 5.8 GHz), because the authorized transmission power levels are much higher than in Europe. Indeed, by contrast with Europe, the American FCC regulatory authority has aimed to stimulate the production and use of wireless products and networks since 1985, by modifying Title 47, Part 15 of the radio frequency spectrum regulations, which governs products using the spread spectrum technique, enabling wireless network applications to operate without a licence (subject to certain maximum power limits) in the ISM frequency bands of 902–928 MHz, 2.4–2.4835 GHz and 5.725–5.850 GHz.

In addition to the higher technical performance offered, this deregulation of the frequency spectrum has eliminated the need to spend time (with the relevant authorities) and money on planning frequency allocations to coordinate radio frequency installations and any possible residual interference. Moreover, when the geographical location of the physical RFID base station is changed, it is no longer necessary to request a new approval or licence for installing the product in its new operating site (when using electronic timing systems for urban sports events such as marathons and triathlons with RFID numbers, for example).

Example
The chosen carrier frequency, 2.450 GHz for example, is conventionally amplitude modulated, by ASK 100% for example (using what is known as OOK carrier modulation) in the narrow band, and is then made to hop to other values such as 2.455, 2.470, 2.425, etc., once every half second. Throughout each 0.5 s, the carrier frequency is modulated by ASK 100% at the start of the data to be transmitted (at 40 kbits s^{-1}, for example). To avoid any confusion, I will say that this frequency hop corresponds to a 'supplementary modulation' of the carrier frequency and causes an overall spread of the radiated spectrum.

Some Definitions and Terminology Relating to SS Families and Similar
Before looking at the details of these SS and similar techniques, we must define some new general terms.

Processing Gain
Processing gain (PG) is one of the principal parameters of SS systems. It is also defined as the value of the spread ratio of the spectrum:

$$PG = \frac{\text{width of the spread band}}{\text{bandwidth of the signal in the baseband}}$$

Regardless of the SS technique used, an increase in this ratio is accompanied by an increase in the bandwidth of the communication channel and the spread of the spectrum, and an improvement in the spreading performance of the system. This parameter therefore plays a part in determining the maximum acceptable number of users in the system, the quality of the reduction effect of multiple routes, resistance to jamming, signal detection quality, and so on. This is why it is so important.

Spectrum and Bandwidth of an SS Channel
In the case of the spread spectrum systems mentioned above, we often use the term 'bandwidth' to denote the width of the band occupied by the spread signal in a specified channel. This width of band may or may not be equal to the space between two adjacent channels, although the space between channels can be equal to, but not greater than, the bandwidth occupied by the channel.

Different Types of SS and Similar Techniques
To conclude this general description of SS techniques, different solutions are commonly used in RFID at UHF and SHF for spreading the spectrum of a radiated signal. The main ones are:

- frequency hopping (FH), known as FHSS, time hopping (TH) and multicarrier CDMA;
- specific modulation using a direct sequence (DS), as in the DSSS system;
- frequency agility systems operating after listening for the presence or absence of another transmitter in the desired transmission channel, known as listen before talk (LBT).

It is also possible to create combinations of these different techniques, giving rise to hybrid systems, which will now be examined in detail.

13.3 Frequency Hopping or Agility Systems for Spreading the Radiated Spectrum of Narrowband Modulated Carriers

In the following section I will describe the details of two systems, namely a frequency hopping system (FHSS) and a frequency agility system (LBT). These two systems, widely used in RFID at UHF and SHF, use carriers whose narrowband modulations are entirely conventional and well known, such as amplitude modulation (AM), frequency modulation (FM) or phase modulation (PM) and their derivatives.

13.3.1 Frequency Hopping Spread Spectrum (FHSS)

The Operating Principle of FHSS

Frequency hopping operates more or less as its name indicates (Figure 13.3(a) and (b)). The FHSS technique is based on the automatic modification ('hopping') of the RF carrier frequency at brief time intervals. The frequency is chosen in a pseudorandom way from a set of frequencies covering a much wider band than the bandwidth required to transmit the signal containing the useful data.

Note

Theoretically, the speed at which the carrier frequency is modified can be faster or slower than the bit rate of the data signal.

This technical of spreading the spectrum of a signal over a wide frequency band is mainly defined, on the one hand, by the frequency at which frequency hops take place periodically as

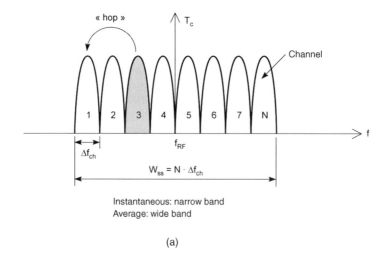

(a)

Figure 13.3 FHSS technique: (a) the principle; (b) three-dimensional representation of the principle; (c) principle of construction of a transmitter operating in the FHSS mode

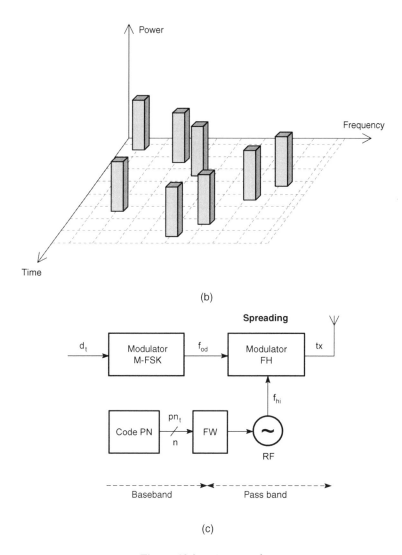

Figure 13.3   *(Continued)*

a function of time (the frequency hop rate) and, on the other hand, by the order or procedure (deterministic, random, pseudorandom, PRBS (pseudorandom binary sequence), etc.) according to which the hops take place (the frequency hop sequence). A more detailed definition of these terms is provided below.

The nature of this technique is such that it reduces interference very considerably, since a signal from a narrowband system will not affect the SS signal unless both are transmitted at the same frequency simultaneously – something that rarely, if ever, happens. The interference rate will therefore be very low and the bit error rate (BER) will be very small or even zero. Figure 13.3 (c) shows the principle of the construction of a transmitter operating in the FHSS mode.

By way of example, the FCC regulations in the USA require manufacturers of equipment operating at 2.45 GHz to use at least 75 frequencies per transmission channel, with a maximum dwell time (occupation time) of 400 ms. With this technique, we can achieve a data rate of about 2 2 Mbits s^{-1}. Faster rates are liable to be affected by numerous errors.

To describe this FHSS technique, we must again define some specific terms.

In Transmission
Frequency hop sequence, or hopping code
The frequency hop sequence, or hopping code, defines:

- the number *n* (or the length) of the chosen frequencies and also
- the temporal sequence in which the carrier frequencies are to follow each other.

This sequence is generally determined by a list or table of pseudorandom frequencies used by the FHSS transmitter to select a specific FHSS communication channel in a given band (see the example below).

Dwell time
This parameter is the time interval for which one of the frequencies of the sequence defined above is occupied or 'lived in' – hence the name of 'dwell time'. This parameter is often expressed in milliseconds, such as 400 ms.

Frequency hop rate
The frequency hop rate, sometimes abbreviated to hop rate, is the rate at which the frequency hopping spread spectrum (FHSS) system changes its transmission carrier frequency. By definition, therefore, the frequency hop rate is the inverse of the dwell time at one of the central frequencies of the FHSS system. It is therefore expressed in hops per second; for example, for a dwell time of 400 ms as mentioned above, the hop rate is 2.5 hops s^{-1}.

Orthogonal hopping code sets
If each of the users uses a different hopping sequence, a number of SS systems can be made to operate simultaneously in the same frequency band without interfering with each other. While one system is using a particular frequency, the others use other frequencies in the same band. Consequently, the FCC requires that the number of transmission frequencies used in the FHSS should be such that a large number of channels can be provided without interference. In each channel, a hopping code set that never uses the same frequency at the same time is called 'orthogonal'.

Detailed Example of a Frequency Hopping Sequence Used in RFID at 2.45 GHz
As mentioned above, RFID base stations operating with the FHSS back scattering method generally use the frequency hopping sequence recommended by the FCC for the Wireless LAN, the IEEE 802.11D standard, as shown below, in order to minimize interference and avoid long periods of collision with systems such as Wireless LAN, WiFi, etc., sharing the same frequency bands.

A frequency hopping pattern, *Fx*, consists of a permutation of all the channel frequencies defined by a table (a 79-channel example is shown in Table 13.1), in which the centre of each channel is spaced from its nearest neighbour by an interval of 0.5 MHz, with the band starting at 2.4225 GHz and ending at 2.4615 GHz.

The IEEE frequency hopping sequences are then used to create a uniform pseudorandom hopping pattern using the whole of the chosen frequency band. For a given number of patterns,

Table 13.1 Example of a frequency hopping table according to IEEE 802.11

Channel number	Frequency (MHz)	Hopping sequence	
		i	$b(i)$
0	2422.5	1	0
1	2423.1	2	23
2	2423.5	3	62
3	2424.1	4	8
4	2424.5	5	43
5	2425.1	6	16
6	2425.5	7	71
7	2426.1	8	47
8	2426.5	9	19
9	2427.1	10	61
10	2427.5	11	76
11	2428.1	12	29
12	2428.5	13	59
13	2429.1	14	22
14	2429.5	15	52
15	2430.1	16	63
16	2430.5	17	26
17	2431.1	18	77
18	2431.5	19	31
19	2432.1	20	2
20	2432.5	21	18
21	2433.1	22	11
22	2433.5	23	36
23	2434.1	24	72
24	2434.5	25	54
25	2435.1	26	69
26	2435.5	27	21
27	2436.1	28	3
28	2436.5	29	37
29	2437.1	30	10
30	2437.5	31	34
31	2438.1	32	66
32	2438.5	33	7
33	2439.1	34	68
34	2439.5	35	75
35	2440.1	36	4
36	2440.5	37	60
37	2441.1	38	27
38	2441.5	39	12
39	2442.1	40	24
40	2442.5	41	14
41	2443.1	42	57
42	2443.5	43	41
43	2444.1	44	74
44	2444.5	45	32
45	2445.1	46	70

(continued)

Table 13.1 *(Continued)*

Channel number	Frequency (MHz)	Hopping sequence	
		i	*b(i)*
46	2445.5	47	9
47	2446.1	48	58
48	2446.5	49	78
49	2447.1	50	45
50	2447.5	51	20
51	2448.1	52	73
52	2448.5	53	64
53	2449.1	54	39
54	2449.5	55	13
55	2450.1	56	33
56	2450.5	57	65
57	2451.1	58	50
58	2451.5	59	56
59	2452.1	60	42
60	2452.5	61	48
61	2453.1	62	15
62	2453.5	63	5
63	2454.1	64	17
64	2454.5	65	6
65	2455.1	66	67
66	2455.5	67	49
67	2456.1	68	40
68	2456.5	69	1
69	2457.1	70	28
70	2457.5	71	55
71	2458.1	72	35
72	2458.5	73	53
73	2459.1	74	24
74	2459.5	75	44
75	2460.1	76	51
76	2460.5	77	38
77	2461.1	78	3
78	2461.5	79	46

x, the hopping sequence can be written thus:

$$Fx = \{fx(1), fx(2), \ldots, fx(i), \ldots, fx(p)\}$$

where $fx(i)$ is the channel number for the ith frequency in the xth hopping pattern and p is the number of channels present in the hopping pattern (79 in the case of FHSS back scattering RFID).

Given the number x of the hopping pattern and the index i of the next frequency (in the range from 1 to p), the channel number is defined by

$$fx(i) = [b(i) + x] \bmod (79)$$

where $b(i)$ is defined in Table 13.1. The base station then uses the second set shown in IEEE 802.11D, where the value of x is defined as

$$x = \{1, 4, 7, 10, 13, 16, 19, 22, 24, 28, 31, 34, 37, 40, 43, 46, 49, 52, 55, 58, 61, 64, 67, 70, 73, 76\}$$

This very unusual set of values was designed to avoid long periods of collisions between different frequency hopping sequences in each set. For example, the sequence of frequency hopping channels created by using $x = 1$ is shown below:

$$F_1 = \{1, 24, 63, 9, 44, 17, 72, 48, 20, 62, 77, 30, 60, 23, 53, 64, 27, 78, 32, 3, 19, 12, 37, 73, 55, 70,$$
$$22, 4, 38, 11, 35, 67, 8, 69, 76, 5, 61, 28, 13, 26, 15, 58, 42, 75, 33, 71, 10, 59, 0, 46, 21, 74, 65,$$
$$40, 14, 34, 66, 51, 57, 43, 49, 16, 6, 18, 7, 68, 50, 41, 2, 29, 56, 36, 54, 24, 45, 52, 39, 31, 47\}$$

The base station starts and runs its normal operating program and then, on receiving a frequency hop trigger signal, it selects a frequency hopping sequence Fj at random from the set shown above. It then makes the RF module switch to the channel described by the selected hopping sequence table and stays there for a maximum period of 400 ms, according to the FCC regulations. On completion of this operation, the base station stops transmitting, starts a timer to count the time for which it remains silent and stores the channel $fj(i)$ that it was using. If a new trigger signal arrives in less than 30 s, the base station continues to use the same sequence Fj and switches to the next channel $fj(i + 1)$. If this is not the case (i.e. if the time exceeds 30 s), the base station selects a new frequency hopping sequence Fk in a random way and restarts the rest of the operation as described above. This procedure results in pseudorandom frequency hops over the whole range of the channels in the band and, on average, equal usage of all the frequencies in the band.

This explanation may appear lengthy and complicated. It may be simpler to imagine a load splitting terminal of a large product distribution organization or a very large logistics company, where there are more than 100 unloading bays in single file each having a base station provided with four antennas operating in multiplexed mode, all transmitting at once and handling every order for hundreds of tags simultaneously. This will enable us to appreciate the true value of these tables and their operating modes.

Spectrum and Bandwidth of an FHSS Channel
Figure 13.4 shows an example of a spectrum of the radiated signal produced by a spread spectrum technique of this type. Clearly, this spectrum occupies a specific bandwidth,

Figure 13.4 Spectrum of the radiated signal produced by an FHSS technique

depending on:

- the initial width of the elementary channel (narrowband);
- the interval (in hertz) between two adjacent channels, allowing for the fact that the channels may be arranged:
  - without a protective space;
  - with a protective space;
  - with a certain permitted degree of overlap between channels.

The bandwidth occupied by the baseband signal in the channel in which it is transmitted can be narrower than the space between channels, to allow frequency tolerances or to provide the necessary protective spacing to enable reliable communication equipment to be deployed.

For FHSS systems and those operating in a narrow band, the bandwidth occupied by the channel must be the maximum authorized bandwidth (measured in Hz) of the modulated signal in the occupied channel. For example, if a single frequency hopping sequence, n, is used and if the channels are adjacent with no overlap, the total bandwidth of the spread signal will be $(n + 1)$ times the bandwidth of the elementary channel. This value $(n + 1)$ will therefore automatically be the same as the processing gain of this type of FHSS with adjacent channels and no overlap.

For FHSS systems, the authorized spacing between adjacent channels is specified by the relevant national authorities (e.g. ARCEP, or Autorité de Régulation des Communications Électroniques et des Postes, in France and the FCC in the USA). In the USA, for example, FCC Part 15, Section 15.247 stipulates that the spacing between adjacent channels must be greater than or equal to the bandwidth of the signal measured at $-20\,dB$ and must also be located within the limits of 25 kHz and 1 MHz.

In Reception

Because of this secrecy, the receiver must know and use the same hopping code as the transmitter, in order to receive the signal correctly. It must then listen to the incident signal at the right frequency, at the right time... not so easy! This is the general theory.

If you have followed my description up to this point, you will certainly be aware that the 'base station/tag' system in RFID at UHF and SHF operates in the 'transponder' mode and uses the back scattering (reradiation) technique. Since the receiver (the base station) of the signal reradiated by back scattering from the tag (where the tag is always tuned to the same frequency as the transmission) is very often physically located in the same enclosure as the transmitter (also the base station) (in fact, 99% of systems are of the monostatic type, or only the antennae are bistatic), knowledge of the hopping code is therefore simple to arrange and does not cost anything! This completes the cycle, and the use of the FHSS technique is therefore clearly suitable for RFID.

Note that, leaving aside the greater resistance to jamming and the assistance with resolving problems of multiple wave paths, back scattering also operates very well in narrowband technology, which can therefore be considered as a degraded form of the FHSS!

13.3.2 Listen Before Talk (LBT)

The LBT (listen before talk) technique also falls within the general category of 'frequency agility' techniques, and in many ways it is related to a simpler form of the FHSS. Furthermore, as explained more fully in the chapters on standards and regulations, Chapters 15 and 16, the FHSS and DSSS modes are difficult or impossible to use in RFID in Europe. These European

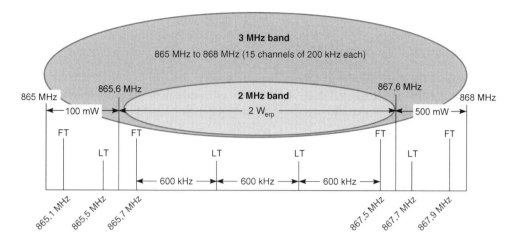

865-868 MHz band, divided into 15 channels of 200 kHz each:

- the 3 lower channels, P_{max} = 100 mW$_{erp}$
- the 10 central channels, P_{max} = 2 W$_{erp}$ (= 3.2 W$_{eirp}$)
- the 2 upper channels, P_{max} = 500 mW$_{erp}$

(a)

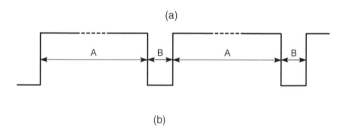

(b)

Figure 13.5   (a) UHF band reserved for LBT. (b) Maximum carrier occupation time

limitations can be partially overcome by using the LBT technique, the operating mode of which is explained below. ETSI EN 302 208-1 and 2 published in September 2004 describe how to use this technique at UHF in a band with a width of 2 MHz (from 865.6 to 867.6 MHz) reserved exclusively for RFID, with a maximum radiated power of 2 W ERP (3.28 W EIRP) during the interrogation of the tags by the base stations, which are referred to as 'interrogators' in this case (see Figure 13.5).

The Operating Principle of LBT

By contrast with the principle used for the FHSS, which can cheerfully hop from place to place in the frequency band allocated to it without considering whether the transmission channel is already occupied, the receivers of the interrogators (base stations) in LBT must be able to minimize interference with other potential users by detecting transmissions from other RF equipment in the frequency band that is used, according to the following criteria:

• If the receiver of an interrogator detects that a channel is occupied by another participant, the interrogator must immediately switch channels before transmitting.

- If the receiver of the interrogator detects that all the channels of the authorized frequency band are occupied by other participants, the interrogator must remain in idle mode. The interrogator must not transmit before its receiver has detected an unused channel.

Operation of the Base Station in LBT

To ensure that the interrogator detects the presence of other equipment with which it might interfere, the receiver of the interrogator must first switch to a band listening mode.

In the Listening Mode

Immediately before each transmission, the receiver of the interrogator (base station) must set itself to the listening mode and check the selected channel for a minimum period of 5 ms, plus a random value (according to 11 possible steps) in the range from 0 to 5 ms, in order to detect the presence of any signal from another participant. In this listening mode, the sensitivity of the receiver of the interrogator must be adjusted to detect a minimum threshold, which is a function of the maximum permitted transmitted power in the band used (Table 13.2).

If the receiver of the interrogator detects a signal with a value above the minimum threshold, this automatically indicates that another device is already occupying the channel. In this case, the interrogator must not transmit, but can scan the band, listening to other channels, for at least 1 ms per channel, until it detects a channel in which the received signal is below the minimum threshold level.

In the Transmission Mode

An interrogator (base station) detecting that a channel is unoccupied can switch to the transmission mode and can then transmit continuously in this channel for a maximum of 4 s. If an interrogator that has started a dialogue with a tag is subject to interference, it can switch to another channel, or switch again, provided that it has rechecked that this channel has been unoccupied for at least 100 ms.

In order to give other users a chance to access the medium, the interrogator must not retransmit on the same channel for at least 100 ms after its transmission. However, the interrogator can immediately listen to all the other channels in the band for a period of at least 1 ms in order to find an unoccupied channel. If the interrogator discovers that another channel is free, it can retransmit continuously for up to 4 s. This process can be repeated for an unlimited number of times.

To avoid unnecessary saturation of the transmission channel, the interrogator parameters must be set to ensure that the transmission period is not longer than that required to read the tags present in the field and to check that there is no additional tag present in the field radiated by the base station.

In conclusion, we can state that the whole LBT system is inherently 'probabilistic' as regards the access to the medium, and therefore it is difficult to establish in a 'deterministic' way the

Table 13.2

Maximum transmission power, ERP, of the interrogator	Detection threshold indicating occupation of a channel (dBm)
Up to 100 mW	−83
From 101 to 500 mW	−90
From 501 mW to 2 W	−96

precise instant at which communication with the tags will be assured or the precise time sequence in which collision management will take place.

Examples

Here are a few brief examples to illustrate all these points. First of all, we must make the assumption (which is not always valid, in view of the cost of the base stations) that the base station is technically well designed and capable of the following:

- switching to listening mode after transmission, and then listening, detecting, changing the reception channel if necessary, listening again, detecting again, and so on, finally switching to transmission mode and re-establishing the carrier to start transmitting in the free channel;
- listening to other channels while transmitting in a specified channel, so that it can rapidly transmit again in a free channel (after 1 ms) – an operation that is easier to describe in a book than to achieve in practice!

In view of the above, as I have already stated, ETSI makes the following specifications:

- The minimum listening time on a channel must be 1 ms.
- Transmission must not be restarted on the same channel until 100 ms have elapsed.
- An interrogator detecting a free channel is authorized to transmit on this channel for a maximum of 4 s.

The operation of the system is inherently 'probabilistic', so let us examine two extreme cases in order to establish its limits. The *best case* is as follows:

- The base station communicates on channel A for x seconds, where $x < 4$ s.
- If the base station is not really an LBT device, but simply intends to use the opportunity to transmit 2 W ERP maximum in the band in question and can only transmit on channel A, it must stop at the end of the maximum period of 4 seconds, for at least 100 ms, which seems a long time, and may be longer if channel A is then occupied by another user.
- If the base station is of the LBT type, with smart software and suitable electronics, it starts to explore the band in which the application is intended to operate, in a systematic, structured, random or pseudorandom way, each step taking 1 ms (or less, if the occupation of a new channel is detected more quickly).
- As a general rule, in a well-designed system (where the base station is the only one, although it does not know this, and is isolated, in a free field and thus without reflections, etc.), the sequence for changing reception channels, listening, detection, etc., and re-establishing the carrier in a new channel takes about 5 to 10 ms. Here is an example. The first channel to be explored, B, is free, and can therefore start transmitting again after about 7 to 8 ms or so. In this case, the maximum overall duty cycle, 'on/(complete period)', therefore becomes $4000/(4000 + 7) = 99.83\%$. Let's not be too particular in this case: we will call it 100%.

In the *worst case*:

- The base station communicates on channel A for x seconds, where $x < 4$ s.
- Then there is an unfortunate incident (a common occurrence).

The base station is installed at an unloading bay of a distribution/load splitting/sorting terminal of a major company (with several base stations per gate and numerous gates in line over a long distance, with reflective floor surfaces, etc., the whole system being officially referred to, because of the number of base stations present, as a *multiple environment* or *dense environment*.

The first channel to be explored, B, is occupied, and so after not more than 1 ms (having ensured that nobody has stopped transmitting in the course of this millisecond) the system goes off to hunt for another. The listening frequency is changed (requiring a few milliseconds of preparation) and listening takes place again for not more than 1 ms. Thus, after about 50 to 60 ms, the system will have explored the whole band (the 10 channels of 200 kHz each in the 2 MHz-wide band; see the relevant chapters on regulations, Chapters 15 and 16). During the same period, all the other base stations on the site will have done the same, modifying their carrier frequencies and thus playing hide-and-seek with our base station. Assuming that the base stations use a random selection method or orthogonal pseudorandom codings (such as those used in the FHSS and DSSS – although this case is more complicated because there are only ten channels available) to determine the transmission frequencies of some and the listening frequencies of others, and since there are many base stations and numerous reflections, it may take some time (about half a second to a second) to find a free channel and to start transmitting, in the hope that no other base station has had the same idea at the same time (leading to a collision of carriers), which would damage our data (data collision). You can expect to have a duty cycle of about, $4/(4 + 0.5) = 88.8\%$, i.e. losses of 11.2%, which, in terms of time, means that a base station will be 'blind' for about 7 minutes in every hour.

Furthermore, experience in these conditions has shown that, depending on the readers used (with or without smart software) and the planned tag management (collision management, reading of different amounts of data, etc.), we could expect to deal with 30 to 100 tags in one second. Anything above this level is the stuff of dreams! (Remember that a forklift truck travelling at $6 \, km \, h^{-1}$ will cover $6000/3600 = 1.66 \, m$ in a second.)

To conclude the discussion of this subject, as you will have noticed, the aim of LBT technology is to make the reader change channel continually, at least every 4 seconds, thus providing a European near-equivalent of the FHSS system, which is impossible or difficult to use because of the local regulations in Europe.

13.4 Spread Spectrum Systems for Spreading the Radiated Spectrum of Wideband Modulated Carriers

13.4.1 A Direct Spectrum Spreading System: DSSS (Direct Sequence Spread Spectrum)

The DSSS technique is a special method for spreading the spectrum of radiated RF signals when the carrier is modulated, by including a 'spreading sequence' in the modulating signal.

Briefly, in the DSSS technique the signal for (phase) modulating the radiated RF carrier is normally produced by 'multiplying' the initial data signal (the data to be transmitted) by a pseudorandom digital signal (call a 'chip sequence', see below), where the chip rate is several times the data bit rate. The result of this technique is that the spectrum of the signal radiated in a carrier having a constant frequency (by contrast with the FHSS) is much wider than that of the initial signal representing the data (Figure 13.6).

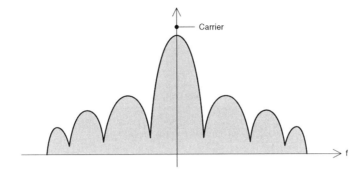

Figure 13.6 Radiated spectrum in the DSSS

Note that, in some cases, phase modulation implies PSK, QPSK or xQAM, and therefore a number of states, as well as concepts of symbols and relations between bit rates, symbol rates and 'randomization' of the energy with an energy dispersion sequence using a special pseudorandom sequence of the kind generally found in DVB (digital video broadcast) transmission in MPEG2.

Let me explain some specific terms used in these modulation techniques.

Chip (in the Context of Digital Radio Communications)

The chip is one of the main elements in the device used by the DSSS technique to prepare for the spreading of the radiated signal spectrum. The chip is the smallest data coding element for spreading the spectrum of the data to be transmitted. Figure 13.7 illustrates the concept of the chip.

The chip is defined by its value, its period and its rate:

- Value: a chip can have one of two values, namely $-1/+1$ in polar notation or $0/1$ in binary notation.
- Its period (t_{chip}) is generally expressed in µs: of course, this signal has its own period t_{chip}. Although this is not obligatory, the period is generally an integer submultiple of the data bit period.
- For the rate, see 'chip rate' below.

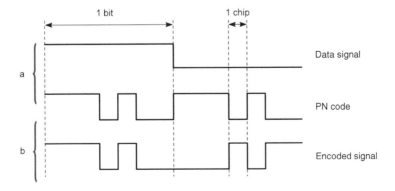

Figure 13.7 Examples of chips

Chip Rate, R_c

If there is a period, there must also be a rate. By definition, the value of this rate R_c, expressed in chips s^{-1}, is called the chip rate and is the inverse of t_{chip}.

Processing Gain in the DSSS

In the DSSS, this parameter is defined by the following ratio:

$$\text{DSSS processing gain} = t_{bit}/t_{chip}$$

Once again, an increase in this ratio is accompanied by an increase in the bandwidth of the communication channel and the spread of the spectrum, and an improvement in the performance of the system.

For information:

- The American regulatory authority FCC does not permit values below 10.
- The IEEE 802.11 (WiFi) working group specifies a processing gain of 11.
- Most commercial systems operate with a value of about 20.

All other things being equal, the DSSS technique can be used to achieve much higher bit rates than 2 Mbits s^{-1}, by contrast with the FHSS system.

Spreading Sequence

The spreading sequence is the pseudorandom sequence (based on a pseudorandom noise code; see the next section) of the data coding elements (chips) used to encode each data bit. To understand how it works, we need to see what it is made of.

The Number of Chips n in the Sequence (or Length)

In theory, n can be any number. In fact, all other things being equal, an increase in the sequence length causes an increase in the desired spectrum spreading effect.

The Period

The period of the spreading sequence is clearly related to the number n of chips in the sequence and the period t_{chip} of each of them: period $= n t_{chip}$. Note that the sequence rate is equal to the chip rate.

However, a problem arises here: for many reasons, which will be discussed later in this chapter, we generally require this period to be equal to the data bit period, but we also want the value of n to be high, leading to rather severe demands in terms of t_{chip} when high data rates (bit rates) are used.

Its Value

The effects of the value of this spreading sequence are such that it cannot be seen as neutral. In fact, this value must be used in such a way that the result of the signal processing is a spread spectrum, but one that meets certain requirements.

To achieve good results, in other words to provide a system resistant to jamming and spying, etc., the resulting spectrum must be as close as possible to white noise, so that all the components of the spectrum are represented and the spread is as perfect as possible. We must therefore consider how its structure is created.

The Structure of Its Value
To ensure that the spreading sequence, composed of chips with values of $-1/+1$ (in polar values) or 0/1 (in binary), has properties similar to those of noise (preferably white noise), there must be a very low correlation between the codes of the sequence, and these must be pseudorandom noise codes (PN codes).

Pseudorandom Noise Codes (PN Codes)
To be usable for DSSS applications, a PN code must meet the following requirements:

- The sequence must be constructed using two-level numbers (which is the case, fortunately).
- The codes must have a low cross-correlation value. The smaller this value, the more potential users can be present in the same band. This must be true of all or part of the PN code. The last comment relates to the fact that, in most RFID applications, the tags enter the electromagnetic field at random times, and two or more PN codes may be superimposed at any time and in any way, which may impede dynamic collision management.
- The PN codes must be 'balanced': in other words, the difference between the number of '1' and '0' must be equal to 1, to ensure a uniform spread of spectral density over the whole frequency band.
- The PN codes must have precise autocorrelation (with the width of a chip) to ensure the synchronization of the code.

In practice, for DSSS applications, PN codes are based on Walsh–Hadamard codes, M-sequences, Gold codes or Kasami codes. In mathematical terms, these sets of codes can be roughly divided into two classes: orthogonal and nonorthogonal codes. Orthogonal codes are those in which there is no interaction between the results (Figure 13.8).

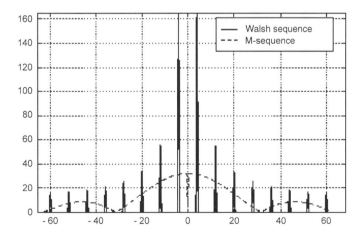

Figure 13.8 Walsh sequences and M-sequence

Walsh sequences belong to the first category, while the others are sequences generated by well-known shift register devices. Thus it is easy to create PN code sequences.

In DSSS systems, if the length of the shift register is n, we can say that the period of the families of codes mentioned above is such that the length of the code is the same as that of the 'spreading factor'. In this case, if each data symbol is combined with a complete PN code, the DS processing gain is equal to the length of the code, and, for all the reasons mentioned above, it is easy to achieve a high DS processing gain.

Example of a spreading sequence
By way of a summary, here is an example of a spreading sequence:

- The DSSS system sends a specific string of bits for each data bit to be transmitted.
- A chipping code is assigned to represent the data bits 1 and 0.
- As the data stream is transmitted, the corresponding code is also transmitted.

For example, the physically transmitted sequence 00010011100 reflects the transmission of a data bit equal to 1.

The chip rate is the rate at which the spreading sequence modulates the carrier frequency.

The Operating Principle of the DSSS
The DSSS system is one of the most widely used SS techniques. Its operating principle is shown in Figure 13.9.

The system has a number of essential steps:

- First of all, the logical data for transmission arrive at a specific data bit rate (Figure 13.10).
- The actual bit coding is then carried out. To avoid excessively complicated explanations, we will consider the simplest form of bit coding, namely NRZ. The spectrum associated with the electrical signal representing these logical data encoded by this bit coding procedure is called the 'baseband spectrum'.

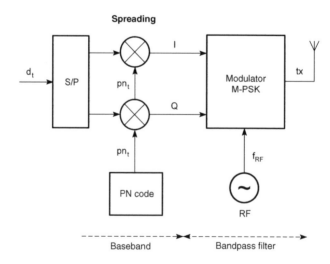

Figure 13.9   The operating principle of DSSS

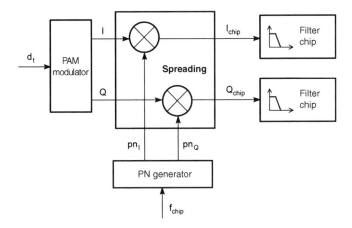

Figure 13.10   Coding and signal input

- A pseudorandom noise code (PN code) is then created, using a shift register to provide a sequence of n chips, each of which has a given period and therefore a specific chip rate. This sequence therefore has a precise total duration, nt_{chip}. This duration is generally equal to the data bit period: $nt_{chip} = t_{data}$.
- A multiplier is used to multiply the instantaneous value of the data bit by the instantaneous value of the PN code. The signal leaving this multiplier has a rate equal to the faster of the two, in other words the chip rate. Its baseband spectrum is therefore larger than that of the initial logical data.
- An HF modulator (AM, BPSK, QPSK, xQAM, etc.) then receives the output signal from the multiplier in order to 'construct' the DSSS signal, called the PN modulated information signal.

Note

The choice of the type of modulation for modulating the final transmitted signal is also very important. One of the benefits offered by the DSSS technique is the fact that, among all the types of carrier modulation, it is possible to use the following forms of phase modulation:

- conventional BPSK (binary phase shift keying) with two phase states, each corresponding to the binary value of the incident signal;
- QPSK (quadrature phase shift keying) with four phase states, in which two successive binary elements of the data form what is known as a 'symbol' (Figure 13.11).

Since it carries two bits per symbol, this modulation therefore enables the real data rate to be increased by a factor of two. The bandwidth of the data signal is now multiplied by a factor of two. On the other hand, the power content remains the same, and therefore the spectral power density is reduced. Of course, this causes an equivalent reduction in the available processing gain. The processing gain is reduced because, for a given chip rate, the bandwidth (used to define the processing gain) is divided by two because of the increased information transfer.

To conclude this note, the above description makes it clear that this form of modulation (QPDK) is to be recommended when we wish to have a high bit rate, provided that the environment in which the RFID system is implemented is not excessively affected by parasitic signals; otherwise the loss of symbols (through errors in the demodulation of the groups of symbols received) may reach disastrous levels.

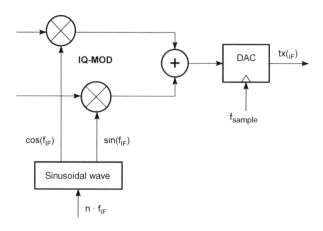

Figure 13.11 The modulation stage

• Finally, a balance double mixer is used to multiply the RF carrier by the PN modulated information signal, in order to transpose the spectrum of this signal to RF.

As a result of this signal processing (DSSS) carried out with the aid of PN codes, the transmitted RF signal occupies a very large bandwidth. Its frequency spectrum is substantially equivalent to that of a noise signal, while its spectral power density falls below the noise level without any loss of information. Figure 13.12 shows a graphic image of this, in the form of the radiated spectral content of a DSSS signal in a case in which the signal is subjected to conventional BPSK phase modulation. The form of this spectrum is therefore of the $(\sin x/x)^2 = \text{sinc}^2 x$ type.

Finally, to sum up:

• DSSS uses the PN code to divide or chop up the 'bit' data for transmission into 'chips'.
• The chip rate is generally of the same order as, or greater than, the data signal.

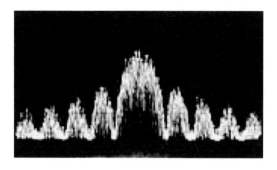

Figure 13.12 Real spectrum of the DSSS

- The chips are then modulated and transmitted.
- The PN code has the function of making the signal spectrum wider.
- The spread of the radiated signal is of the $\mathrm{sinc}^2 x$ type.

In view of the above concepts, we must define some new parameters.

Bandwidth of a DSSS Channel

For systems using DSSS, we generally define the bandwidth occupied by a spread channel as the value of the frequency band lying within the two 'nulls' limiting the main lobe of the radiated spectral density (i.e. the 'null-to-null band' or the 'frequency difference between the main lobe nulls') of the DSSS signal in the occupied channel, shown as $2R_c$ ($R_c = $ chip rate) in Figure 13.13(a) and (b).

Receiver

In the receiver, the reverse operation of 'despreading' is the same as the operation for spreading the spectrum. The received signal is again multiplied with the same (synchronized) PN code.

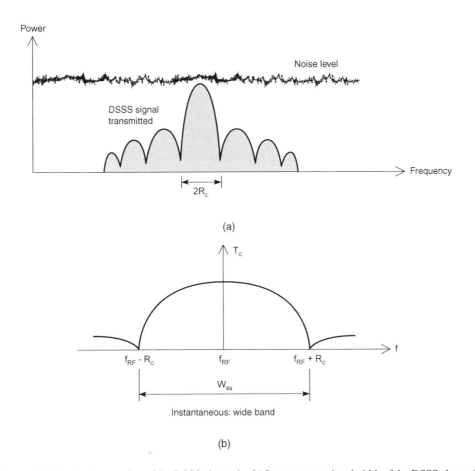

Figure 13.13   (a) Occupation of the DSSS channel. (b) Instantaneous bandwidth of the DSSS channel

Table 13.3

	FHSS	DSSS
Obtaining a high processing gain	Difficult	Easy
Possibility of a high data rate	Medium	High
Tolerance of interfering signals	Very good	Good
Presence of near (or very near) interfering transmitters (near/far effect)	Very little effect	Major problem
Ease of integration into existing systems	Easy	Complicated
Operating distance/range	Good	Better
Fast PLL required for the hop sequence?	Yes	n.a.
Bit rates at which best compromise is achieved	$<2\,\mathrm{Mbits\,s^{-1}}$	$>2\,\mathrm{Mbits\,s^{-1}}$
Cost	Low	High
Power consumption	Low	High

Since this operation is based on the values $+1$ and -1, it completely deletes the coding of the received signal and enables the original signal to be restored. Consequently, any jamming signal present in the channel will be dispersed before the data are detected, thus reducing the effects of jamming.

For example, in the case of BPSK RF modulation, demodulation is carried out by mixing/multiplying the same PN modulated carrier with the incident signal. At the output of this operation, we obtain a signal that reaches a maximum level when the two signals are strictly equal, or when they are correlated. The correlated signal is then filtered and sent to the BPSK demodulator.

Minimum Bandwidth of the Receiver
This is the minimum range of frequencies (all frequencies or the only frequency) that the receiver must be capable of receiving.

For a first approximation to its value, note that the power contained in the main lobe (equal to $2R_c$) of the radiated signal is equal to about 90% of the total power. You should also note that the bandwidth that includes half of the power of the main lobe is equal to $1.2R_c$, i.e. 45% of the total power. If the latter value is considered to be sufficient, this means that we can produce receivers whose input bandwidth is smaller than the whole radiated spectrum, causing only a slight rounding of the rising and falling edges of the signals in the time domain.

Comparison Between FHSS and DSSS
This lengthy description is summarized in Table 13.3, which shows the main structural differences between systems using the FHSS and DSSS.

13.5 'Hybrid' Spread Spectrum Techniques: DSSS and FHSS

Of course, we can always try mixing different systems together in the hope that something useful will emerge. Sometimes, mixtures, or marriages, or hybrids, are created, with the aim

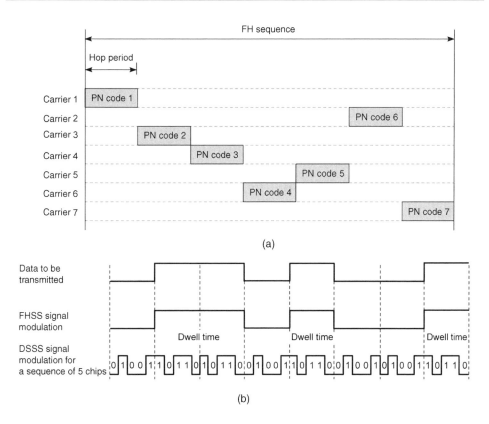

Figure 13.14   (a) Time multiplexing of PN codes. (b) Generation of the signals to be transmitted

of benefiting from some of the best features of both the FHSS and DSSS. The DS/FH spread spectrum technique is one combination of the FHSS and DSSS.

A data bit is distributed over several frequency hop channels (carrier frequencies). In each frequency hop channel, a PN code of integer length is multiplied with the data signal (Figure 13.14(a) and (b)).

Since the FH sequence and PN codes are coupled, an address is a unique combination of FH sequences and PN codes. To minimize the probability of collisions (the probability that two users will share the same frequency at the same instant), the frequency hopping sequences are chosen in such a way that the two transmitters having different FH sequences do not share more than two frequencies at the same time. This is often called the time-shift random technique.

13.6 Back to the Future

Table 13.4 shows the choices made by the experts of ISO SC 31/WG 4/SG 3 (in other words the RFID air interface subgroup) for the ISO 18000-x family of standards governing the RFID

Table 13.4

18000-x	Frequency		Narrow band	Spread spectrum		
				DSSS	FHSS	LBT
18000-2	<135 kHz		x	n.a.	n.a.	n.a.
18000-3	13.56 MHz	Mode 1	x	n.a.	n.a.	n.a.
		Mode 2				
				Option	Option	
18000-4	2.45 GHz	Mode 1	x	US	US/Europe	n.a.
		Mode 2				
18000-5	5.8 GHz	On ice!				
				Option	Option	
18000-6	860 to 960 MHz	Mode A	x	n.a.	US	Europe
		Mode B	x	n.a.	US	Europe
		Mode C	x		US	Europe
18000-7	433 MHz			n.a.	n.a.	n.a.

air interface layer for RFID processing of objects, for different frequencies and fields of application.

Note

The operation of the physical layer and the communication protocols in ISO 18000-4 Mode 1 (at 2.45 GHz) and ISO 18000-6 Mode B (860 to 960 MHz) is very similar, making it possible to use exactly the same integrated circuit, regardless of the operating frequency. The back scattering return path also makes it possible to provide the forward channels, and consequently the return channels, regardless of whether spreading techniques (FHSS or DSSS or hybrid) are used or not (narrowband).

Here are some specific examples of commercially available systems.

13.7 Examples at SHF

13.7.1 Detailed Example 1: FHSS, ISO 18000-4 Mode 1 (and 18000-6 Mode B)

Here, in Tables 13.5 and 13.6 are some specific examples of commercially available systems.

13.7.2 Detailed Example 2: Hybrid: DSSS Forward Link, FHSS Return Link (Example: the Intermec System)

Tables 13.7 and 13.8 are more examples of commercially available systems.

Table 13.5 Forward link, from the base station to the tag

Bit coding	Manchester
Duty cycle	50%
Bit rate of the transmitted data	20–40 kbits s^{-1}
Nominal carrier frequency	2.45 GHz
Type of carrier modulation	ASK 100%, called OOK (on/off keying)
On/off duty cycle	50% (due to 50% Manchester bit coding)
Ratio of on/off carrier levels	40 dBc
Carrier spreading	Yes
Type of spreading	FHSS
Spread across channels	Yes, over 79 channels with central points varying from 2422.5 to 2461.5, in steps of 0.5 MHz
Hop rate	Essentially dependent on the regulations in different countries: for the USA (FCC Part 15, Section 15.247) it is 400 ms
Frequency hop sequence	The pseudorandom hopping sequence table is shown in Table 13.1
Table	IEEE 802.11 hopping sequence
Spectral occupation of channel	Bandwidth at -20 dB $= 0.5$ MHz (conforming to FCC Part 15)
Maximum power, EIRP	36 dBm, made up of:
	- maximum amp. output level: 30 dBm
	- gain of base station antenna: 6 dB

Table 13.6 Return link, from the tag to the base station (in the back scattering mode)

Forward carrier frequency	2.45 GHz, unmodulated
Carrier spreading	Yes
Type of spreading	FHSS; for details, see below
Modulation of antenna impedance	OOK (on/off keying): back scattering is provided by modulation of the tag antenna impedance
Data bit coding (format of the signal in the baseband)	FM0 (frequency modulation), also known as bi-phase space
Duty cycle of the FM0 bit coding	50%
Data rate of the transmitted data	20–40 kbits s^{-1}

13.8 FHSS, LBT, DSSS... and RFID

Clearly, although the attenuation due to the medium varies with $1/r^2$, the signal received at long distance varies with $1/r^4$ and its value is often at the same level as the ambient industrial noises, all these techniques are used in RFID at UHF and SHF because the communication distances are long.

Admittedly, as I have already emphasized, it seems to be difficult to make the tag follow the transmitter's frequency hops, since the circuits on the tag do not know the hopping sequence. However, in RFID this is unimportant, since the system operates in the back scattering mode and therefore the tag reflects some of the incident wave, which is consequently at the same frequency as that of the radiation reaching the tag. Furthermore, the receiver of the base station is generally located in the same unit as the transmitter (regardless of whether the systems have monostatic or bistatic antennas); consequently, it is easy for the receiver to know and constantly follow the same sequence during reception. Alternatively, it is possible to use a wideband input stage, which is less satisfactory as regards noise, but much simpler in terms of circuitry.

Table 13.7 Forward link, from the base station to the tag

Data bit coding	Manchester (50%)
Bit rate of the transmitted data	307.7 kbits s^{-1}
Nominal carrier frequency	2.442 GHz (fixed frequency)
Type of carrier modulation	ASK 100%, called OOK (on/off keying)
Ratio of on/off carrier levels	40 dBc
On/off duty cycle	50% (due to 50% Manchester bit coding)
Carrier spreading	Yes
Type of spreading	DSSS
Spreading sequence	31 chips maximal length pseudorandom noise (PN)
	Sequence 001 1010 0100 0010 1011 1011 0001 1111
	A logical '1' bit is a complete cycle of noninverted PN
	A logical '0' bit is a complete cycle of inverted PN
Chip rate	9.54 MHz ($=31 \times 307.7$ kbits s^{-1})
Maximum power, EIRP	36 dBm, made up of:
	- maximum amp. output level: 30 dBm
	- gain of base station antenna: 6 dB

Table 13.8 Return link, from the tag to the base station (in the back scattering mode)

Central forward carrier frequency	2.45 GHz, not modulated by the interrogator
Carrier spreading	Yes
Type of spreading	FHSS
Spread across channels	75 channels in three bands of 25, with central points at 2442, 2418 and 2465 and 0.4 MHz channel spacing
Hop rate	Depends on the regulations of different countries For the USA (FCC Part 15) it is 0.4 s
Frequency hop sequence IEEE 802.11 hopping sequence	Pseudorandom hopping
Spectral occupation of channel	Bandwidth at -20 dB $= 0.5$ MHz (conforming to FCC Part 15)
Bit coding (baseband signal)	NRZ – space (NRZ, invert on zero)
Bit rate of the transmitted data	149 kbits s^{-1}
Carrier modulation	Multiplication of a modulated subcarrier with the carrier
Subcarrier frequency	597 kHz $= 4 \times$ data rate $= 4 \times 149$ kbits s^{-1}
Subcarrier modulation	BPSK

Final note

One common question relates to the time for which the carrier frequency remains present, relative to the wave propagation time on the forward and return paths between the base station and tag.

The speed of light is about $300\,000\,\text{km s}^{-1}$, giving a propagation constant of about $3.5\,\text{ns m}^{-1}$ in round numbers. The usual maximum operating ranges for remotely powered RFID are of the order of $10\,\text{m}$, giving a round trip length of $20\,\text{m}$ or a travel time of $75\,\text{ns}$. If we assume that this propagation time must be negligible with respect to the other similar values in RFID applications, it must be at least 10 times smaller than the smallest maximum time value of the system. In the present case, this time would be $75 \times 10 = 750\,\text{ns}$. We can therefore consider using systems in which the shortest time element signifying the presence of a frequency is of the order of a microsecond (μs), implying frequency hops at a hop rate equal to a frequency of 1 MHz. As you will have seen in the preceding examples, the values found in practice are much greater!

14

Interactions and Conclusion

14.1 Relations, Interaction and Performance: How They Are Affected by the Choice of Bit Coding and the Types of Modulation Used

When choosing a bit coding method, and consequently a decoding method, we must always bear in mind that it needs to be efficient in all respects, but also as simple as possible, to keep the chip area and the power consumption of the tag and base station (possibly a portable reader) as small as possible.

This choice, as well as the choice of bit rates, carrier modulation principles and power requirements (also related to the bit coding, see above), has a direct effect on the sidebands of the transmitted signal spectrum, and therefore on whether or not the system will comply with the radiation compatibility and sensitivity standards and regulations (FCC, ETSI, etc.) (see Chapter 16).

14.1.1 Effects of the Choice of Bit Coding and Modulation Type

The following paragraphs provide some notes on the effects and consequences of the choices made concerning the bit coding and types of carrier modulation used. These notes are based on long experience, including both theoretical (protocol and spectral research) and practical on-site study of, for example, types of noise and parasitic phenomena due to industrial environments. This comment is aimed especially towards newcomers in this field, who often tend to make choices based on general principles, only to find that the resulting designs prove to be disastrous on the ground.

Let me remind you once again that, in RFID contactless communication, the technical solution adopted must be uniform in all respects, including the principle and type of modulation, protocol, collision management, noise sensitivity, net bit rate, bit coding, permitted transmitted power, range, sideband width and level, radiation standards – and much more besides!

RFID at Ultra and Super High Frequencies: Theory and Application   Dominique Paret
© 2009 John Wiley & Sons, Ltd

The following section will provide a detailed examination of the following matters:

- the efficiency of the power source;
- the immunity to noise;
- bit counting and bit error problems;
- frame synchronization and format error problems.

14.1.2 Forward Link

Efficiency of the Power Source and Energy Efficiency
Let us see how energy efficiency is affected by the type of modulation.

In ASK 100% or OOK
The ASK 100% modulation, or OOK, the principle is a very special case of the ASK modulation type, because, depending on the bit coding (e.g. Manchester encoding), the carrier is completely (or almost) interrupted for short periods of time, known as pauses (Figure 14.1(a)).
We can therefore distinguish two operating phases in the forward link:

- Phase 1: during the presence of the carrier. During this phase, the reader sends the full power level to the tag, which is free to use it.
- Phase 2: during the pause and the absence of the carrier.

Since almost all tag integrated circuits available on the market are currently made in CMOS technology, known as a 'static' technology, regardless of their operating modes (which may be synchronized with the presence of the carrier, or, more commonly at UHF and SHF, not

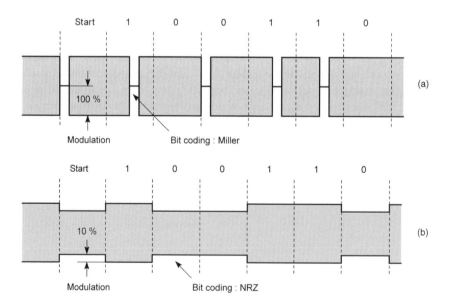

Figure 14.1 (a) ASK 100% modulation Miller bit coding. (b) ASK x % modulation and NRZ coding

synchronized with it, simply because of the power consumption of HF dividers), the clock signal supply to the logical circuits (such as wired or micro-controlled logic and memories) stops when the carrier is cut off (during the pause). Consequently, since there is no clock, the circuits momentarily cease operating and switch to idle mode.

Since the logic circuits of the tag are no longer consuming energy, the filtering and power supply capacitor in the transponder integrated circuit remains fully charged, thus making it possible to carry out other similar operations, which consume very little power, to ensure the correct operation of the link and provide the internal security and protection. In this case, this type of modulation is as efficient as FSK modulation in energy terms, since 100% of the energy is delivered only when necessary, regardless of the binary content of the bit.

Moreover, this carrier cut-off ('pause') forms an integral part of the signal representing the bit information, and therefore has no effect on transaction time. Clearly, all other things being equal (regarding synchronization quality, frame loss, etc.), a decrease in the carrier cut-off time with respect to the bit period increases the capacity for establishing high communication rates. To summarize the last two sentences, we can say that the base station provides the whole amount of energy when the tag needs it, but provides no energy when it is not needed, and the tag can use the maximum available power to supply itself, or, all other things being equal, can operate at a longer range.

In ASK x % (Modulation Index m = x %)

This type of carrier modulation and the associated bit coding both contribute to the supply and quality of the energy source available at the tag.

In ASK x % modulation (Figure 14.1(b)) and in LF and HF RFID, where the tag operates in synchronization with the carrier, the tag integrated circuit consumes power constantly because of the permanent presence of the carrier (and therefore the clock signal). If the base station carrier is modulated in ASK x % mode by the data (with the same rectification/filtering capacity and power consumption as those of the tag), the power available for the operation of the tag is reduced proportionally to the square of the modulation index, thus reducing the mean level of energy available to supply the transponder.

By contrast with the previous case, the break due to the pause described above is no longer applicable with ASK x %, and the bit coding has a direct effect on the time for which the carrier amplitude is reduced to transmit information. The shorter the bit time, the more energy is available at the transponder.

Now let us look at the effects of this, using two examples of bit coding.

With NRZ Bit Coding

Consider a generic example used in contactless RFID, which can easily be adapted to the UHF parts of ISO 18000-6 A, B and C. Thus, in ASK 10% (nominal modulation index $m = 10\% \rightarrow V_{min}/V_{max} = 81.8\%$; see the section above on the modulation index), using NRZ coding, a '0' signifies an 18.2% reduction in the carrier amplitude and, given that the modulation index is often defined as 8 to 14%, this gives in the worst case a maximum reduction in carrier amplitude of 24.5% ($m = 14\% \rightarrow V_{min}/V_{max} = 75.4\%$) throughout the whole period of the '0' bit or, in an even worse case, during a long sequence of '0' (Figure 14.2). During this time, the integrated circuit, which is undersupplied by the presence of a reduced carrier, continues to operate, and therefore consumes power as it does in the presence of a '1'. Moreover, the transmitted power, and therefore the possible operating range, varies not directly as a function of the recovered voltage but as a function of its square (V_2/R); in other words, all

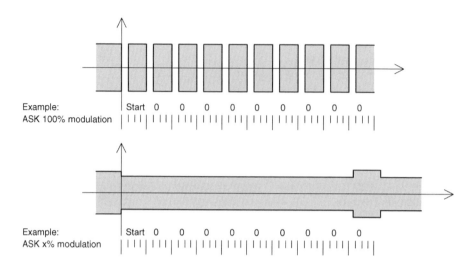

Figure 14.2   The worst case of remote supply of the tag in ASK x %

other things being equal, the correct useful operating range is reduced by 0.754×0.754 56.8%, a value limited by the under-supply of the transponder, which can occur during the transmission of momentary sequences of '0' (which is necessary and always possible).

With Manchester or Miller Bit Coding
Using the Manchester coding only reduces the carrier amplitude during half of the bit time, regardless of the bit value, so in statistical terms it would be expected to halve the efficiency of the energy supply. Miller or modified Miller bit coding would be a much better solution for minimizing the possible carrier reduction, while retaining the 'self-clocking' aspect of Manchester coding.

By Way of Conclusion
For the forward link (from the base station to the tag), the energy efficiency of ASK 100% modulation is better than that of ASK 10% modulation, and, if it is permitted by local regulations, it results in greater operating flexibility because:

- The transaction range is greater.
- The processing power capacity is greater for a given range.
- A less powerful base station can be used for the same range.

On the other hand, we must remember that, in terms of bandwidth occupation, the sideband level is greater in ASK 100% than in ASK 10%.

Reliability of Communication, Immunity to Noise, Signal-to-noise Ratio
and Incident Signal Detection
Reliability of communication and immunity to noise are important factors for the correct operation of a system, especially in the case of difficult electromagnetic environments in which most long range (i.e. UHF) RFID devices have to operate. The higher the energy in the signal representing the useful information, the greater will be the reliability of communication and immunity to noise.

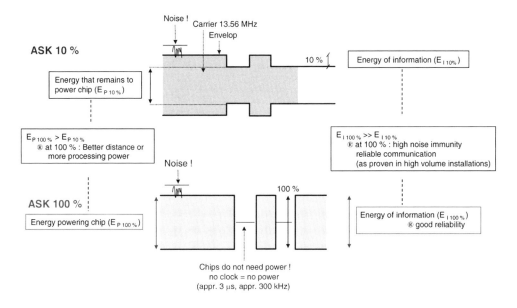

Figure 14.3   Example of comparative immunity to noise

It is evident (Figure 14.3) that, compared with ASK x % modulation, ASK 100% modulation is:

- more energy-rich (the sidebands of the spectrum are higher);
- more reliable as a result;
- less disturbed by the presence of noise.

Moving Tags
We must also consider that the tag is generally moving while in operation relative to the base station, even if the movement is small; this creates variations in the PD received at the terminals of the tag antenna. These levels of PD are of the same kind as those produced by carrier modulation created intentionally by the base station, and may interfere with the proper detection of the incident signal.

Since the tag cannot recognize that an unintended signal is superimposed on the intended signal, this significantly reduces the acceptable *S/N* ratio. This is even more noticeable in the case of ASK x % modulated signals, which are already smaller compared with those modulated by ASK 100%. Figure 14.4 clearly shows the problems that may arise.

Effect of the Type of Bit Coding of the Forward Link
To make matters quite clear, in radio frequency communications the only really reliably detectable effects are deliberate changes or variations in the signal, where the variations in amplitude are consequently more significant.

NRZ
In an NRZ encoded signal, there is no transition in the transmitted and received signal during sequences of consecutive '0' or '1'. A variation in the amplitude of a reference level (due to the movement of the tag, for example) therefore causes a considerable reduction in the *S/N* ratio.

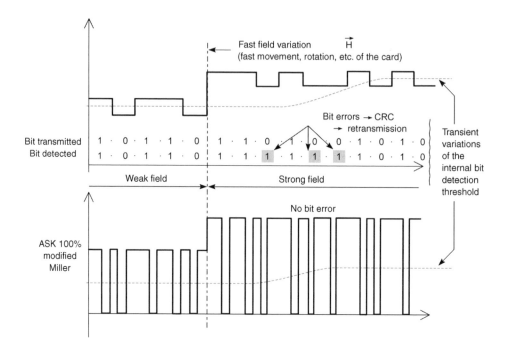

Figure 14.4   Effect of tag movements on the quality of the link

Manchester or Miller and Modified Miller
If Manchester, Miller or modified Miller bit coding is used, there are regular changes of state, making it easy to refresh the reference levels.

At this point, we may conclude that the 'worst' system is a combination of ASK x % and NRZ, and the 'best' is the combination of ASK 100% and Manchester or modified Miller.

Bit Counting and Bit Error

NRZ
As I have shown, NRZ bit coding cannot be used to provide a range of logical states such as '0', '1', 'no data', etc. It is therefore difficult to detect an end-of-transmission frame in a reliable way in the lower layers.

Manchester or Miller and Modified Miller
Conversely, Miller, modified Miller or Manchester encoding can be used to encode '0', '1' and 'no data', thus enabling the tag to use a counting method to verify and check the conformity of the number of bits in the received frame. This enables possible transmission format errors to be identified quickly, without the need for parity bits or CRC, which are major consumers of supplementary bits and consequently of communication time. The main advantage of these types of bit coding is that the probability of error detection in the lower layers is greatly increased, thus allowing more reliable and rapid error recovery.

   In terms of the communication protocol (the upper layers of the OSI model), this means that a device must be implemented for retransmitting data if the receiver detects an error. The initial error detection must be carried out with a very high probability, to avoid their detection at higher levels and more serious problems. This can be achieved by using parity checking, CRC, bit counting, etc., as described above, at the cost of a reduction in the net bit rate or an unnecessary or unjustifiable increase in the rate. When these errors have been detected, the communication protocol is responsible for retransmission, which by its very nature reduces the effective transaction speed between the base station and tag. The bit coding chosen for use must therefore be such that it does not reduce the bit rate, or reduces it as little as possible. This is the case with Manchester, Miller and modified Miller coding, but not with NRZ.

Frame Synchronization and Format Errors

It is all very well not to have bit counting error during a frame, provided that we have also succeeded in marking the start of the frame correctly. It is also essential to avoid wasting time in detecting the start of frame.

NRZ

NRZ bit coding is based on the assumption of correct synchronization on the start bit. Consequently, there is a high probability of incorrect synchronization in a noisy environment. We must therefore design an upper-layer communication protocol with an overhead (header, preamble, etc.) to provide retransmission and resynchronization protocols, which greatly increase the transaction management time.

Manchester or Miller

Miller and Manchester bit coding carry out synchronization on a start bit, but additionally, because of the self-clocking bit coding, the synchronization redundancy ensures correct frame synchronization, even in environments where the electronic noise is considerable, e.g. by averaging the bit clock pulses to achieve final synchronization. The efficiency of this coding even makes it possible to send frames (data, commands, etc.) without using start and stop bits. For commands of not more than one byte (8 bits), this saves 20% of the time (by comparison with symbols composed of 10 bits, i.e. start bit + 8 data bits + stop bit), and provides a much faster effective net data rate.

14.1.3 Return Link

Energy Efficiency of the Bit Coding

Figure 14.5 again shows the two types of bit coding that can be used for the return link, from the tag to the base station, operating on the load modulation principle. These are the Manchester or Miller encoded subcarrier and BPSK types. In terms of their noise resistance, both have many transitions and can therefore be strongly recommended.

As mentioned above, for simple reasons of synchronization, BPSK coding usually requires a clock run in sequence and a PLL in the receiving system of the base station, but, apart from that, all other things being equal, the operating range of a system using BPSK coding is inherently shorter than that of a system using the Manchester or Miller encoded subcarrier method. This is because, as Figure 14.5 shows, the time interval in which the power supply capacitor on board the transponder can be recharged while the transponder is communicating is 75% in the Manchester encoded subcarrier system but only 50% in BPSK, making it possible to communicate over a longer range with the former system.

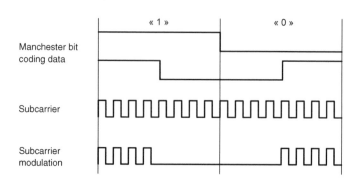

Figure 14.5   Bit coding using a subcarrier

14.1.4 Concluding Comments on Bit Codings and Modulation Types

As shown above, if we use bit coding including transitions, such as Manchester or Miller and modified Miller, we can recover the bit clock and improve the effective data rate by comparison with NRZ, achieving much better results in terms of transaction time in a high-noise environment.

In the Forward Link, from the Base Station to the Tag
The advantages of Manchester or modified Miller bit coding are:

- long frames without start and stop bits at the bit synchronization level, and therefore a higher data rate;
- a difference between 'data' and 'no data', providing options for bit counting;
- a change of state with each bit, giving greater immunity to noise.

ASK 100% amplitude modulation provides:

- the highest energy when it is needed, and therefore a longer communication range;
- no reduction in data rate;
- greater structural immunity to noise.

An example of a chosen system is ISO 18000-4 for 2.45 GHZ and 18000-6 for UHF from 860 to 960 MHz.

In the Return Link, from the Tag to the Base Station
In the return direction, from the tag to the base station, using load modulation, the coding used in the system must ensure that the load modulation does not draw too much power and must permit easy and rapid recognition, even in the presence of noise, and must also be such that it enables a possible collision or collisions between transponders to be identified.

An example of a chosen system is ISO 18000-4 and -6, as cited above.

14.1.5 With or Without a Subcarrier?

An odd heading, you may think. What has this got to do with interactions between bit coding, modulation type, transmission and demodulation quality, etc.?

However, you should know that this relates to one of the main keys to the performance of RFID systems, and communication techniques using 'subcarrier' frequencies have been well known and widely used for many years in numerous applications (such as the SECAM and PAL television systems). Therefore, this is not exactly breaking news.

Operating Principle
Let us briefly survey the intrinsic qualities of this method.

Spectrum of a Signal in the Baseband with and Without a Subcarrier
In RFID, we often make a specific change, or changes, to the existing standard bit coding, by introducing a subcoding including a new signal at a higher frequency. The baseband spectrum of the initial bit coding is thus modified and transposed to either side of this new frequency.

Spectrum of the Composite Signal Transposed to RF
When the RF carrier of the radiated signal is modulated (at either the base station or the tag), the whole of the bit coding, including its subcarrier, is transposed to RF. Figure 14.6 shows a general example that summarizes the two sections above.

Clearly, this spectrum occupies a bandwidth that depends on:

- the subcarrier frequency, where an increase in this frequency causes the sidebands containing the energy due to modulation to move farther from the principal carrier;
- the initial bit rate of the data to be transmitted, which directly determines the width of the signal bands located on each side of the central lines with respect to the idle value of the subcarrier.

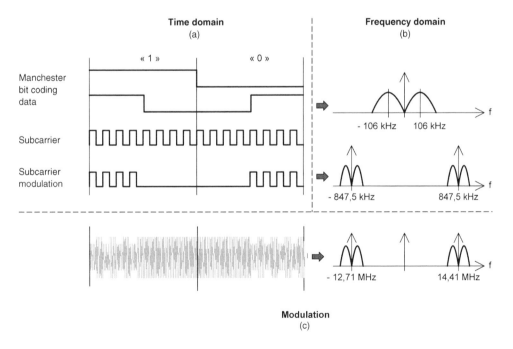

Figure 14.6   (a) Signal in the time domain. (b) Signal in the frequency domain. (c) The resulting RF signal

These two direct dependencies (which are, however, independent of each other) can be used, if necessary, to juggle between the data rates, transmission channel widths, quality/ease/ flexibility of reception of the signals from the tags, collision management rates, etc. Let us take a look at these areas.

During the Forward Link Phase
The RF carrier sent by the base station (the main carrier) to provide the forward link is modulated (by ASK or another method) by a binary signal, which is simply encoded with a standard bit coding whose spectrum is conventionally located on each side of the carrier frequency.

During the Return Link
The main carrier, still supplied by the base station, provides the remote power supply (for remotely powered tags) and also a physical support for the content of the signal forming the return link (the back scattering signal produced by the tag antenna load modulation).

Ease and Selectivity of Reception
To make it easier for the receiver part of the base station to recover, demodulate and interpret the very weak signals from the tag, it is useful to be able to supply the input stage of the receiver with a signal in which the useful spectral response is not too difficult to isolate. This is easily done by creating a bit including a subcarrier frequency for the return link. This has several advantages, as follows:

- If the subcarrier frequency is high, then the resulting frequency difference between the carrier frequency and the values of $f\_carrier \pm frequency\_subcarrier$ makes it easier to isolate the content (spectral density) of the part corresponding to the modulation, thus decreasing or eliminating the need for main carrier rejection devices based on expensive components such as circulators or bidirectional couplers (see Chapter 19 on base station technology) and facilitating demodulation.
- Since we can create several possible values (multiples and/or submultiples) of subcarrier frequencies, it is possible to meet the requirements of different transmission channel widths whose values can be adjusted to comply with current local and/or worldwide regulations. We will consider these important subjects in detail in Chapters 15 and 16, which are concerned with the options of the ISO 18000-6C protocol and the constraints imposed by RF regulations. We can then optimize the system in many ways in respect of the channel widths and/or number, the data rate of the forward link and the speed of processing and management of a high number of collisions that may occur in the return link.
- It facilitates the reception, extraction and validation of the signals received in the presence of a large amount of noise, because of the presence of numerous transitions in the received signal.

14.2 General Conclusion of Part Three

By way of a general conclusion to this lengthy part of the book, I would emphasize that we must simultaneously consider all the parameters and their numerous interactions: this certainly does not make things any easier! However, if we omit even one of these parameters from our calculations, we may finish by choosing an unsatisfactory compromise that fails to meet the overall requirements of the proposed application.

Depending on the type of application (in terms of slow/fast rate, short/long range, collision management, etc.), this work may produce results that often differ radically from each other, but in which the essential requirement of the forward link is that it must provide an optimal remote power supply for a given operating range, must have the best spectrum in terms of local regulations for a decoding rate suitable for the proposed application and must have a receiving and decoding circuit on each tag that is as simple and reliable as possible, with the lowest possible power consumption.

Note that, in certain cases, it is possible to consider various compromises that, while differing from each other, all permit satisfactory operation. In such cases, the choice of a final design is based on other parameters that we have not yet considered, such as the availability or otherwise of a generic family of components, their interoperability, their cost, etc.

Part Four

Standards and Regulations

This fourth and penultimate part is one of the most important and essential elements of this book. The reader is asked to read it with particular care, because any system must not only operate correctly but also comply with numerous national and international standards, local regulations and similar rules, which have a profound effect on the technical principles chosen and/or used. However, before continuing, let me avoid any misunderstandings by drawing a distinction between standards and regulations.

A *standard* is a document or set of documents defining the considered and coherent technical choices that make it possible to design products with a high level of interoperability. Many national institutions and organizations (AFNOR in France, DIN in Germany, ANSI in the USA, BSI in the UK, etc.) and international organizations (ISO, CENELEC, ECMA, ETSI, etc.) are authorized to draw up standards.

Theoretically, compliance with standards is not compulsory. Every designer can decide whether or not to comply with them. However, he may also ask a manufacturer to deliver products conforming to a specific standard, if he so wishes. As a general rule (there are many exceptions), a standard and compliance with it is only mandatory if a state expressly stipulates the application of the standard in a law, a decree or any kind of ministerial order.

A *regulation* is an official document or constraint issued by an authority directly connected to the state or to a group of states, and compliance with it is stipulated by laws, decrees or orders. For your information, regarding RFID in France at a particular date, the figure and table below provide some details on the position of France in the European standardization system and on the mechanisms of the state and its regulatory authorities.

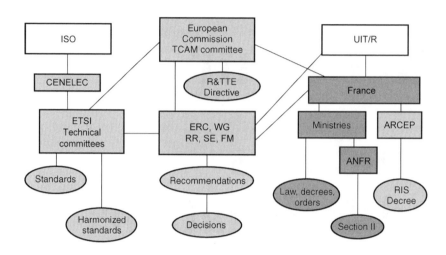

Figure Example of French structure standardization system

Table Distribution of responsibilities of ARCEP, ANFR and the ministries concerned

Ministry of Telecommunications	Telecommunications Regulation Authority (ART)[a]	National Frequencies Agency (ANFR)
General regulatory powers	Issue of ARP licences	Planning and forecasting of spectrum
Approval of the decisions of ART that have general application	Issue and delivery of RI licences	Management of certain ART networks (convention)
Approval of pricing	Allocation of associated frequencies Planning of ART spectrum Establishment of general conditions of authorization	Control, coordination, dealing with interference

[a]ART has now been renamed as ARCEP (see the next chapter).

Those of you who have read my earlier books and have already had dealings with RFID at LF (below 135 kHz) and HF (at 13.56 MHz) will certainly be surprised by my emphasis on the importance of regulations in general and local regulations in particular. The reason is simple. As you will discover in this part, whereas regulations for frequencies below 135 kHz and 13.56 MHz are approaching worldwide harmonization, this is certainly not the case for UHF and SHF, where the development of RFID systems in different geographical areas requires considerable technical planning.

15

Standards for RFID at UHF and SHF

15.1 The Purpose of the Standards

One of the main aims of the standardization authorities and the standards that they issue is to ensure interoperability of equipment and convenience for users, while also helping to reduce the end costs of products by standardization and working against the use of protectionism, proprietary solutions, licences, patents, etc.

Obviously, this is more effective when applied to widely used products such as consumable labels, which will all become contactless eventually. Given the benefits of existing contactless smart cards, the idea of applying the same contactless technology to these products is attractive.

As with their predecessors covering contactless smart cards, the goal of standards for item management in RFID is:

- A tag according to the standard must be readable by all the base stations according to the standard.
- A base station according to the standard must read all tags according to the standard.

This has the merit of being clear, at least.

15.2 Users and Providers of Standards

The major consumers of standards are the users of open-loop systems based on electronic labels, as found in the labelling and supply chain management (SCM) markets. Because of the worldwide circulation of the products, this demand for standardization reflects a basic necessity for these markets, comparable with bar codes.

In the case of industrial identification systems of the type in which products are monitored along a local production line in a factory, known as closed-loop systems, most applications are custom-built and can, if necessary, use proprietary systems without any proper standardization.

RFID at Ultra and Super High Frequencies: Theory and Application   Dominique Paret
© 2009 John Wiley & Sons, Ltd

15.3 The ISO/OSI Layer Models

Figure 15.1 shows the overall structure of the OSI model.

As already described in detail, the procedure for an exchange between two or more RFID (i.e. contactless) elements is based on a medium (air), signal processing (bit coding, carrier modulation, etc.) and protocols that must be complied with to ensure correct operation and an understanding of the messages that are exchanged.

It is often useful to relate these concepts to a specific structured communication model such as that described by the ISO in its well-known document ISO 7498 of November 1984, which specifies the communication model known as OSI (open systems interconnections). Without going into the details of the layers of the OSI model, we need to know that they are arranged in this way so that we can divide up the resulting structure and develop each layer independently. The structure is such that each layer must only communicate with the layer immediately above or below it. The only exception to this rule occurs if there is no adjacent layer.

The structure of the OSI model is such that the level of abstraction increases as we rise towards the upper layers. Here is a brief summary of the contents of each layer and their relationship with RFID applications.

Layer 0: Media

Strictly, this layer does not form part of the ISO/OSI standard. It has been given its name by custom and habit, because it defines the elements below layer 1.

This layer defines the media in which the communication takes place and the associated connected points, i.e.:

- the media, which in the case of RFID may be air, liquids, metallic materials or combinations of these;
- the network topology, which may be of the bus, star or ring type or, for RFID applications, in the broadcast mode;

Figure 15.1   Conventional representation of the OSI model

- the network termination/adaptation impedances;
- the types of sockets used for connecting the elements to the network, etc.

All are described in purely concrete terms.

Layer 1: Physical Layer (PL)

According to the ISO/OSI, 'The purpose of the physical layer (PL) is to provide mechanical, electrical, functional and procedural means to activate, maintain and de-activate physical connections for bit transmission between data-link entities.'

This layer specifies the way in which the signal is transmitted and has the role of ensuring the physical transfer of bits between the different nodes, according to all the properties (electrical, optical, radio frequency, etc.) of the system. Clearly, the physical layer must be the same for each node (tag) within a single network (such as a stack of tags on a pallet).

This layer has the task of describing:

- the representation of the bit, in other words the type of bit coding and its temporal properties (period, position in time, etc.); in the case of RFID, these codings may include Miller, modified Miller, NRZ, Manchester, Manchester coded subcarrier, BPSK, etc.;
- the general definition of the type of transmission medium (e.g. single wire, differential pair); in our case, RF waves at UHF or SHF and their respective modulations;
- the definition of the electrical, optical, radio frequency and other levels of signals, e.g. the minimum and maximum electromagnetic field strengths (E and H), etc.;
- the bit synchronization.

To assist with the definition of these parameters, the physical layer (PL, Layer 1) is divided into a number of sublayers:

- the physical signalling sublayer (PLS) (bit coding, bit timing, synchronization);
- the physical medium attachment sublayer (PMA) (characteristics of the command stages (drivers) and reception stages);
- the medium dependent interface sublayer (MDI) (connectors).

To sum up, this layer defines the bit, in terms of its coding type (e.g. NRZ, bi-phase, Manchester) and its duration, but has nothing to say about the way that the bits are, or may be, arranged and interconnected to form bytes, words, frames, etc.

Layer 2: The Data Link Layer (DLL)

According to the ISO, 'The purpose of the data link layer (DLL) is to provide functional and procedural means to establish, maintain and release connections between network entities and to transfer data link data units. A data link connection is established using one or more physical connections between a single pair without an intermediate node.'

Layer 2 is one of the most important layers in the model, because if it is well written its contents can be used for the physical design of logical controllers (hardware or software). The data link layer (Layer 2, DLL) is composed of two sublayers:

- the logical link control sublayer (LLC);
- the medium access control sublayer (MAC).

The MAC Sublayer
Moving from the lowest to the highest level, the MAC sublayer has the following functions:

- In reception, it accepts messages to be transmitted to the LLC sublayer.
- In transmission, it presents messages from the LLC layer to Layer 1.

   Its tasks are therefore as follows:

- framing the message and organizing trains of bits, e.g. by serialization in transmission and deserialization in reception;
- managing access to the physical layer (CSMA/CR, CMxx, etc.);
- conflict arbitration (at the bit level) (e.g. bit collision management in RFID);
- acknowledgement of reception of messages;
- detection of transmission errors (e.g. CRC with or without error detection);
- signalling transmission errors.

The LLC Sublayer
The tasks of the LLC sublayer are as follows:

- to check the integrity of the format of transmitted and received frames;
- if possible, to provide frame structure error correction (in relation to the format, etc.);
- the error recovery and retransmission procedure, etc.;
- frame identification;
- message filtering, etc.;
- overload warning;
- invitations to transmit or receive.

   In RFID, Layer 2 covers the encoding of transmitted frames, including parity bits or CRC, collision management for conflict arbitration, etc.

Remarks on Layers 1 and 2
These two layers form what is commonly known as the 'lower' or 'equipment' layers of the model, since the fixed and known functionality specified by them for a given protocol makes it easy to provide this functionality in physical terms in the form of chips (special-purpose integrated circuits or dedicated micro-controllers). Note that the use of these methods in system design is very efficient in terms of signal processing time and has only a slight effect on the overall protocol management time.

Note
Numerous working groups in the ISO are examining these layers, one protocol at a time. The family of 'contactless smart card' standards (ISO 14443, 15693) and RFID (ISO 18000-x) are typical examples of description of Layers 1 and 2, and form what is commonly called the 'air interface'.

Layer 3: Network
According to the ISO, 'The purpose of the network layer is to provide the functional and procedural means of exchanging data units of the network layer (in packets) over network connections between entities of the layer above (transport).'

Put more simply, this means that the task of this layer is to provide interconnection commands, path finding (i.e. routing) and sorting for the data units presented to it. Its role is to send data over the available links to the following network function, until the desired destination is reached. Its main function is therefore to provide:

- data packet forwarding;
- packet and data circuit routeing;
- multiplexing (of data packets, etc.);
- error checking for the functionality relating to this layer;
- flow control (prevention of transmission bottlenecks).

In the case of RFID, part of the collision management and the phase of selecting the tag with which a link is to be established is included in Layer 3.

Layer 4: Transport

According to ISO, 'The purpose of the transport layer is to provide transparent data transfer between session entities, thus relieving these entities from any concern with providing reliable and cost-effective data transfer (optimal transfer). The transport layer optimizes the use of the available network services in order to provide the requisite performance at the lowest cost for each of the session entities.'

This layer provides interconnection between systems, and consequently also provides:

- flow control (ensuring that addressees are not overloaded);
- splitting and reassembly of messages into packets;
- error checking (in respect of loss or duplication of packets, modification, alterations, etc.);
- message sequencing (delivery in the order in which they were presented);
- end-to-end control;
- end-to-end optimization of the data transport.

In RFID, the combination of Layers 4, 5 and 6 is called the transmission layer, or communication protocol, covering the UID or type request exchanges according to ISO 18000-3 (HF) and ISO 18000-6 (UHF), the selection commands, the data rate selection commands, etc.

Layer 5: Session

According to the ISO, 'The purpose of the session layer is to provide the means necessary for co-operating presentation entities to organize and to synchronize their dialogue and to manage their data exchange. To do this, the Session Layer provides services to establish a session connection between two presentation entities and to support orderly data exchange interactions.'

This layer provides the functions for supporting the dialogue between processes, such as the initialization, synchronization and termination of the dialogue, and it also makes the constraints and characteristics of implementations in the lower layers transparent to the user. For example, in RFID, after collision management, the selection of one of a group of tags is one example of a Layer 5 function. Because of this, the references to the different systems consist of symbolic names instead of network addresses. It also includes elementary synchronization and recovery services for exchanges.

By way of illustration, let us assume that at the end of the collision management procedure (the point at which all the UIDs of the different tags in the electromagnetic field are known), it is decided to establish a special link with one or more of them; a special session is then established to 'provide . . . to manage their data exchanges'. Here again, parts of the initialization and collision management protocols, and the communication protocol in some cases, are included in Layer 5.

Layer 6: Presentation

According to the ISO, 'The purpose of the presentation layer is to represent data which are communicated to each other by application entities, or to which they refer in the course of their dialogues.'

In a heterogeneous environment, this layer is used to describe the data in a coherent way and encode them in a unique form for transfers via the network (air, in the case of RFID). The service provided is a kind of transcoding, making the application processes independent with respect to the representation of the transmitted data. This layer is responsible for handling the problems associated with the representation of the data to be exchanged or manipulated by the applications. In other words, this layer is concerned with the syntax of the exchanged data, thus enabling the application entities to deal with the semantic aspects of the data only.

Note

To avoid any misunderstandings, here are two brief definitions from the *Collins English Dictionary*, 21st Century Edition, HarperCollins, 2000, ISBN 0-00-472529-8 ©HarperCollins Publishers 1979, 1986, 1991, 1994, 1998, 2000:

- Semantics: *n.* the branch of linguistics that deals with the study of meaning, changes in meaning, and the principles that govern the relationship between sentences or words and their meanings.
- Syntax: *n.* the branch of linguistics that deals with the grammatical arrangement of words and morphemes in the sentences of a language or of languages in general.

Layer 7: Application

According to the ISO, 'This layer is the only one in the model that provides services directly to the application process. The application layer must provide all the OSI services that can be used directly by application processes.'

This layer is at the highest level of abstraction. Its purpose is to enable decisions to be made at the application level, without considering the means used for the data exchange. It contains the semantics of the application. This layer can offer services such as:

- resource allocation;
- cooperative synchronization;
- user interface;
- user program for network services;
- special communication drivers for user programs, etc.

15.4 ISO Standards for Contactless Technology

The general standards for contactless RFID proposed by the ISO (ISO 14443, 15693, the 18000-x family, etc.) mainly describe OSI Layers 1 (PL, or physical layer) and 2 (DLL, or data link layer), in other words what the professionals commonly refer to as the air interface or the lower layers, relating to the organization of the logical data contained in the memory fields.

Most of the standards directly relating to RFID are currently available from the ISO/IEC (International Standards Organization/International Electrotechnical Commission, based at Geneva). Each of these has been discussed at length, commented on and voted on by the technical committees (TC) or joint technical committees (JTCS), subcommittees (SC), working groups (WG) made up of groups of experts (GE) meeting in special task forces (TF), subgroups (SG) or ad hoc groups, in which the technical contributions are listed according to their filing dates under the numbers xxx. The examples below show the structure of the document references:

- ISO/CEI (or IEC) JTC 1/SC 17/WG 8/TF2/no. xxx or
- ISO/CEI (or IEC) JTC 1/SC 31/WG 4/TF2/no. xxx.

All these documents are debated (hotly!) in the national bodies such as AFNOR (France), DIN (Germany), BSI (UK), ANSI (USA), etc., during the development of the obligatory sequence of documents (WD, working draft; CD, committee draft; FCD, final committee draft; FDIS, final draft international standard), which leads up to the final stage, namely the official and authentic IS (international standard), the Bible for whole generations of RFID system designers.

15.4.1 Who does what in RFID?

Figure 15.2 is a summary flow diagram of the different ISO bodies working on the principles of contactless/RFID operation, for all applications.

Within JTC1, two major bodies are involved with contactless technology. In SC17, one of the branches is concerned with all matters relating to contactless smart cards and personal identification, while in SC31 one branch is dedicated to all matters relating to item management via AIDC (automatic identification data capture), in other words the field known to all professionals as RFID.

In RFID, TC23/SC19/WG4 is also concerned with 'electronic identification methods for animals and agriculture', TC104/SC4/WG2 is concerned with 'remote labels of freight containers' and TC204/WG4 is concerned with 'identification equipment for motor vehicles'. Everything else is shown in the diagram.

15.4.2 The Difference Between Contactless Smart Cards and Labels

To avoid any confusion, let us now draw an important distinction between the different branches of these applications.

JTC 1/SC 17 – Cards and Personal ID

The first of these, handled by SC17, is the field of tags/transponders made in the well-known physical and mechanical format of smart cards, regardless of their applications. The basic ISO standards issued to date for contactless smart cards are as follows:

- ISO/IEC 10536 – close coupled cards, for very short distances;
- ISO/IEC 14443-x – proximity cards, for short distances;
- ISO/IEC 15693-x – vicinity cards, for 'vicinity' distances;

and their associated base stations. These applications operate at 13.56 MHz.

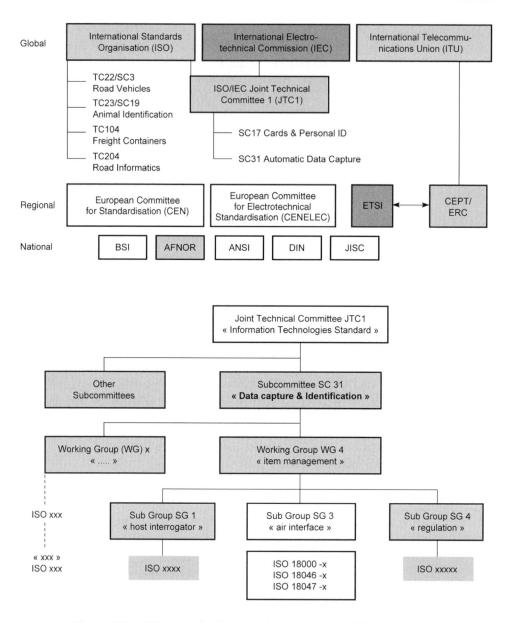

Figure 15.2 ISO standardization committees involved in RFID applications

JTC 1/SC 31 – Automatic Identification Data Capture

The second branch of the applications relates to automatic identification (Auto ID) and radio frequency labelling applications (RFID). In these applications, the physical format of the transponder is not necessarily the conventional smart card format, but may be a key fob, a paper label, a seal or any other kind of common or uncommon shape, of small or large size, regardless of the reading/writing ranges. In this context, subcommittee SC31 is concerned with the large family of standards numbered ISO 18000-1 to -x (x has reached 7 at the date of writing) and other associated standards.

The aim of these few lines is to make it clear that, regardless of the mechanical format of a card, certain vicinity standards such as ISO 15693-x, mentioned in the preceding section, are equally applicable to labels, if this meets the requirements of the planned application.

TC 23/SC 19 – Agriculture

The third area is concerned with contactless identification of animals (sheep, cattle, domestic pets, horses, etc.). This area is covered by ISO 11784 and 11785 and ISO 14223-x, operating mainly at 134.2 kHz.

A general remark on all these standards
For further information, you should consult the official documents available from the ISO.

15.4.3 Standards for Contactless Identification of 'Items'

Subcommittee SC31 of JTC1 of the ISO/IEC, AIDC (automatic identification and data capture techniques), has requested its working group WG4 to investigate 'RF item ID' at all operating frequencies. WG4 was instructed to work on the basis of 'evidence' (rather than theoretical speculations) that other experts were able to examine, together with the many new working drafts supplied to the standardization committees.

The very numerous applications of RF item ID (baggage tracing, product identification in major stores, traceability systems, library catalogues, pharmaceuticals, etc.) and the financial implications of these standards (the predicted daily consumption ranges from several million to tens of millions of units) have only added to the complexity of the work of the standardization team. It was therefore necessary to produce more than one standard to cover all the applications. By way of illustration, the following lines are intended to give the reader an idea of the minimum performances and requirements of a tag according to the targets of these applications:

- It must be programmable without contact (OTP at least).
- It must have writing and reading capabilities.
- It must allow bidirectional communication.
- It must operate at a minimum range of about 60–70 cm.
- It must include a collision management device capable of handling more than 100 tags.
- It must be able to handle at least 30 to 50 tags per second.
- It must be inexpensive.
- It must be as robust as possible in respect of the environment.
- It must be reliable.

In parallel with its own work, WG4 has tasked four main subgroups to deal with specific areas, as follows:

- SG1: data syntax;
- SG2: unique ID for RF tags;

- SG3: air interface/communication protocol
- SG regulatory issues: relations with regulations and problems arising.

We shall now examine the results and implications of the work done by these subgroups.

Why so many Standards? or, the Harsh Facts about the Life of a 'Label'

RFID applications are not at all simple. Let us briefly run through the different stages of the lifetime of a single tag, as shown by one example:

(a) When produced in the USA, a product such as a pair of jeans is labelled with a tag. Short-range readers operating at slow or fast rates are placed along the production line to monitor the manufacturing process. At the end of the production line, the product is packed in a small individual box on which another tag is applied to follow its logistical route.

(b) At the end of the line, the box is placed with other boxes in a large case, which is then placed on a pallet. At the factory gate, the whole contents of the pallet are checked by a system that reads all the tags of all the (small) packages. This is a case of long-range reading, with a device that must be able to manage numerous collisions quickly.

(c) The pallet reaches the American airport for export to Europe. On departure, the whole pallet is scanned by a long distance reading procedure (according to the American FCC standards and regulations) in a noisy environment. This is done by both the airline, to check its destination, and by the customs, to detect and prevent any counterfeiting or parallel markets. On arrival in Europe, long range reading takes place again (according to the European ETSI standards and regulations this time), for the same reasons. (Of course, the ETSI standards are different from the US standards and regulations – as described in a later part of this chapter – and the Japanese ARIB standards and regulations are not the same either!)

(d) Consolidation, breaking up, pallets, boxes, packets, etc., together with supply chain management (SCM), are essential features of distribution, all with their own requirements in terms of traceability, resupply, mixed pallets, etc.

(e) The last stage: before you can take this wonderful product home, your supermarket assistant will carry out a final proximity or short range reading, without collision management, using a manual reader (pistol-grip reader) at the check-out (because it was a little too big to be recorded on the medium distance reader with automatic collision management located under the rubber belt). Finally, there is a brief deactivation procedure to comply with privacy requirements in view of the problems of individual liberties (Data Protection Act), and all we need to do is carry out a little recycling, to maintain our Green credentials, using RFID for automatic waste sorting, and the life cycle of our little friend the tag is complete! So much for this brief exposé of the numerous stages in the life of one tag: as you will have noticed, the tag will be subject to short and long range reading and writing operations, transmission of powerful, weak or medium signals, fast and slow data rates, collision management devices, noisy environments, American and European standards, etc., at different points in its career – a bewildering succession of events that the tag must be able to cope with throughout its life cycle.

> **Note**
> No doubt you have become aware of the very large number of stages in the life of a product in which tags are used for reading and writing, and the large number of people who are benefited by this. If we simply divide the initial cost of the tag (very small) by this large number, you will see that each elementary operation costs practically nothing. If we take this line of reasoning to its limit, we could even say that users should pay tag manufacturers for their assistance in these applications! (After so many pages of technical material, we can dream a little, can't we?)

The ISO 18000-x Family of Standards for Contactless Identification of Items

As I have mentioned, a generic group of standards, ISO 18000-x, has been developed for RFID for item management (air interface), on the basis of the work done by SG3 of WG 4 and following a review of existing systems. These standards cover the possible and authorized operating frequencies, i.e. those below 135 kHz, 13.56 MHz, 433 MHz, the band from 860 to 960 MHz, 2.45 GHz and 5.8 GHz, each of which has its own virtues and merits, although none of them can claim to cover the whole range of RFID applications to which they relate.

The field of applications covered by the ISO 18000-x family is clearly defined by its generic title: *Information Technology Automatic Identification and Data Capture Technics – Radio Frequency Identification for Item Management – Part x: Parameters for Air Interface Communications at xxx (Frequencies)*, i.e. the techniques of capturing data for items identified by radio frequency methods.

In view of the environmental constraints and characteristics of each application, this set of standards does not claim to identify the best operating frequency or frequencies, but in its 'Part *x*' it describes all the parameters concerned with communication via the air interface at all the frequencies authorized around the world for RFID (see Chapter 16 on regulations). At present, the ISO 18000-x set of texts has seven parts (the ones in bold type below are those that relate to this book; the ones in italics have been described in my earlier books, References 1 and 2 mentioned in the Preface):

Part 1[1] – Generic parameters for air interface communications for globally accepted frequencies
Part 2 – Parameters for air interface communications below 135 kHz
Part 3 – Parameters for air interface communications at 13,56 MHz
Part 4 – Parameters for air interface communications at 2.45 GHz
Part 5[2] – Parameters for air interface communications at 5.8 GHz
Part 6 – Parameters for air interface communications at 860–960 MHz
Part 7 – Parameters for air interface communications at 433 MHz

The content of ISO 18000-x

In general terms, these standards describe the physical and data link layers (Layers 1 and 2 of the OSI model).

---

[1] This part defines the parameters and functions used in this family of standards, and states the general physical characteristics that must be supported by a tag according to the ISO 18000 family. Be careful not to confuse ISO 18000-1 (above) with ISO 18001, which describes in a generic way all the possible applications of RFID in item management.
[2] Note that Part 5 was put 'on ice' in February 2003 because of a lack of participants. It may well be resurrected in due course, but at the present date (2008) it is resting in peace!

Physical layer

- The accepted methods of transferring energy/power from the base station, called the 'interrogator', to the tag, using radio frequency waves
- The carrier frequencies f_c accepted by the standards
- The communication signals that must be provided between interrogators (base stations) and tags, and between tags and interrogators: bit representation and coding, bit rate, type of carrier modulation

DLL (data link layer)

- Data encoding
- Structures of frames and communication protocols between interrogators and tags
- Methods of collision management
- Organization of memory plan management
- CRC calculation, etc.

Note that a number of parameter options are often stated in order to comply with the different local radio frequency standards and regulations around the world and to satisfy certain special requirements of applications.

To enable communications to be established between interrogators and tags and to facilitate the independence of the protocols, the decision was taken to structure the communication between base stations and tags using frames conforming to the principles of the OSI model (which is why the model was described at the start of this chapter). Each frame is generally delimited by a start of frame (SOF) and an end of frame (EOF), which are generally implemented using the principle of code violation.

Standards Supplementing the ISO 18000-x Family
This basic set of standards, *Information Technology – Automatic Identification and Data Capture Technics,* is supplemented by other important standards relating to:

- *The vocabulary and definitions of RFID terms* used in the ISO 18000-x family of standards: ISO/IEC 19762-3 – RFID harmonized vocabulary.
- *Methods for testing 'performance' and 'conformity'.* These are:
  - ISO/IEC 18046-1 to -3 – RFID device performance test methods;
  - ISO/IEC 18047-x – RFID device conformance test methods. The lower case x here is directly related to the lower case x representing the frequencies used in the different parts of ISO 18000-x.
- Finally, the *intrinsic structure of the unique tag number*:
  - ISO/IEC 15962 – RFID for item management – unique identification for RF tag;
  - ISO/IEC 15963 – RFID for item management – unique identification for RF tag.

In view of the specialist nature of this boom, which is concerned with the principles and applications of RFID at UHF and SHF, I will limit my discussion to ISO 18000-4, -6 and -7, which relate to these frequencies. For ISO 18000-2 and -3, relating to frequencies below 135 kHz and 13.56 MHz respectively, the reader should consult my two earlier books (References 1 and 2 mentioned in the Preface) in which these matters were dealt with comprehensively.

15.4.4 The ISO 18000-4, -6 and -7 Standards for RFID at UHF and SHF

Rather than describe these standards in the official numerical order, I shall tackle them in the sequence ISO 18000-6, -4 and -7, dealing with the UHF band at 860–960 MHz, then the 2.45 GHz band and finally the 433 MHz band. The reasons for this approach are very straightforward:

- The UHF frequency band at 860–960 MHz is and will be the most widely used band for inexpensive tags operating at long range (from 1 to 10 m). This is why the ISO decided to standardize it first.
- The decision was then made to develop ISO 18000-4, for RFID in the 2.45 GHz band, by reusing large parts of the ISO 18000-6 protocols used in the UHF band at 860–960 MHz, in order to guarantee upward interoperability of RFID systems, compatibility of chips and low cost. Note that one of the operating modes at this frequency is exactly identical to ISO 18000-6 mode B for the 860–960 MHz band.
- Finally, specific dedicated applications at 433 MHz were added a little later.

So much for the history of the sequence in which these standards were presented.

ISO 18000-6 – Frequency Band from 860 to 960 MHz

The ISO drew up the ISO 18000-6 standard with the aim of harmonizing the various projects and proposals developed by different companies and the technologies used by the main chip manufacturers (in alphabetical order: NXP/Philips Semiconductors, STμe, TI and several others). This standard currently specifies three types, namely A, B and C, which differ mainly in the methods of communication and tag selection:

- Type A, mainly used by Texas Instruments, Bistar and RAFSEC, manages collisions with a time slot method of the slotted ALOHA type.
- Type B, promoted by NXP/Philips Semiconductors, Intermec and TAGSYS, manages collisions and selects tags using a 'binary selection tree'.
- An amendment of the initial standard, entitled Amd1, introduced type C to allow the use of the code numbers of the EPC C1 G2 family from EPCglobal, described in a later part of this chapter.

Figure 15.3 provides a summary of the structural architecture of the three options A, B and C.
Note that 18000-6 type B (860-960 MHz) and 18000-4 mode 1 (2.45 GHz) are similar in terms of the communication protocols and collision management, and can easily be supported by a single chip, for example.

Forward Link

Let us look at the special features of the physical layer of the forward link of options A and B of this standard.

ISO 18000-6 type A

Bit coding. Type A bit coding is provided by the PIE (pulse interval encoding) method, the principle of which has already been fully discussed in Chapter 11. I will not describe it any further here.

Carrier modulation. Before the carrier modulation, the electrical signals representing the PIE symbols are first shaped by what is known as a 'negative-going raised cosine' filter process (Figure 15.4).

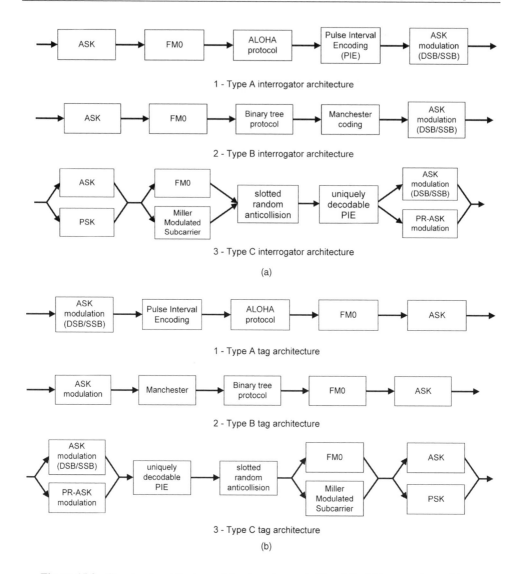

Figure 15.3   Structural architecture of the three types, A, B and C: (a) base stations; (b) tags

The carrier is modulated by the ASK modulation procedure, for which the modulation depth
D was defined in Chapter 12. Figure 15.5 shows this modulation again and Table 15.1 shows the
corresponding values.

ISO 18000-6 type B
Now let us examine the type B physical layer.

Bit coding. Type B bit coding uses conventional Manchester encoding and does not require
any further description.

Carrier modulation. The carrier is again modulated by the ASK procedure. To enable the
planned RFID applications to comply with different local regulations around the world, the
carrier modulation supports two options with different values of the modulation index

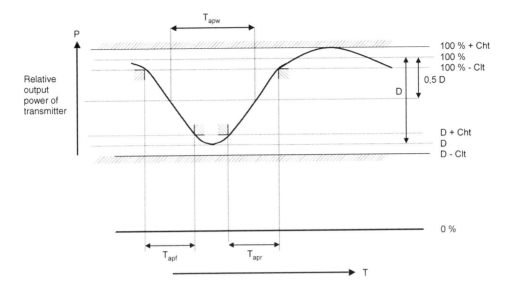

Figure 15.4 Shaping of the PIE signal

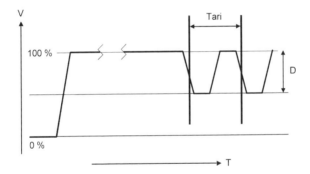

Figure 15.5 Modulation depth

$m = (A - B)/(A + B)$, namely 15% nominal (from 18 to 20%) and 90 to 100%. Figure 15.6 shows the envelopes of these two forms of modulation.

At the base station, the Manchester encoded data modulate the carrier as shown in Figure 15.7.

Table 15.1 Modulation parameters

Parameter	Minimum	Nominal	Maximum
T_{apw}		$10\,\mu s$	
D	27%		
T_{apf}		$4\,\mu s$	
T_{apr}		$4\,\mu s$	
C_{ht}			$0.1D$
C_{lt}			$0.1D$

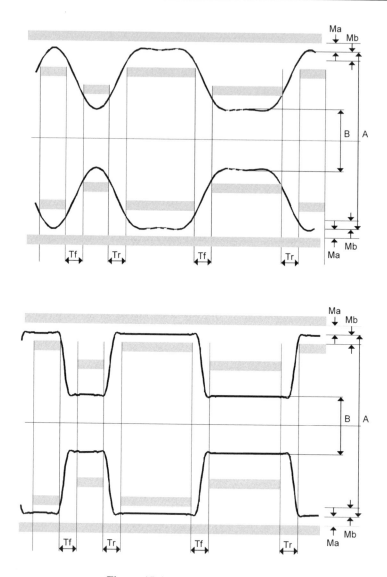

Figure 15.6 Modulation envelopes

Regardless of which type, A or B, is used, the carrier must be interrupted in order to move or hop from one frequency to another where the use of FHSS techniques is authorized by local regulations (mainly in the USA and Canada). To limit the transient disturbance of the radiated spectrum, ISO 18000-6 specifies the carrier cut-off and establishment periods, and also the minimum time for which the carrier must be present. Figure 15.8 shows the corresponding shapes of these signals and Table 15.2 gives the values of these parameters.

Return Link (Common to Types A and B)
The procedures governing the return link signals are common to types A and B.

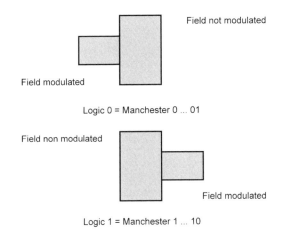

Logic 0 = Manchester 0 ... 01

Logic 1 = Manchester 1 ... 10

Figure 15.7   Example of carrier modulation by Manchester bit coding

Physical layer

In the return link phase, type A and B tags transmit data to the base station by reradiating some of the incident energy towards the base station by the back scattering technique, which has been fully described in this book. To do this, the tag switches the value of its 'reflectivity' (using the value of ΔRCS) between two states, called low and high reflectivity. Note that the standard makes no stipulation about the way that this change of reflectivity is provided in physical terms. Figure 15.9 simply shows the physical state of the reradiated signal in the medium used:

• The *space* state (also known as the nonmodulated state) represents the normal condition in which the tag is (remotely) powered by the base station and can receive and decode the data in the forward link (in 99% of applications, this is the low-reflectivity state of the tag).
• The *mark* state (also called the modulated state) represents the other condition created by changing the load/matching configuration of the tag antenna during the back scattering phase (as mentioned in the preceding chapters, this corresponds to the high-reflectivity state of the tag in 99.9% of applications).

Bit coding. In the return link, from the tag to the interrogator, ISO 18000-6 specifies the use of FM0 (bi-phase space) coding which was described in Chapter 11. Transmission of the most significant bit first (MSB first) is also specified. Note that this bit coding was chosen to facilitate demodulation on arrival at the base station.

Bit rate. The bit rate of the return link is 40 kbits s^{-1} $\pm$ 15% (i.e. the period is approximately 25 μs). The tolerance of the bit period is due to the fact that, for obvious technical reasons, the internal clock of the tag is local and asynchronous with respect to the incident wave (no internal divider operating at UHF and consuming energy, very wide frequency ranges from 860 to 960 MHz, etc.).

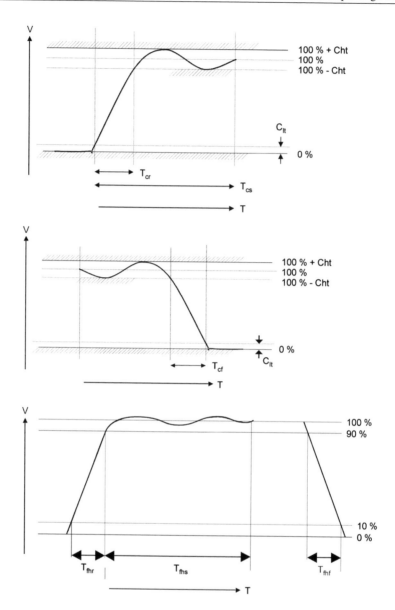

Figure 15.8 Shapes of the establishment and cut-off times in the FHSS

Table 15.2 Carrier rise and decay times in the FHSS

Parameter	Minimum	Maximum
T_{fhr}		$30\,\mu s$
T_{fhs}	$400\,\mu s$	
T_{fhf}		$30\,\mu s$

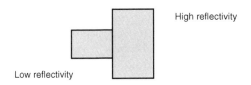

High reflectivity

Low reflectivity

Figure 15.9   Example of states of reflectivity of tags in the two back scattering phases

DLL layer
I will not provide a paraphrase of the protocol section of the standard, which you will have to read for yourselves sooner or later when developing your own applications. I suggest you read it now for its description of the frame structures of the forward and return links and the collision management methods, whose operating principles have already been explained in detail in my earlier books. I will simply mention a small peculiarity that arises from the virtually obligatory asynchronism of the return link from the tag to the base station.

Message format. The return link message consists of *n* data bits, preceded by a 'preamble'. The data bits are then sent with the MSB (most significant bit) first. What is this preamble for?

Return link preamble
As shown in Figure 15.10, the return link preamble consists of 16 bits, containing multiple code violations under the FM0 encoding rules, which act as frame markers/delimiters and provide a transition between the end of the return link preamble and the start of the useful data being transmitted. At any rate, this is the official version according to the standard! In fact, for various reasons mentioned above concerning the power consumption of the tag and the asynchronous clock of the incident wave, one of the principal functions of the preamble is to enable the base station circuits to lock on to the (asynchronous) clock of the data sent from the tag and then start to decode the message. This is an example of how to read between the lines of a standard!

In this diagram, during the bit period, the low coding state represents the low reflectivity of the tag and the high state represents the higher reflectivity, as shown in Figure 15.9. Now it is your turn to track down all the subtleties hidden in the official protocols and documents of the ISO.

For a little diversion, let us look at proprietary systems that are not covered by ISO standards but are designed to operate in the same frequency band of 860–960 MHz; as certain restaurant guides might say, this subject is 'worth going out of your way for' (although they don't say whether the aim is to avoid or enjoy!).

Proprietary Systems not Covered by ISO Standards
We must admit that there are many other devices used in RFID air communications systems in the UHF band from 860 to 960 MHz that do not operate in accordance with ISO 18000-6 typesA

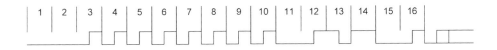

Figure 15.10   Format of the return link preamble

and B as described above. You may be astonished, or at least surprised, to discover this. I will not keep you in suspense: here are the main reasons for this state of affairs. I have arranged them in four groups under headings that are deliberately exaggerated but indicate the nature of the situation.

Forerunners and Antecedents
Obviously, standards did not spring into being after the Flood; in the years before they existed, some companies had already developed RFID systems operating at UHF and SHF, thus acting as the forerunners of this market and enabling it to flourish, which in turn made it necessary to draw up standards in order to avoid a state of (total) anarchy. Some of these companies are long-established, often serve well-defined markets and have found it unnecessary to include 'standard-compliant' products in their ranges.

The Ousted Rivals
Absolutely anybody can participate in the ISO standardization process, via his or her national organization (such as AFNOR in France or ANSI in the USA), thus contributing to the development of the official documents. Of course, contributions from participating private or public companies, government bodies, universities, research laboratories and the like are discussed by national bodies before a decision is made as to whether they should be supported at the national level. This provides a first level of filtering for proposals. This is followed by votes at the international level in the ISO, together with commentaries, commentary resolutions (the famous ballot resolution meetings, or BRMs) based on a consensus, forming a second stage of the process. Some technical proposals are accepted as they are, or after a greater or lesser degree of amendment – and some are not accepted at all. So there are some ousted rivals or, to be blunt, losers – and losers may be good or bad.

'Good losers' are those who accept that their technical proposals or solutions were not suitable in their original state, but go back to develop components or systems that will conform to the established standards, thus increasing the influence of the market, contributing to its expansion due to the effects of standardization and interoperability, and sharing in the profits.

'Bad losers', who may be piqued or envious, will continue to claim loudly that their solutions are technically or economically excellent or even the best of all (which may be true!), and who then go on to spread despondency and hold back the development of the market.

The Diehards
These characters will have nothing to do with standards, whether before, during or after the process. They refuse to participate in the standardization process, are quite unconcerned by this and remain firmly on the margin. Such companies generally serve closed or niche markets that have no particular need for standards and are happy to support proprietary systems.

The 'Not Invented Here' Syndrome
This syndrome, expressed as 'because it wasn't invented by us', or 'we are the best, the others are useless', or 'let's reinvent the wheel', or in many other versions, is very much with us. The champions in this field are undoubtedly our friends across the Atlantic, closely followed by many countries in the Far East. Therefore, we must learn to live with it. In the field of RFID, our colleagues, friends and competitors across the Atlantic were slow off the starting blocks. In 1998–1999, however, seeing the Europeans making all the running, they first attempted to hold back the progress of the ISO's work to give themselves time to catch up, and then

diversified their solutions on the 'not invented here' principle, finally proposing what was, give or take a few technical tweaks, exactly the same thing – but seen from their viewpoint. After this little story, which may be condemned as politically incorrect but which is only the historical truth (sceptics should note that I can supply the dates and documentary evidence), let us go on to the sequel.

The same principle could be seen at work in the tangle of organizations created by MIT (Massachusetts Institute of Technology), including Auto-ID Center, EPC, EAN, UCC GenCode (later GS1), EPCglobal, etc., which joined the RFID family in 2002–2004. To summarize, therefore, on 1 October 1999 MIT responded to requests from American clients (suppliers and distributors) in the field of mass production, by starting work on the new opportunities for the use of RFID. Accordingly, MIT, acting jointly with the Uniform Code Council, the Gillette Company and Procter and Gamble, set up a dedicated centre with major technical and financial support, called the Auto-ID Center (for automatic identification), which had the task of developing the electronic product code (EPC) system. The EPC specifications produced by the Auto-ID Center combine RFID technology with a special-purpose IT architecture for storing and accessing data (history, characteristics, etc.) on a product via a network (the Internet).

EPC Global and Auto-ID Labs
Having developed the EPC system, the American Auto-ID Center gained a foothold on other continents with the aim of gathering an increasing number of users and experts (lobbying, lobbying, lobbying,. . .) to develop the system according to market requirements. Thus, in June 2003, UCC-EAN International (the world leader in conventional bar codes) decided to approach Auto-ID to 'adapt' the widely used EPC bar code coding to the existing EAN/UPC coding on the bar code medium. This rapprochement between Auto-ID Center and UCC-EAN International became official with the creation of the EPCglobal and Auto-ID Labs networks and the specification of the functions of each participant. According to them, 'Auto-ID Labs established on different continents will pursue the technical development of solutions. EPCglobal will validate the specifications developed by Auto-ID Labs and market them in the form of technical standards. For this purpose, this structure must bring together, form and inform the community of users and achieve implementation on the ground.' Figure 15.11 shows a flow diagram of the EPCglobal structure.

The Auto-ID Center/EPCglobal specifications are intended to establish:

- a unique code for each EPC product;
- an open, multisectorial, global system;
- a special-purpose IT architecture;
- a more comprehensive database;
- simple and inexpensive tags.

All of this has a curious similarity to the aims of ISO 18000-x as described above – except as regards the unique code for each item identified. To make matters clearer, here are the major differences between the planned EPC solution and the ISO solution mentioned above.

ISO 18000-x
The ISO 18000-x standard family, and 18000-6 in particular, is based on the fact that the integrated circuit of the tag has a unique identifier (UID) incorporated in it. ISO 18000-x

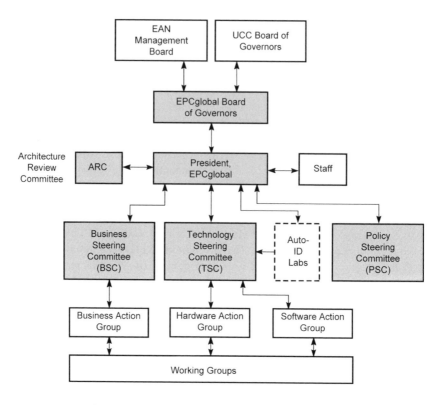

Figure 15.11 Flow diagram of the EPCglobal structure

defines the binary size of this as 40 or 64 bits (for information, a 64-bit UID corresponds to (only!) 18 billion billion combinations). This UID, etched directly on the chip by the integrated circuit maker to guarantee its uniqueness, is naturally only accessible by reading. Because of its structure, this unique UID number is directly related, not to the physical item (product or article) to which it is attached but to the number of the integrated circuit contained in the tag.

Note that all information relating to the item itself (type, reference, individual number, traceability, etc.) can easily be stored in an E2PROM memory incorporated in the integrated circuit of the tag, and its content can be read without necessarily having access to a special database, thus avoiding unnecessary increases in communication time.

A final point is that the algorithm used for collision management and listing the tags present at a given instant in the electromagnetic field radiated by the base station is, by its very nature, directly related to the UID of the chip described above and to its size.

EPC

The EPC (electronic product code) system is based on a numbering scheme designed to provide and allocate a unique identification number to each of the physical items, sets and systems. Information on the physical nature of the item is not stored directly in the EPC, which acts rather as a reference element (or address, or path) for accessing the information, which can be found via a network such as the Internet – when network access is operating correctly, which is not always the case! In other words, the EPC, rather like a bar code, is an address that shows a remote computer where it can find the information on the item in a database, via Internet access.

The EPC is allocated by an appropriate authority. Because of the nature of the application, it cannot be (or should not be) programmed, marked or inscribed into the integrated circuit of the tag except by the manufacturer or user of the item during the final initialization of the electronic label (by programming on the production line of the item, for which the manufacturer or user is fully responsible in law, as in the case of the bar code). This EPC is therefore what is known as a user data element and is not the UID of the tag's integrated circuit as specified in ISO 18000-6 types A and B. In the simplest form of the EPC system (see the different classes below), only this number is finally etched and/or stored in the label (or in its PROM/WORM memory).

Similarly, if we wish to use the EPC for collision management (when reading multiple items in a packaging unit, for example), this means that the uniqueness of the EPC that is issued must be guaranteed by a specific authority. This is an important criterion for distinguishing between a bar code (where the value is identical for a whole family of products) and EPC/RFID applications. This is because the use of bar codes at present requires an 'item after item' identification process, whereas the RFID technique enables many items to be identified simultaneously, using collision management devices.

The Different Classes of EPC
Independently of the operating frequencies specified by EPCglobal (13.56 MHz at HF, and the 860 to 960 MHz band at UHF) and the current local regulatory constraints, EPCglobal has set out classes of products in accordance with their expected functionality. Figure 15.12 shows these classes, nested in a kind of 'Russian doll' system.

I will expand this summary by providing a brief outline of these classes.

Class 0
Class 0 tags are remotely powered, with a 64-bit EPC code. Originally, Class 0 was strongly promoted by the Matrics Company, with a read-only component.

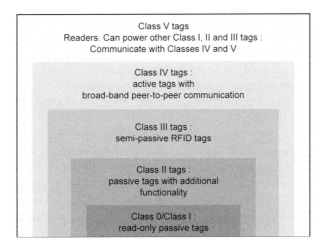

Figure 15.12   EPCglobal classes of product families

Class 1 identity tag
Tags in Class 1 (Version One) are also remotely powered and must provide the following minimum functionality in addition to the characteristics of Class 0:

- The EPC is encoded in 96 bits. Note that the second generation, called Class 1 G2 (see below) provides for a 128/256 bit OID (object identifier).
- A read-only memory, or a write-once memory (fusible link PROM), which amounts to the same thing in the present case.
- The presence of an OID.
- A device for inactivating the tag when necessary ('kill feature').

Originally, this class was strongly promoted by Alien Technology Corporation, using a WORM (write once read memory).

The technical content of Class 1 V1, in a few words
For the forward link, from the interrogator to the tag:

- Manchester bit coding, such as that specified in ISO 18000-6 type B, with an adjustable rate of 40, 80, 160 kbits s^{-1};
- ASK carrier modulation.

For the return link, from the tag to the interrogator (operating, of course, on the back scattering principle, which has been fully described in this book):

- Two types of bit coding can be used: either FM0 bit coding as in ISO 18000-6 or Miller coded subcarrier coding.
- The rate can be adjusted between 40, 80, 160, 320 and 640 kbits s^{-1} by using the available subcarrier frequency options.

When the bit coding has been carried out, the radar cross-section of the tag antenna (the value of ΔRCS) is varied, followed by reradiated carrier power modulation of the ASK and/or PSK type. Note that, in the latter case, we must take into account not only the actual scalar value of RCS but also its complex value, modulus and phase (refer back to the beginning of Chapter 9 if necessary).

A last important point is that EPC Class 0 and Class 1 V1 (with no real on-board write memory) can only be operated via a computer network to provide the link between the tag's OID and the significant and specific features of the item. To make matters quite clear: if the network fails (due to an accident, breakage, strikes, etc.), nothing will happen!

Class 1 generation 2 and the origins of ISO 18000-6 type C
At the start of 2003, for very different reasons, the US Department of Defense (US DoD) and the Wal-Mart chain – the biggest consumer goods distribution chain in the world – asked their supplies to provide packages, pallets and boxes with RFID EPC tags including 256 bits of read/write memory (E2PROM) with effect from 1 January 2005. This is because the structure of read/write memory tags is such that we can avoid the constraints and effects of network accidents, mentioned above, by enabling data to be read locally without using a computer network (using a small handheld RF reader, for example), and the user can mark the tag with

its own 256 bits of data in any way he wishes (i.e. on the same physical substrate). These two benefits give a huge advantage to this new class, Class 1 G2, which is likely to supersede Class 0 and Class 1 V1, or to be much more commonly used in any case. Moreover, since the handling of the UHF air interface in Class 1 G2 clearly shows the influence of ISO 18000-6 (simply because the same companies were involved in the development of both standards!), it is very similar to the specifications of ISO 18000-6 and has been incorporated in it since 2006 (after the inevitable discussions) in the form of Amendment 1 and Part C, joining the existing Parts A and B. Note that a strong consensus has been achieved between the companies supporting this project (TI, NXP/Philips SC, Intermec, ST, μEM, Alien Technology Company, Impinj, etc.).

The technical content of Class 1 G2
The technical details of Class 1 G2 can be summarized as having adopted the best parts of existing RFID protocols, with the addition of some new ideas. For the forward link, from the interrogator to the tag:

- A PIE bit coding like that of ISO 18000-6 type A, but inverted, as mentioned above, and with a period that can be adjusted between 6.25 and 25 μs to comply with local regulations.
- Carrier modulation based on different derivatives of ASK modulation, i.e. the well-established double sideband (DSB ASK) or single sideband (SSB ASK) modulation, and also phase reversal (PR ASK), all three of which were described in Chapter 13.

Since it is impossible to know which areas of the world a tag will travel to, all these options must be implemented in the tag, because they enable the spectrum radiated by the base station and the reradiation by the tag to be optimized, to ensure compliance with the various local regulations around the world (FCC, ETSI, ARCEP, ARIB, etc.).

For the return link, from the tag to the interrogator (operating, of course, on the back scattering principle, which has been fully described in this book), two types of bit coding can be used: either FM0 according to ISO 18000-6 or Miller coded subcarrier coding (according to the guidelines in ISO 14443 for smart cards), the value of which can be varied from 25 to 640 kbits s^{-1}.

When the bit coding has been carried out, the radar cross-section of the tag antenna (the value of ΔRCS) is varied, followed by reradiated carrier power modulation of the ASK and/or PSK type.

The comments made in the previous section are applicable here as well, for the same reasons.

Figure 15.13 shows the range of solutions offered by the ISO 18000-6 type C architecture. Because of this range of possibilities, we can:

- produce systems whose bit rates and performance can be optimized according to local regulations;
- handle a high theoretical number of collisions per second on a probabilistic time slot principle, easily derived from what was described above (up to 1600 tags s^{-1} in North America and only 800 tags s^{-1} in Europe (2 W ERP LBT), according to EPCglobal, because of the limits imposed by ETSI by comparison with the FCC);

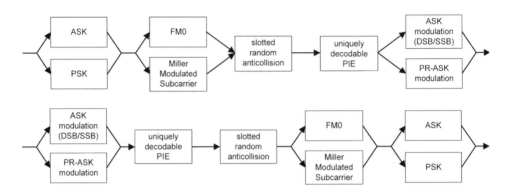

Figure 15.13   Technical solutions offered by the ISO 18000-6 type C architecture

- have more robust functionality, which can operate in environments subject to a high noise level and radio frequency pollution;
- provide better resolution of certain problems inherent in near and far fields;
- use numerous base stations on the same site ('dense environment'), subject to certain limits imposed by local regulations (FCC, ETSI, etc.). We will return to this point later on.

Summary and comparison of ISO 18000 types A, B and C (adaptable to/supporting EPC Class 1 G2)
Table 15.3 summarizes the main properties of ISO 18000-6 types A, B and C

Class 2 higher functionality tag
In addition to the features of Class 1 described above, tags in Class 2, also of the remotely powered type, have these properties:

- the presence of a tag ID (TID);
- an optional read/write user memory;
- an optional 'packetized' communications protocol.

Class 3 semi-passive tag
Like Class 2, in that it also operates by back scattering between the tag and the base station, this class uses a local power source (battery, accumulator, etc.) on board the tag to supply its logical functionality; in other words, a Class 3 tag is a battery-assisted Class 2 tag. This makes it possible to design tags incorporating sensors of physical quantities (temperature, pressure, acceleration, pH, etc.).

Class 4 active tag
This is the top of the EPC range. A system in this class must:

- allow active communication (according to ISO 18000-7, for example; see below);
- provide direct tag-to-tag communication;
- allow tags to be networked with each other.

Finally, I have provided a brief survey of the hierarchy of different EPC entities in Appendices 1, 2 and 3 to this chapter.

Table 15.3 ISO 18000-6 types A, B and C

Parameter	Type A	Type B	Type C
Forward link (base station to tag)			
Bit coding	Pulse interval encoding (PIE)	Manchester	Pulse interval encoding (PIE) (inverse)
Type of modulation	ASK	ASK	ASK (DSB, SSB, PR)
Modulation index	15 to 100%	18 or 100%	90% nominal
Modulation depth	27 to 100%	30.5 or 100%	80 to 100%
Bit rate	33 kbits s^{-1} (mean) (to facilitate compliance with local regulations)	10 or 40 kbits s^{-1}	26.7 to 128 kbits s^{-1}
Return link (tag to base station)			
Bit coding	FM0	FM0	FM0
Subcarrier			Miller coded subcarrier, 40 to 640 kHz
Radar cross-section modulation (back scattering)	ASK	ASK	ASK or PSK
Bit rate	40 to 160 kbits s^{-1}	40 to 160 kbits s^{-1}	40 to 640 kbits s^{-1}
Tag UID	64 bits (40 bits SUID)	64 bits	16 to 496 bits
Type of collision management	ALOHA (time slotted)	Binary tree	Random slotted bit arbitration
Linearity of collision arbitration procedure	Up to 250 tags	Up to 2^{256}	Up to 2^{15}
Memory addressing	Blocks up to 256 bits	Byte blocks	1, 2, 3 or 4 byte writes
Forward link error detection	5 bit CRC for all commands (with 16 bit CRC for all long commands)	16 bit CRC	16 bit CRC
Return link error detection	16 bit CRC	16 bit CRC	16 bit CRC

Other Products

We cannot ignore the fact that there are many other proprietary products on the market, operating outside the ISO standards and having their own benefits and drawbacks. Generally, we can say that one of their main benefits is that their design is simple (sometimes they are called 'ET', for elementary tags, which is yet another different concept), and thus their applications are limited, but their prices are often attractive, or appear to be. The simplicity of their design obviously means that the silicon chips can be smaller.

This simplicity is generally due to the fact that they use simplified communication protocols (such as read-only, with automatic loop reading, very few data, no collision management facility, etc.); the simplicity may be such that there is no protocol at all, and there's the rub. In fact, if there is no area of the chip dedicated to the communication protocol to be handled, all the tag has to do is to be remotely powered as soon as it enters the electromagnetic field, reset itself correctly and then send its UID to the base station by back scattering. This well-established technique is known as TTF (tag talk first); in other words, the tag starts to respond without waiting for an interrogation request from the base station. One example of this is one of the forerunners of the RFID market, namely the chip originally designed by the South African IPICO Company and then taken over by μElectronique Marin. However a system will only operate correctly in TTF subject to two important conditions:

• On the one hand, when the system is used, it must be ensured that only one tag is present (or is always present) at any one time in the field; otherwise, it is difficult or even impossible to manage the collisions of the signals returned in a completely asynchronous way by the tags.
• On the other hand, if one or other of the tags (of the same kind or of different kinds) is already communicating in the field, the appearance of a TTF (tag talks first) tag wreaks havoc in the ongoing communication. This is one of the main reasons why the ISO committees dealing with RFID have only standardized the RTF (reader talks first) mode, with the aim of providing all the possible and necessary interoperability for industrial RFID solutions, regardless of whether they are open- or closed-loop types, and regardless of the number of tags present simultaneously in the field.

One way of tackling the problems of TTF has recently been proposed, in the form of the TOTAL (tag only talks after listening) concept, in which the tag listens to see if another base station is talking before it starts to communicate on its own behalf.

All these systems – TTF, TOTAL, etc. – have the avowed aim of minimizing the price of the circuit by reducing the size of the chip. Unfortunately, as I will go on to show you, the current sizes of the chips – even the EPC C1 G2 type – are so small ($0.4\,mm \times 0.4\,mm$) that they are already very difficult to deposit and position correctly on antennas/inlets at a rate of several million components per day with good enough reproducibility to be able to achieve rejection rates of only a few ppm – if the price is to be kept realistic.

Now that you are fully informed, it is up to you to choose the best systems for your applications.

ISO 18000-4 – RFID Operating at 2.45 GHz

This standard applies to passive tags (no on-board transmitter, therefore using back scattering for the return link) operating at 2.45 GHz. I will describe it at this point in the book, immediately after dealing with ISO 18000-6, because historically it was drawn up after the latter. Admittedly, there are some differences, due to the physical phenomena occurring between the 860–960 MHz band and the 2450 MHz band, but they are not too important. This meant that large sections of the ISO 18000-6 standard could be, and were, simply pasted into ISO 18000-4!

ISO 18000-4 comprises two very distinct parts, corresponding to markedly different operating modes, as follows:

• Mode 1, for tags operating in the passive–remotely powered mode.
• Mode 2, orientated more towards tags that can operate in passive–battery-assisted mode.

Note that the first part was mainly supported by Intermec (USA) and NXP/Philips SC (Netherlands/Austria). The second part was mainly supported by NEDAP (Netherlands) and Siemens Automatisme (Germany and Austria).

Mode 1

Mode 1 closely resembles ISO 18000-6 Part B, simply because it has been 'cut and pasted' from the latter. I will therefore not go into the technical details of this part of the standard. Consequently, an integrated circuit designed according to ISO 18000-6 can work equally well at 2.45 GHz, subject to a few provisos:

- The attenuation in air is greater at 2.45 GHz than at 900 MHz (see Chapter 6), and therefore, 'all other things being equal' (i.e. given the same constraints in terms of regulations), the communication range will be shorter.
- The power consumption of a conventional CMOS integrated circuit increases with its operating frequency, so it consumes more at 2.45 GHz than at 900 MHz. Consequently, in the same conditions, there is a direct effect on the communication range.
- The width of the authorized frequency band around 2.45 GHz is generally such that FHSS and DSSS techniques can be used.

Given the first two provisos, we can say that, all other things being equal, the operating range at 2.45 GHz is shorter than at 900 MHz for remotely powered tags. The use of these tags, justified by other factors, is therefore dependent on other parameters, including, of course, the local regulations.

You may be aware that the use of UHF (860–960 MHz) is problematic, or even prohibited, in certain Asian countries. In such cases, 2.45 GHz is the only option! All necessary measures must be taken to compensate for the two factors mentioned above, by making it possible to use higher EIRP power levels and antennas with higher gain (and therefore stronger directivity). Given all this, the use of the 2.45 GHz frequency will again become 'feasible' and 'viable'.

The question will then be, 'But what do we do locally, if we can't provide a higher radiated EIRP?' The answer is simple: read on!

Mode 2

If the power emitted by the base station is insufficient, it may be difficult to power the tags remotely. To overcome this, we can fit batteries to the tags, so that they remain passive but become 'passive–battery assisted'. This is what is described as the 'Mode 2 long-range high data rate RFID system' in ISO 18000-4.

In this operating mode, the gross bit rate can reach $384 \, \text{kbits s}^{-1}$ in the case of a read/write (R/W) tag. For a read-only (R/O) tag, the bit rate is $76.8 \, \text{kbits s}^{-1}$. Note that this mode does not specifically require the use of a battery-assisted tag.

The whole system consists of a base station (interrogator) and at least one of the following three types of tag:

- a read/write (R/W) tag;
- a read-only tag (R/O);
- a special version of the R/O tag that can interpret a notification channel so that it can be used in high bit rate applications.

The standard was also designed to provide for combined systems using the different types of tags listed above. Obviously, the base station must at least operate with standard R/O tags.

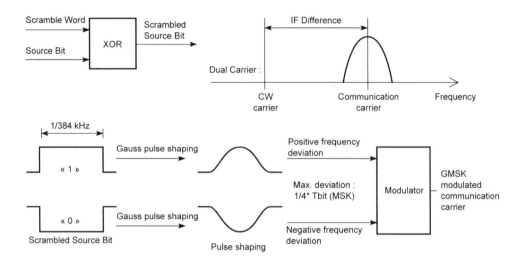

Figure 15.14 Forward link block diagram of the system

To allow all the above tags to operate at the same time, a tag talk first (TTF) principle is used. All the tags must also reradiate (back scatter) a fixed sequence, called the notification sequence, which starts with synchronization information and the tag data (including the 'user tag ID').

Forward link
Figure 15.14 is a block diagram of the forward link system proposed by the ISO.

Return link
As already mentioned, the tag responds in the TTF mode by back scattering, based on load modulation of the antenna impedance, including the presence of a subcarrier frequency of 153.6 kHz in the notification phase and 384 kHz in the communication phase. During the notification sequence, the subcarrier modulation is as follows:

- DBPSK for R/W tags;
- DBPSK or OOK for R/O tags;
- Manchester DBPSK in the communication phase (R/W only).

Figure 15.15 is a block diagram of the coding and modulation schemes according to the ISO documents.
Table 15.4 shows the main differences between modes 1 and 2 of ISO 18000-4.

ISO 18000-7 (433 MHz)
This standard was strongly promoted by the American authorities after the events of 11 September 2001, as a way of facilitating the identification and/or tracing of the routes, contents, etc., of shipping containers arriving on their territory. This standard, based on a proposal by the American SAVI Company, closely associated with the DoD (Department of Defense), is strongly orientated towards battery-assisted tags and is designed to provide communications where remote powering is difficult, in other words mainly over 'very' long ranges, also using pure radio transmission methods (transmission and reception circuits at each

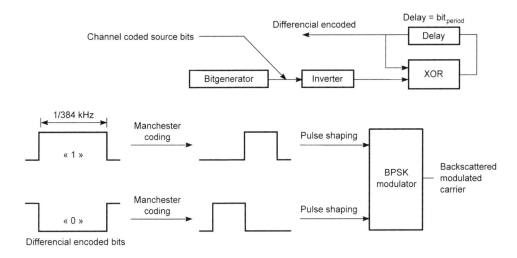

Figure 15.15 Return link block diagram of coding and modulation schemes

end, with base stations and tags operating in a true 'active–battery-assisted' mode). Figure 15.16 shows the essential reasons for the choice of the 433 MHz frequency for these applications.

Some European countries, like France, abstained from voting for the use of this frequency for RFID at the ISO, because it was already widely used for remote control systems for access control, vehicle doors, garage doors and tyre pressure measurement systems (TPMS) (see my book *Multiplexed Networks for Embedded Systems*, 2007, available from John Wiley & Sons, Ltd, Chichester), all systems that are undergoing rapid development. The fear was that using this frequency for RFID would only add to the existing levels of radio interference.

A Few Details
According to ISO 18000-7, RF communication between the interrogator (base station) and the tag uses a narrow UHF band with the following characteristics:

- Carrier frequency: 433.92 MHz $\pm$ 20 ppm
- Type of modulation: FSK
- Frequency deviation: $\pm$35 kHz

Table 15.4

ISO 18000-4	Mode 1	Mode 2
Protocol Characteristics	RTF Passive – back scatter	TTF Passive – back scatter Battery assisted Long range High speed
Bit rate	40 kbits s^{-1}	76.8 or 384 kbits s^{-1}
Collision management	Yes, determined by the base station in the collision management sequence	Yes, defined when the system is installed in the tag

Band	303 MHz	315 MHz	418 MHz	433 MHz	868 MHz	915 MHz	2400 MHz
	302 - 305 MHz	314,7 - 315 MHz 42 dBuA/m @10 m	418,95 - 418,975 MHz 10 mW ERP	433,050 - 434,790 MHz 10 mW ERP 10 %	868 - 868,6 MHz 25 mW ERP 1 %	902 - 928 MHz	2400 - 2483,5 MHz
USA	✓	✓	✓	✓		✓	✓
Canada	✓	✓	✓	✓		✓	✓
Great Britain				✓	✓		✓
France				✓	✓		✓
Germany				✓	✓		✓
Netherlands				✓	✓		✓
Singapore		✓		✓	✓	✓	
Taiwan	✓	✓	✓	✓			✓
China/ Hong Kong		✓		In process		Limited	Limited
Australia				✓		Limited	Limited
Summary	Limited acceptance	Limited acceptance	Limited acceptance	Better Choice	Limited duty cycle	Limited acceptance	Poor technical performance

Figure 15.16   Comparison of different possibilities according to SAVI

- Low symbol: $f_c - 35\,\text{kHz}$
- High symbol: $f_c + 35\,\text{kHz}$
- Bit rate: 27.7 kHz
- Wake-up signal: 30 kHz

The wake-up signal is a subcarrier, called a tone (constant frequency) at 30 kHz, with a duration of 2.5 to 2.7 s, transmitted for the purpose of waking up all the tags present in the communication areas. In response to this signal, the tags that were asleep (i.e. in sleep mode/power down mode) wake up and switch to the ready state where they wait for a command from the interrogator.

Communication between the interrogator and tag is of the master–slave type, in which the interrogator always initializes communications and then listens for the response from the tag. If there are multiple responses from several tags, they are managed by a special 'collection' algorithm.

15.4.5 Standards, Measurement and Tests for Conformity and Performance

You will have to wait for Chapter 20 to discover the details of standards, measurements and methods of testing for conformity and performance. Before we get there, I will be describing the physical construction of tags and base stations in Chapters 18 and 19. So please bear with me for now.

This is the end of our (fairly) brief survey of standards directly relating to contactless RFID systems at UHF and SHF. Let us continue with a review of other relevant documents, regulations and general-purpose standards.

15.5 Appendix 1: Hierarchy and Structure of the EPC System

Although it is a slight digression from the theme of this book, I would like to provide an outline of the structure of the EPC system hierarchy. This consists of two levels, hardware and software entities, defined as follows by EPCglobal.

Hardware

Tags means the physical system consisting of the 'integrated circuit + its antenna', which is attached to an object (box, pallet, etc.) described by the generic term 'item'. It is the element that carries the digital data.

EPC is the unique identity number (the 'item' code) applied to this object. It is programmed into the memory of the tag's integrated circuit. Overall, this value is the unique pointer for enquiries and investigations concerning the article associated with the EPC.

The *reader*, or rather the interrogator or base station, is the portable or fixed element having the functions of detecting the presence of a tag in the operating space of the system and capturing the data contained in the tag. It is connected to a network via software layers of the EPC middleware or EPCglobal network type, which I will now briefly describe.

Software

EPC middleware is the intermediate software layer between the hardware and the application. It is designed to process data flows from the tags or elements that capture data on events from one or more readers. It has the function of filtering, aggregating and counting the data from the tags and reducing their quantity before sending them to the higher-level application layers.

The *interface* denotes the protocol for transferring the EPC middleware data to EPC IS (information service).

EPC IS (known previously as PML server) makes the data available in XML in a PML format to the other services, such as the 'read data' and 'data and object class-level data' for the tags. Its purpose is to store all the data relating to the EPC in question (static value, time and instance level).

PML is the product markup language, or physical markup language; like XML, it uses an XQL request structure for interrogating and establishing structured relations for access control, authorization and authentication of EPC numbers.

ONS is the object naming service, a distributed resource that knows the storage location of the EPC data. It provides a global search service for translating an EPC into one or more URLs (Internet uniform reference locators) where more information about the object can be found, subject to authentication/authorization procedures for security protection, making it possible to obtain the data structure, attributes, physical parameters and log for the whole supply chain, etc.

SAVANT is the name given to the servers, which act as warehouses for EPC data and associated parameters and which support sophisticated and flexible middleware for meeting PLM requests.

15.6 Appendix 2: Structure of the EPC Number

To illustrate all these concepts, here is a summary of the structure of an EPC number. It consists of several data fields (Figure 15.17).

Header

The header field defines everything that follows it. In this case it shows:

- the number of bits in the EPC message (e.g. 96);
- the type (class 2);
- the version (G2);

Element	Header	EPC Manager	Object Class	Serial Number
Bits	8	34	20	34
Values$_{10}$	0-256	0-17, 179, 869, 183	0-1, 048, 575	0-17, 179, 869, 183 (inc 0-9)

Class 0	64 bits
Class 1	96 bits
Class 1 G2	128/256 bits
Class 2	Class 1 with larger memory and read/write
Class 3	Class 2 with sensors (semi-passive)
Class 4	

EPC type	Header size	First bits	Domain manager	Object class	Serial number	Total
96 bit	8	00	28	24	36	96
64 bit type I	2	01	21	17	24	64
64 bit type II	2	10	15	13	34	64
64 bit type III	2	11	26	13	23	64

Figure 15.17 Structure of EPC 96 coding

- the length of the following fields;
- other information such as 'locked', 'killed', 'hidden', password.

EPC Manager
The EPC manager field specifies the company or entity responsible for assigning the codes (and their uniqueness) in the following fields; e.g. Coca-Cola, Ricard, etc.

Object Class
The object class field specifies the item, i.e. the stock keeping unit (SKU) or consumer unit; e.g. 25 cl bottle, decaffeinated, etc.

Serial Number
The serial number field specifies the unique number allocated to the items in a given class; e.g. 123 456 789.

Example
ISO 18000-6 mode C specifies that the TID (tag identifier) memory reserved for EPCglobal applications is structured as follows:

- From 00h to 07h, the memory must contain the value E2h (1110 0010$_2$) to indicate that an EPC is being used.
- From 08h to 13h, the memory must contain a 12-bit encoded mask-designer identifier (obtained from EPCglobal).

- From 14h to 1Fh, the memory must contain a 12-bit number associated with the vendor of the product.
- Beyond 1Fh, the memory may contain data specific to the tag and/or the vendor (e.g. the serial number of the tag, a code of any type, the date, etc.).

15.7 Appendix 3: Some Facts about the Everyday Performance of ISO 18000-6 mode C – EPC C1 G2

EPC C1 G2 was launched in the USA, with much razzmatazz and numerous presentations (papers, slides, presentations, conferences, etc.), as 'the best of the best' in RFID. Admittedly, the concept has many benefits, mainly in relation to the numerous bit coding and subcoding options, which facilitate adaptation to the various local standards and regulations in different countries. However, without being too negative, it is sometimes necessary to assess matters more objectively. Briefly, we should not confuse hardware, software, paperware and slideware!

Out of the plethora of parameters and terms used to promote the EPC C1 G2 concept, there are two in particular that are often emphasized, namely 'management of high numbers of collisions per second' and 'operation in dense environments', which I have already mentioned.

Managing High Numbers of Collisions Per Second

The best figures announced by EPCglobal were 1600 tags s^{-1} (see the upper left-hand corner of Figure 15.18).

Given the binary lengths of the different commands required to run the collision management procedure of the random-slotted collision arbitration type, namely Query = 22 bits (including the session number, CRC, turn-round time, etc.), ACK = 18 bits, etc., and the minimum values of the mean bit time (with as many '0' as '1' in the message, i.e. a bit time of

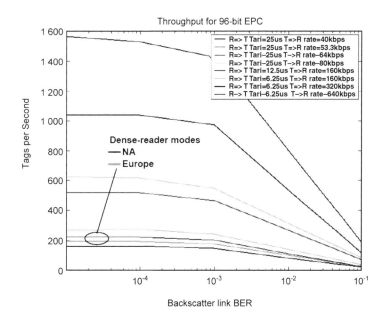

Figure 15.18 Tag read rate according to the EPCglobal presentation

about 1.25 Tari) with a minimum Tari value of 6.25 μs (i.e. $T_{min\ bit\ mean} = 7.8\ \mu s$), it is true that the minimum time required by the protocol may be only 1/1600 of a second (i.e. about 625 μs, meaning that about 80 bits are exchanged during the procedure) for managing a collision, provided that there is only one tag in the field!

However, when there are many tags in the field, a complete collision management cycle (Figure 15.19) requires a two-step procedure, with an initial wake-up (using the Query

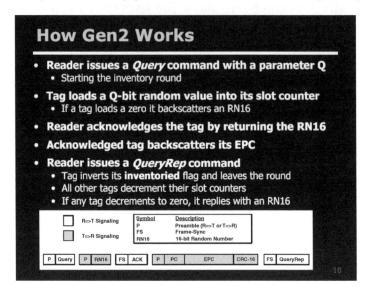

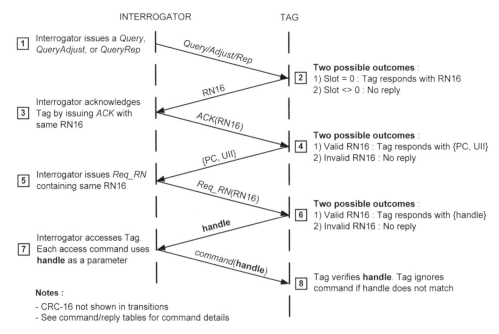

Figure 15.19   'How Gen2 works' according to EPCglobal

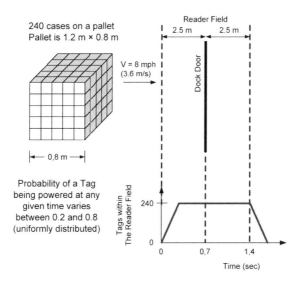

240 cases on a pallet
Pallet is 1.2 m × 0.8 m

V = 8 mph
(3.6 m/s)

|← 0,8 m →|

Probability of a Tag
being powered at any
given time varies
between 0.2 and 0.8
(uniformly distributed)

Reader Field
2.5 m   2.5 m

Dock Door

Tags within
The Reader Field

240

0

0   0,7   1,4

Time (sec)

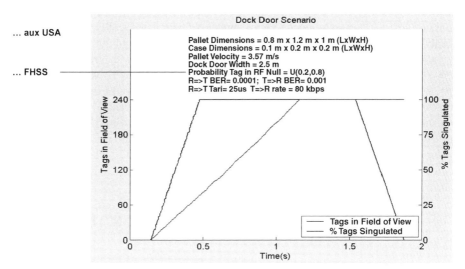

... aux USA

... FHSS

Dock Door Scenario

Pallet Dimensions = 0.8 m x 1.2 m x 1 m (LxWxH)
Case Dimensions = 0.1 m x 0.2 m x 0.2 m (LxWxH)
Pallet Velocity = 3.57 m/s
Dock Door Width = 2.5 m
Probability Tag in RF Null = U(0.2,0.8)
R=>T BER= 0.0001; T=>R BER= 0.001
R=>T Tari= 25us T=>R rate = 80 kbps

Tags in Field of View: 240, 180, 120, 60, 0
% Tags Singulated: 100, 75, 50, 25, 0
Time(s): 0, 0.5, 1, 1.5, 2

—— Tags in Field of View
—— % Tags Singulated

Figure 15.20 Door dock scenario according to EPCglobal

command), after which reading is restarted (using the Query Adjust command), requiring two times 1/1600 of a second, so the actual maximum rate is 800 tags s^{-1}. This is still quite good, but only half of the stated level! It would not be so acceptable if the numbers were on your payslip.

Another original EPCglobal document (Figure 15.20) indicates (and may lead us to hope) that we can easily read 240 cases on a pallet with a base measuring 1.2 m × 0.8 m and a height of 1 m, passing through a door dock at a speed of 3.6 m s^{-1}.

The volume of the loaded pallet is about one cubic metre, i.e. 1000 litres. If we assume that the cases have identical dimensions at the end of the production line (i.e. the pallet is 'uniform'), each case has a capacity of about 1000/250 = 4 litres, and forms a cube with a side measurement of about 16 cm. If we fit these with UHF labels with $\lambda/2$ antennas, in other words with lengths of approximately 16 cm, all the cases must be spaced at intervals of about $\lambda/2$, and a very large amount of interference will occur from one label to the next because of reradiation. This destructive interference will hide or mask some of the cases, preventing them from being read by the base station. The figures that have been announced are therefore simply the most enticing results achieved by a 'paper' simulation for slide presentations (by the way, paper hardly reradiates at all – otherwise we would have heard about it!), but these do not in any way represent the daily reality of reading all labels with a high and known success rate. You may of course reply that this may, or must, be true, because these presentations originate from the USA, implying that FHSS techniques are used, which may help. True, but it will not resolve all the difficulties (see the full discussion of this matter in Chapters 8 and 9), and anyway, we are operating in Europe! Therefore, be careful when developing your applications.

Dense Environment

This point has already been mentioned in this chapter, in a lengthy consideration of the possibilities offered by FCC 47 Part 15 and ETSI 302 208. The original slides show (Figure 15.18, reading from the bottom upwards) that, in a dense environment (10 to 50 base stations on one site), the maximum theoretical number of 'simulated' collisions per second falls markedly in the USA and Europe (ETSI 302 208), to about 210 and 190 respectively.

In reality, as shown by actual measurements on uniform pallets, with a reading rate of more than 90–95%, these numbers fall to 160 for the USA (FCC 47 Part 15 – 4 W EIRP – FHSS), 80–90 in Europe (ETSI 302 208 – 2 W ERP – LBT – unsynchronized base stations, multiple multiplexed antennas) and about 50 in France (according to the current ARCEPT regulations, with a single channel at 250 kHz, 500 mW ERP, even without allowing for the unfavourable effects of a low authorized duty cycle for band occupation, with operation at 10%!). Figure 15.21 is a simple illustration and summary of the hard facts of this daily reality, disregarding all publicity, paperware and slideware.

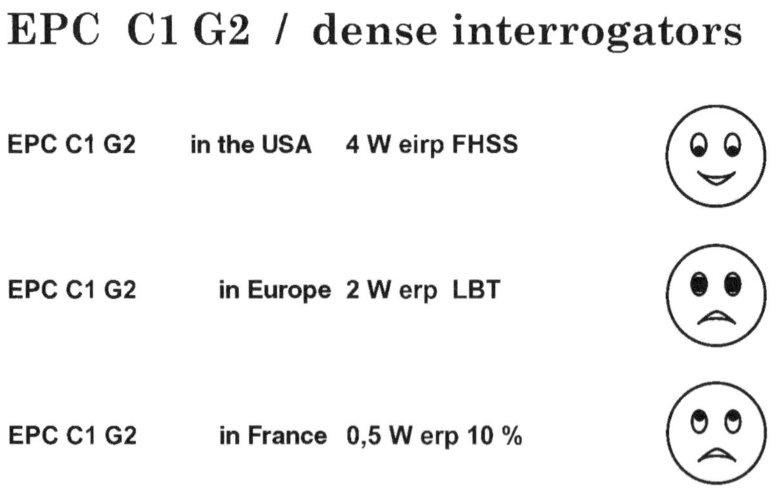

EPC C1 G2 / dense interrogators

EPC C1 G2	**in the USA**	**4 W eirp FHSS**	🙂
EPC C1 G2	**in Europe 2 W erp LBT**		🙁
EPC C1 G2	**in France 0,5 W erp 10 %**		🙁

Figure 15.21 Number of collisions per second as seen by EPCglobal, aka the siren song...

In view of this, ETSI (and many other working groups) have gone back to work and have now provided some supplements and responses for the original ETSI 302 308, adding specific operating modes for accessing the UHF channels in which transmission at 2 W in the LBT mode is permitted. For anyone interested, the documents in question are ETSI TS 102 562 and TR 102 649-1, 102 463 and 102 313.

16

Regulations and Human Exposure

By their very nature, radio frequency systems cannot exist in isolation and must also allow for the external environments in which they are embedded, which are themselves regulated by many other standards, such as the RF pollution and radiation standards and regulation, as well as those relating to (hypothetical) health problems and the problems of safety (electrocution, etc.). These are the matters that I will now examine.

> **Note**
> In a book of this kind, I can only give an overview of the state of these standards and their implications. For further information, you should consult the official authorities (Figure 16.1) and texts (the addresses of the suppliers of these documents are listed at the end of the book). The aim of the following sections is to make the reader aware of the many daily problems of RFID applications and their relationships with some of the parallel documents, regulations and standards.

All the conventional systems described in this book have transmitting antennas located at the base stations. There are numerous standards and regulations issued by government bodies that stipulate, in the form of authorized radiation levels and specific ranges, the constraints and restrictions (for radiation, pollution, susceptibility, etc.) on equipment used for RFID applications.

16.1 Survey of Standards and Regulations

The problem of acceptable radiation levels and tolerated RF pollution is far from simple, because of the wide variety of regulations, laws and exemptions that have developed, often for historical reasons, in different countries. In short, this is a complex issue.

A certain number of worldwide, American, European and French authorities lay down and regulate these parameters.

16.1.1 Worldwide

The International Telecommunication Union (ITU), based at Geneva, has a Radiocommunication Sector (ITU-R), which draws up recommendations for the technical characteristics and

RFID at Ultra and Super High Frequencies: Theory and Application Dominique Paret
© 2009 John Wiley & Sons, Ltd

Figure 16.1 The main regulatory authorities around the world

operating procedures of radiocommunication services and systems, following the well-known Conference Preparatory Meetings (CPM). Note that the ITU has divided the world into three 'regions', plus a fourth one, 'Space' (Figure 16.2(a)).

Figure 16.2(b) and (c) summarize the UHF and SHF options allocated by regulators for RFID.

16.1.2 In the USA

In the USA, for all matters relating to RFID systems of the SRD (short range device) type, at UHF and SHF, the 'Bible' for RFID telecommunications applications is the document issued by the US Federal Communications Commission (FCC), under the aegis of the American National Standards Institute (ANSI), which has drawn up the equally well-known reference text US Code of Federal Regulations (CFR) Title 47, Chapter I, Part 15 – *Radio Frequency Devices*, which I have already mentioned many times.

16.1.3 In Europe

With the aim of standardizing the test and measurement methods and following the ITU recommendations, the European Conference of Postal and Telecommunications Administrations (CEPT) comprises around 50 member countries, the experts meeting under the auspices of ETSI (European Telecommunications Standards Institute), which operates in partnership with the ISO. The latter body has the task of developing recommendations via the ERO (European Regulation Organisation), which includes the ERC (European Radiocommunications Committee).

One of the reference documents is the well-known CEPT/ERC recommendation ERC 70 03, *Relating to the Use of Short Range Devices (SRD)*, which you are strongly advised to obtain

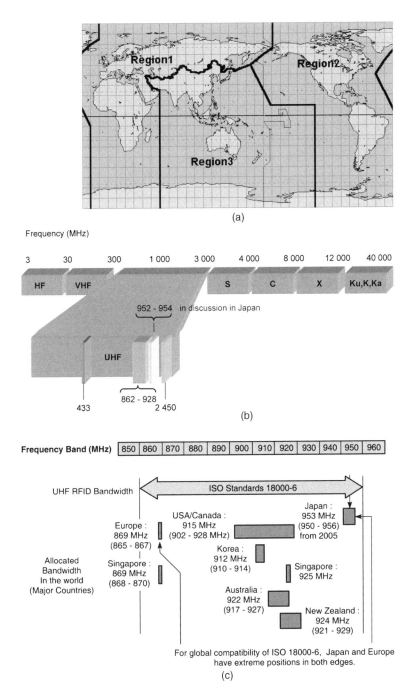

Figure 16.2 (a) Regions of the world according to the ITU. (b) Worldwide RFID options for UHF and SHF. (c) Worldwide RFID options for the 860–960 MHz UHF band

(from the www.ero.dk website, based in Denmark). I shall refer to this document regularly in this book. Figure 16.3 shows the values listed in this document.

The main current European standards relating directly to RFID applications are as follows:

- The ETSI documents of the EN 300-xxx family, Electromagnetic Compatibility and Radio Spectrum Matters (ERM); Short Range Devices (**bold type** indicates those applicable to UHF and SHF):

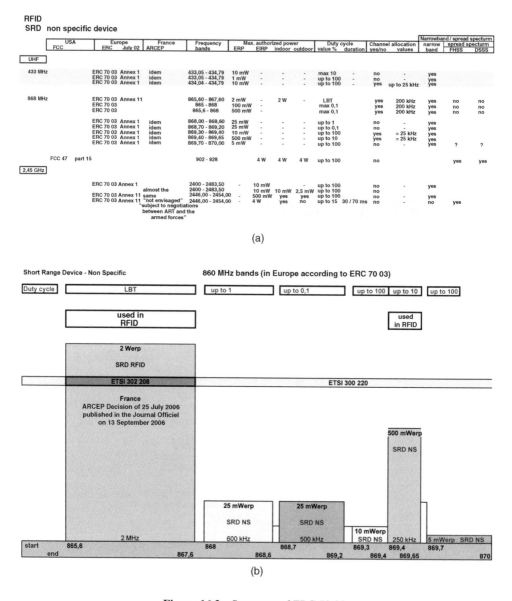

(a)

(b)

Figure 16.3   Summary of ERC 70 03

- EN 300 330 – frequency range 9 kHz to 25 MHz, i.e. covering 125 kHz and 13.56 MHz
- **EN 300 220 – frequency range 25 MHz to 1 GHz, i.e. covering 433 and 860/960 MHz** which is divided into two main parts:
  - ETSI EN 300 220-1 – Part 1: Technical characteristics and test methods
  - ETSI EN 300 220-2 – Part 2: Harmonized EN under Article 3.2 of the R&TTE Directive
- **EN 300 440 – frequency range 1 GHz to 40 GHz, i.e. covering 2.45 and 5.8 GHz** which is also divided into two main parts:
  - ETSI EN 300 440-1 – Part 1: Technical characteristics and test methods
  - ETSI EN 300 440-2 – Part 2: Harmonized EN under Article 3.2 of the R&TTE Directive
  Standards dealing specifically with RFID UHF applications:
- **ETSI 302 208-x – Electromagnetic compatibility and radio spectrum matters (ERM) – Radio-frequency identification equipment operating in the band 865 MHz to 868 MHz with power levels up to 2 W:**
  - Part 1 – Technical characteristics and test methods
  - Part 2 – Harmonized EN under Article 3.2 of the R&TTE Directive
- Finally, ETS 300 683 for EMI

> **Note**
> I hope that you will have noticed, at the beginning of this book, that the official designation of the UHF band covers frequencies from 300 MHz to 3 GHz and that the top of this band is covered by two standards, ETSI 300 220 and 300 440.

In France
These parameters are set by two authorities:

- ANFR (National Frequencies Agency);
- ARCEP (Postal and Electronic Communications Authority, formerly known as ART).

These authorities, working on the basis of European recommendations wherever possible, produce documents for use in developing French standards and regulations for short range devices (SRD), nonspecific systems and RFID, with which the RFID branch is concerned.

16.2 Summary of Regulations in the USA, Europe, France and the Rest of the World Relating to RFID at UHF and SHF

Here is a very brief summary of the present situation worldwide at a given date. It is provided subject to the usual reservations and does not claim to be exhaustive.

16.2.1 UHF Frequencies in RFID

The authorized frequencies for RFID applications are in the vicinity of 433 MHz, in the 860 to 960 MHz band, around 2.45 GHz and finally at 5.8 GHz, depending on the country. Here are a few examples of the major regulations in force on 1 January 2009.

433 MHz

USA		
USA/Canada	433.92 MHz	$4.4\,\mathrm{mV\,m^{-1}}$ at 3 m (approx. 7 μW)
USA	433.92 MHz	Under discussion $11.0\,\mathrm{mV\,m^{-1}}$ at 3 m (approx. 33 μW) + improved duty cycle
Europe		
Europe	433.05–434.79 MHz	$P_{\mathrm{ERP}} = 10\,\mathrm{mW}$; maximum 10% duty cycle
	433.05–434.79 MHz	$P_{\mathrm{ERP}} = 1\,\mathrm{mW}$; up to 100% duty cycle
	434.04–434.79 MHz	$P_{\mathrm{ERP}} = 10\,\mathrm{mW}$; up to 100% duty cycle
Asia		
Japan	433.xx MHz	$35\,\mathrm{\mu V\,m^{-1}}$ at 3 m, not authorized as yet, likely to change
China	430.0–432.0 MHz	$6\,\mathrm{mV\,m^{-1}}$ at 3 m • for wireless equipment for intruder alarms • for short distances for vehicle, garage, emergency call, intruder and alarm applications

The 860 to 960 MHz Band

Use of the frequencies in this band is very idiosyncratic, because of the lack of harmonization in the regions concerned.

USA		
USA/Canada	902–928 MHz (for information, the central value is 915 MHz)	$P_{\mathrm{EIRP}} = 4\,\mathrm{W}$ (without licence) Note: 1 W conducted maximum + antenna with gain of + 6 dB FHSS authorized $50\,\mathrm{mV\,m^{-1}}$ at 3 m (single frequency systems) $P_{\mathrm{EIRP}} = 30\,\mathrm{W}$! (with licence) FCC Part 90, LMS (note: 3 W conducted)
Europe		
Europe	869.4–869.65 MHz	$P_{\mathrm{ERP}} = 500\,\mathrm{mW}$ (dc = 10%)
	865.6–867.6 MHz	$P_{\mathrm{ERP}} = 2\,\mathrm{W} \sim P_{\mathrm{EIR}} = 3.28\,\mathrm{W}$ LBT (listen before talk) Off time = 100 ms, on time = 4 s (see also the ERC 70 03 table in Figure 16.3)
Asia		
Japan	952–954 MHz	$P_{\mathrm{EIRP}} = 4\,\mathrm{W}$, as for the USA
China	917–922 MHz	Under consideration; $P_{\mathrm{ERP}} =$ between 2 and 4 W with LBT
Korea	908–914 MHz	20 LBT channels
Rest of the world		
Australia	918–926 MHz	$P_{\mathrm{EIR}} = 1\,\mathrm{W}$
	864–868 MHz	$P_{\mathrm{EIR}} = 4\,\mathrm{W}$
	921–929 MHz	$P_{\mathrm{EIR}} = 1\,\mathrm{W}$
New Zealand	864–868.1 MHz	$P_{\mathrm{EIR}} = 4\,\mathrm{W}$

South Africa	869.4–869.65 MHz	$P_{EIRP} = 500\,mW$ (dc = 10%) (in progress)
	915.2–915.4 MHz	$P_{EIRP} = 8\,W$ (RFID)
		(proposed by SABS WG1 – TC 74)
	915.3–915.6 MHz	15 W EIRP (note: 5 W conducted)
	860–930 MHz	Divided into 15 channels of 5 MHz each

16.2.2 RFID Frequencies at UHF (2.45 GHz) and SHF (5.8 GHz)

The same applies to the frequency band around 2.45 GHz, not in relation to its value, which is clearly defined, but as regards the subbands and power levels that can be radiated and the local restrictions, possible exemptions and their conditions, and the reading and writing distances in the case of passive transponders without on-board batteries. These frequencies are also used by other applications such as Bluetooth, Wireless LAN and WiFi.

2.45 GHz

USA		
USA/Canada	2400–2483 MHz	$P_{EIRP} = 4\,W$
		FHSS; no duty cycle or $50\,mV\,m^{-1}$
		at 3 m (single frequency systems)
Europe		
Europe	2446–2454 MHz	$P_{EIRP} = 500\,mW$ indoors/outdoors,
		no duty cycle
		$P_{EIRP} = 4\,W$ indoors duty cycle = 30 ms/170 ms
France		ARCEP does not accept 4 W P_{EIRP}
		(see also Figure 16.3)
Asia		
Japan	2400–2483.5 MHz	$3\,mW\,MHz^{-1}$: $P_{EIRP} = 1\,W$ with the
		same restrictions as in USA/Canada
China	2400–2425 MHz	$250\,mV\,m^{-1}$ at 3 m ($P_{EIRP} = 21\,mW$)

5.8 GHz

USA		
USA/Canada	5725–5850 MHz	$P_{EIRP} = 4\,W$
		FHSS; no duty cycle
Europe		
Europe	5725–5875 MHz	$P_{EIRP} = 25\,mW$; no duty cycle for nonspecific
		short range devices
France	Not authorized	
Asia		
Japan	Not allotted	($P_{EIRP} = 6\,W$ for toll collection, but not for RFID)
China	No allocation	

16.2.3 A Last Comment

Remember that it is not enough to comply with the maximum authorized values; we must also make sure that all the spectra radiated by the base stations and reradiated by the tags fall within the ranges concerned. The shape of the transmitted spectrum depends mainly on the type of bit coding, the bit rate used for the communication and the collision handling principle. Here is a little supplementary information that should help you to find your way.

16.2.4 Special Remarks on the use of UHF and SHF Frequencies

Many debates have raged about the UHF and SHF frequencies that should or should not be allocated to tag and RFID applications, and many national and international authorities are being subjected to serious questioning about the allocations of frequencies and associated bandwidths, the maximum authorized power and permitted duty cycles. In fact, as I have already pointed out, the currently authorized bands are not the same around the world, and, furthermore, without going into details and getting embroiled in interminable arguments, the EIRP and/or ERP power levels allocated to short range devices (SRD), whether nonspecific (NS) or specifically RFID devices, may be 10 mW ERP, 100 mW ERP, 500 mW ERP, 2 W ERP or 4 W EIRP, depending on the country, which is not at all helpful when it comes to the development of RFID products for worldwide use.

Let us now look at some other matters that are directly related to the signals radiated by base stations.

16.3 Standards for Magnetic and Electrical Fields in a Human Environment: Human Exposure

16.3.1 History

The International Commission on Non-Ionizing Radiation Protection (ICNIRP) was set up in 1992, and its work has been supervised by the World Health Organisation (WHO).

The first European preliminary standards/recommendations appeared in 1995, and then in 1997, in the form of these documents from the TC 211 committee of CENELEC – UTE (Union Technique de l'Electricité):

- ENV 50166-1:1995 – Human exposure to electromagnetic fields – low frequency (0 Hz to 10 kHz)
- ENV 50166-2:1995 – Human exposure to electromagnetic fields – high frequency (10 kHz to 300 GHz)
- ES 59005:1997 – Human exposure to electromagnetic fields for mobile telecommunication equipment in the range of 30 MHz to 6 GHz

In 1998, these documents, which are very hard for the ordinary reader to understand, were translated into the form of the 'ICNIRP guidelines for limiting exposure to time-varying electric, magnetic and electromagnetic fields up to 300 GHz', published in *Health Physics* in April 1998, volume 74, no. 4, and in 1999 all these documents were used as the basis of the

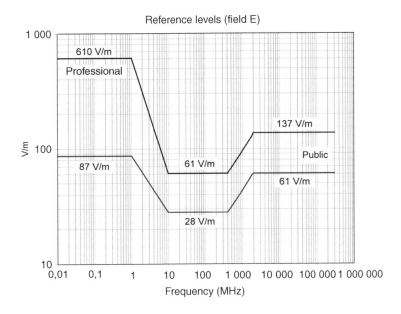

Figure 16.4   Example of RF reference levels in an electrical field

reference document: R&TTE Council Directive 1999/519/CE Requirement for RFID Device: Article 3 (a), 'The protection of the health and safety of the user and any other persons', 1999/519/CE, Council Recommendation of 12 July 1999, for limiting public exposure to electromagnetic fields (from 0 Hz to 300 GHz), on which a standard and measurement methods were to be based. An example of RF reference levels in an electrical field is given in Figure 16.4.

Working on the basis of the ICNIRP Council Recommendation 1999/519/EC, but also on other existing standards (ANSI, IEEE 95.1, DIN/VDE 0848, ICNIRP, AS/NZS 2772.1, etc.) of different countries (France, UK, USA, etc.), in order to prove conformity, TC 106 developed the measurement standard EN 50357, which was published in October 2001, under the title 'Procedure for evaluation of human exposure electromagnetic fields (EMFs) from devices used in Electronic Article Surveillance (EAS), Radio Frequency Identification (RFID) and similar equipments'.

For example, this was ratified in France by Decree 2002-775 of 3 May 2002 issued under Section 12 of Article L.32 of the Posts and Telecommunications Code relating to the 'limited values of exposure of the public to electromagnetic fields emitted by equipment used in telecommunications networks or by radio installations'.

That concludes our summary of the history.

16.3.2 *MPE and SAR*

EN 50357 mentions two essential terms: MPE (maximum permissible exposure) and SAR (specific absorption rate). The latter is defined as follows: 'The specific absorption rate (SAR) is the (mathematical) time derivative of the incremental energy absorbed by an incremental mass:

$$SAR = \frac{d}{dt}\left(\frac{dW}{dm}\right)$$

which can also be written as follows, using the electrical parameters normally encountered in contactless applications:

$$SAR = \frac{\sigma|E|^2}{2\rho} = c\frac{\Delta T}{\Delta t}\text{ expressed in mW g}^{-1}$$

where σ is the conductivity of human tissue in S cm^{-1}, ρ is the density of human tissue in g cm^{-3}, c is the specific heat capacity of the tissue in J g^{-1} °C^{-1}, E is the peak electrical field strength (which is why the value 2 appears in the denominator) in V cm^{-1} and ΔT is the temperature difference in °C over a time interval Δt in s.

Looking at these parameters, you will clearly see how difficult it is to assign specific values to them. Let me give some examples:

- Are we considering local exposure? or general?
- What kind of tissue are we looking at – face, hand, man, woman, child, ...?
- On the subject of tissue, the conductivity of an 'average muscle at rest' is 0.66 S cm^{-1}. What if it is neither average nor at rest?
- The same applies to the conductivity and permittivity, and therefore to the thermal impedance, of human tissue. To help us to quantify all this, the standard lists more than 800 different values, according to the frequencies used (10 Hz to 10 GHz) and the human organs concerned (liver, heart, skin, etc.).

In any case, since we cannot be expected to 'cook' anybody for experimental purposes, we are concerned here with the modelling (never perfect) of human tissue (using special viscous liquids). The results are finally obtained by measuring the received power in the equivalent liquid, using a calorimeter. Unfortunately, these methods are difficult to use and have low reproducibility, except in relation to standard human figures, known as anatomical models, which are specially designed and very expensive (Figure 16.5). You may like to know that the leading stars in this field have been given the names of Visible Man, MEET Man, Hugo, Norman, etc., and that the International Commission for Radiological Protection has defined a 'standard man' with a height of 1.76 m ± 5% and a weight of 73 kg ± 5%, to represent the nonuniform structure of the human body. His tissues have realistic dielectric properties and the resolution of the quantities used is better than or equal to an interval of 10 mm.

Returning to specific details, we find that, by way of example, at 13.56 MHz ($\sigma = 0.51$ S cm^{-1} and $\rho = 1.04$ g cm^{-3}) a conventional base station for the proximity reading of contactless smart cards (approximate power applied to the antenna, $P = 0.6$ W, i.e. a radiated power of $P_a = 20\,\mu$W) has an SAR of about 40 to 50 mW kg^{-1}, and that the maximum value authorized by the FCC in the USA is 80 mW kg^{-1} for the whole body, or 1.6 to 2 W kg^{-1}, i.e. 1.6 mW g^{-1} for partial exposure (of the head or the trunk for six minutes). Therefore, this system is well within the permitted limits.

If we estimate that the density (= mass/volume) of human tissue is close to unity (we are almost entirely made of water), and therefore 1 g of human tissue represents $1 \, cm^3$ of human tissue (1 kg of water is equivalent to 1 litre $= 1 \, dm^3$ and therefore 1000 g is equivalent to $1000 \, cm^3$, 1 g is equivalent to $1 \, cm^3$ and $1.6 \, mW \, g^{-1}$ is equivalent to $1.6 \, mW \, cm^{-3}$), we can

Antenna	Normative dimensions (in cm)			Informative dimensions (in cm)		
	a/b/c	X	Z	Height	Width	Depth
Wall mounted unit	15	20	-	20-100	20-55	5

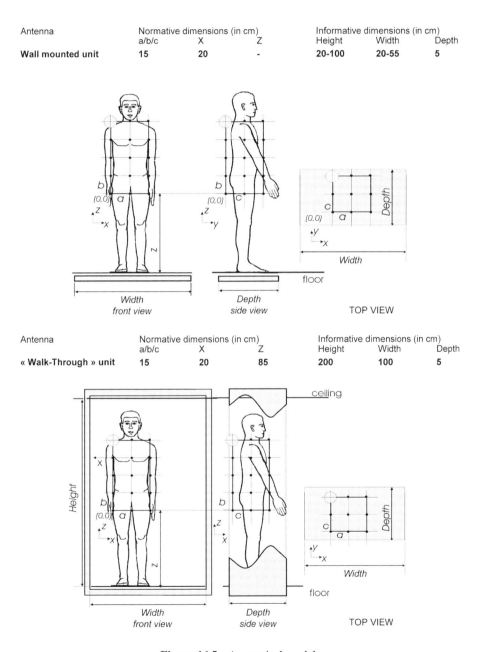

Antenna	Normative dimensions (in cm)			Informative dimensions (in cm)		
	a/b/c	X	Z	Height	Width	Depth
« Walk-Through » unit	15	20	85	200	100	5

Figure 16.5 Anatomical models

then extrapolate these values for the volume of a finger, a hand, and so on, without allowing for thermal inertia of course. In other words, for long-term exposure, we obtain a value of 16 mW for a volume of $10\,cm^3$ (e.g. a small cube of meat measuring 2.15 cm on each side). Very similar values have been found with RFID devices operating at UHF and SHF (see References 1 and 2 mentioned in the Preface, which provide very full details of these matters).

Following these technical/culinary considerations, we shall now look at the test and measurement methods for contactless devices.

16.3.3 An International and European Problem

Most of the energy levels, magnetic fields, etc., that are used are a long way below the authorized maximum values, but it is true that there is a lack of harmonization of the authorized maximum values that would enable a true maximum to be defined.

Before leaving the discussion of this subject, interested readers may wish to consult the document (which admittedly is rather old) summarizing the papers presented in May 2000 at the UTE Conference on 'Exposure to Electromagnetic Fields: Problems of Standardization in the Context of the European Recommendation', while anticipating the next issue of EN 50475 in June 2009.

16.4 Other Requirements to be Met

By way of conclusion to this chapter, I must point out that there are many other generic requirements to be met when working in RFID, although they are almost entirely outside the scope of this book. Let me just mention one aspect, that of privacy and all the matters relating to individual liberties that are a source of intense interest for CNIL (National Commission for IT and Civil Liberties) in France, Article 29 of the Brussels authorities, and Katherine Albrecht, who leads the famous American lobbying group CASPIAN (Consumers Against Supermarket Privacy Invasion and Numbering), but seem to receive little attention when it comes to the permanent use of credit cards, mobile telephones and the numbers printed on banknotes. I will not spend too long on these matters, which are regularly chewed over in the press, nor will I be drawn into any major arguments. Since the contents of this book are intended to be technical, I will simply draw your attention to some points that are already well established, as follows:

- Numerous devices have already been implemented in chip form for managing, blocking or locking access to specific data, or for 'killing' some or all of the functionality of a tag (according to state requirements), using special commands and a bit, which is appropriately called the 'kill bit', already included in ISO 18000-6C (be careful not to confuse this with the EAS anti-theft function). So the problem is not inherent in the tag and its circuitry, but in the choices made by users about what they wish to do with it. Everyone has his own problems!
- Fraudulent interception of RF communication is always possible (using 'stray waves', based on the principle of radiation of electromagnetic waves) if the resources are available. Any espionage or counterespionage service around the world will confirm

this! However, the resources used to process these intercepts may or may not be authorized (high power transmitted against the regulations for reading at long range, etc.), legal (fraudulent decryption, etc.), difficult to design and manufacture (requiring advanced technology and engineering skills, special components, etc.) or hard to deploy (heavy, bulky equipment, etc.). Therefore, is it all really worthwhile? But that is another story.

17

The Effects and Repercussions of Regulations on Performance

We will now examine the details of the parameters specified in RFID standards, the texts of the regulations, the constraints in terms of human exposure and their direct and indirect effect on the operating ranges of tags using the 433 MHz, 860–960 MHz and 2.45 GHz frequencies.

First of all, I must provide a few further details concerning the realities concealed behind the small print and subtleties of the regulations. We can talk about 'transmission' in the abstract, but we need to know where it takes place, in what bands, at what frequencies, by what method, etc. These are the matters that I will now examine.

17.1 Frequency

As a first step, let us look at the parameters relating to the permitted transmission frequencies.

17.1.1 Frequency Bands

As mentioned above, the local, regional and international regulations specify the bands that can be used in broad categories of applications (broadcasting, business band radio, etc.) in different countries.

The field of RFID applications with which this book deals is governed directly by the regulations on SRDs (short range devices). Essentially, the concept of an SRD is based on the principle of 'my freedom stops where the freedom of others begins'. Because of the short distances covered, users must not interfere with their neighbours and (theoretically) should not be affected by them. The great class of SRDs is also divided into two groups, the first comprising 'specific' devices relating to dedicated applications (such as medical FM broadcasting), while the second incorporates 'non-specific devices', covering everything else, including RFID devices if they are not specified elsewhere. Consequently, many RFID applications are classed as nonspecific SRDs.

Having made this clear, let us return to the definition of the frequency bands. For ease of reference, specific names are often given to these bands, e.g. 'UHF band' or 'the 2.45 GHz

RFID at Ultra and Super High Frequencies: Theory and Application Dominique Paret
© 2009 John Wiley & Sons, Ltd

band', a practice that can lead to considerable confusion, because these abbreviations may imply a range of bandwidths that is rarely the same from one application to another (e.g. the UHF band used for television in Europe varies from around 470 to 860 MHz, but the UHF band allocated to RFID in Europe occupies just a few MHz around 868 MHz); above all, it varies between countries (in the USA, for example, the UHF band for RFID extends all the way from 902 to 928 MHz). If it is any consolation, matters are even worse when we come to specify the actual values for the so-called 2.45 GHz band. Therefore, it is always advisable to adopt a reserved attitude while requesting more precise information about the frequency band in which your application is to operate – and if this is not forthcoming, then be very wary indeed.

17.1.2 The Value of Frequency (Carrier Frequency and Band)

When the true value of the usable frequency band is known, it will be necessary to specify the carrier frequency, or frequencies, before transmission takes place. These values are clearly important, if only for the purpose of calculating the attenuation (in dB) of the propagation of the wave, which depends directly on this parameter (see Chapter 6), and for helping to gain an initial idea of the operating range.

17.1.3 Band Occupancy or Division Modes

I have deliberately written 'the frequency or frequencies' because there are several ways of occupying a given frequency band.

Band without Channel Allocation

If the band occupancy mode is 'unrestricted', any value of carrier frequency can be chosen within the band in question. In this case, the band is 'without channel allocation'. However, it is not authorized to 'wander' all over the band, and in many cases the maximum bandwidth authorized for transmission is strictly specified.

Band with Channel Allocation

By contrast with the previous mode, some regulations (FCC, ETSI, ARIB, ARCEP, etc.) stipulate precise values of carrier frequency and/or specific values of subbandwidth in some bands and/or application types, to avoid a potentially anarchic use of frequencies. The total band is therefore divided into a number of transmission 'channels' and is thus known as a band with channel allocation.

Note
Theoretically, it is not essential for the channel widths to be strictly identical over the whole frequency band.

17.1.4 Modulation, Frequency Hopping, Spread Spectrum and Frequency Agility

As I pointed out in Chapter 12, an unmodulated carrier tends to be of little use, and I have mentioned numerous possible methods of modulation.

Type of Modulation

Leaving aside the presence or absence of channel allocation, as mentioned above, any properly drawn-up regulation will spell out, on many occasions, the type or types of carrier modulation (ASK, FSK, etc.) that can be supported, thus implying a specific radiated frequency spectrum. If nothing is specified, then any method is permitted – which is not uncommon!

Spread Spectrum

That is not all. In addition to the types of modulation listed above, there are techniques for occupying a wider spectrum, such as FHSS, DSSS, LBT, etc., as discussed at length in Chapter 13. Depending on whether they are locally authorized (for different countries), the global performance (mainly in terms of operating range) of the planned application may change radically, since these techniques can theoretically be used to operate (i.e. to detect signals) even below the surrounding noise level and to overcome some or all of the problems of parasitic reflection. When comparing one system with another, therefore, or one country with another, you should be careful only to compare what is truly comparable (see the section below that is wholly concerned with this point) and pay no heed to siren voices, which will be of no help to you.

One last remark can be made on this subject. Depending on the allocated bandwidth and the country involved, it may not be possible to have the same maximum number of permitted hops in a band or the same number of hops per second, and this will obviously have direct effects on the signal (or carrier)-to-noise ratio, and therefore on the communication range as well.

17.2 Transmission Level

Transmission is a simple enough idea, but we also need to know what the maximum levels are, whether it takes place indoors or outdoors, and for how long. Let us look at the parameters relating to transmission levels.

17.2.1 Band Occupancy Duty Cycle

First of all, we must understand that we are not allowed to transmit a carrier continuously, in sustained mode ('continuous wave', or CW). This would give rise to problems of disturbance, RF pollution or simultaneous occupancy of a single band by many users. We therefore define a duty cycle Δ (or rate) of temporal band occupancy, which obviously varies from $0.x\%$ (hardly any transmission time at all) to 100% (permanent occupancy).

At the frequencies that can be used for RFID, and especially at UHF and SHF, we often encounter band occupancy rates ranging between 0.1, 10, 15 and 100%. Moreover, this occupancy rate is defined with respect to a reference time base, which varies from a second to a minute, or more commonly to an hour, etc., according to the standards and regulations. Once again, then, you should carefully examine the specific definition of the duty cycle of the particular frequency band in which the application is to operate.

Clearly, this band occupancy limit may be very troublesome in applications in which numerous tags move rapidly past the base station antenna (with a high passage rate) and there is insufficient time to read them. This also strongly affects the choice of dynamic collision management methods (because there are many elements arriving and departing simultaneously, and constantly!).

To provide a specific example, ARCEP in France specifies a maximum radiated power of 0.5 W ERP at UHF, with a utilization duty cycle of 10%, in other words only six minutes of activity per hour (seriously!), or 6 seconds per minute, or 60 out of every 600 ms. Now it is up to you to find an intelligent way of dealing with the tags entering and leaving the electromagnetic fields. In the 2 W ERP LBT mode, this usually requires a duty cycle of nearly 90% for operation. As for 4 W EIRP, or 4 ERP, broadband systems using options such as the FHSS, this looks like a pipedream at present – but who knows? (I know, but I'm not saying. . .). If we are thinking of using UWB (ultra wide band), impulsed radio and Chirp SS systems, we must be crazy!

17.2.2 Indoor Versus Outdoor

We may speak of maximum transmitted power, but in what surroundings? Indoors or outdoors? Here is another troublesome parameter to consider when comparing permitted power levels or resulting operating ranges.

Depending on the countries involved, the authorized frequencies, the local regulations, etc., we may have to draw a further distinction between transmitters located 'physically' (or pseudo-physically) inside and outside a building. In fact, as mentioned in Chapter 7, at the UHF and SHF frequencies used for RFID, a conventional wall thickness of 22 cm causes an attenuation of about 15 to 20 dB, depending on the materials used (bricks, breezeblocks, concrete, reinforced concrete, etc.). Clearly, then, if we make conventional measurements of RF pollution at 10 or 30 m in a free field for a given pollution rate, using the methods proposed by the ETSI or FCC, we will have to accept a higher level of power radiated by the base station when it is inside a building than when it is outside.

Theoretically, everything will be just fine if 'indoor' is clearly distinguished from 'outdoor'. Remember the old riddle, 'When is a door not a door?' Unfortunately, the harsh reality of RFID applications means that we must answer, 'When it is ajar!' To make this problem clearer, let us consider two specific examples:

- Our first example is an access control system for a garage, showroom, etc., where the base station is physically located inside the building, just behind the folding entry door.
- In the second example, the base station is located inside a storefront, by the entrance, within a shopping mall.

In the first case, the door is always open during the day, for obvious reasons concerning the nature of the business. Therefore, is the base station in the building considered to be 'indoor' or 'outdoor'? This situation is commonly encountered in warehouses or depots, where the base stations are located just inward from the loading bays and are separated from the bays by plastic strip doors exposed to all weathers.

In the second case, the whole shopping mall may be seen as a private space belonging to a single business that rents its premises to the traders, so that the combination of stores is interpreted as a single building and the RFID installations are described as 'indoor'. Alternatively, the inside of the mall may be treated as a shopping street by the local authority, for reasons of security and access by public protection agencies, in which case the RFID installations are considered to be 'outdoor'.

In fact, the precise answers to these questions can be found in standards, regulations, by-laws, etc., but you will have to examine all of the small print very thoroughly![1]

If you are finding this material rather indigestible, please take some time to review Chapters 6 and 7.

17.2.3 Maximum Radiated Power Levels and/or Electrical Fields

In Chapter 6, I spent some time in examining the many different definitions of power levels (EIRP, ERP, r.m.s., peak, etc.). I did this for a very good reason.

It was because international and national standards and regulations happily play with terms and definitions that are all technically correct, but that tend to be politically slanted, in that they show lower or higher values when it is expedient to do so. So be very careful not to confuse the terms EIRP and ERP (no I!). Theoretically, in Europe, according to the ETSI standards, radiated power is measured in ERP below 1 GHz and in EIRP above this level.

However, you should also know that, for many frequency bands (such as HF), there is no definition of maximum radiated power – but that does not mean that everything is permitted! In fact, since the next section will describe the maximum levels of radiated fields, the system that is used must conform to this and limit the radiated power, instead of transmitting freely at any level.

There are also some standards that specify a maximum 'number of base stations per km^2' (i.e. a concentration of systems per unit of area), in addition to authorized maximum power levels.

17.2.4 Antenna Directivity and Selectivity Angle

Here is another example of hypocrisy, related to the previous section:

- The term EIRP includes the concept of 'isotropic', and therefore cannot take account of lobes and directivity of radiated electromagnetic fields.
- The term ERP implies an underlying $\lambda/2$ dipole antenna, and therefore a known gain and directivity. As I showed in Chapter 5, the directivity pattern of this antenna is about $\pm39°$. Consequently, if the power flux density is insufficient, it will be impossible to read bulky objects (such as the whole contents of a pallet) outside this lobe, particularly in the proximity of the base station antenna (Figure 17.1). In order to read these objects, it will be necessary to move them farther away and provide either a longer operating range or more radiated power.

17.2.5 Maximum Levels of Radiated or Interfering Fields

The main authorities responsible for issuing these standards or recommendations are the ETSI in Europe, the FCC in the USA, and the ARIB in Japan. For information, Figure 17.2 shows the relationships between the different organizations in Europe.

---

[1] By way of example, here is an extract from Annex 11 of ERC 70 03 dated September 2005, concerning RFID: 'When measured outside a building, at a distance of 10 metres, any emissions due to RFID devices must not exceed the field level equivalent to one which would be produced by a 500 mW device located outside the building, the measurement being made at the same distance. If a building is composed of a number of premises, such as stores in shopping malls (or the like), the measurements must relate to the stores located at the boundaries of the inside of the building.'

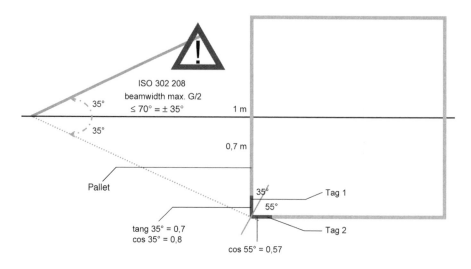

Figure 17.1   Example of ISO 302 208 – UHF 2 W ERP LBT

For UHF and SHF, depending on the country and the local regulations, maximum radiated electrical field strengths E (in $V\,m^{-1}$) and magnetic field strengths H (in $A\,m^{-1}$) are specified, at stated distances, for large groups of frequency bands. Here again, close attention must be paid to the associated parameters, such as measurement distances, measurement methods, units of measurement and associated masks, as well as the specified values.

Methods of Measurement
The measurement methods described in these standards and regulations are sometimes very specific, if not 'specialist' (such as r.m.s., peak, quasi-peak, peak to peak, etc.), and in some cases,

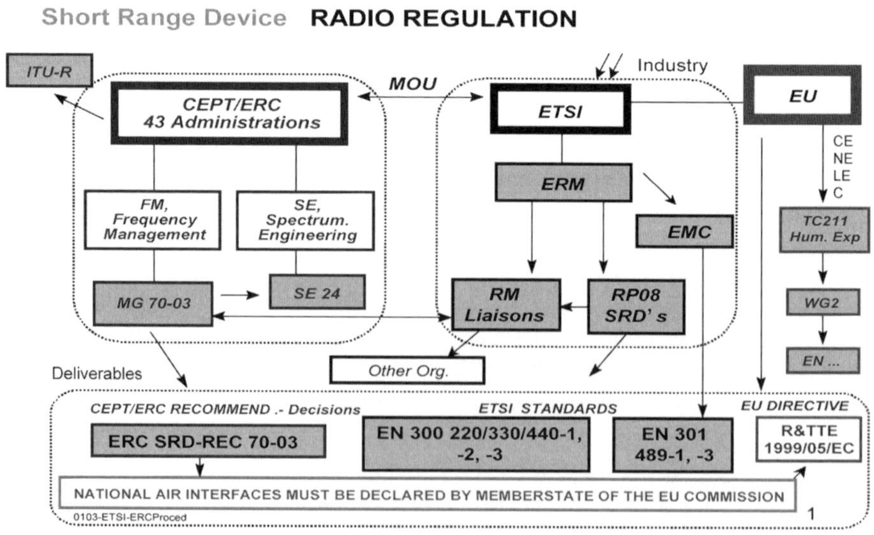

Figure 17.2   Organization of SRD regulators

surprisingly enough, they provide the criteria for the final choice of a solution, particularly as regards the choice of collision management systems (described in a later section).

An Observation (and Not a Random One)

Note that most of the measurements relating to ETSI and FCC standards are to be made according to the CISPR 16 standard, which describes the apparatus and methods to be used (Figure 17.3).

This is not just a random observation, because most of the specified values mentioned above have to be measured in quasi-peak levels (Figure 17.4). Clearly, this tells you nothing at all, presented in this form. That is understandable, but here is a way to read between the lines.

COMMISSION ÉLECTROTECHNIQUE INTERNATIONALE

INTERNATIONAL ELECTROTECHNICAL COMMISSION

CISPR 16-1

Première édition
First edition
1993-08

COMITÉ INTERNATIONAL SPÉCIAL DES PERTURBATIONS RADIOÉLECTRIQUES
INTERNATIONAL SPECIAL COMMITTEE ON RADIO INTERFERENCE

Spécifications des méthodes et des appareils de mesure des perturbations radioélectriques et de l'immunité aux perturbations radioélectriques

Partie 1:
Appareils de mesure des perturbations radioélectriques et de l'immunité aux perturbations radioélectriques

Specification for radio disturbance and immunity measuring apparatus and methods

Part 1:
Radio disturbance and immunity measuring apparatus

© CEI 1993 Droits de reproduction réservés — Copyright – all rights reserved

Aucune partie de cette publication ne peut être reproduite ni utilisée sous quelque forme que ce soit et par aucun procédé, électronique ou mécanique, y compris la photocopie et les microfilms, sans l'accord écrit de l'éditeur.

No part of this publication may be reproduced or utilized in any form or by any means, electronic or mechanical, including photocopying and microfilm, without permission in writing from the publisher.

Bureau Central de la Commission Electrotechnique Internationale 3, rue de Varembé Genève, Suisse

IEC

Commission Electrotechnique Internationale
International Electrotechnical Commission
Международная Электротехническая Комиссия

CODE PRIX
PRICE CODE XE

Pour prix, voir catalogue en vigueur
For price, see current catalogue

Figure 17.3 Extract from CISPR 16 specifying the apparatus and methods to be used

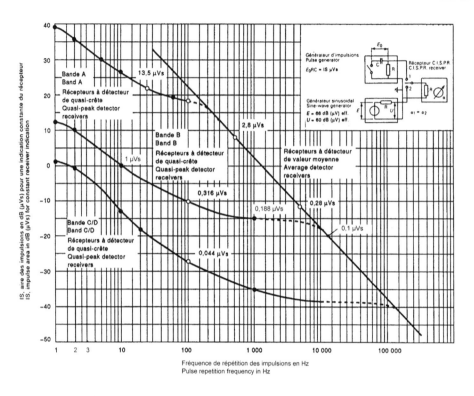

Figure 17.4   Quasi-peak measurement method to be followed

A quasi-peak detection measurement system operates, as its name indicates, on a quasi-peak basis! What this means is that we measure a value nearly equal to the peak of the signal. That is true, but we also use a diode and a capacitor C to carry out a simple rectification of the incident alternating signal (to detect the peak of the signal), and we also connect a resistor R with a specified value in parallel with this capacitor, which integrates the received and measured signal. Evidently, the capacitor C discharges slightly, as a function of the time constant $\Theta = RC$ of the circuit constructed in this way, thus providing quasi-peak detection – and this is where the 'quasi' becomes particularly important.

This is because the base station is constructed so as to radiate the most powerful signal possible (especially when a longer range is required), and therefore if it transmits less often the capacitor will be discharged more fully and the measured quasi-peak value will decrease, and the system will conform more easily to the permitted maximum levels. Thus this quasi-peak measurement is sensitive to the command repetition cycles used in exchanges between base stations and tags. The choice of the collision management method used is therefore not at all independent. Is this strange? Not at all. If we use probabilistic collision management, such as an ALOHA slotted method (see my earlier book, Reference 1 mentioned in the Preface), the base station only transmits commands occasionally (thus tending to decrease the quasi-peak value) and frequently listens for data returning from the tags. However, in the case of deterministic collision management systems such as the 'bit-to-bit' type, the base station takes over more frequently (thus recharging the capacitor more frequently), and so is rather less favourable. This is one of the technical reasons behind the inclusion of different collision management modes in the ISO 18000-x standards, including fast 'bit-to-bit' methods for short operating ranges (and therefore lower transmitted power levels) for handling applications on fast production line conveyors using the same label applied to a packet and also (and equally) 'time slot' collision management methods for overcoming the problems of long range applications, e.g. in the case where largely immobile pallets have to be read while complying with the values measured by the quasi-peak method.

17.2.6 Notes on the Use of ETSI 300 220 for 433 and 860–960 MHz, and ETSI 300 440 for 2.45 GHz

It is always worth remembering that, for HF RFID (at 13.56 MHz), the FCC in the USA often specifies maximum values of these levels in an electrical field E in μV m^{-1} at a certain distance (30 m), while the ETSI in Europe specifies levels in a magnetic field H in dBμ A m^{-1} at another distance (10 m), which is a great help in understanding the figures!

Throughout this book, I have also pointed out that RFID applications at UHF and SHF operate in the far field, where $d > \lambda/(2\pi)$ – let us say, more than a few metres. In this case, we must bear in mind that the electromagnetic wave produced by the base station antenna is propagated, and, if we only consider the amplitudes of the electrical field E and magnetic field H of this wave on the axis of the antenna, we should remember that the relation between these fields in the far field is

$$E = Z_0 H$$

where Z_0 is the impedance of the air ($Z_0 = 377\,\Omega$).

It is a simple matter to use this equation to convert the value of the radiated electrical field E (expressed in dBμ V m^{-1}) in a magnetic field H (expressed in dBμ A m^{-1}), using the relation E (dBμ V m^{-1}) $= H$ (dBμ A m^{-1}) $+ 51.5$ dB, as shown in Chapter 4.

How to Comply with the Standards and Regulations

Now we have established the limits that each system must comply with in order to operate legally without exceeding the FCC or ETSI values, in other words the maximum power levels and/or the power levels in far electromagnetic fields that satisfy the equation in the preceding section above a few metres. As I have also shown in Chapters 4 and 5, there is a direct relation between the effective value of the radiated electrical field and the radiated power EIRP in the far field:

$$E \approx 7 \frac{\sqrt{P_{\text{radiated}}}}{r}$$

where

$$P_{\text{radiated}} = P_{\text{radiated}} I^2$$

in which equation P_{radiated} is the equivalent isotropic radiated power P_{EIRP} and $R_{\text{radiation}}$ is the radiation resistance of the base station antenna.

For a given system, the designer can easily use these equations to find the maximum level that the current I_{max} flowing in the base station antenna must not exceed if it is to conform to the maximum value permitted by the regulations and/or standards, thus:

$$I_{max} = \frac{Er}{7\sqrt{R_{radiation}}}$$

Masks and Templates

A maximum level of a radiated field E or H having all the attributes mentioned above is meaningless without its associated template, or mask, within which the transmitted spectrum is required to fit.

17.2.7 Local Regulatory Authorities

In each country, the authorities concerned with problems of regulation act independently. Depending on their technical and political links with other countries, and depending on the local possibilities, they may or may not follow the 'Recommendations' issued by international bodies (UIT, CCIR, CEPT, etc.) in order to harmonize the bands, power levels, templates, etc., as far as possible. The options are often limited because of the earlier history of radio frequency communications (e.g. where a band is already occupied by military applications, by GSM mobile telephony, by television, etc.). In France, as I have already mentioned, the National Agency for Frequencies (ANFr) and the ARCEP (formerly ART, Radio and Telecommunications Authority) are responsible for assigning frequencies and the use and limits of these frequencies.

17.2.8 Unlicensed Operation, Exemptions and Site Licences

Note that, if we wish to transmit in an unauthorized band, or one already occupied by other users, or at a higher power level than that specified for an authorized band, etc., it is always possible to apply to the authorities for an exemption. Requesting an exemption is one thing, but obtaining it is quite another! However, there is no reason not to try. Clearly, acting in combination with others (in associations, industrial groupings, etc.) will add weight to your arguments and may bring results, although even a solitary approach has been known to succeed. If exemptions are granted, they may be subject to temporary conditions or requirements for use (e.g. operation on a trial basis, or for six months, or for a renewable stated period, etc. – all kinds of stipulations may be made).

At the time of writing (end of 2008), this is still the case in some countries – particularly throughout Europe – in the case of UHF for the use of the 2 MHz wide band dedicated to RFID applications (865.6–867.6 MHz) subject to the conditions of 2 W ERP and LBT specified by ETSI 302 208 and recommended by ERC 70 03, because the military are currently using parts of this band. In such cases, the problem is still very much with us and unfortunately it can often only be resolved by a tactful approach to the ministries concerned (Interior, Defence, Industry, etc.).

While on this subject, I will mention the subtleties of exemptions and other discreet transactions that may lead to the following compromises, even if it is impossible to 'release frequencies' for good and all from their present occupants:

- allocating narrower frequency bands than those requested;
- allocating the band but only if lower power levels are used;
- allowing the use of less power in a narrower band than that requested;
- allocating the band and allowing the use of the desired power, but only for indoor operation, not outdoor;
- special exemptions dependent on local factors that are closely specified (to the nearest metre);
- geographical 'zoning' of operating sites or locations in cities and/or regions;
- and so on... useful, if not perfect!

In short, these compromises do nothing to promote technological advance, and often leave local industrialists lagging behind other countries.

17.2.9 Populations of Interrogators: Single, Multiple and Dense

Here is another form of classification that is extremely important. Simply stated, this is a matter of evaluating problems of cohabitation between base stations (interrogators) present on the same site, where their density is measured in numbers per square kilometre. The aim is to avoid interference between them.

Single
This is the best-known and simplest case, in which there is a single isolated base station. In this case, everything I have said above is applicable, subject to the restrictions described earlier, of course (in terms of the duty cycle, etc.).

Multiple
If the number of base stations present on the same site is 10 or less, the base station density is described as 'multiple'. If we wish to avoid problems, we need to place more precise limits on parasitic signal reflection and out-of-band or out-of-channel signal interference, by using more effective templates; we must also consider the use of more appropriate techniques (time division, frequency and spread spectrum multiplexing, FHSS, LBT, etc.).

To give an idea of the problems, at 910 MHz in free space the attenuation of a signal due to propagation in air alone is as follows (this was demonstrated in Chapter 6):

$$\text{att (dB) at } 910\,\text{MHz} = 31.68 + 20 \log r \quad \text{where } r \text{ is in m}$$

The attenuation at 1 km is therefore

$$\text{att (dB) at } 910\,\text{MHz} = 31.68 + 20 \log 1000$$

$$\text{att (dB) at } 910\,\text{MHz} = 31.68 + 60$$

$$\text{att (dB) at } 910\,\text{MHz} = 91.68, \quad \text{i.e. approximately } 92\,\text{dB}$$

By way of example, here are some commonly encountered cases. If a base station in the USA radiates a power of 4 W EIRP (+ 36 dBm), then the strength of the radiated signal still present at 1 km will be $36 - 92 = - 56$ dBm. Similarly, in Europe, at the authorized level of 2 W ERP, i.e. 3.28 W ERP ($33 + 2.14 = 35.14$ dBm), approximately -57 dBm will remain. You may remember (if not, see Chapter 10 again) that in remotely powered applications (with an operating range from 7 to 8 m), the return signal from a tag is at about -65 dBm, well below the level received from another, interfering transmitter located 1 km away.

Dense

Now the stakes are raised. This term is used when there are 10 to 50 base stations on the same site. This situation is commonly encountered, for example, with RFID applications at the loading bays of major carriers (which may have as many as 60 to 90 gates, often with two to four antennas per gate). This is not the place to consider the huge extent of the total radiated spectrum – or the reflections of all the signals involved. However, we must look at the details, because they are extremely important and need to be examined closely. To enable a system to operate correctly, we must be able to manage everyday operations. We must therefore look very closely at the different regional environments, governed by the CEPT/ETSI in Europe and by the FCC in the USA and Canada.

US and Canadian Environments

First of all, let us consider RFID applications in the USA and Canada. Since they are regulated by the FCC, everything is simpler, because the usable frequency band is much wider and the FHSS principle can be used.

Multichannel operation

The frequency band from 902 to 928 MHz (bandwidth 26 MHz, maximum EIRP 4 W) can be divided into 50 channels of 500 kHz each, and the transmissions from the base station can use the FHSS principle, which makes matters much simpler. Figure 17.5 shows an example of this operating mode, according to ISO 18000-6 type C, using PR-ASK modulation, Tari = 25 µs and a tag response at 64 kbits s^{-1} carried by Miller subcarrier coding with an M value of 4, i.e. a subcarrier frequency of 256 kHz.

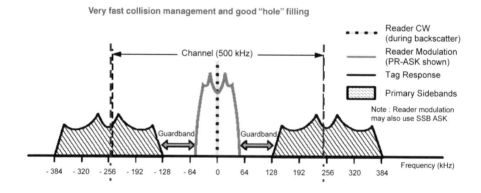

Figure 17.5 Multichannel operation

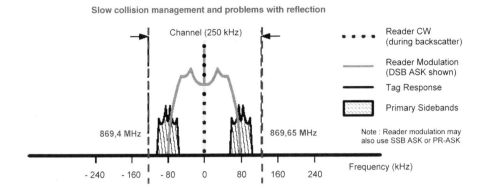

Figure 17.6   Operation in the single-channel mode

European Environments
The European context, governed by the CEPT/ETSI, is much more complex. We must consider two different approaches.

Operation in the single-channel mode
In a long range single-channel RFID, the transmissions take place in the 869.4–869.65 MHz band (bandwidth 250 kHz, maximum ERP 500 mW, maximum duty cycle 10%) used in the single-channel mode with a channel width of 250 kHz to provide sufficient bandwidth and the highest possible bit rate. In this case, the transmissions from the base stations and the responses from the tags are separated in time, by using the half-duplex communication principle of the protocol used, and also by providing for the synchronization of the base stations co-located on the same site to operate their tags sequentially. Finally, all the base stations transmit a permanent signal in order to listen for the responses from the tags. Figure 17.6 shows an example of this operating mode according to ISO 18000-6 type C (near the optimum in terms of speed of collision management), with DSB ASK modulation, Tari $= 25$ µs and a tag response at 20 kbits s^{-1} carried by Miller subcarrier coding with an M value of 4, i.e. a subcarrier frequency of 80 kHz.

Operation in the multichannel mode
According to ETSI 302 208 (for a 3 MHz wide band at UHF) including frequencies from 865.6 to 867.6 MHz, a 2 MHz wide band divided into 10 channels of 200 kHz each (Figure 17.7), it is permissible to use a maximum power of 2 W ERP in the LBT mode. I will show that the differentiation between the base station transmissions and those of the tags must take place at the level of the spectra radiated by the base station–tag systems.

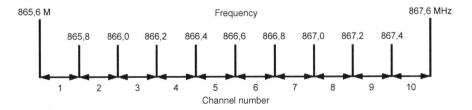

Figure 17.7   Operation in the multichannel mode

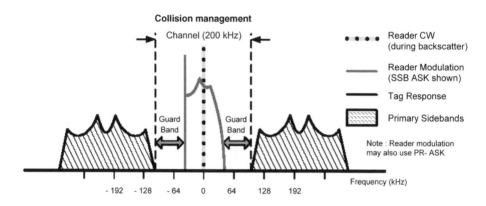

Figure 17.8   Example of the multichannel mode

Because of the bandwidth of about 200 kHz required for communication (because of the speed of the forward and return links required for fast collision management), the necessary reserved spaces between transmission channels, etc., according to ISO 18000-6 type C, the spectra produced by transmission from the bases stations (with DSB or SSB-ASK modulation) and by the tag responses occupy more than the normal width of a channel. One way of keeping within the channel limits specified by ETSI 302 208 is to lock the base station transmission carrier frequencies to the frequencies of certain channels in the band and adjust the system parameters (speed, Tari, xxx ASK, M, etc.) so that the frequency spectra of the tag responses fit exactly, or as nearly as possible, within the permitted bands for adjacent channels. Figure 17.8 shows an example of this operating mode, using SSB-ASK modulation, Tari $=$ 25 μs and a tag response at 53.3 kbits s^{-1} carried by Miller subcarrier coding with an M value of 4, i.e. a subcarrier frequency of 213.3 kHz. Clearly, this frequency separation is a further constraint on the possibilities offered by the LBT principle, preventing the use of some potential transmission channels and some options for resolving 'black holes' due to reflection, and also making listening to the return channels more specific because of the possible collisions of spectra in channel $n + 1$ due to simultaneous transmissions from base stations in channels n and $n + 2$.

In view of the serious problems arising in dense environments, the ETSI improved its proposals in 2007 by publishing a number of technical reports recommending some methods of using the 10 channels in question. These reports, TS 102 562 and TR 102 313, 436 and 649, are required reading for everyone, because, as is usually the case, they are full of important footnotes in small print.

Operation in the 'multiple–single' channel mode
In the band described above, authorized by ETSI 302 208, there is nothing to prevent users from operating both in the LBT mode and in the 'single narrowband channel' mode in a specific channel; in other words we can use the ten 200 kHz channels for different communications. This means that the spectra of the signals sent by the base station and the responses reradiated by the tags alternate and are included in a single channel. Even in the case of a base station operating at the slowest bit rate (with the highest Tari value of 25 μs), the spectrum radiated by the base station is a close fit in the channel width. Given that the tag responses must also be located in the same channel, they must appear slightly later in time, meaning that we can only use time multiplexing, not frequency multiplexing. In order to achieve the widest possible

spectrum reradiated by the tag (for the fastest collision management), this spectrum must be locked to the carrier and symmetrically distributed, and must therefore be based on a response not requiring the presence of a subcarrier frequency (using FM0 modulation, for example), making it necessary to use circulators or bidirectional couplers at the inputs of the reception stages of the base station, in order to isolate the return link from the forward link (see also Chapter 19).

Concluding remarks

Let us remain in Europe for a while. The strict ETSI 302 208 standard setting out the specifications of the LBT system permitting a power level of 2 W ERP around 866 MHz for specific RFID applications states that the threshold indicating the occupation of a channel during the listening process is −96 dBm. Since 2 W ERP (+ 33 dBm) is equivalent to 3.28 W EIRP (33 + 2.14 = + 35.14 dBm), the total possible discrepancy due to the signal attenuation is 35.14 + 96 = 131.14 dBm.

As I mentioned in Chapter 6, the attenuation of a signal due solely to its propagation in air is as follows:

$$\text{att (dB)} = 32.5 + 20 \log f \ 20 \log r$$

where f is expressed in GHz and r is expressed in m. Therefore

$$\text{att (dB)} = 32.5 + 20 \log(0.866) + 20 \log r$$

$$\text{att (dB) at 868 MHz} = 31.27 + 20 \log r$$

Now let us calculate the value of r such that the attenuation in air is 131.14 dBm:

$$20 \log r = 131.14 - 31.27 = 99.87$$

where $\log r = 4.99$, i.e. $r = 98$ km, in an ideal free space!

Even if the reality is quite different (at about 10 km) because of attenuation due to many other objects (buildings, trees, etc., always encountered over this distance), this means that the LBT receiver, which always listens before talking, will almost always detect a weak signal, but one above the significant threshold of −96 dBm, in the RFID band, and this will prevent everything from operating correctly. Indeed, some troublemakers, who oppose the state of European regulations, refer to this system as LNT (listen never talk!).

To sum up, when an RFID system operates in the LBT mode in a multiple or dense environment (such as the loading bays of stores or warehouses with numerous gates, each having several antennas), if we wish to avoid excessive detection of signals above −96 dBm that prevent any 'talking' in the UHF RFID band because of excessive channel occupancy, it is useful (i.e. 'highly advisable') to provide time synchronization between the base stations (interrogators) of all types, in order to achieve the best possible operation in the LBT mode.

By Way of Conclusion

In dense environments, if we wish to manage collisions as speedily as possible:

- there are 50 FHSS channels available in the USA
- and 5 LBT channels in Europe.

			Region I EUROPE		Region II ASIA	Region III USA	Effects
		EXAMPLE		France			
Bandwidth							Selectivity of transmission / reception
Division of band into channels							Signal/noise level
Spread spectrum, FHSS / DSSS							Detection in noise, distance
number of hops							Detection in noise, distance
Carrier frequency							Applicable worldwide or not
Radiated level							
max. radiated power							erp/ eirp ==> distance
max. duty cycle							Dynamic collision management
indoor / outdoor							Limitations on applications
radiated fields H, B at d							Max. operating distance
methods of measurement							Principle of collision management
shapes / values of templates							Max. communication speeds
							Remote collision management
Selectivity angle of the antenna							Communication volume / distance
Local regulatory authorities							Everything

Figure 17.9   Effects of regulations on the performance of an RFID system

17.3 Summary

The skeleton table in Figure 17.9 provides a summary of the above text, showing the main direct effects of the various parameters on the operating ranges and performance of RFID tags at UHF and SHF, according to the regulations. You will also find this table useful as a way of making precise comparisons between the operating ranges of different systems and in different countries.

17.4 Comparison Between Europe and the USA

In the UHF band from 860 to 960 MHz in Europe, based on the specifications of ERC 70 03, we can envisage several cases, according to the subband frequencies:

(a) The maximum ERP is limited to 10 mW at present with a duty cycle of 100%. In these conditions it is only possible to read and write tags at relatively short range, if we are dealing with passive tags (remotely powered tags without any on-board power source). Furthermore, the requirement to use the existing narrow channels means that, in practice, we have to use ASK carrier modulation with low bit rates in order to fit the spectrum templates.

(b) Provided that we use a narrowly specified low duty cycle (dc = 10%), we can use a power level of 500 mW ERP ($P_{\text{EIRP}} = 820$ mW). We can then achieve practical ranges of about 2.50 m for reading and about 2 m for writing. These ranges can also be achieved with conventional gate systems operating with a duty cycle of 100% (constant operation) at 13.56 MHz, without any exemption for standard RF interference measured at 10 m at 42 dBµ A m^{-1} (ETSI 300 330) (see the paragraph on this subject below).

(c) As mentioned above, it is now possible to use a power level of 2 W ERP (3.28 W EIRP) in LBT conditions (with a duty cycle of about 95% in the case of multiples devices, but lower in a dense environment) and achieve an operating range of 4–5 m for reading and 3–4 m for writing, similar to those shown below for the USA.

In the USA and Canada, because of the completely different allocation of frequencies – there is no cordless or mobile (GSM) telephony in these bands – the authorized radiated power can reach a maximum of $P_{EIRP} = 4$ W with a duty cycle (dc) of 100%. Furthermore, because the bandwidth is much larger than in Europe, the FCC authorizes the use of carrier frequency hopping, using the FHSS (in the 860–960 MHz band) or the DSSS (at 2.45 GHz), and it is possible to achieve practical operating ranges of about 4–5 m for reading and 3–4 m for writing. This is all summarized in Figure 17.10.

We can only hope that, in the near future, i.e. in about three to five years – by which time the authorities concerned, namely the UIT, CEPT/ERC and ARIB plan to harmonize the allocations of UHF frequencies and the associated channel widths – it will be possible to make tags operate in a virtually identical way in the 860 to 960 MHz band in all three regions of the world.

17.5 UHF or 13.56 MHz Around the World and in Europe

I have often been asked, 'Which is the best frequency, 13.56 MHz or UHF?' or 'I've heard that the performance levels are better at UHF than at 13.56 MHz. Is this true? What do you think?' and so on. Leaving aside the protocols, bit rates, etc., you will have noted that one of the key problems in RFID at UHF and SHF is to satisfy (comply with) the regulations (in terms of power, templates, radiation levels, appropriate measurement methods, etc.) in force in different countries – and we can be quite sure that they will differ widely between frequencies and between countries. (Review the detailed sections above if necessary.)

17.5.1 UHF

As I have stressed, the available UHF frequency bands vary considerably around the world (902 to 928 MHz in the USA, around 868 MHz in Europe and around 952 to 954 MHz in the Far East):

1. The maximum authorized levels differ widely, as follows:
   - USA: 4 W EIRP (FHSS and DSSS permitted), dc = 100%, indoor and outdoor
   - Europe: 500 mW ERP = 820 mW EIRP, dc = 10%, indoor and outdoor
                        = 2 W ERP = 3.82 W EIRP, LBT in door and outdoor
   - Spain, Italy, Turkey and Russia: 500 mW ERP = 820 mW EIRP, dc = 10%, indoor and outdoor
2. The authorized on/off duty cycles are also different:
   - USA: 100% at 4 W EIRP
   - Europe: 100% at 10 mW ERP
   - Europe: 10% at 500 mW ERP
   - Europe: LBT at 2 W ERP (from 99.9% to a much lower level in dense environments)
3. Because the transmission is at UHF and the antennas are often tuned at $\lambda/2$, the transmission is inherently directive and the lobe has an aperture of about ± 30 to $40°$ measured from the principal axis of the antenna. This makes the base station 'blind' to part of the space in front of its UHF transmission antenna and it is often necessary to move objects away to make them readable in the angle of the directivity cone. This effect is enhanced when an antenna with a high gain is used, since directivity increases with gain.

Example in UHF

	Region I EUROPE				Region II ASIA varies according to country	Region III USA	Main technical and application parameters affected
Applicable in France	x	x		x			
Band (in MHz)	868.00 - 868.60	869.40 - 869.65	865.6 - 867.6	868.00 - 868.60	952 - 954	902 - 928	Selectivity of transmission / reception
Channel division / narrowband	no	yes - 25 kHz	yes - 200 kHz	no	yes	yes	Signal/noise level
FHSS / DSSS / LBT	no	no	LBT	no	~ LBT	FHSS	Detection in noise, distance
permitted number of hops	none	none	10	none		50 / 75*	Detection in noise, distance
Central value of frequency	868.xx	869.55	n.a	868.3		915	Applicable worldwide or not
Signals radiated power	25 mW ERP / 41 mW EIRP	500 mW ERP / 820 mW EIRP	2 W ERP / 3.28 W EIRP	25 mW ERP / 41 mW EIRP	4 W EIRP / 1 W conducted	4 W EIRP / 1 W conducted	erp/ eirp ==> distance
duty cycle	up to 100%	up to 10%	(LBT -99%)	up to 100%	up to 100%	up to 100%	Dynamic collision management
indoor / outdoor	outdoor	outdoor	outdoor	outdoor	outdoor	outdoor	Limitations on applications
fields H, B	n.a	n.a	n.a	n.a	n.a	n.a	Max. operating distance
methods of measurement templates	CISPR 16 / ETSI 300 - 220	CISPR 16 / ETSI 300 - 220	ESTI 302 208 / ESTI 302 208	CISPR 16 / ETSI 300 - 220		FCC / FCC	Principle of collision management / Max. communication speeds
Dense environments	poor	poor	poor	poor		good	Remote collision management
Angle / directivity of the antenna	approx. +/- 40°	approx. +/- 40°	approx. +/- 40°	approx. +/- 40°	approx. +/-30°	approx. +/- 30°	Limitations on applications / Communication volume / distance
Local regulatory authorities documents	CEPT / ERC 70 03	CEPT / ERC 70 03	CEPT / ERC 70 03	ARCEP (ex ART) / ERC 70 03	ARIB / ARIB 81	FCC / FCC 47 part 15	Everything

Figure 17.10 Example of the effects of regulations on the performance of an RFID system

Table 17.1

UHF		Distances (theoretical, not operational)
	In the USA	About 7–8 m (with 4 W EIRP, 100% + FHSS)
	In Europe	About 0.6 m (with 10 mW and dc = 100%)
		About 4 m (with 500 mW ERP and dc = 10%)
		About 6–7 m (with 2 W ERP, LBT)

4. Where the transmission and reception circuits are concerned, in the USA and in other countries it is possible to use systems operating with spread spectrum techniques, either the FHSS or DSSS, which greatly improves the chances of extracting a signal from the noise and increases the available operating ranges. In Europe and France, this is not permitted, and systems are required to operate in a 'divided' band mode, called the narrowband mode, or in the LBT mode in the countries where this is permitted.
5. Mandatory regulations concerning strengths of radiated polluting fields are not usually an obstacle or a critical or sensitive area for RFID applications.

Summary

These parameters make it possible to achieve the theoretical distances given in Table 17.1 at a given date (2007), for a given component technology and for the same functionality (reading or read/write), according to the current regulations. These distances do not allow for operating tolerances and are based on operation on the axis of the antenna, so they will be reduced according to the directivity of the antenna used, the tolerances, reflection and absorption (see Chapters 5, 6 and 7).

17.5.2 13.56 MHz

This frequency has been used for a long time around the world. The pollution limits are approximately the same in the USA (according to FCC 47 part 15), in Japan (set by ARIB), in Europe (set by ETSI 300 330) and in France (set by ARCEP), at about $+42\,\text{dB}\mu\,\text{A m}^{-1}$ at 10 m or an equivalent level in $\text{dB}\mu\,\text{V m}^{-1}$ at 30 m worldwide. The same applies to the 100% duty cycle and the template. This is a major advantage, making it possible to produce and supply a product that can circulate freely throughout the world with the same operational performance. Furthermore, since January 2004 it has been possible to increase this limit to $+60\,\text{dB}\mu\,\text{A m}^{-1}$ in Europe, according to ERC 70 03 (see the next section).

Let me swiftly put an end to all the foolish ideas that are commonly put forward in the nonspecialist press and/or by the uninitiated, by pointing out that, by contrast with UHF/SHF, no maximum power levels are specified for HF transmission at 13.56 MHz (except for those that would result in the maximum SAR values advocated by the WHO). It is only the RF pollution limits that are specified, for a given range.

We can comply with these limits by using products according to ISO 15 693 and ISO 18000-3 in continuous operation (with a duty cycle of 100%: this is important) at ranges of about 90 cm to 1.2 m with simple circular or rectangular antennas or 1.2 to 1.5 m with figure-of-eight wound antennas. We can cover areas from 2 to 2.5 m wide with systems located in gates and even much larger areas (several tens of m^3) with other systems in arch structures, without excessive directivity or unwanted reflections, thus covering a large volume, which is clearly defined in physical terms.

2004: A New Development in Europe, Thanks to ETSI: + 60 dBμ A m⁻¹ at 10 m at 13.56 MHz

2004: A New Development in Europe, Thanks to ETSI: $+ 60\ dB\mu\ A\ m^{-1}$ at 10 m at 13.56 MHz

At last, there is something new from CEPT/ERO/ETSI!

Constraints Imposed by ETSI 300 330 at 13.56 MHz

As mentioned above, the ERO (European Regulation Organization) sets certain limits on the use of 'nonspecific' SRDs (short range devices) such as those used in RFID applications in the ISM band at 13.56 MHz. The most important of these constraints at present is the one relating to prevention of pollution in the radio frequency spectrum. This is set by ETSI 300 330, which has specified for a long time that, in specific measurement conditions (in terms of measurement methods, conformity with templates, etc.), regardless of the radiated power, the magnetic field radiated by the base station and measured at 10 m must not exceed $H_{max\ ETSI\ 10m} = 42$ dBμ A m⁻¹ or, in terms of the equivalent electrical field strength, $E_{max\ ETSI\ 10m} = 47.4$ mV m⁻¹. As we have seen, the relation between the far electrical field and the radiated power in the far field (10 m is the beginning of the far field at 13.56 MHz) is approximately as follows:

$$E = \frac{7\sqrt{P_a}}{r}$$

and therefore

$$P_{a\ max} = [(rE_{max\ ETSI\ 10\ m})/7]^2$$

$$P_{a\ max} = [(10 \times 47.4 \times 10^{-3})/7]^2 = 4.585\ \text{mW}$$

Since January 2004, however, ERC 70 03/ETSI has revised the maximum radiated magnetic field level from 42 to 60 dBμ A m⁻¹, measured at 10 m as before. This is an increase of 18 dB or, alternatively, 3×6 dB. Since the value of 6 dB in terms of voltage is equivalent to a numerical ratio of 2, this global increase amounts to $(2 \times 2 \times 2) = 8$, and therefore

$$E_{max\ ETSI\ 10\ m} = 47.4 \times 10^{-3} \times 8 = 379.2\ \text{mV m}^{-1}$$

On the other hand, the equation

$$E = \frac{7\sqrt{P_a}}{r}$$

indicates that, at the same distance d, E is proportional to the square root of the power radiated by the base station antenna, P_a. Consequently, the maximum power radiated by the base station, $P_{a\ max}$, for a given antenna is allowed to increase in a ratio of $8^2 = 64$. Therefore

$$P_{a\ max} = 4.585 \times 64 = 293.44\ \text{mW}$$

All other things being equal for a base station operating at 13.56 MHz (number of turns N_1, radius r_1 and section s_1, and therefore with the same radiation resistance R_a), I have shown that the radiated power $P_{a\ max}$ is proportional to I_1^2:

$$P_a = R_a I^2$$

and, given that (see References 1 and 2 mentioned in the Preface)

$$I = \sqrt{\frac{P_1 Q_1}{L_1 \omega}}$$

the power radiated by the base station antenna P_a must be proportional to the electrical power P_1 applied to it (the conducted power) and P_1 must be increased by a factor of 64 in order to obtain the maximum benefit from this in a specified system.

On the other hand, if we replace I_1 with its value:

$$P_a = R_a \left(\frac{H_0 \times 2r_1}{N_1} \right)^2$$

we see that, for a given base station antenna with a radius r_1 and number of turns N_1, we can have a magnetic field H_0 in the centre of the antenna that is eight (8) times stronger than before, and consequently we can achieve a longer operating range in the near field.

In a previous book on RFID using inductive coupling (Reference 1 mentioned in the Preface), I showed that the value of the magnetic field H in the near field (the area at a distance of less than $\lambda/2\pi$; in this case, $\lambda = 22$ m) could be written thus:

$$H(a, r) = \frac{1}{(1 + a^2)^{3/2}} H(0, r)$$

where $a = d$/ret $H(0,r)$ with field H in the centre.

At a distance d that is long (but not enormously long) with respect to the radius of the antenna r (for example, $a = 2, 3, 5, \ldots$), we will therefore still be in the 'near field' of the transmission antenna. The value of a^2 is therefore large with respect to 1, and the equation is reduced to the form

$$H = \frac{1}{(d^2/r^2)^{3/2}} H(0, r) = \frac{1}{(d^3/r^3)} H(0, r)$$

and therefore

$$H(d, r) = \frac{r^3}{d^3} H(0, r)$$

For a specified and constant value of r, this equation takes a form of the type constant/d^3. The magnetic field $H(d)$ then varies with $1/d^3$, and therefore

$$H \approx \frac{\text{constant}}{d^3}$$

and therefore

$$d = \sqrt[3]{\frac{\text{constant}}{H}}$$

Thus, everything else being equal as regards the base station antenna (same antenna, L, radius, Q, etc.), if we multiply the value of P_a by 8, then for the same field value H operating the same tag, the new operating range d_2 will be equal to the root of 8 times the former value d_1, i.e.

$$d_2/d_1 = \sqrt{8} = 2.82$$

In previous books (References 1 and 2 mentioned in the Preface), I showed that we could obtain a theoretical operating range, on the principal axis of the antenna, of the order of a metre, when using a standard tag (5×5 cm and six turns) with a threshold magnetic field of $40\,\mathrm{mA\,m^{-1}}$, while complying with the ETSI limit of $42\,\mathrm{dB\mu\,A\,m^{-1}}$. With the new authorized value of $+60\,\mathrm{dB\mu\,A\,m^{-1}}$, d becomes approximately 2.8 m (with a single antenna) or about 4 m in a gate structure.

We can therefore conclude that, in Europe, the operating range at 13.56 MHz (about 2.8 m) with a single antenna (with a duty cycle of 100% and very low directivity) is roughly equivalent to what can be achieved at UHF (reckoning on the basis of a theoretical free space distance of about 4.10 m and much more directivity) with $P_{\mathrm{ERP}} = 500\,\mathrm{mW}$ and a duty cycle limited to only 10%.

What about the Problem of Sideband Levels?
It is only the maximum size of the template 'straitjacket' that has increased: the maximum permitted values of the sidebands according to ETSI 300 330 have not changed. This means that there are still some restrictions on the use of this $+60\,\mathrm{dB\mu\,A\,m^{-1}}$ level. Let us take a closer look.

To ensure that the sidebands of the signal radiated by the base station continue to fit the permitted template, the first thing to do is to reduce the communication speed. Unfortunately, ISO 15693 and ISO 18000-3 provide no official specifications; if we reduce the bit rate, we find ourselves 'outside the ISO'. On the other hand, the return channel from the tag to the interrogator can continue to operate at higher ISO speeds, because the signals sent by tag load modulation are very weak. So how can we use this feature? The answer is simple:

- At short range with high bit rates, we can identify products on a production line.
- At short range with low bit rates, we can identify items on shelf display units (with clearly defined magnetic field shapes).
- At medium to long range, with high bit rates, we can monitor pallets and primary and secondary packaging with large volumes (in $\mathrm{m^3}$) according to ISO 15693/18000-3 at $+42\,\mathrm{dB\mu\,A\,m^{-1}}$.
- At long range, with low bit rates, outside the ISO specifications, using the $+60\,\mathrm{dB\mu\,A\,m^{-1}}$ level, we can provide EAS (electronic alarm surveillance) at check-outs.

To sum up, in Europe, according to these regulations, 13.56 MHz gate systems and UHF systems with monostatic base stations can provide substantially identical results in terms of operating range.

> **Important**
> We must adjust these values according to the type of material (air, plastic, water, metal, etc.) forming the objects in the field.

SHF (2.45 and 5.8 GHz)
For practical purposes, the same considerations apply to the frequency band around 2.45 GHz, not in relation to its value, which is clearly defined, but as regards the subbands and power levels

Table 17.2 Characteristics of RLAN transmitters

EIRP	Omnidirectional	$+20\,\text{dBm}$
FHSS	Bandwidth at 3 dB	$<0.35\,\text{MHz}$
	Number of channels	79
	Hop sequence increment	1 MHz
DSSS	Bandwidth at 3 dB	15 MHz
	Null-to-null bandwidth	22 MHz
	Number of channels (overlapping)	14, every 5 MHz, can be selected by the user
Duty cycle	From 1 to 99%	

that can be transmitted and the restrictions for different countries, the possible exemptions and their conditions, and also the reading and writing distances in the case of passive transponders without on-board batteries. These frequencies are also used by many other applications such as Bluetooth, Wireless LAN, WiFi, etc.

The Combination of Power and Bandwidth Requirements for Fixed and Moving Tags at SHF

In order to meet the power and bandwidth requirements stated above, we can consider various alternatives for systems operating at 500 mW EIRP and above.

If the system uses the FHSS mode over the whole band from 2.4000 to 2.4835 GHz, transmitting a power of 4 W EIRP (indoor) and 500 mW EIRP (outdoor), there may be unacceptable interference with other radio communication services. One way of avoiding this is to reduce the frequency spread to operate inside a limited spectrum mask where systems radiate at these higher levels.

Note the three very narrow frequency bands in which we can transmit a maximum power of 36 dBm (4 W EIRP), or at least a level in excess of 27 dBm (500 mW EIRP), and the much wider band in which the power is limited to only 10 dBm (10 mW EIRP).

Comparison Between RLAN and RFID

For your information, I have listed below the main features of RLANs (radio local area networks) and RFID systems that operate in the same frequency bands, for the simple reason that both are included in the category of SRDs (short range devices).

RLAN Applications

The frequency hopping (FHSS) and direct sequence spread spectrum (DSSS) techniques are used in RLAN systems (IEEE 802.11B Wi-Fi) and the receivers have characteristics of this type (see Tables 17.2 and 17.3).

Table 17.3 Characteristics of RLAN receivers

FHSS	Sensitivity of receiver	$-90\,\text{dBm}$ or better
	Noise bandwidth	1 MHz
	Number of channels	79
DSSS	Sensitivity of receiver	$-90\,\text{dBm}$ or better
	Noise bandwidth	15 MHz
	Number of channels (overlapping)	14, every 5 MHz, can be selected by the user

Note

Most of the applications use ominidirectional antennas with a typical maximum gain of 2 dBi (see Tables 17.4 and 17.5).

Table 17.4 RFID applications

EIRP	In the USA	+36 dBm (4 W) (indoor)
	In Europe (depending on the country)	+27 dBm (500 mW) (indoor/outdoor) +36 dBm (4 W) (indoor) and dc = 15%
	Antenna gain	> +6 dBi
	Beamwidth	<90°
FHSS		
	Bandwidth at −3 dB Tx	<0.35 MHz
	Number of channels	79 (optionally 20)
	Hop sequence increment	1 MHz (optionally 0.35 MHz)
DSSS		
	Bandwidth at −3 dB	15 MHz
	Null-to-null	22 MHz
	Number of channels	every 5 MHz, can be selected by the user
	Number of channels with overlapping	14
NB (narrow band)		
	Number of channels	3 (can be positioned anywhere)
	Bandwidth at −3 dB	<0.01 MHz
	Channel spacing	0.6 MHz

Table 17.5 Typical SRD applications

EIRP	+10 dBm (10 mW)
Antenna gain	0 dBi
Beamwidth	360°
Bandwidth of channel at −3 dB	1 MHz
Narrowband	Anywhere in the 2400–2483.5 MHz band

17.6 Appendix: The Main Standards and Regulations

The lists below show the main ISO/IEC and ETSI standards and regulations that are most commonly used in the field of contactless devices, and especially for identification cards (13.56 MHz), radio frequency identification of animals (<135 kHz) and item management (all frequencies, mainly 13.56 MHz and at UHF/SHF).

Note

For further information, readers should contact the ISO, ERO, UIT and ETSI, whose addresses are given at the end of this book.

17.6.1 Standards for Contactless Smart Cards (at 13.56 MHz)

ISO/IEC 10536	**Contactless integrated circuit(s) cards – Close coupled cards**	
ISO/IEC 10536-1	Part 1	Physical characteristics
ISO/IEC 10536-2	Part 2	Dimensions and location of coupling areas
ISO/IEC 10536-3	Part 3	Electronic signals and reset procedures
ISO/IEC 10536-4	Part 4	Answer to reset and transmission protocols

ISO/IEC 14443	**Contactless integrated circuit(s) cards – Proximity cards**	
ISO/IEC 14443-1	Part 1	Physical characteristics
ISO/IEC 14443-2	Part 2	Radio frequency power and signal interface
ISO/IEC 14443-3	Part 3	Initialization and anti-collision
ISO/IEC 14443-4	Part 4	Transmission protocols

ISO/IEC 15693	**Contactless integrated circuit(s) cards – Vicinity cards**	
ISO/IEC 15693-1	Part 1	Physical characteristics
ISO/IEC 15693-2	Part 2	Radio frequency power and signal interface
ISO/IEC 15693-3	Part 3	Anti-collision and transmission protocols

ISO/IEC 10373	**Test methods**
ISO/IEC 10373-1	Test methods – generalities
ISO/IEC 10373-4	Test methods for contactless close coupling cards (ISO 10536)
ISO/IEC 10373-6	Test methods for contactless proximity cards (ISO 14443)
ISO/IEC 10373-7	Test methods for contactless vicinity cards (ISO 15693)

17.6.2 Standards for Animal Identification (at 132.4 kHz)

ISO/IEC 1178x	**Radio frequency identification of animals**
ISO/IEC 11784	Radio frequency identification of animals – code structure
ISO/IEC 11785	Radio frequency identification of animals – technical concept

ISO/IEC 14223-x	**Radio frequency identification of animals**	
ISO/IEC 14223-1	Part 1	Radio frequency interface
ISO/IEC 14223-2	Part 2	Command code structure (preliminary stage)
ISO/IEC 14223-3	Part 3	Applications

17.6.3 Standards for Item Identification (LF, HF, UHF, SHF)

Information technology – Automatic identification and data capture – Radio frequency identification (RFID) for item management (at all frequencies)

ISO/IEC 19762	**Harmonized vocabulary**	
ISO/IEC 19762-1	Part 1	General terms relating to AIDC
ISO/IEC 19762-2	Part 2	Optically readable media (ORM)
ISO/IEC 19762-3	Part 3	Radio frequency identification (RFID)

ISO/IEC 18000-x   **RFID for item management – Air interface**

ISO/IEC 18000-1	Part 1	Generic parameters for air interface
ISO/IEC 18000-2	Part 2	Communications below 135 kHz
ISO/IEC 18000-3	Part 3	Communications at 13.56 MHz
ISO/IEC 18000-4	Part 4	Communications at 2.45 GHz
ISO/IEC 18000-5	Part 5	Communications at 5.8 GHz
ISO/IEC 18000-6	Part 6	Communications at 860–960 MHz
ISO/IEC 18000-7	Part 7	Communications at 433 MHz

ISO/IEC 1804x   **Test methods**

ISO/IEC 18046-x	Part 1	Conformance (one per frequency)
ISO/IEC 18047-x	Part 2	Performance: Three parts: (a) tag, (b) interrogator, (c) system
ISO 15961		Data protocol: application interface
ISO 15962		Data protocol: data encoding rules and logical memory functions

ISO 15963-x   **RFID for item management**

ISO 15963	Unique identification for RF tags
ISO 15963-1	Host interrogator – tag functional commands and other syntax features
ISO 15963-2	Data syntax
ISO 15963-3	Unique identification of RF tag and registration authority to manage the uniqueness

17.6.4 Regulatory Standards (ETSI, FCC, ARIB)

ETSI 300 xx0   **Radio equipment and systems; short range devices**

ETSI 300 330	Technical characteristics and test methods – 9 kHz to 25 MHz
ETSI 300 220	Technical characteristics and test methods – 25 MHz to 1000 MHz
ETSI 300 440	Electromagnetic compatibility and radio spectrum matters (ERM) – short range devices – radio equipment to be used in the 1 to 40 GHz frequency range
Part 1	Technical characteristics and test methods
Part 2	Harmonized EN under Article 3.2 of the R&TTE Directive

ETSI 302 208-x   **Electromagnetic compatibility and radio spectrum matters (ERM) – Radio-frequency identification equipment operating in the band 865 MHz to 868 MHz with power levels up to 2W ERP**

Part 1	Technical characteristics and test methods
Part 2	Harmonized EN under Article 3.2 of the R&TTE Directive
ETSI TS 102 562, 102 649, 102 463, 102 313	These technical reports supplement the text of ETSI 302 208 and facilitate its application
RCR STD-1	RFID equipment for premises radio station
RCR STD-29	RFID equipment for specified low power radio station
CEPT/ERC 70-03	Relating to the use of short range devices (SRD), Annex 11 – Recommendation 70-03

FCC   US Code of Federal Regulations (CFR) Title 47, Chapter I, Part 15.
                                 Radio frequency devices, US Federal Communications
                                 Commission
ARIB   ARIB STD-T81 – RFID equipment using frequency hopping
                                 system for specified low 'informative references'

17.6.5 EPC Standards

EPCglobal™   EPC™ radio frequency identity protocols, Class 1 Generation 2 UHF RFID,
                    Protocol for communications at 860 MHz–960 MHz, Version 1.0.9

Part Five

Components for Tags and Base Stations

The last section of this book provides a detailed description of the main electronic and other production methods used to manufacture tags and base stations for UHF and SHF (Chapters 18 and 19), and also shows how to develop and produce a commercial electrical data sheet for a finished tag (Chapter 20).

We shall start by considering the structure of a tag.

18

RFID Tags

18.1 Some General Remarks

Virtually all the UHF and SHF tags available on the market operate in a return link mode (from the tag to the base station) using the back scattering principle, and are therefore 'passive'. These have been described at length in this book. In physical terms, their components are as follows:

- For inexpensive, remotely powered tags ('passive batteryless' tags): an antenna, an element (which may or may not be an integrated circuit) providing the identification number and other functionality, and optionally a substrate (paper, plastics, etc.).
- For 'passive battery-assisted' tags: an antenna, an integrated circuit, a battery (cell, accumulator, paper battery, etc.) and optionally a substrate (paper, plastics, etc.). In a tag with an integrated circuit (silicon chip), this may take the form of mini-packages (such as the TSSOP 8), micromodules or bare chips, connected to the antenna with microwire or directly mounted in a flip chip arrangement in contact with the antenna.

18.2 Summary of Operating Principles

The following sections provide a brief overview of the back scattering function, with the aim of defining the various electrical and electronic functions expected in a commercial tag.

18.2.1 The Operating Principle of the Exchange between Base Stations and Tags

The exchange between a base station and a tag normally takes place by a half duplex (HDX) procedure managed by the base station, according to a communication protocol known as 'reader talk first' (RTF), whereby the base station initializes the dialogue, and the tag, which may be locally or remotely powered, remains silent up to that point.

RFID at Ultra and Super High Frequencies: Theory and Application   Dominique Paret
© 2009 John Wiley & Sons, Ltd

I will briefly run through the different phases of this exchange, which were detailed in Chapter 9.

Forward Link

In the first phase, the base station sends a carrier frequency, which is generally amplitude modulated (by ASK) or, less commonly, frequency modulated (by FSK), to carry the desired commands to the tag. The tag antenna is tuned to the carrier, and receives and absorbs the signal from the base station.

If the tag incorporates an electronic component, this then has to carry out a number of functions:

- Detection of the incoming signal is performed by rectifying it, so that it can be used as the tag's (remote) power supply, and also by carrying out the amplitude (or frequency) demodulation of the incoming wave.
- The electronic component of the tag decodes and interprets the command received from the base station and is then able to operate (e.g. by responding to the request, changing the contents of its memory, etc.).

Return Link

In a second phase, the base station transmits an unmodulated sustained carrier to enable the tag to communicate. The tag then switches to its response phase (for the return link). The principle of this operation is based on the modulation of a reradiated/reflected part of the incoming (unmodulated) wave received from the base station (see Chapter 9 again if necessary). This operating mechanism is known as 'back scattering' (scattering of the return wave).

For this purpose:

- The internal physical or electronic components of the tag modulate the load impedance of the antenna in accordance with the variation of the data to be transmitted to the base station.
- This detunes the tag's receiving antenna and thus changes the value of its impedance in accordance with the binary variation of the data to be transmitted.
- As a result of this deliberate detuning, some of the incoming wave is not absorbed by the tag, but is reflected, and its phase varies with the variation of the digital data to be transmitted by the tag. In other words, the reflection coefficient is modulated according to the variations of the data. In this way the tag varies the power of the wave reflected towards the reader.

The back scattered return signal therefore consists of part of the incoming wave modulated by shifts, xSK, or more precisely by binary shifts, BxSK, generated by the variations in the rate of the data to be transmitted by the tag.

Note

It is difficult, or even impossible, to use these techniques at lower frequencies, because of the size of the antennas that have to be incorporated in the tags.

18.3 The Technology of Tags

Despite this lengthy list of electronic functions to be provided, their technological implementation is a relatively simple matter. It is possible to use a single diode or a simple integrated circuit in the tag to provide the numerous functions described above, namely:

- powering the tag (remotely if appropriate) (using rectification to provide this if possible);
- demodulating the modulated incoming wave;
- modulating to provide the return link.

18.3.1 Remotely Powered Passive Tags (Batteryless)

These exist in many types, with and without integrated circuits, with greater or lesser degrees of complexity. To make matters clear, I will divide them into the following categories:

- chipless (i.e. no integrated circuit);
- with an on-board semiconductor component such as a diode or an integrated circuit.

Chipless Surface Acoustic Wave (SAW) Tags

These tags, which have no integrated circuits and are therefore known as chipless, account for a small part of the market at present (a few per cent), and in some quarters it is expected that they will make up 10 to 12% by 2015. These tags are read-only, and therefore also operate without a power supply. Their operation is based on the well-known piezoelectric effect, which is used to construct a reciprocal electromechanical transducer (see below).

Operating Principle

Figure 18.1 introduces the principle of chipless surface acoustic wave tags. The electrical field of the waves radiated by the base station is converted to a potential difference by means of the

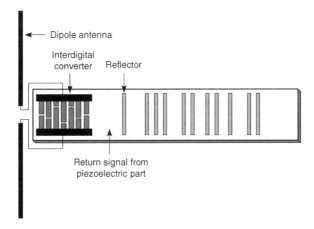

Figure 18.1 Principle of chipless surface acoustic wave tags

tag antenna. The piezoelectric effect of the substrate used, acting as a mechanical electrical transducer (converter), causes the chip to be deformed mechanically in accordance with the variations of the received electrical signal. This mechanical deformation produces a mechanical/acoustic wave that is propagated over the surface of the chip (which is made by planar microstrip technology, for example), which is why we use the term 'surface acoustic wave' (SAW). The velocity of mechanical propagation of this acoustic wave in the material is about 3000 to 4000 m s^{-1}, which is very low by comparison with the speed of light and electromagnetic propagation in air or in a vacuum.

'Reflectors' in the form of 'fingers' are often placed on the chip or substrate (by etching notches or depositing aluminium to form interdigital transducers, or IDTs) in order to filter and reflect a small proportion of the incident mechanical wave at precise instants, which differ according to their relative mechanical positions. Using the reciprocity of the piezoelectric effect, these reflections are reconverted into electrical signals by the transducer/converter and modulate the tag's retransmission to the base station (using the same principle of back scattering by modulation of the phase and/or power and antenna load). The mechanical and geometrical position of the reflectors on the chip can be used to encode the reflections to form a binary data element (a number that is specified once and for all in the factory – see below).

One advantage of this technique is that the propagation of the mechanical wave in the substrate is much slower that the propagation of an electromagnetic wave in air, and therefore the wave retransmitted by the tag and containing its response will reach the base station some time after all the reflections caused by the electromagnetic environment of the site of the RFID system, thus providing better immunity to the application environment. It is therefore relatively easier to recognize this signal in the ambient spectrum.

Example of the Performance of Chipless SAW Tags
Unique ID and Read-Only Devices

Because of the mechanical implementation (dimensions) of the fingers and the maximum substrate area, which can be achieved at a reasonable price, the unique tag number (unique ID) is encoded in 16 to 64 bits in practice, which limits the range of applications to some extent. For your information, 16 bits = 66 564 ID numbers.

Generally, all the fingers are placed on the substrate at the start of the industrial tag manufacturing process, and those that are not required are then removed or modified when the unique ID is created. The customization operation generally requires the use of expensive machinery, and can therefore only be carried out in the substrate or tag production plant, not at the point where the tags are applied to the final products. This imposes structural limits on applications such as those envisaged by EPCglobal, and means that these tags are intrinsically suitable for a read-only device, read-only tag and proprietary types of applications.

It is worth noting that the global SAW tag (GST), for example, is currently attempting to resolve the problem of unique numbers encoded in 128 bits ($= 3.4 \times 10^{38}$ unique codes) by using 32 reflectors and 64 QAM and pulse modulation AM methods, in order to increase the spectral efficiency (in bit s Hz^{-1}) of the bit rate per width of spectrum occupied and to achieve a high number of reads per second.

Reading Distance

Since this type of tag has no integrated circuit ('chipless'), there is no need to provide any remote power supply for the tag or fit a battery in it. The only problem is that of knowing whether the base station will be capable of recovering enough of the return signal from the tag for correct operation of the system. Given the sensitivity of the base station receivers available at present, we find that the reading distances of the tags are about 7 to 10 m when the power emitted by the base station is 10 mW ERP, either at UHF (300 to 900 MHz) or at SHF (2.45 GHz).

Bit Transfer Rate

I have mentioned that the acoustic wave is propagated mechanically in the substrate at about $3000 \, \text{m} \, \text{s}^{-1}$. By a simple proportionality rule, therefore, the wave must travel 1 cm in 3 μs. Given the actual dimensions of the substrates, and therefore the mechanical distances travelled by the wave, the possible rates are about 300 to 500 kbits s^{-1}, making it feasible for the tags to be in rapid movement.

Operating Temperature

Since the interdigital comb structures can be formed on a ceramic substrate, it is quite feasible to use tags of this type for operation at high temperatures (200 to 300 °C), which is impossible for tags fitted with integrated circuits.

18.3.2 Semiconductor Tags

Tags with Capacitor Diodes (for EAS Applications, for Example)

Tags with capacitor diodes are '1 bit' passive tags. The bit indicates whether or not a tag is present in the irradiated field. These tags are widely used in applications such as EAS (electronic automatic surveillance), for theft prevention, for example (they may be attached to garments), and they are encapsulated in a plastic casing that is strong enough to resist mechanical wear and tear. They are very easy to construct. The tag consists of two antenna strands, forming a $\lambda/2$ dipole antenna, and a capacitor diode.

Operating Principle

The operating principle of these tags is as follows:

- The base station sends a pure wave.
- The receiving dipole (antenna plus diode) of the tag is intrinsically nonlinear, and therefore creates harmonics, generally of low amplitude.
- Capacitor diodes have the property of using the energy stored in the capacitor to amplify the harmonic frequencies and reradiate them.
- If the base station receiver detects a frequency that is a multiple of the transmission frequency, the alarm is triggered.

To improve the performance of the systems in the presence of noise, the signal from the base station is modulated. The harmonics reradiated by the tag are therefore also modulated, so that

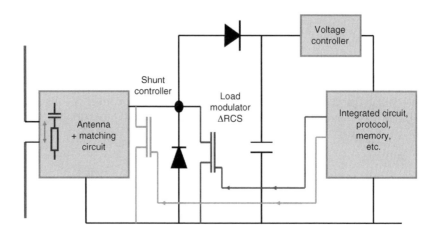

Figure 18.2   Schematic diagram of a UHF tag

the tag's signal can be distinguished from noise, parasitic signals or signals from other transmitters sharing the same frequency band.

Integrated Circuit Tags

This is the most common type of tag on the market. Apart from the antenna, it contains an integrated circuit that may be more or less complex. I will describe one of the fullest configurations here, but of course you are free to disregard the options that you will not need.

The 'generic' integrated circuit includes the following functions (see Figure 18.2):

- An impedance matching circuit. This input circuit has the task of adapting the antenna impedance so that the whole system forms a 'conjugate match' in order to recover as much power as possible, at a given frequency or in a given frequency band. This circuit often takes the form of a simple tuning capacitor, although, as we have seen in Chapter 6, this is often completely inadequate.
- A device for rectifying the incident wave. This has the task of reconstructing a continuous voltage and making it available to power the tag locally. The device used for this purpose is generally a full-wave rectifier bridge (consisting of diodes or transistors) or a voltage doubler circuit, in which the diodes have low operating thresholds, and a filter capacitor. This capacitor usually has a very low capacitance, and is therefore small, because of the UHF or SHF operating frequencies and the low power levels (a few tens of μW) required for the integrated circuit.
- A power on reset (POR) device. When the integrated circuit has been supplied with power via the carrier frequency, it must go through the POR sequence to prepare itself for communication.
- A local clock regeneration system. In order to decode the incoming signals, it may (or may not) be necessary to regenerate a clock that is (more or less) linked to the incident frequency or the bit rate received. Clearly, if these stages operate at a (high) incident frequency, they will consume a lot of energy. These problems can be overcome in a number of clever ways (all patented, of course!).

- Protocol decoding via a state machine. The core of the circuit is a wired logic for appropriately decoding and encoding the signals, according to the communication standards and modes used. This part may be more or less complicated and bulky, according to the standards followed and their (numerous) options. For further information, see the chapters in Part Three, where I have identified some very simple devices (not covered by the ISO) and other more complicated ones (according to ISO 18000-6 part C, compatible with EPC C1 G2, for example) for implementing applications in accordance with different local regulations.
- A low-consumption local oscillator. The rate of the data exchanged in the return link (from the tag to the base station) generally ranges from tens to hundreds of kbits s^{-1}. In order to keep the power consumption of the tag to a minimum, while also increasing the possible operating range in the remotely powered mode, it is generally preferable to recreate a local clock in the tag, of the order of magnitude stated above, using a special circuit that is completely independent of the incident frequency, instead of providing input dividers, which operate at high frequencies and consume a large amount of energy.
- A return link encoder (bit coding). One part of the integrated circuit must handle the signals (frames) returned by the tag and provide the special bit coding for the return link.
- An antenna impedance modulation control stage. This stage has the function of modulating the antenna impedance of the tag in accordance with the variation of the data to be returned to the base station, thus creating the back scattering signal.
- A setting stage (near field/far field) and its setting element.
- An E2PROM memory and the usual access, protection and/or correction devices for it.
- A circuit called the charge pump circuit. The function of this circuit is to provide a special power supply to the programming/writing device of the E2PROM memory.
- Optionally, a device for encrypting the data that are carried.

Remotely Powered Passive Tags (Batteryless)
These are the most commonly used type, and their construction essentially follows the principles described in earlier chapters of this book. Each tag comprises an integrated circuit, as described above, and an antenna. When the functionality of the tag has been specified, it is simply necessary to choose the type of integrated circuit to be included in it and then move on to the design of the antenna (see below).

Battery-Assisted Tags
As already mentioned, a local battery may be added if the power recovered by the tag is insufficient for its functionality. The tags are usually divided into two classes (see Chapter 2 if necessary).

Battery-Assisted Passive Tags
Battery-assisted passive tags do not have on-board RF transmitters. They include:

- Tags operating at short or medium range but having a lot of on-board circuitry (requiring a large power supply), such as large memory plans, powerful microprocessors or microcontrollers for the rapid operation of these memories, etc.

- Tags used for automatic toll functions or for automated traffic control on specific lanes of motorways, which can be usefully provided with batteries to vary the value of Γ from $+1$ to -1 in order to obtain the widest possible variation of their radar cross-section.

Battery-Assisted Active Tags

Battery-assisted active tags have on-board RF transmitters to provide the return link from the tag to the base station. This family of tags includes those operating over (very) long distances, such as tags fitted to shipping containers, which must be identified at long range and where the conventional back scattering technique cannot provide a sufficient power supply or an adequate return signal. Note that ISO 18000-4 (for 2.45 GHz) and 18000-7 (433 MHz) currently describe suitable communication protocols for these situations.

By providing a true transmitter function on board the tag, it is also possible to consider the use of fully bidirectional communication (full duplex, which is faster) between the base stations and tags (rarely used at the present time) without the need to design special communication protocols including orthogonal carrier modulation, in other words modulation without interaction between the signals of the forward and return links.

Note that, in the two examples of battery-assisted operation (passive and active) described above, the local circuits are generally woken up by detection of the presence of the forward wave from the base station, in order to achieve a longer on-board battery life and minimize the local energy consumption of the tag.

18.4 Antennas for Tags

First of all, I would remind you that this book is not concerned with antenna design. The reader is advised to consult the many books dealing with these matters in detail. However, I have tended to describe standard $\lambda/2$ antennae in the preceding sections of this book, to facilitate an understanding of this subject, since these antennas are particularly easy to produce. These antennas are very widely used in RFID and serve their purpose very well, on the whole. However, if applications require any special functionality (small dimensions, specified directivity, special environments, etc.) it is necessary to design completely different antennas, while always bearing in mind that the antenna of an RFID tag is used for multiple purposes, as follows:

- recovering the RF power radiated by the base station;
- transferring the power to the integrated circuit with minimal losses, in order to optimize the operating range;
- supplying this power under the output impedance, which is conjugated (resonance) with that of the integrated circuit;
- enabling the data transfer (amplitude modulation of the forward link);
- reflecting/reradiating some of the power of the incident signal during the return phase (back scattering technique);
- supporting the structure(s) of the rectifier circuit(s) of the integrated circuit.

The problem here is one of producing a low-cost antenna that is small enough to be applied to small objects, while still having sufficient gain and not too much directivity, and that is

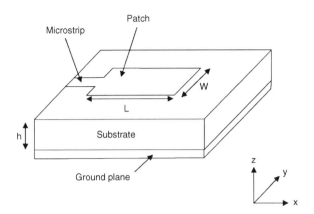

Figure 18.3 Example of a patch antenna

reproducible and can withstand the proximity of other tags without too many problems (optimal forward/backward effect), etc.

Many industrial companies, universities, engineering faculties and research laboratories around the world are able to provide assistance with the design and development of these antennas. Designers often use simulation tools that may be more or less 'heavyweight' (in terms of both physical weight and financial burden!), as follows:

- Three-dimensional electromagnetic fields, three-dimensional technologies (to create structures of all kinds). The best-known examples are Ansoft HFSS, IMST Empire, CST Microwave Studio, etc.
- Electromagnetic fields in 2.5 dimensions, 2.5-dimensional technologies (for creating structures limited to flat objects). The best-known examples are Sonnet EM Suite, Agilent Momentum (formerly HP), Ansoft Ensemble, etc.

These will help to establish the rough outlines of the antenna design. The next step generally requires the employment of certain persons with white hair (or no hair at all!) who have many years of experience in RF analogue techniques and especially in UHF, SHF and other radar applications. Why this is so, nobody knows. ...

Note that one of the main methods used at present is that of designing 'patch' antennas (see Figure 18.3), because many laboratories working in the mobile telephony field (mostly using the same frequency bands) have designed this type of antenna, owing to the miniaturization of telephones. However, you should be aware that the special features required for RFID antennas are very different from those used for GSM.

18.4.1 General Problems of Antenna Design for UHF/SHF RFID Tags

The antennas designed for UHF and SHF tags have special features required by RFID applications (see Figure 18.4). For many reasons, which I have already mentioned, it is often (or always) necessary to solve several problems simultaneously, as follows:

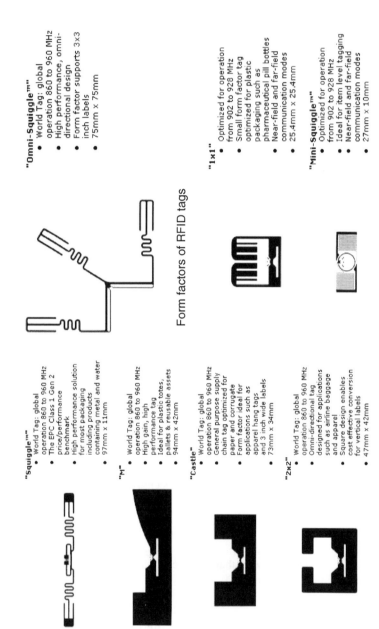

"Squiggle"™
- World Tag: global operation 860 to 960 MHz
- The EPC Class 1 Gen 2 price/performance benchmark
- High performance solution for most packaging including products containing metal and water
- 97mm x 11mm

"M"
- World Tag: global operation 860 to 960 MHz
- High gain, high performance tag
- Ideal for plastic totes, pallets & reusable assets
- 94mm x 42mm

"Castle"
- World Tag: global operation 860 to 960 MHz
- General purpose supply chain tag optimized for paper and corrugate
- Form factor ideal for applications such as apparel hang tags and 3 inch wide labels
- 73mm x 34mm

"2x2"
- World Tag: global operation 860 to 960 MHz
- Omni-directional tag designed for applications such as airline baggage and apparel
- Square design enables cost effective conversion for vertical labels
- 47mm x 42mm

"Omni-Squiggle"™
- World Tag: global operation 860 to 960 MHz
- High performance, omni-directional design
- Form factor supports 3x3 inch labels
- 75mm x 75mm

"1x1"
- Optimized for operation from 902 to 928 MHz
- Small form factor tag optimized for plastic packaging such as pharmaceutical pill bottles
- Near-field and far-field communication modes
- 25.4mm x 25.4mm

"Mini-Squiggle"™
- Optimized for operation from 902 to 928 MHz
- Ideal for item level tagging
- Near-field and far-field communication modes
- 27mm x 10mm

Form factors of RFID tags

Figure 18.4   Examples of UHF tags for different applications

- There is a mechanical and dimensional problem concerning the tag antenna, which must of course be as small as possible.
- We need an antenna whose physical design is simple and as close as possible to a true $\lambda/2$ dipole antenna, while being largely independent of the frequency.
- We must allow for the metallic environment, providing screens (back planes) if necessary.
- The dielectric materials must be of high-performance but inexpensive (obvious enough!).
- The antennas must have little or no loss. This is affected by the quality of the conductors, and therefore the maximum thickness of the conducting layers, which must be reduced to a minimum (for reasons of cost) while allowing for skin effects, etc.
- The tag antenna must be inductive, since the tag must always have an input capacitor.
- The radiation resistance of the antenna must match the level due to the technology of the integrated circuit, which is not obvious when the values of R_{ic} are examined.
- The conjugate matching ($LC\omega^2 = 1$ and $R_{ant\,t} = R_l$) between the impedance of the tag antenna and that of the load has to be maintained over the whole of the 860–960 MHz band in order to comply with ISO 18000-6 (well, we can always dream. . .), so that the maximum available energy can be recovered regardless of the received frequency.
- The same applies to the quality factor $Q = X/R$.
- The nominal gain and directivity must be known (easy!) and reproducible in the production process (not so easy!).
- It must be possible for the antenna to be either folded or unfolded, thus creating (or not creating) a continuous ohmic short circuit at the input of the integrated circuits according to the composition of their input stages (voltage doubler, etc.).
- The mechanical design of the antenna must be such that the integrated circuits can be easily located and deposited on it by a precise and reproducible industrial process (see below).
- The assembly consisting of the antenna and the integrated circuit must be able to accept the industrial production tolerances in terms of the bonding and covering of the flip chip assembly, in order to achieve the desired tolerances in the operating range, especially in the case of those operating in a narrow band (according to local regulations) and therefore having a high value of Q, etc.

If we fail to consider even one of these points, the whole system will fail! It will be no use to dream of the reproducibility, performance, quality, ppm and wonderful operating ranges measured in anechoic chambers and proudly quoted by marketing departments and vendors. In short, it will be a disaster (to look on the bright side).

18.4.2 Problems of Mounting the Integrated Circuit (Chip) on the Inlet

Perhaps you thought these matters had already been dealt with? No! Let us take a look at some of the commonest problems encountered in the industrial production of UHF and SHF tags, especially the problem of the reproducibility of their functional performance (to guarantee the 'min/typ/max' part of the tag data sheet; see Chapter 20), which is concerned with the art and craft of mounting the chip on its antenna in a reproducible way. This is far from simple, especially for a manufacturer aiming to be a market leader with an output of

Figure 18.5   Examples of commercially available tags

several million per day! (For information, an output of 1.4 million tags per day means that 16 tags per second must be produced, working round the clock, and, of course, they must all operate correctly and essentially in the same way after 100% individual functional testing. Of course, if 16 tags per second seems a lot for one machine to handle, you can provide 16 machines in parallel, each producing 1 tag per second – but that entails a lot more expense!)

As I have described at length in Chapter 6, the conjugate matching of the source (tag antenna) to the load (integrated circuit) is one of the main quality parameters of a tag. This means that it is necessary to know, and guarantee, the imaginary and real components of the assembly produced in this way. However, when the chip (integrated circuit) has been bought and delivered as a 'bare chip' or miniature casing, we must eventually connect it to the antenna that has been so carefully designed – and it is at this point that new problems arise.

Effects of Mounting the Integrated Circuit on the Tag Substrate
The types of casing (bare chip, miniature casing, etc.) and their physical implementations in tags have very considerable effects on the values of the real and imaginary components of the antenna load impedance. The positioning on the x, y and z axes of the chip and its precision with respect to the antenna are crucial and can greatly detract from the performance of the completed assembly if care is not taken (mainly concerning the operating range). See Chapters 5 and 6 again if necessary.

Each producer is responsible for simulating and calculating in detail the effects (impedance mismatching, detuning, reduction of operating range, etc., as described in Chapters 5 and 6) of the available industrial processes and tools for chip deposition, according to the form of the finished tag antenna (produced by etching, screen printing, ink deposition, inkjet, offset, gravure printing or other methods) and the dimensions of the connection pads (surfaces) of the chips in relation to those of the antennas.

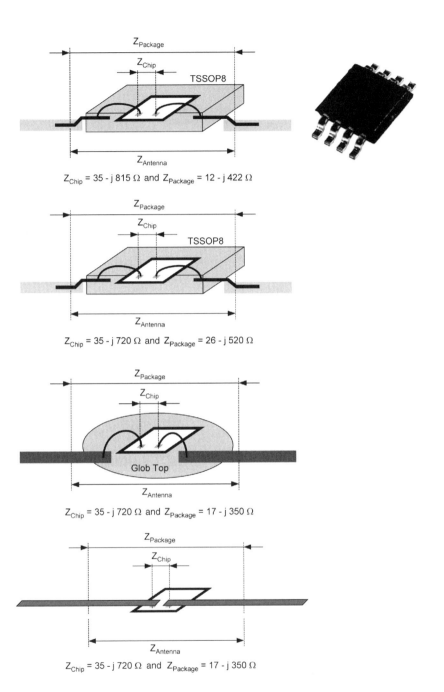

Figure 18.6 Effects of packaging

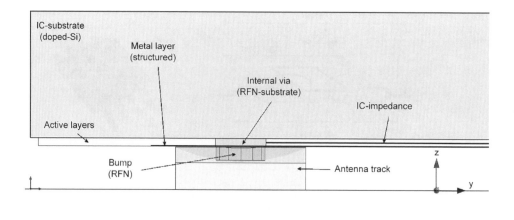

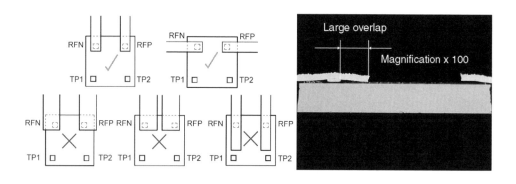

Figure 18.7   Effects of mounting the integrated circuit on its substrate (documents from NXP)

Without wishing to be defeatist in any way, I would point out that the parameters mentioned above and others relating to the manufacturing process, etc., are key elements that must be resolved in order to ensure the reproducibility and satisfactory day-to-day performance of tags, and often make the difference between 'good' and 'less good' tag producers.

Figures 18.5 to 18.7 give some examples of package and mounting incidences on electrical values of the chip and the tag.

Now let us take a look at the technical and technological aspects of the design of base stations.

19

The Base Station

This chapter is not intended to provide detailed instructions for building a UHF and SHF base station, but rather to point out some of the common problems facing developers of these systems. In fact, the design of these systems is a complex matter and requires considerable experience of the UHF and SHF fields; the methods used, and the engineering skills involved, are both numerous and excellent. In this chapter I will therefore simply outline the different designs and the benefits, properties and major constraints of the proposed solutions and architectures.

Perhaps I should say 'we', rather than 'I'. The reason is that I owe a great deal to Christian Ripoll[1] and François de Dieuleveult,[2] both well-known experts (and friends!) in the field of radio frequency technology, who have given me considerable assistance in writing this chapter and suggesting ideas and solutions to offer to my readers. My warmest thanks to them.

19.1 Introduction

Let us start by considering the basic problem of an RFID base station operating at UHF or SHF. As described in the previous chapters, the base station radiates a level of power (P_{EIRP}) that may vary from several hundred milliwatts (820 mW EIRP in Europe) to 4 W EIRP (in the USA), depending on the current local regulations.

I have shown that, when a tag receives enough energy (several tens of μW, more precisely 35 μW in the case of the NXP/Philips Semiconductors U_code), it reradiates part of the

---

[1] Christian Ripoll is a researcher and lecturer in the Signal Processing Department at ESIEE (Higher Institute of Electrical and Electronic Engineering). The ESIEE is a member of the Chamber of Commerce in Paris and is located in the inner suburbs of the capital, at the Cité Descartes, Noisy-Le-Grand. Among its other activities, the ESIEE maintains a dedicated team of lecturers and researchers and a laboratory working on the processing of analogue and digital signals, as well as a highly specialized radio frequency engineering laboratory, managed by Christian, for joint development of industrial projects with external companies.

[2] François de Dieuleveult is a 'Technology Consultant' specializing in signal processing at the Technological Research Department of the CEA (French Atomic Energy Commission) at Saclay. Over many years, François has written a large number of technical articles in the specialist press and has also produced books on radio frequency and associated technologies (published by Dunod).

---

incoming wave to the base station by back scattering. The power recovered at the base station is commonly of the order of a few hundreds of nW or even pW, which the base station circuits must be capable of dealing with! The sensitivity of the receiver must therefore be considerably below the nW level, in order to be able to detect and process the weak return signal in the correct way, bearing in mind that this signal is present at the same time as the powerful signal transmitted at the same frequency or in the same band. The scene is thus set for a conflict between the two leading characters – one at several watts and one at a few nanowatts. This sketch of the situation should be enough to make it clear that the problem is essentially one of keeping the return wave sent by the tag separate from the forward wave sent by the power amplifier of the base station.

Within this general field of the back scattering techniques, which are widely used in RFID at UHF and SHF, I shall now consider some of the different options for designing base stations, which may be fixed, for long distance communication, or may be the portable, light, compact and preferably of the low-energy type known as 'handheld', as well as the various associated methods for 'isolating', or 'decoupling', the forward and return links.

19.1.1 The Base Station

The base station receives a very weak return signal from the tag. It must then:

- receive the signal;
- amplify it;
- process the signal (filtering, etc.);
- detect the signal;
- demodulate it;
- and finally carry out the digital processing of the signal.

To do this, the base station receives on its antenna (which may also be used to transmit the carrier frequency) the wave reflected by the transponder, separates it from the transmitted signal by means of a directional coupler, a bidirectional coupler or a circulator (selective for the signal direction), and sends it to the receiving part (Figure 19.1).

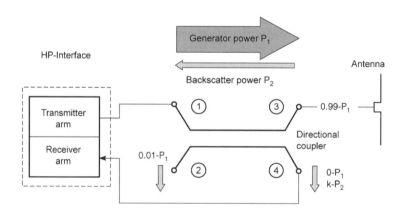

Figure 19.1   Block diagram of a base station

19.1.2 Base Station with Two Antennas ('Bistatic')

This is the most basic solution of all! If we want to be sure of separating the transmission of the forward wave from the reception of the return wave, this is the system that will (almost certainly) provide the desired result.

This kind of system with two separate antennas is called a 'bistatic' or 'bistatic antenna' system (see Chapter 8). In order to separate the forward wave (transmission) from the return wave produced by back scattering from the tag, a fixed base station can use two separate antennas. In this way, the signal carrying the energy is mechanically and physically isolated by placing its transmission antenna at a 'sufficiently great' distance from the antenna receiving the return signal, thus preventing overloading and/or blinding the receiver by part of the transmitted wave, given that, as mentioned above, the transmitted signal is often billions of times stronger than the received signal (with a power of several watts, as against 1 nW) and both signals (in static back scattering with an unmodulated return signal) are on exactly the same frequency at all times (or very close in terms of frequency in dynamic modulation according to ISO 18000-6 part A or B) or when the bit coding of the tag requires a subcarrier, according to ISO 18000 part C, for example, as a result of the use of the back scattering principle at UHF and SHF.

In order to increase this separation, use may be made of antennas having high forward/back ratios in order to optimize the solutions by simple mechanical arrangements, and it is also possible to ensure that the radiation sublobes or auxiliary lobes of the transmission antenna do not interfere with reception by the receiving antenna. Should we require further refinements, it is even possible to transmit the signals with a given linear polarization and receive signals with an orthogonal linear polarization, with the tag's circuitry and antenna producing or contributing to the appropriate rotation.

Of course, such solutions are more costly, bulkier and heavier than others that are described below.

19.1.3 Base Station with a Single Antenna ('Monostatic')

This kind of system, called a 'monostatic' or 'monostatic antenna' system to distinguish it from the bistatic type, is obviously cheaper, since it uses a single antenna for transmission and (simultaneous) reception to perform all the transmission and reception functions on the same frequency (or almost the same). Such a system is designed to ensure that none, or very little, of the large amount of energy present on the transmitter path towards the antenna is deflected towards the receiver, where it would swamp the tiny received signal. This poses some technical problems, but also provides some benefits.

Let us start with the technology. This function can be provided using components known as circulators and bidirectional couplers, which are familiar to RF specialists. I will provide more details of their operating principles further on. Unfortunately, however, the separation (or associated rejection level) between paths is usually only about 20 to 30 dB, so unless other measures are taken, a monostatic system cannot be as sensitive as a bistatic one, because a part (even if only a small part) of the powerful transmitted signal is reinjected into the receiving section of the system. Consequently, the operating distance is intrinsically limited. To give you an idea of the orders of magnitude involved, if all other things are equal (especially the transmitted power level), we can achieve an operating distance of about 12 m with a bistatic system, compared with 8 to 9 m for a monostatic system, simply because it is easier to separate the received tag return signal from the transmitted signal in the bistatic system.

Now let us consider the benefits of monostatic systems. A single antenna will please users of portable (handheld) systems, where the problem is always how to achieve a balance between the volume, weight, battery life and operating distance. This makes it easier to identify tags in locations where a fixed base station cannot be installed (e.g. where tags are used on items placed far away from computer networks, as in tree health monitoring in forestry, location and feeding of animals in fields, stocktaking in warehouses, etc.). Subsequently, it is a simple matter to use PDA, RS 232 or GSM interfaces and GPS location systems to communicate with the central computer base and download data, without any transcription errors due to manual intervention.

19.1.4 Notes on the Electronic Components Used in an RFID Base Station

The techniques used for the design of RFID systems at UHF and SHF and the associated base stations are very different from those used at LF and HF. They make use of devices typically operating at radio frequency,[3] sometimes including novel electronic 'beasts'. Examples of these are bidirectional couplers, circulators and strip lines.

I shall provide a brief description of these devices before going on to describe their use in UHF and SHF base station architectures, to avoid any uncertainty.

Circulator

A circulator is a (radio frequency) junction with three ports (inputs/outputs) arranged in the shape of a Y (Figure 19.2). This system is magnetically coupled to a polarized ferrite material, which has the function of guiding the main power between two consecutive ports. Like the coupler, which is examined in the next section, the circulator forms a symmetrical junction, but, unlike the coupler, it is not reciprocal.

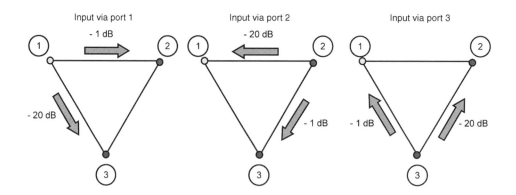

Figure 19.2   Principle of the circulator

[3] J. D. Gerdeman, *RF/ID – Radio Frequency Identification – Application 2000*, Research Triangle Consultants, 2000.

Operating Principle of a Circulator

To understand how a circulator works, we should imagine it as a box containing three transmission lines at 120° to each other. These transmission lines are placed between two ferrite discs. On the other side of the ferrite, there is a nonferrous ground plane and then a magnet, followed by a piece of iron, which is placed there to screen the whole assembly from external magnetic fields and also to act as an efficient dissipator of the heat that it may generate.

When a radio frequency is introduced at one of the three ports, two rotating magnetic fields, of equal strength but opposite phase, are induced in the ferrite. Assuming that the magnetic field produced by the magnets is strong enough (as it is carefully designed to be), the opposing rotating fields are cancelled out on the adjacent transmission line and are reinforced on the other two lines. The result is that the radio frequency radiation passes without attenuation into the adjacent transmission line, but does not pass into the third line (this is the clever part!).

Thus a circulator allows waves to pass in one direction only, e.g. from port 1 (input/source) to port 2 (output/load), or from port 2 (formerly the output, now the input) to port 3 (load), or, finally, from port 3 (formerly the load) to port 1 (formerly the input). To sum up, a circulator is a system such that a signal applied to one of its ports is transferred to the next adjacent port, without the need for any special action. It is therefore possible to transfer (or 'circulate') RF radiation from one port to the next in a clockwise direction on the three points of the Y connection. Also, when one of the ports is set to its matching condition, the other two are isolated in the opposite direction (as an example, port 3 matched → communication from port 1 to port 2, but not from port 2 to port 1). Therefore, when somebody sells you a circulator with a matched load connected to port 3, it is called an isolator from port 2 to port 1. This simultaneous phenomenon of circulation and isolation is the origin of the name 'circulator'.

Clearly, this will only work well (simultaneous circulation and isolation) if the ports are terminated on their matched impedances, since in this case there will be no parasitic radiation reflected, as the standing wave ratio (VSWR) will be as close as possible to 1 : 1. As usual, this can never be completely achieved, for numerous physical and practical reasons (width of the band, losses, temperature variations, nonlinearity of the ferrite components, etc.). In practice, a circulator provides isolation of about 20 to 30 dB in the reverse direction. It may be necessary to provide up to three circulators in series in order to improve the isolation factor of a connection. As regards the insertion losses of the forward paths, these generally range from 0.25 to 1 dB (10 to 15% loss in insertion) and are mainly due to heat dissipation in the ferrite and magnetic materials and detuning of the transmission lines across the bandwidth of the transmissions required.

To conclude this brief description of circulators, it is worth noting that the intrinsic nonlinearity of the ferrite components often gives rise to harmonic frequencies of the transmitted signals, resulting eventually in intermodulation products of every order. As a general rule, the use of circulators implies the simultaneous use of filter networks.

Summary

If the power enters at port P_1, the power at port P_2 will be equal to the power injected at port P_1, subject to the insertion losses P_I (in the region of 1 dB):

$$P_2(\text{dBm}) = P_1(\text{dBm}) - P_I(\text{dB}) \text{ dB}$$

Port P_2 is then called the coupled port. The power P_{3i} collected at port P_3 (called the isolated path) will be the power injected at P_1, subject to the isolation (and therefore theoretically equal to zero). This isolation I is generally about 20 dB in the UHF band and above:

$$P_{3i}(\text{dBm}) = P_1(\text{dBm}) - I(\text{dB}) \quad \text{dB}$$

Port P_3 is then called the isolated port. Figure 19.3 summarizes the operation of the circulator.

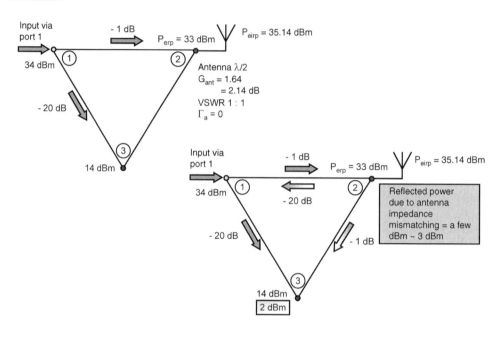

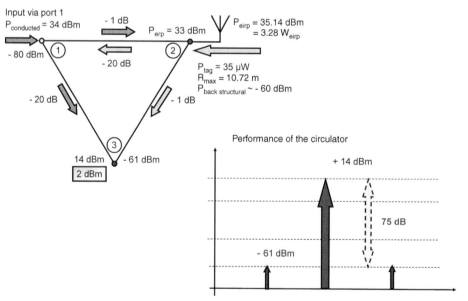

Figure 19.3 Example of the use of a circulator in Europe, at 2 W ERP (33 dBm) in the LBT mode

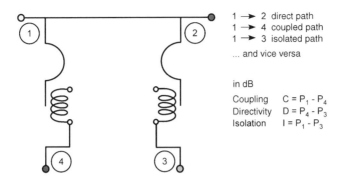

Figure 19.4   Bidirectional coupler

Directional Coupler

A directional coupler is a radio frequency junction with four ports (inputs/outputs). Figure 19.4 shows an example of the implementation of a bidirectional coupler, which has the function of sampling part of the signal (1/10 to 1/100, i.e. 10 to 30 dB of the main power, for example) circulating in a given direction. Here the direction of propagation of the wave is of primary importance, and the properties of the coupler are highly dependent on this.

Let us consider two possible cases:

- When the main energy flows from port P_1 to port P_2 (Figure 19.5), path 1 is strongly coupled to path 2, weakly coupled to path 4 and completely decoupled (theoretically) from path 3. In this case, path 2 is called the forward path, path 3 is called the isolated path and path 4 is called the coupled path.
- If the propagation is from port P_2 to port P_1 (Figure 19.6), path 2 is strongly coupled to path 1, weakly coupled to path 3 and completely decoupled (theoretically) from path 4. In this case,

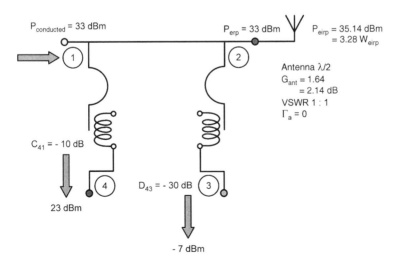

Figure 19.5   Example: Europe, 2 W ERP (33 dBm), LBT

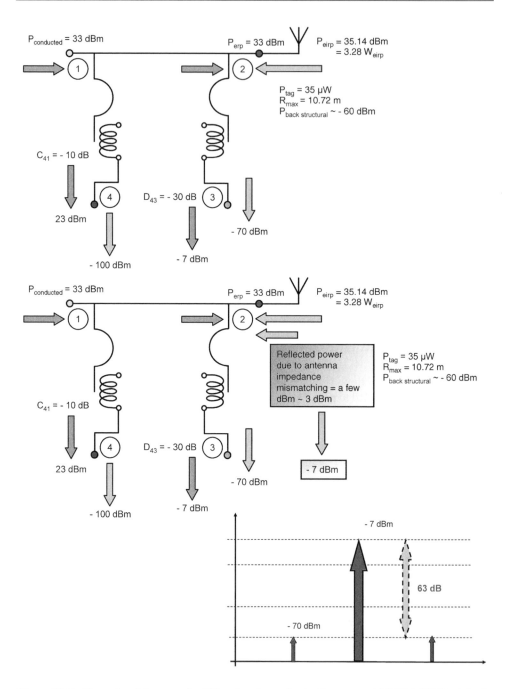

Figure 19.6 Example of the use of a bidirectional coupler in Europe, at 2 W ERP (33 dBm) in the LBT mode

path 1 is called the forward path, path 4 is called the isolated path and path 3 is called the coupled path.

The definition of the coupled and isolated paths therefore depends on the port at which the incident energy is injected, which gives the coupler its directional properties.

Coupling, Directivity and Isolation
The performance of a coupler is generally assessed in terms of three parameters, namely the coupling, the directivity and the isolation (in addition to the standard reflection coefficients), which will now be defined.

Coupling
The coupling factor, C, is defined as the ratio between the incident power P_1 at port 1 and the power P_4 collected on the coupled path, at port 4:

$$\text{In linear terms, } C = P_1/P_4$$

$$\text{or, expressed in dB, } C(dB) = P_1(\text{dBm}) - P_4(\text{dBm})$$

Directivity
The term directivity, D, is defined as the ratio between the coupled power P_4 and the unwanted power P_3 on the isolated path (with a value that may range from 20 to 30 dB, or even below):

$$\text{In linear terms, } D = P_4/P_3$$

$$\text{or, expressed in dB, } D(\text{dB}) = P_4(\text{dBm}) - P_3(\text{dBm})$$

Theoretically, the power P_3 collected at P_3 should be zero. Owing to the technological constraints on the construction of bidirectional couplers, this is only true to a certain extent – hence we require the value of D! Directivity thus provides a measure of the quality of energy coupling as a function of the direction of the incident power levels concerned.

Isolation
The isolation factor, I, is defined as the ratio between the incident power P_1 at port P_1 and the power P_3 collected at P_3 on the isolated path:

$$\text{In linear terms, } I = P_3/P_1$$

$$\text{or, expressed in dB, } I(\text{dB}) = P_1(\text{dBm}) - P_3(\text{dBm})$$

Note
The equation for I can be formulated differently, as a function of C and D. In linear terms, $I = P_1/P_3$:

$$I = \frac{P_1}{P_4} \times \frac{P_4}{P_3} = C \times D$$

$$\text{or, expressed in dB, } I(\text{dB}) = P_1(\text{dBm}) - P_3(\text{dBm}) = C(\text{dB}) + D(\text{dB})$$

19.2 Examples of Base Station Hardware Architecture

Having introduced the bidirectional coupler and the circulator, the two essential elements for the separation of the signals transmitted and received by the same antenna in RFID when back scattering is used, I shall now go on to examine some examples of RFID base station structures, with the benefits and drawbacks of each type.

By way of example, and without any claim to be exhaustive, here are three possible base station hardware architectures that may be considered, depending on the type and class of application required. For simple reasons of cost and size, I will describe single-antenna base station designs, in other words monostatic systems. In some cases I have indicated a bistatic version where this may be required for better performance. The systems in question are:

- a circulator system, using Schottky diode demodulation;
- a bidirectional coupler system, using synchronous I/Q demodulation or synchronous I/Q demodulation with measurement and control of the carrier level.

Note that this does not in any way determine the hardware and software systems required for the use of spread spectrum methods such as FHSS, DHSS and LBT to comply with other provisions of the current standards and/or regulations.

It may also be of interest to note that it is quite easy to model and simulate the following:

- The circuits proposed in these examples of base stations, using simulation tools such as ADS (Agilent Development Software), MWO (MicroWave Office), Touchstone, etc.
- The fields produced and radiated by the antennas: in two dimensions, using Sonnet EM suite, Agilent Momentum, Ansoft Ensemble, etc., or in three dimensions, using the HFSS tools produced by Ansoft or the products of IMST, CST Microwave Studio, etc. These products are generally expensive, but their results can be used to predict the final performance of a device with a confidence level of about 85 to 90%, which is quite satisfactory, and may even be better than that!

19.2.1 Important Note

(a) In the following section, you should note that the values of P_{ERP} that are mentioned are expressed (according to the measurement methods in the current standard CISPR 16) in terms of peak or quasi-peak power, and are therefore independent, or almost independent, of the fact that the transmitted carrier is on/off modulated (ASK 100%, with a duty cycle of 50%, and Manchester bit coding) in the forward link and with a constant pure carrier in the return link (back scattering from the tag to the base station).

(b) To facilitate the comparison of the relative performances of these different hardware architectures, all the examples are based on the same assumptions as regards operation. Assuming that we are in Europe, we shall comply strictly with the European RFID regulation, namely ETSI 302 208-1 (for further details, see Chapters 15 and 16 on standards and regulations), which was implemented in Europe in 2006 by ERC 70-03, using a monostatic base station transmitting a maximum power of

2 W ERP in the LBT mode, with a $\lambda/2$ antenna (beamwidth 70°) and using a tag with a $\lambda/2$ antenna and an NXP/Philips Semiconductors U_code circuit whose typical minimum consumption is 35 μW, giving a theoretical remote power distance of 10.72 m.

The global forward and return link budget is shown in Table 19.1. The return power

$P_{\text{back tag unmodulated}}$ is therefore -62.18 dBm.

Table 19.1

Transmission		
	Base station	$P_{\text{ERP}} = 2\,\text{W max} = 33\,\text{dBm}$ $G_{\text{ant bs transmission}} = 1.64 = 2.14\,\text{dB}$ $P_{\text{ERP}} = 3.28\,\text{W} = 35.14\,\text{dBm}$
Air		
	Air	att at 867 MHz at 10.72 m = 51.87 dB
Tag		
	Tag	Received signal $= -16.73\,\text{dBm}$ $G_{\text{ant t}} = 1.64 = 2.14\,\text{dBm}$ $P_{\text{t}} = -14.59\,\text{dBm} = 35\,\mu\text{W}$ $P_{\text{s structural}}$ (when load matching is present) $= -12.45\,\text{dBm}$
Air		
	Air	att at 867 MHz at 10.72 m = 51.87 dB
Reception	Base station	Received signal $= -64.32\,\text{dBm}$ $G_{\text{ant bs recept}} = 1.64 = 2.14\,\text{dB}$ $P_{\text{back bs structural}} = -62.18\,\text{dBm}$

For your information, Figure 19.7 shows the values obtained by a comparison between transmitted power levels of 500 mW ERP and 2 W ERP. Note that:

- This example relates to a bistatic base station such that

$$G_{\text{ant bs}} \text{ in transmission} = 2.14\,\text{dB}$$

$$G_{\text{ant bs}} \text{ in reception} = 4\,\text{dB}$$

- Also, even though the transmitted power level of 500 mW ERP is lower, the power received by the base station is slightly greater than in the case of 2 W ERP, because the maximum remote power supply distance is shorter.

Table 19.2 shows some final information on the maximum power that the amplifier can, or should, deliver according to the local regulations for UHF (860–970 MHz).

Global link budget		Europe ERC 70 03	Europe ETSI 302 208	Notes
Frequency ERC 70 03	MHz	869.50 500 mW 10%	867.60 2 W LBT	max. frequency of the band concerned for att_ max
P_ ERP_ bs_ max	dBm	27.00	33.00	i.e. P_ ERP max. authorized
G_ ant_ bs @ transmission	dB	2.14	2.14	λ/2 dipole antenna with G_ ant_ bs = 1.64 ; angle < 70"
P_ EIRP_ bs max	dBm	29.14	35.14	(i.e. P_ ERP + 2.14 dBm)
r_ max remote powering	m	5.39	10.79	Friis equation ... in the best possible case! for the tag concerned, P_ t = 35 μW = -14.5 dBm
att @ of F and r	dB	45.92	51.93	at the frequency concerned and r_ max_ remote powering
G_ ant_ t	dB	2.14	2.14	λ/2 dipole antenna with gain G ant_ t = 1.64 dB
P_ t	dBm	−14.56	−14.56	
	μW	35.00	35.00	Philips Semiconductors U_ code at UHF
r_ max_ useful	m	2.70	5.40	the harsh reality: reflection, detuning, etc.
att @ of F and r	dB	45.92	51.93	at the frequency concerned and r_ max_ remote powering
G_ ant_ t	dB	2.14	2.14	λ/2 dipole antenna with gain G ant_ t = 1.64 dB
G_ ant_ bs´ @ reception	dB	2.14	2.14	different receiving antenna with G_ ant_ bs = 4
P_ back_ structural	dBm nW	−56.27	−62.29	at the input of the base station demodulator

Figure 19.7   Link budget

Clearly, therefore, at UHF, with a slightly lower radiated power EIRP (3.28 W EIRP instead of 4 W EIRP), the RF output amplifiers of European RFID base stations must deliver a conducted power to the (useful) load of the antenna, which is greater (2 W as opposed to 1 W) than that of base stations in the USA. Allowing for the antenna connecting wires, the losses in these, the possible mismatch between the output impedances of the amplifier and of the base station antenna (VSWR = 1.5), the manufacturing tolerances, the variations of components over time, etc., we find that the amplifier must deliver a gross output of 2.8 W sinusoidal with almost complete spectral purity, thus complying with the specified levels for pollution and other spurious signals that must not interfere with the adjacent paths of the permitted frequency band. To achieve this, we must design a power amplifier operating as a class A linear device, in other words dissipating at least as much power as it has to supply to its load when at rest (with no

Table 19.2

		Maximum radiated power	Maximum conducted power	
USA	FCC 47 part 15	4 W EIRP	1 W max	Therefore $G_{bs\ min} = 4$
Europe	ETSI 302-208	2 W ERP, LBT	2 W max	Therefore 3.28 W EIRP (maximum beamwidth 70°) and therefore $G_{bs\ min} = 1.64$
	ERC 70-03	0.5 W ERP/dc = 10%		Therefore 0.82 W EIRP

carrier present). In other words, the total power is at least 2×2.8 W, making it necessary to use a power stage that can handle about 6 W and is therefore often fitted with a heat sink and/or a fan as appropriate.

19.2.2 Base Station with Circulator and Schottky Diode Demodulation

Figure 19.8 is a schematic diagram of a monostatic architecture based on a circulator.

Details of the Structure of this Base Station

- The power supplied by the power circuits of the base station flows to the antenna via a circulator.
- This power is therefore present, subject to the insertion losses, at the antenna (note that no filter element is shown in the diagram, but in practice one must be provided to comply with the standards for parasitic emission levels).
- The antenna converts the conducted electrical power to radiated electromagnetic power P_{EIRP} (effective isotropic radiated power) with a value (in dB) of $P_{\text{EIRP}} = P_{\text{cond}} + G_{\text{ant bs}}$.
- Using its antenna, the tag, located at a given distance, collects an amount of power that is a function of the attenuation of the medium in free space (on a matched load to optimize the received power; see the Friis equation reproduced in an earlier part of this book in Chapter 6).
- As explained previously, a power level $P_{\text{s back}}$ is reradiated and modulated by the variation of the load impedance of the tag antenna as a function of the variation of the data.
- This reradiation takes the form of the modulation of the amplitude of the power flux density and the phase of the incident carrier, which is retransmitted to the base station.

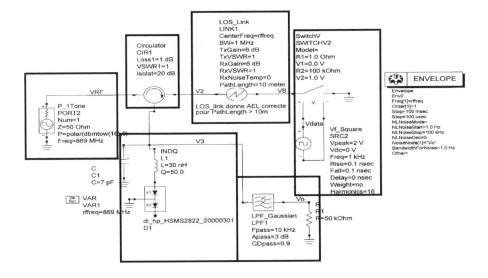

Figure 19.8   Base station with circulator and Schottky diode demodulation

- When it reaches the (single) antenna of the (monostatic) base station, the return signal is directed by the circulator (in which it flows from port P_2 to port P_3, always with an insertion loss P_1) towards the mixer stage, which consists of a Schottky diode.

To be more specific, let us provide some numbers for this example for a circulator having the following values of insertion loss P_1 and isolation I:

$$P_1 = 1 \text{ dB and } I = 20 \text{ dB}$$

During the Forward Link, from the Base Station to the Tag
For the carrier to be transmitted, port P_1 is the input port of the circulator. Given that $P_1 = 34$ dBm $= \sim 2.5$ W (the injected value of $P_1 = 34$ dBm is due to the fact that, in order to obtain a real level of $P_{\text{conducted}} = 33$ dBm at the output to drive the base station antenna, the insertion loss of 1 dB in the forward path of the circulator must be compensated). Then:

- By coupling, the power level at port P_2 (the coupled path) will be

$$P_2 = P_1 - P_I$$

Therefore $P_2 = 34 - 1 = 33$ dBm conducted, giving a maximum permitted level of 2 W ERP radiated with a $\lambda/2$ antenna.
- Due to incomplete isolation at port P_3 (isolated path)

$$P_3 = P_1 - I$$

$$P_3 = 34 - 20 = 14 \text{ dBm} = 25 \text{ mW}$$

During the Return Link Phase
The back scattering signal from the tag
During the return link phase, with no modulation of the load impedance of its antenna (tag antenna matched to its load), as mentioned above, the tag returns a power of $P_{\text{back tag unmodulated}} = -62.18$ dBm $= \sim 1$ nW. For this signal, port P_2 becomes the input port of the circulator.

1. The transponder signal travels from port P_2, where it appears as $P_{\text{back tag unmod}}$, to port P_3, which is connected to the input of the mixer, which is assumed to be matched by a load of $50\,\Omega$. The signal power is then as follows:

$$P_{\text{in mixer}} = P_3 = P_2 - P_I = P_{\text{back tag unmodulated}} - P_I$$
$$P_{\text{in mixer}} = (-62.18) - 1 = -63.18 \text{ dBm}$$

In fact, the load formed by the diode is never perfectly matched, since there are always some losses due to reflection or mismatching, of the order of 3 dB. The actual mixer input power to be taken into consideration is therefore -66.18 dBm.

> **Note**
> During the phase of back scattering response from the tag when the load impedance of the tag antenna is modulated, the collected power $P_{\text{back tag mod}}$ will be, as shown in Chapter 8, four times greater than that of the phase without modulation, i.e. $P_{\text{back tag unmodulated}} + 10\log 4 = -62.18\,\text{dBm} + 6\,\text{dB} = -56.18\,\text{dBm}$.

2. At port P_1, due to the imperfect isolation of the circulator, we find

$$P_1 = P_2 - I$$
$$P_1 = -62.18 - 20 = -82.18\,\text{dBm} \approx 10\,\text{pW}$$

Power derived from the base station carrier

(a) Since the carrier is always transmitted at full power to act as the support signal for the return data, as in Case 1, there will be a residual quantity from port P_2 at port P_3 because of the imperfect isolation, as follows:

$$P_{\text{in mixer}} = P_3 = P_2 - I$$
$$P_{\text{in mixer}} = 34 - 20 = 14\,\text{dBm} = 25\,\text{mW}$$

In this case also, with the Schottky diode as the load, the power will be $+11\,\text{dBm}$ instead of the $+14\,\text{dBm}$ level that was expected, since the diode is not perfectly matched and the reflection or mismatching losses are about 3 dB.

(b) In many cases, the impedance of the base station antenna is not at all well matched (if only because of the wide frequency band to be covered) to the output impedance of the circulator. We therefore have to allow for the presence of reflected waves and the associated reflection coefficient and VSWR. This means that the return signal from the tag at port P_2 is joined by the carrier power, which is reinjected into the circulator through this port. In our case, if the load represented by the antenna is well designed, the reinjected power will be about 3 dBm (2 mW), and this will of course immediately flow to port P_3 by coupling.

Global Budget of Signals Present at Port 3 (see Table 19.3)
To sum up, we find the following at P_3:

- from the high-level carrier $= 14\,\text{dBm} + (2)\,\text{dBm} = 25\,\text{mW} + (1.6\,\text{mW})$;
- from the low-level back scattering signal $= -63.18\,\text{dBm}$.

Note that, even if the antenna is well matched, the carrier power level ($+14\,\text{dBm}$) reaching the mixer input is much greater than the return signal ($-63.18\,\text{dBm}$), which carries the data from the tag.

Now let us estimate the peak voltage at the diode input:

$$P_{\text{in mixer}} = U_{\text{rms}}^2/R = U_{\text{peak}}^2/(2R)$$

Table 19.3

	Power (dBm)		Carrier (dBm)	Back scattering signal (dBm)
	P_1		34	
	P_2			− 62.18
	Input port	Output ports		
By coupling	$P_1 \rightarrow$	P_2	33	
	$P_2 \rightarrow$	P_3		− 63.18
	$P_3 \rightarrow$	P_1		− 64.18
By isolation	$P_1 \rightarrow$	P_3	14	
	$P_2 \rightarrow$	P_1		− 82.18
	$P_3 \rightarrow$	P_2		
By reflection to P2 (depending on mismatching)			(3)	
Coupling	$P_2 \rightarrow$?	P_3	(2)	

Therefore

$$U_{\text{peak mixer}} = \sqrt{P \times 2R}$$

$$U_{\text{peak mixer}} = \sqrt{25 \times 10^{-3} \times (2 \times 50)} = \sqrt{2.5} = 1.58 \text{ V peak}$$

This is the case of synchronous demodulation with a sufficiently high carrier level to make the mixing effective.

Note

In order to cover the whole range of RFID applications, let us briefly consider the case of a portable base station whose transmission power P_{ERP} is deliberately reduced to $+ 15$ dBm (approximately 30 mW, theoretical distance 1.30 m) to avoid excessive energy consumption. A calculation similar to that shown above would give us a mixer input power of -4 dBm (400 μW) instead of $+ 14$ dBm. This would result in the following peak voltage:

$$U_{\text{peak mixer}} = \sqrt{P \times 2R} = \sqrt{0.4 \times 10^{-3} \times (2 \times 50)} = \sqrt{4 \times 10^{-2}} = 200 \text{ mV peak}$$

This value is not sufficient to reach the opening threshold of the diode (0.7 V), and therefore the diode will perform its mixing by using only the quadratic portion of its current–voltage (I–V) transfer function. This will result in significant conversion losses, which we shall now calculate.

Now let us calculate the conversion losses of the mixer in the two cases above, for a load of 10 kΩ at the output of the filter:

- The input power is -82 dBm.

- The output power and the conversion losses for a local oscillator of $-4\,\text{dBm}$ are

$$P_{\text{out mix}} = \frac{V_{\text{out}}^2}{2R_{\text{out}}} = \frac{(35\,\mu\text{V})^2}{2 \times 10\,\text{k}\Omega} = -109\,\text{dBm}$$

$$G_{\text{mix}} = \frac{P_{\text{out mix}}}{P_{\text{in mix}}} = -109 + 82 = -27\,\text{dBm}$$

- The output power and the conversion losses for a local oscillator of $+6\,\text{dBm}$ are

$$P_{\text{out mix}} = \frac{V_{\text{out}}^2}{2R_{\text{out}}} = \frac{(95\,\mu\text{V})^2}{2 \times 10\,\text{k}\Omega} = -100\,\text{dBm}$$

$$G_{\text{mix}} = \frac{P_{\text{out mix}}}{P_{\text{in mix}}} = -100 + 82 = -18\,\text{dBm}$$

To carry out the synchronous detection at the input of this single Schottky diode (unbalanced mixer), the carrier required for this mixing is actually the residual signal from paths 1 and 3 of the circulator, coupled by imperfect isolation. This has a number of consequences:

- The phase of the carrier is not necessarily synchronous with the phase of the return signal, which may cause the cancellation of the signal if the relative phase difference is $90°$:

$$s(t) = a(t)\cos[\omega_0 t + \varphi(t)]$$
$$p(t) = \cos(\omega_0 t)$$
$$d(t) = 0.5 \times a(t)\cos[\varphi(t)]$$

- The level of the carrier is not easily optimized, unless an amplifier or an attenuator is added.

A more complex demodulator structure, such as a diode bridge with a transformer on the RF and local oscillator paths, would enable the residual oscillator and RF elements at the output to be kept much smaller, thus improving the signal-to-noise ratio. In theory, the low-pass filtering removes these HF components without any problems, but in practice the electromagnetic coupling between the input and the output of the filter prevents satisfactory rejection.

The demodulated signal is then filtered to limit the thermal noise band, reject the RF and local oscillator residual components, and block the double frequency produced in the mixing. The choice of filter type is important because of its amplitude and phase transfer function. A Gaussian filter is generally used, because it preserves the phase linearity. The signal is modulated in amplitude and phase, but since we are dealing with a small bandwidth system ($\pm 1\,\text{MHz}$ when the bit rate is $256\,\text{kbits s}^{-1}$), it is more important not to distort the phase of the modulated signal.

Limitations of this Architecture
The main quality of this architecture is its simplicity. However, it has technical limitations due to its low sensitivity. This is because the sensitivity of a receiver (at an ambient

temperature of 17 °C) is given by the following formula (see also the detailed example in Chapter 10):

$$\text{Sensitivity} = [(kT)\text{BWdr}] \times \text{NF} \times \text{LBM}$$

where $kT = -174$ dBm at 17 °C. Given that BWdr is the noise band of the receiver, allowing for the bit coding and bit rate (1 MHz in this example, allowing for a maximum bit rate of 256 kbits s^{-1}), NF is the noise factor of the receiver, S/N is the signal-to-noise ratio at the output of the low-pass filter, and LBM is the safety margin (including tolerances, etc.), then in terms of dBm, at 17 °C, we find that

$$\text{Sensitivity} = [-174 + 10 \log \text{BWdr} + 10 \log \text{NF} + \text{LBM}] + 10 \log(S/N)$$

A First Difficulty
The noise factor of the receiver is very unfavourable, being, to the first order of approximation, equal to that of the mixer, which also causes considerable losses. Since the noise factor of a mixer is greater than its conversion losses, it is very degraded in this case.

Another Difficulty
We cannot add a low-noise amplifier to mitigate the first problem, because the carrier level, which is coupled due to the imperfect isolation of the circulator, is too high and the power of the amplifier would have to be overdesigned, which would be very costly.

This base station architecture is therefore useful in the following cases:

- where a single antenna is used for a very compact reader;
- where the components are very few in number and inexpensive (except for the circulator).

19.2.3 Base Station with a Bidirectional Coupler and IQ Demodulation

Now let us examine the schematic diagram of a monostatic base station architecture based on a bidirectional coupler (Figure 19.9) designed to separate the transmission and reception paths from each other (by direct application of the directivity properties determined by the direction of propagation of the wave). In this case, I have chosen a three-way directional coupler (in other words, the fourth path is terminated indirectly on its matched load of 50 or 75 Ω).

Examination of the Structure of this Base Station

The power generated by the base station passes through a bidirectional directive coupler, which has the function of guiding the incident power (from the generator) and reflected power (from the tag) to two separate access points, enabling each path to be processed appropriately. The power generated by the base station is therefore present, subject to insertion losses (of 0 to 1 dB), at the antenna. In practice, filtering must also be provided to comply with the standards for parasitic emission levels. The part of the circuit that sends back the data is identical to the low-cost base station described above, and will not be described again here.

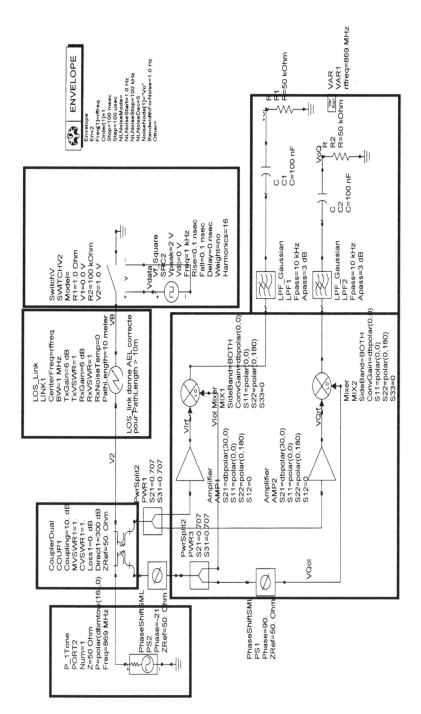

Figure 19.9 Base station with a coupler and IQ demodulator

The quadrature demodulation (known as I and Q demodulation) of the return signal from the tag is carried out as follows:

- by dividing the signal into two equal parts, using a power splitter;
- by amplifying the resulting two paths with low-noise amplifiers with a gain of 30 dB;
- by mixing each of the two paths separately with a local oscillator obtained from the transmitted carrier by means of multipliers, after carefully phase shifting one of the two paths of the local oscillator through 90° to provide the necessary quadrature for demodulation;
- finally, by filtering the I and Q signals resulting from the demodulation.

Let us examine this in more detail. We shall initially assume that the directivity of the coupler is ideal, but its effect will be assessed later on. Since the junction is symmetrical and reciprocal, the properties of the power arriving from the antenna will be the same, as we have seen:

- The return signal from the tag, present on the associated coupled path (port P_2) of the coupler, is divided into two by a power splitter, thus creating two identical paths for the demodulation.
- These two signals can then be amplified, since only the weak return signal from the tag is amplified in these two paths. The presence of the carrier is (theoretically) zero as a result of the isolation due to the operation of the bidirectional coupler. Thus there is no risk of saturating the amplification by the simultaneous presence of the carrier. Consequently, it is at this point in the circuit that the sensitivity will be improved.
- The noise factor of a sequence of amplifiers consisting of elements in series is given by the formula of Friis:

$$F_{tot} = F_1 + \frac{F_2-1}{G_1} + \frac{F_3-1}{G_1 G_2} + \cdots$$

where (F_1, G_1) are the noise and gain factors of the first stage, (F_2, G_2) are the noise and gain factors of the second stage, and so on.

If the gain G_1 of the first stage, the low noise amplifier (LNA), is sufficient, then according to the above equation the noise factor of the demodulator will be equal, to a first approximation, to that of the LNA, i.e. F_1, which must therefore be kept as small as possible (at about 1 dB). This will make it possible to use less expensive mixer stages.

To obtain the local oscillator signal for the demodulation, a small part of the output power of the generator (10% in the case of a 10 dB coupler) is sampled from the other path (port P_4) of the bidirectional coupler (by using the coupling provided by the coupler). In the present case of a 33 dBm generator (2 W ERP LBT), this corresponds to a 23 dBm signal (200 mW) acting as the local oscillator signal. This signal is also separated into two identical paths, using a power splitter. To provide the I and Q demodulation, one of the branches of this splitter is connected directly to a demodulator, while the other is connected via a 90° phase shifter to the other demodulator.

Clearly, we must ensure that the multiplier circuits used for the I and Q demodulation can withstand this level. For information:

$$P_{eff} = \frac{U_{peak}}{2R}$$

where $U_{peak} = U_{rms} \times \sqrt{2}$, $P_{rms} = U_{rms}/R$ and so

$$P_{rms} = U_{peak}^2/(2R)$$

As $P_{peak} = 200\,\text{mW}$ then $P_{rms} = 0.2/1.414 = 0.141$, and therefore $U_{peak} = 3.76\,\text{V}$.

If it is necessary to attenuate the signal, a simple $50\,\Omega$ resistor connected in parallel with the input may suffice. For an integrated circuit multiplier, the level of the local oscillator is commonly expressed in mV (often 200 mV peak to peak), whereas for a diode multiplier the level depends on the chosen structure (7, 17 or 27 dBm), giving different grades of performance in terms of conversion loss and isolation between paths. It is worth noting that it is useful to operate with a high local oscillator injection level to achieve greater linearity in the mixer (even if this is not immediately obvious), which means fewer intermodulation products at the output, although the pollution of the base station card (due to local oscillation retransmitted by the antenna, for example) will be greater unless special measures are taken.

The I and Q quadrature demodulator makes it possible to avoid the recovery of the carrier phase during the synchronous demodulation that takes place in the two multipliers. Thus, with operation in two paths, this is equivalent to the projection of any vector (Fresnel plane) on the horizontal and vertical axes. Over a period of time, one of the two projections will therefore have a value other than zero, and the following equations will be true:

$$s(t) = a(t)\cos(\omega_0 t) + E(t) = a(t)\cos[\varphi(t)]\cos(\omega_0 t) - a(t)\sin[\varphi(t)]/\sin(\omega_0 t)$$
$$p_I(t) = \cos(\omega_0 t), \; p_Q(t) = \sin(\omega_0 t)$$
$$d_I(t) = 0.5 \times a(t)\cos[\varphi(t)], \; d_Q(t) = 0.5 \times a(t)\sin(\varphi)$$

We then simply need to process the two paths by Gaussian filtering (for the same reasons as those mentioned above for the first, low-cost, base station) and ensure that the processing is correct (by finding the maximum).

Given the same initial conditions of compliance with the European regulations and ETSI 302 208-1 (2 W ERP in LBT mode), and assuming that we are using a bidirectional coupler in which $P_2 = P_1$, we can draw up a complete power budget for the different ports. With a coupling C of 10 dB, a directivity D of 30 dB, an injected power P_1 of 33 dBm, corresponding to 2 W, we find:

$$P_2 = 33\,\text{dBm} = 2\,\text{W ERP}$$
$$P_4 = P_1 - C$$
$$P_4 = 33 - 10 = 23\,\text{dBm} = 200\,\text{mW}$$
$$P_3 = P_4 - D$$
$$P_3 = 23 - 30 = -7\,\text{dBm} = 200\,\mu W$$

During the Forward Link, from the Base Station to the Tag

Let us start by considering the different origins of the signals present in the coupled path P_3:

(a) First of all, there is a 'parasitic' (unwanted) signal from the forward part of the base station which appears in the coupled path P_3 because of imperfect isolation. Since

$$I = P_{3i}/P_1 = D + C$$
$$I = 30 + 10 = 40 \, dB$$

then, in our case, if $P_1 = 2 \, W \, ERP = 33 \, dBm$, the signal P_{3i} originating from the base station will be

$$P_{3i} = P_1 - C - D = 33 - 10 - 30 = -7 \, dBm = 200 \, \mu W$$

(b) Now let us consider the carrier signals that have passed through the coupler in the normal way:

- If the load terminating port P_2 is very well matched, there is no power reflection, and the only energy present at P_2 is the energy P_{2i}, which comes from P_1 because of imperfect isolation (as shown in section (a) above).
- If the load terminating port P_2 is not well matched, because of imperfect matching of the antenna load (actually, all the components following the coupler), there will be wave reflection, and a certain amount of power P_{2r} also returns to port P_2, due to the reflection on the load *and* the coupling provided by the bidirectional coupler. Assuming that the insertion losses of the coupler are zero (forward path), $P_2 = P_1$, if $|\Gamma_a|$ is the reflection factor of the antenna load, the power returned/reflected by the antenna P_{2r} towards port P_2 will be $P_2 |\Gamma_a|^2$, and the coupled power P'_{2r} recovered at P_2 will therefore be $P_2 |\Gamma_a|^2 C$.

Example
Where $P_2 = 33 \, dBm = 2 \, W$, if $|\Gamma_a| = 0.315$ (VSWR = 2), then $|\Gamma_a|^2 = 0.1$

$$P_{2r} = P_2 |\Gamma_a|^2$$

or

$$P_{2r}(dBm) = P_2(dBm) + 10 \log(|\Gamma_a|^2)$$
$$P_{2r}(dBm) = 33 - 10 = 23 \, dBm = 200 \, mW$$

In this case, the power P_{2r} at the coupled port P_2 will be

$$P'_{2r} = P_2 |\Gamma_a|^2 C$$

or

$$P'_{2r}(dBm) = P_{2r} + C$$
$$P'_{2r}(dBm) = 23 - 10 = 13 \, dBm = 20 \, mW$$

Note
Even if the matching was virtually perfect and the VSWR was 1.1 (i.e. $\Gamma_a = (1.1 - 1)/(1.1 + 1) = 0.048$), the power P_{2r} would still be $-7 \, dBm = 200 \, \mu W$.

During the Return Link, from the Tag to the Base Station
The problem in receiving the signal retransmitted (by back scattering) from the tag is that the level of this useful signal, which will now be calculated, is much lower than that of all its 'parasitic' companions described above.

Let us re-examine the numbers in the second case.

For the carrier
In the return link from the tag to the base station, the carrier is constantly maintained, to support the return of the tag data. We have assumed that the power supplied to the coupler is 33 dBm (less if there are any insertion losses – which will be ignored), thus obtaining the level of $33 + 2.14 = 35.14$ dBm $= 3.28$ W EIRP $= 2$ W ERP maximum in the LBT mode in Europe on a $\lambda/2$ antenna. Thus, with $P_1 = P_2 = 33$ dBm, the part of the power reaching P_2 due to imperfect isolation is the same as in the previous section, i.e.

$$P_{2i} = P_1 - I = 33 - 40 = -7 \text{ dBm} = 200 \,\mu\text{W}$$

to which we must add the part of the reflected power arriving by coupling, if the antenna impedance is mismatched ($|\Gamma_a| = 0.315$, i.e. VSWR $= 2$), as follows:

$$P'_{2r} = \left[P_2 - 10 \log\left(|\Gamma_a^2|\right)\right] - C = (33 - 10) - 10 = 13 \text{ dBm} = 20 \text{ mW}$$

giving a total carrier power of $(20 \text{ mW} + 200 \,\mu\text{W}) = 20 \text{ mW} = 13$ dBm.

For the signal arriving from the tag
The useful signal received by the base station from the tag, $P_{\text{back tag unmodulated}}$, has been calculated to be -62.18 dBm (see the global link budget a few pages above) at the output of the base station antenna (in reception). For a circuit consisting of the coupler followed by a filter with losses of $P_{if} = 2$ dB and the antenna, therefore, this corresponds to the following usable power level received from the tag:

$$P''_{2r} = P_{\text{back tag unmodulated}} - P_{if} - C = -62.18 - 2 - 10 = -74.18 \text{ dBm}$$

The global level
At the input of the reception part of the base station, i.e. at port P_2, the spectral representation of the RF signals present will therefore show a pure carrier at -7 dBm (up to $+13$ dBm, depending on the antenna impedance mismatching), together with a signal returned by the tag and received at the base station demodulator at an equivalent level of -74.18 dBm! The receiver must therefore have a dynamic of approximately 100 dB!

Let us return to the directivity D of the coupler and evaluate its effect when it is equal to 30 dB. We shall assume a coupling of 10 dB and a generator power of 33 dBm. The signal level on the return path is given by the following equations:

$$P_{2i} = P_1 - C - D = 33 - 10 - 30 = -7 \text{ dBm} = 200 \,\mu\text{W}$$

which is a much higher power level than that of the modulated signal (-74.18 dBm). This means that the low-noise amplifiers must have power compression points greater than $-7 + 30 = 23$ dBm, but this limits us to a gain of 30 dB, which may be insufficient for long distance reading. This problem is tackled by using the third proposed structure.

Limitations of this Architecture

This architecture is also limited in terms of sensitivity. It is better than in the previous case of detection by diode, but is still low, owing to the excessively weak coupler directivity, and the weak modulated signal cannot be amplified as much as we might wish, because this would saturate the low-noise amplifier. This architecture is therefore useful in the following applications:

- where a single antenna is used for a very compact reader;
- where the reading distance is medium, because the amplification is limited to the receiver;
- medium-cost solutions.

One Possibility...

To overcome this problem, we could consider using a second antenna for reception, thus making the system bistatic (and increasing the cost), as shown in the schematic diagram in Figure 19.10, which omits the *I* and *Q* amplification and local oscillator signal generation circuits. Note than only one path of the coupler is used.

In this case, there is obviously no injection of the (strong) carrier signal into the path of the (weak) signal received from the tag. Should you choose this system, you will be able to review the calculation procedures in the preceding sections and modify them accordingly.

In this case, theoretically at least, there is no 'pollution' of the weak return signal by the strong carrier signal; however, you should note that this is only true of the designer who carefully specifies his installation (in terms of printed circuit designs, choice of components, etc.) in order to minimize any electrical and electromagnetic coupling. To help you avoid a number of problems, the following section will summarize the standard recommendations for implementation that should be followed for RFID applications at UHF.

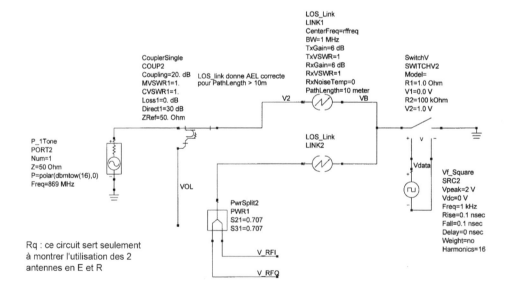

Figure 19.10   Circuit with bistatic antenna systems

Some Practical Recommendations
To reduce electrical coupling

- Ensure that the isolation between the local oscillator and RF paths of the mixers is sufficient (40 dB is often the best achievable level). If this condition is not met, the residual local oscillator signal will appear on the RF path and will be mixed with it, producing an incorrect continuous (DC) component. This is a well-known problem in homodyne receivers (where the local oscillator has the same value as the incident carrier frequency). There are not many solutions to this problem; however, it is possible to reduce the continuous (DC) component at the mixer output, using a high-pass filter. In this case, we must ensure that the modulated signal does not contain too much DC energy, in order to avoid excessive degradation of the signal-to-noise ratio.
- Try to provide the best possible match for all the units connected by 50 Ω lines, to avoid the presence of reflected power leading to the creation of standing waves, which would result in points of excess voltage and excess current, promoting electromagnetic radiation.

To reduce the electromagnetic coupling, ensure that the design of the printed circuit tracks (the mask) is such that electromagnetic coupling is minimized, as follows:

- Avoid long parallel lines carrying strong signals.
- Use perpendicular intersections where possible (and orthogonal ones in multilayer circuits).
- Do not place lines carrying weak signals in the proximity of strong signal lines and power supply lines, unless the power supply decoupling is excellent, in order to avoid varying the DC levels and generating fluctuations that would degrade the phase noise of the oscillators.
- Shorten the local oscillator lines.
- Enclose the different parts of the base station in a compartmented metallic casing to avoid electromagnetic coupling.
- Ensure that the isolation sources between the local oscillator and RF paths of the mixers are sufficient (40 dB is often the best achievable level). If this condition is not met, the residual local oscillator signal will appear on the RF path and will be mixed with it, producing an incorrect continuous (DC) component.

19.2.4 Base Station with a Bidirectional Coupler and IQ Demodulation, with Carrier Level Control

Figure 19.11 is a schematic diagram of a monostatic architecture that is also based on a bidirectional coupler, but in which the carrier level is measured in order to improve the *I/Q* demodulation and the performance of the system.

Here is a brief explanation of the distinctive features of this new architecture. The main difference from the previous types is that a loop for controlling the carrier level is included on the modulated return signal path, but this modification of the architecture is only justified if a single antenna is used.

The aim of this new architecture is to add the carrier (which arrives on the return path due to the imperfect directivity of the coupler), defined as 'residual', to the carrier, which is reconstituted and then weighted with respect to amplitude and phase.

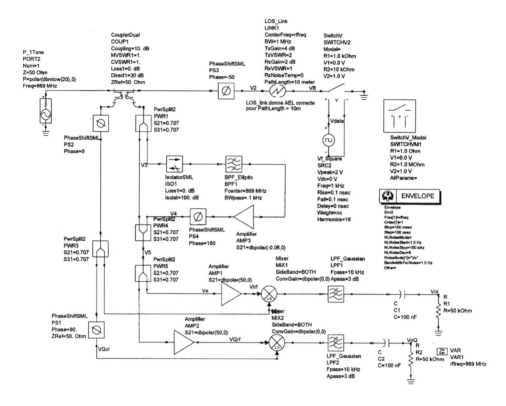

Figure 19.11 Monostatic base station with bidirectional coupler and carrier measurement

This loop must include Figure 19.12:

- A phase locked loop (PLL) for recovering the carrier phase. The bandwidth of this loop must be as small as possible (this may be a limitation on the range of application if the PLL is maintained or the locking band), but this can be envisaged for this kind of application if the frequency is known (transmission in a specified channel), and therefore the voltage controlled oscillator (VCO) can have a very small range of frequency variation to remove the modulation band and retain the carrier in the purest possible form. We may note that this system is much more effective if we use a return signal including a subcarrier (ISO 18000-6 part C) as in HF RFID systems operating at 13.56 MHz, since the modulated signal is not degraded and the adjustment of the loop is facilitated.
- A variable gain amplifier (VGA), to ensure that the carrier level corresponds to the residual carrier level. This means that the residual signal must be sampled (by a 10 dB coupler, for example), since a detector provides an image of the envelope level in order to control the gain of the VGA with respect to a set point. Note that this is possible because the level of the residual signal is fixed.
- A phase shifter in the loop to provide phase opposition of the signal with respect to the residual signal (assuming that the phase–frequency detector has a phase shift of 0°).

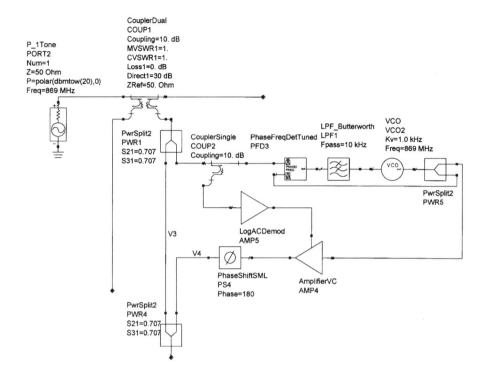

Figure 19.12   Electrical circuit

Limitations of this Architecture

The advantage of this structure is its excellent sensitivity, since the residual carrier is attenuated by about 60 dB (beyond this level, electromagnetic coupling takes priority over this electrical isolation). The principal disadvantage is that its adjustment is relatively complicated, because the two essential components, namely the phase locked loop (for controlling the phase) and the variable gain amplifier (for controlling the amplitude), are highly sensitive to the dynamic variations of the loop.

The limitations of this architecture arise from the fact that back scattering modulation using subcarrier frequencies is very advisable. Note that some options in RFID at UHF and SHF according to ISO 18000-4 or -6 part C permit the use of subcarrier frequencies in the return signal.

19.2.5 Base Station with a Bidirectional Coupler and IQ Demodulation, with Carrier Level Control and Cancellation

Figure 19.13 is a global schematic diagram of a commercial base station. Let us briefly survey this proprietary system. It can be divided into a number of separate units.

Local Oscillator

The local oscillator unit is based on a VCO (voltage controlled oscillator), controlled via a phase locked loop (PLL) so that the frequency of the VCO can easily be changed to comply with local regulations or to permit the use of frequency agility solutions such as FHSS or LBT.

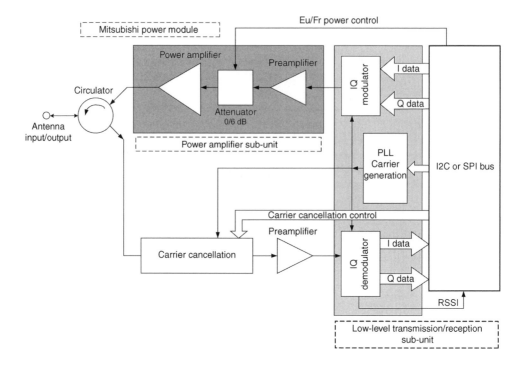

Figure 19.13   Base station with circulator and demodulation by carrier cancellation

Modulator

The carrier modulation (ASK type) provided by the binary stream of the forward link (from the base station to the tag) is carried out, in accordance with the systems described in ISO 18000-6 parts A, B and C, by means of a modulator with I and Q inputs (vectors in quadrature). This type of system facilitates the modification of the carrier modulation index, enabling optimal use to be made of the possibilities offered by the spectrum templates, subject to local regulations according to the communication bit rates used.

Amplification Circuit and Power Amplifier

The amplification circuit is composed of several stages connected in series, making it possible to achieve the maximum 1 W conducted power permitted in the USA (i.e. 4 W EIRP with a 6 dB antenna) or the 2 W ERP permitted in Europe (i.e. 2 W maximum conducted power applied to a $\lambda/2$ antenna (with a gain of 1.64), equivalent to $2 \times 1.64 = 3.28$ W EIRP). To achieve spectral purity of the transmitted wave, the power amplifier is commonly of class A (meaning that at least as much power is dissipated in the load as in the amplifier, so that a global dissipation of 6 to 7 W maximum is required), or if necessary class AB, often based on a circuit or module such as those made by Mitsubishi. Note that a programmable attenuator is connected in the middle of the chain, so that the transmitted power can be adjusted according to the specific requirements of local regulations.

Circulator or Isolator

This enables the antenna to be driven in the forward direction, while helping to separate the outward and incoming signals, as described earlier in this chapter.

Antenna Diversity

The purpose of the antenna diversity function is to control the time multiplexing of an antenna set with the ultimate aim of filling all, or most, of the 'black holes' (field cancellations due to single and multiple reflections of the wave determined by the paths), and thus improving the tag reading and writing distances and coverage.

Antenna(s)

The set of antennas is specific to each application, according to the chosen gain, bandwidth, polarization, forward/backward ratios, forms, etc.

Reception Circuit

This is often the area that conceals the well-kept secrets of proprietary systems! Briefly, the aim is to suppress as much of the carrier as possible (by isolation, subtraction, etc.) before the demodulation of the return signal, in order to amplify only the microscopic signals that arrive from the tag and that reflect its presence via the dynamic variation of its RCS.

Demodulator

These demodulators are often of the I and Q type, because the use of this technology makes it possible (among other things) to achieve a much lower BER (bit error rate).

RSSI Device

Although this is not a prerequisite for conventional RFID applications at UHF, it becomes virtually essential for systems according to ESTI 302 308 – LBT, because decisions must be made according to whether or not any transmissions are present in the desired transmission channel, where the received signal level exceeds a specified threshold.

High-Sensitivity Threshold Receiver

To comply with ETSI 302 208, we must use a high-sensitivity amplifier capable of detecting signals in the desired channel above or below the threshold of -96 dBm and triggering or preventing the talk phase of the LBT procedure.

Synchronizing Device

At UHF, when RFID systems are located in a 'dense environment' (see Chapter 16), it is often useful to synchronize the bases stations with each other in respect of the operation of the frequency hop sequences provided in FHSS and/or LBT, in order to avoid constant and unregulated occupation of the spectrum either by pure transmission or by multiple reflections that are not fully controlled, which in LBT would make it impossible to talk at any time (giving rise to the pejorative term LNT, i.e. 'listen never talk'). Note that ETSI, being fully aware of this problem, has recently added a supplement to ESTI 302 208 in the form of a technical report (TR) on this matter, to avoid total anarchy among the suppliers of base stations.

Microcontroller + ASIC or FPGA or...

Obviously, the microcontroller is designed to manage the whole base station and offer the user all the functionality that may be needed. Rather than describe this in purely abstract terms, let us consider a base station designed to comply with European regulations. In this case, the base station must comply with the following:

- at least the handling of air interface communication protocols such as those specified in ISO 18000-6 types A, B and C;

- the local regulations in force at a given date, namely ETSI 302 208, i.e. 2 W ERP (3.28 EIRP) with operation in the LBT mode for Europe, or ERC 70 03, i.e. 0.5 W ERP (0.820 W EIRP) with a duty cycle of 10%.

These requirements mean that we must provide dedicated I/O ports (2 × 2) for microswitches allowing us to select (a) the desired protocols and (b) the local regulations.

In terms of software, the microcontroller must control the following elements, in addition to the pure handling of the ISO 18000-6 types A, B and C (EPC C1 G2) protocols mentioned above:

- The instantaneous value of the frequency transmitted by the base station. This command, sent via an I2C or SPI serial link, provides the PLL (outside the microcontroller) with the numerical value to be entered into the register dedicated to its frequency divider.
- The PLL operation, which will take place according to the time sequence specified in ETSI 302 208 (conformity with the on and off times between two possible accesses to the medium – air).
- The frequency change command, which is entirely controlled by the value of the RSSI (received signal strength information), which may be analogue or may have been digitized in advance, depending on the resources of the microcontroller used, and which is sent from the reception part, making it possible to determine whether or not another carrier is present in the channel concerned (according to the levels specified in ETSI 302 208) during the L phase of the listen before talk procedure.
- The change in the transmitted power level, according to local regulations, using an I/O port operating a 'programmable' attenuator present in the amplification circuit.
- The operation of the antenna diversity multiplexer (a total of four antennas). This is time multiplexing, because the antennas are operated one after another, and the on-board software must be able to handle singletons and doublets when reading in order to determine the actual number of tags present in the radiated field. Note that the target of reading 200 objects passing through the antenna field at a speed of $1.5 \, \mathrm{m \, s^{-1}}$ is to be assessed according to the volume of the field, the maximum on time, the estimated statistical on/off ratio of the LBT procedure in a single channel (remember that $\mathrm{on_{max}} = 4 \, \mathrm{s}$) and the field switching time. For example, if we assume that the volume is approximately $3 \, \mathrm{m} \times 2 \, \mathrm{m}$ ($1.5 \, \mathrm{m}$ on each side of the arch supporting the antennas over a width of $2 \, \mathrm{m}$, giving a total volume of $6 \, \mathrm{m^3}$), then if the 200 objects travel at $1.5 \, \mathrm{m \, s^{-1}}$, each of them will remain in the field for $4 \, \mathrm{s}$. The antennas must therefore be multiplexed with a minimum switching cycle of approximately $1 \, \mathrm{s}$.
- The operation of an RS 232 or 485 link for control by a host (PC, network, etc.).

19.3 Examples of Products

Before ending this chapter, here are some examples of components found in proprietary base stations.

19.3.1 Protocol Handling Circuits

By way of example, Figure 19.14 shows the base station structure recommended by Watkins and Johnson (WJ), using standard integrated circuits from their catalogue. Naturally, this includes all the functions described above. Note that many other circuits are also on the market (some examples are: R1000/Indy from INTEL/Impinj; AS 3990 from AMS).

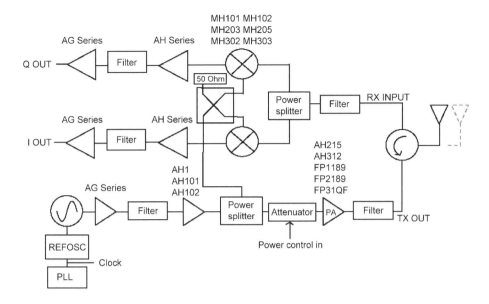

Figure 19.14 Base station: as proposed by WJ Semiconductors

Figure 19.15 is a block diagram of a dedicated circuit (ASIC) made by the same company, for handling most variants of the standard RFID communication protocols operating at UHF (ISO 18000-6 types A, B, C, etc.). Note that the output amplifier of this circuit delivers a power of only + 20 dBm (100 mW), which is quite adequate for small handheld readers, but generally too low for long distance readers.

19.3.2 Power Components

As explained above, what we still need is an adjustable gain amplifier in which the transmitted power levels can easily be adjusted to comply with the various local regulations around the world. By way of example, Figure 19.16 shows the Mitsubishi RA 13H 8891 MA circuit, which meets these requirements.

19.3.3 A Commercial Base Station

Finally, Figure 19.17 shows an example of a commercial base station produced by Samsys (photographs taken by the author).

19.4 Antennas for Base Stations

The subjects discussed below could fill entire books – but these already exist. My aim here is therefore not to go over the same ground, but rather to outline the key points of the antennas used in RFID applications at UHF. Many of the general features of these antennas have already been described in Chapters 4 and 5 but, in the light of the preceding sections on regulations and other ISO standards, I shall summarize the crucial points relating to the specification and choice of base station antennas.

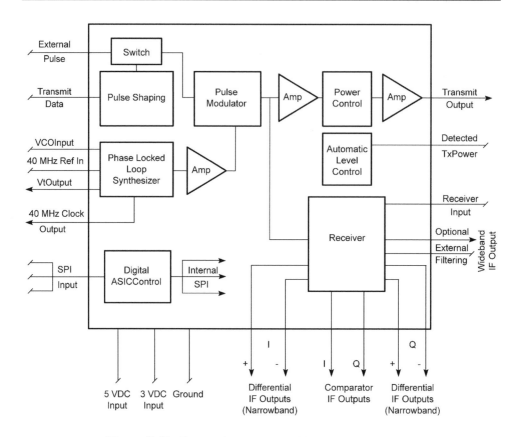

Figure 19.15   Base station: as proposed by WJ Semiconductors

19.4.1 Power Gain

Normally, the more gain, the better. This is not always true in RFID, because as I have shown the gain and directivity of an antenna are related, and we generally look for the best compromise between the gain and low forward directivity with a low backward/forward ratio to achieve the largest possible reading volume. This generally means that we must use a gain of about 2 to 6 dBi, and rarely more than this, except in special cases (e.g. logarithmic antennas).

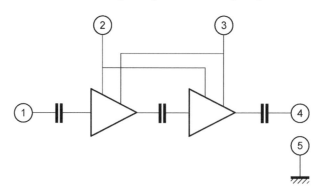

Figure 19.16   Example of an integrated circuit for a power amplifier

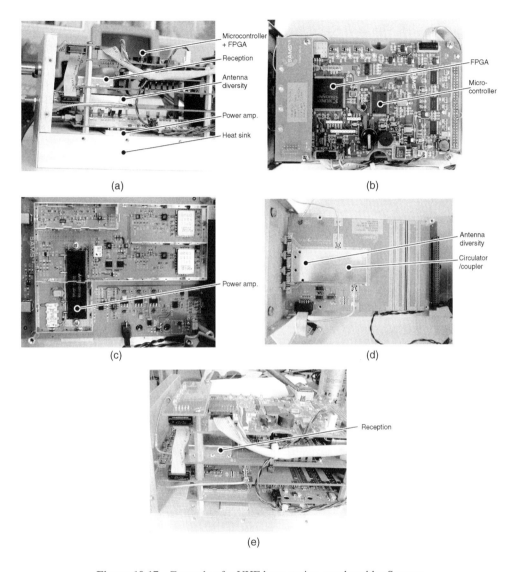

Figure 19.17 Example of a UHF base station, produced by Samsys

19.4.2 Power

The main aim is to adjust the conducted power levels, in ERP and EIRP, permitted by local regulations, and combine them in an intelligent way with the gain and directivity of the antennas to be used for the applications.

19.4.3 Directivity

This parameter covers a multitude of sins, including the associated beamwidths, which may be specified by standards (e.g. ETSI 302 208), the presence or absence of secondary lobes that strongly affect the propagation paths of unforeseen waves that may cause constructive reflection, resulting in hot spots, or destructive reflection producing 'black holes'. The choice of the forward/backward ratios is extremely important where other base stations are present (which is often the case) and/or where systems operate in the LBT mode in 'multiple' and 'dense' environments, or when base stations with bistatic antennas are used.

19.4.4 Polarization of the Wave

Linear, circular or elliptical – that is the (big) question! Here again, the choice between types of polarization is often guided by the characteristics required for the application, as well as the regulations to be observed.

19.4.5 The Mechanical Shape of the Antenna

Should the antenna be flat (circular, square, rectangular) or not (Yagi type)? Once again, our choice must be guided by the application. For example, a logarithmic Yagi antenna, with high gain and directivity, is preferable if we wish to know the plane in which the wave will be propagated. For example, this may be flat, at ground level, for reading the last tags located near the ground at the bottom of pallets, to avoid problems of parasitic reflections and black holes.

19.4.6 Antennas Integrated into the Base Station, or not, and Vice Versa

If the antenna is remote, and therefore not integrated into the base station, there must be a cable between the base station and the antenna. The presence of a cable inevitably means problems of length, losses, mismatching, VSWR, reproducibility, degradation of performance, etc. – in short, it spells trouble! If the antenna is integrated into the base station (or vice versa), in what is known as a 'one base station per antenna' structure, this will avoid the above problems – but at a price!

19.4.7 Multiple Antenna Devices

If the application requires the coverage of a large volume, we must consider using a number of antennas to achieve this coverage. Let us look at two conventional solutions:

- If the antennas are not integrated into the base station (or vice versa), we must run cables of the same length to each of them; otherwise we will encounter different losses, mismatching, VSWRs, etc. – meaning more trouble! We will also have to provide time multiplexing for this whole arrangement and ensure that the systems are synchronized with some of their companions.
- If the antennas are integrated (one circuit per antenna), we will escape the problems noted above, but will incur costs and difficulties in providing the synchronization between all these systems, especially if they are not all from the same supplier. We should also note that a standard specifying a standardized synchronization mechanism where there are time differences between antennas is being developed (the ETSI is very interested in resolving these problems and those of dense environments), but generally each user follows his own path; in other words, he uses proprietary solutions.

19.5 Some Concluding Remarks

We have summarized the main conventional architectures that can be used for developing the hardware of UHF/SHF base stations. There are many other solutions that are often variants or subvariants of those described above and many proprietary products (whose contents cannot be disclosed), often featuring patented technical and technological devices, which are, as usual, subject to licences and royalties.

To conclude this chapter, let me simply point out that the same evaluations and preliminary theoretical calculations have to be carried out in all cases, regardless of the forms of coupling used at the terminals of the base station antenna. Now that you know the general outline, you can fill in the details as you please.

Finally, should you require further information about these matters, you should consult the many specialist books on radio circuitry, especially those describing radar techniques, which will provide full details of all the specific points relating more or less closely to the general principles of back scattering.

20

Conformity, Performance and Methods for Evaluating Tags and systems

Now that we are approaching the end of the book, we will conclude our detailed examination of the theory, methods and technology of this UHF and SHF field of RFID by looking at the procedures for estimating, measuring and guaranteeing performance.

There are still all too many tags (in the form of discontinued commercial products) and base stations that are sold without technical data sheets providing detailed descriptions of the purpose, conformity and expected performance of these systems. This situation often leads to insurmountable problems in the installation of systems, due to a concealed lack of interoperability. Having experienced this kind of trouble all too frequently in the field, I will now attempt to overcome these difficulties by, firstly, providing a brief survey of conformance and performance standards, and, by way of a conclusion to the book, a (reasonably) straightforward description of the measurements to be made and the associated test methods, to give you a fairly accurate idea of a commercial end product (tag), including of course its integrated circuit, its antenna and its specific package (paper, plastics material, etc.).

20.1 Official Measurement and Test Methods

20.1.1 ISO 18047-x Conformance Tests and ISO 18046 Performance Tests

It is all very well to say that products conform to an ISO standard. However, what matters is the proof. To summarize:

- Claims of conformity can only be enforced against the person making the claims – even if a statement is signed to this effect.
- Proving conformity is a matter of producing measurements made by the manufacturer or by a recognized independent laboratory. This is better, but there is still no guarantee of the reliability of the measurements or of the appropriateness of the measuring instruments used.

- It is possible to obtain the 'stamp' of an independent laboratory that is certified/approved/ accredited (by an Accreditation Committee), which has made measurements with the measuring equipment specified in the test standard, to remove all doubt as to the veracity of the claims made. This is the perfect situation.

It is for this purpose that the ISO 18047-x set of standards was drawn up.

20.1.2 The ISO 18047-x Conformance Tests

As with the ISO 18000-x series, the 'x' of the ISO 18047-x standards corresponds to the frequency covered. These standards provide a detailed description of the measuring equipment and methods that must be used in all cases to ensure that the resulting values have the meaning given to them in the ISO 18000-x documents. You should consult the official texts of these standards in order to learn about the many details of their contents. It is not the aim of this book to examine and comment on the whole of their wording. However, it is useful to know that these conformance test standards are important, because it is not at all simple to measure tiny signals among a host of much more powerful ones, and the results can mislead even the most expert professionals unless great care is taken. Those who are particularly unsure or easily confused should carefully and thoroughly re-examine the fine mathematical and physical distinctions between the FFT (fast Fourier transform) (often used on all kinds of occasions without consideration of its deeper significance) and its predecessor, the DFT (discrete Fourier transform). They will then realize that these methods are fundamental to an evaluation of the measurements obtained for back scattering signals received in RFID.

Meanwhile, in order to remove some areas of doubt concerning the conformance criteria specific to ISO 18000-6, I will point out that:

- The base station (interrogator) must:
  - support types A, B and C of the standard,
  - be able to switch from one to the other and also, depending on the application; control the sequence for allocation of the time ratio between the two types.
- As for the tag, this must:
  - support at least one of the three types and, optionally, the other two;
  - remain silent when it receives a modulated signal from a base station that it does not support or recognize;
  - operate over the whole frequency range from 860 MHz to 960 MHz;
  - show a minimal variation of the delta radar cross-section.

While on this subject, the operating performance may vary according to the exact operating frequency within the 860–960 MHz range, depending on the antenna and the tuning of the tag.

20.1.3 ISO 18046 Performance Test Methods

Once you have undergone all the conformance tests described above and received the coveted award of 'passed', rather than the shameful 'failed', you will have your papers stamped by an organization certified by the duly accredited authorities, confirming that your product conforms to the standard. There may be a fly in the ointment, however. Your strongest competitor may have received the same certificate and, everything else being equal, he is bound to claim that his products offer a better communication range than yours! This will be repeated constantly, if that is what it takes. Of course, this is annoying, or even infuriating. If we want a 'tie-breaker', we must make measurements so that we can evaluate (or claim that we have evaluated) the performance of the two products in an impartial way, in order to put a stop to all the negative publicity. This is the purpose of ISO 18046-x, which can be briefly described as the 'performance' standard. Essentially, it offers reliable measurement methods, with reproducible results, which anyone can use to evaluate, measure and publicize recognized performance levels and thus provide a firm basis for comparisons between products and between systems. The standards in question are:

- ISO 18046-1 – System performances (including the interrogator and tags system)
- ISO 18046-2 – Interrogator performances (performance of the base station)
- ISO 18046-3 – Tag performances

Each of these standards contains specific sections on inductive coupling systems or those operating in any propagation mode, and covers all the frequencies used for RFID applications.

This concludes our rapid overview of the standardizing of the measurements and methods of the conformance and performance tests provided by the ISO for RFID at UHF and SHF. Beyond this, all manufacturers of base stations, tags and systems who sell products are required to supply the specifications along with these products, as in the case of the specifications, data sheets, etc., supplied with integrated circuits. These are the matters that I will now examine.

20.2 Required Parameters

In addition to the communication protocols (governing the transmission frame formats, collision management techniques, etc.) relating to the chosen operating standards and/or the integrated circuits used in the tags, I will now provide a list of the major parameters that, seen from the outside and in a given environment (air or special environment such as water or metallic material, etc.), ambient temperature, etc., are the main factors to be considered by an end user when he is specifying and implementing his system:

- The specifications that the tag must meet (standards, etc.):
  - minimum and maximum values of the mandatory modulation index M_i of the forward wave;
  - the nominal tuning frequency of the tag, f_0;
  - the value of the bandwidth B_p (or the quality factor Q of the tag);
  - the radiation pattern of the tag, showing the performance of its antenna in terms of directivity.
- To enable the maximum possible operating ranges to be determined, the values of the threshold electrical fields E above which the tag memory can be read ($E_{\text{thres read}}$) or written to ($E_{\text{thres write}}$) are required.
- In order to avoid certain kinds of damage (possibly fatal) to the tag or its immediate environment (e.g. burning or initiation of fire in the packaging box to which it is applied), we must be able to specify the authorized minimum distance for correct operation (i.e. the

closest the tag may come to a strongly transmitting base station). For this purpose, we must know and specify:

- the value of the maximum electrical field E_{max} (absolute maximum rating), which must not be exceeded;
- the maximum temperature rise of the tag package at the authorized value of E_{max} (or, more precisely, its thermal resistance R_{thtag}; see below).

- In order to confirm the minimum and maximum operating ranges mentioned above and ascertain that the (standardized?) base station can interpret the return signal (back scattering), the range of electrical field values E_{min_rcs} and E_{max_rcs} over which there is a guaranteed value (standardized, if possible) of dynamic variation of the radar cross-section (ΔRCS) of the tag is required.

- Some time parameters, such as:
  - the minimum and maximum rise times of the field E tolerated/supported by the tag;
  - the tag's reset time (including the reset time of the integrated circuit, but also the effects due to its tuning circuit), etc.

20.3 Simple Methods of Measurement

You should find the following section helpful. It provides a brief description of some simple measurement methods that can be used to evaluate these essential parameters of applications.

20.3.1 Measuring Electrical Fields

Let us begin with the measurement of electrical fields, E_{thres}. We must start with a commercial base station that can deliver a power of the order of one watt EIRP and an antenna with known gain and directivity. On this subject, it is worth noting that many suppliers in the RF field can provide reference antennas, or 'measuring antennas'. However, if you are unable to obtain one of these, it is fairly simple to calibrate an existing antenna, using the theory set out in Chapters 5 and 6, and this works very well.

Using the arrangement shown in Figure 20.1, for a given frequency:

1. The base station delivers a known and constant conducted power. After connecting an attenuator in series in the link to the antenna of known gain, it is then a simple matter to vary the radiated power EIRP. It is worth making a few comments at this point. Clearly, we must ensure, on the one hand, that the attenuator impedance (often 50 Ω) is indeed in the correct ratio with the cable used for the antenna link and, on the other hand, that the whole base station/cable/antenna is correctly matched and that the VSWR (standing wave ratio) of the system is similar.

2. The tag for measurement is then placed in the equatorial plane of the base station antenna and oriented in the direction of its maximum directivity (i.e. in the direction of its maximum sensitivity) at a permanently fixed distance, which you are free to set at about three to five times the wavelength of the measurement frequency, thus establishing far field conditions. As I will go on to demonstrate, this arrangement, without moving the tag towards the base station in order to discover the distance at which it starts to operate, avoids

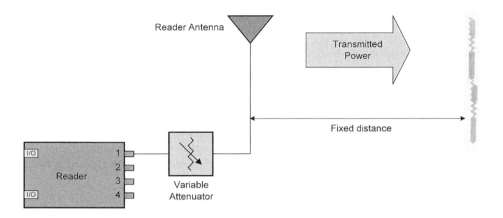

Figure 20.1   Example of an arrangement for measuring electrical fields used for operation

the expense of using an anechoic chamber, which absorbs all the waves that may be reflected. This is shown in Figure 20.2, and is needed for measurements made at UHF and SHF according to ETSI 300 220 and 302 208.

In fact, at a fixed measurement distance (i.e. for a given frequency, at a position where the propagation paths of the radiated waves will always be the same), the potential reflection and absorption of the measurement room will always be identical for all the measurements, and therefore the effects of these will be constant and will not affect the differences between the results. On the other hand, if we were to follow the common practice of varying the distance between the base station and tag during the measurement, without using an anechoic chamber, the measurements would be strongly affected by parasitic reflections, leading to the creation of 'black holes' and hot spots (see Chapters 7 and 8), which would seriously distort the results. Above all, this would adversely affect the reproducibility of the measurements from one site to another.

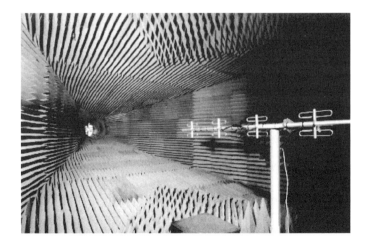

Figure 20.2   Anechoic chamber

3. Note that, if we place the reference antenna in the same location as the tag whose performance is to be measured, we can measure the electrical field present when the base station is transmitting its maximum power, without any modulation of the carrier.

4. The tag is put in position and the carrier is modulated according to the communication protocol associated with the integrated circuit used, sending the simplest possible command, such as the 'identifier read command', and a spectrum analyser with an additive antenna (of the $\lambda/2$ type for example) located near the tag is used to check that the tag is sending a return signal, indicating that it is being remotely powered and is operational. Consequently, we can operate independently of the sensitivity of the base station and we do not need to know whether the variation of the radar cross-section conforms to any particular value.

5. Starting at the highest level of power transmission by the base station, we use the calibrated attenuator to reduce the power until the tag return signal disappears from the spectrum analyser. We then use the reference antenna to measure the corresponding value of the threshold field strength.

> **Note**
> Using this procedure, we can operate independently of the sensitivity and demodulation performance of the base station receiver, and measure only the performance relating to the tag.

If we then continue the measurements throughout the band in which the tag is claimed to operate, we can determine its bandwidth with respect to the threshold electrical field strength, and therefore also with respect to power, given the simple relationship between the field E and the power P (see Chapter 5). The same measurements are made for the electrical field in the write procedure and then the tag can be rotated about its principal axes to find its radiation pattern, using the same methods.

20.3.2 Measuring ΔRCS

To avoid unnecessary repetition, I would ask you to refer back to the method shown in Chapter 8 for measuring and estimating the value of ΔRCS.

20.3.3 Maximum Electrical Field Strength and Maximum Temperature for Operation of the Tag

These two parameters are strongly correlated with each other. Seen from the outside, it is clear that the maximum electrical field strength that a tag can withstand is an essential parameter for the proper specification of a project. This value is defined by two principal procedures.

By The Design of the Tag by Its Manufacturer

When a tag is designed, its designer has to take into account the maximum current that can flow in the input circuit of the integrated circuit used, in order to ensure that the product operates correctly. Note that this current cannot be seen or measured by the end user of the tag. For any given tag antenna design, the current strength is always directly related to the electrical field present around the tag. Since the tag manufacturer is responsible for choosing the integrated circuit, determining its correct use and designing the tag antenna, he must also specify and state the maximum electrical field strength that the tag can withstand 'by design'.

In Relation to the Outside World

Sometimes, a maximum authorized electrical field strength for the tag that is different from that stated above may be specified, although the previously stated maximum current will not be exceeded. Why is this? The explanation is as follows.

RFID applications such as supply chain management (SCM) often use tags formed on paper substrates, which are applied to packages of the cardboard box type. Consequently, the problem is not the maximum current as mentioned in the preceding sections, but rather the general temperature rise in the tag, which itself is determined by the current flowing in the tag and therefore by the local electrical field surrounding the tag. Before we light the fuse, figuratively or literally (imagine a case of explosives labelled with a UHF tag, halted in front of a reader with the tag close to the base station), let us see what we must do to avoid any disasters.

The dissipation of the tag system that is purchased by the end user, consisting of the integrated circuit and the substrate on which the antenna is deposited (paper, polyester, etc.), together with the antenna technology (etching, screen printing, ink deposition, inkjet, epitaxial copper growth, gravure printing, offset printing, etc.), the connections between the integrated circuit and the antenna (bonding, flip chip, bumps, etc.) and the packaging (paper, plastics material and/or film, adhesive, etc.), must be quantified using conventional thermal parameters such as the maximum authorized operating temperature and the global thermal resistance of the tag, which in the last analysis is equivalent to specifying a maximum electrical field strength for the application. Note also that the production technology for the tags described above are often characterized by poor heat dissipation, since they have high levels of thermal resistance (expressed as $°C\,W^{-1}$). In any case, these values must be measured and published; otherwise we cannot be responsible for the consequences in terms of communication by smoke signals and fireworks!

As for the temperature measurements, these are easily carried out using infrared sensors or even, to a first approximation, with adhesive heat sensor pads, which can be stuck to the tag to indicate the temperature reached in the presence of the maximum authorized electrical field. For further information, you should review the end of Chapter 3 where these values are listed in detail.

20.4 By Way of Conclusion

Although I have no wish to act as a righter of wrongs or a technical legal expert, I hope that the foregoing sections will have made it clear that, if you are to make proper use of a commercial tag, a data sheet as shown in the example below must be provided with the tag. Table 20.1 shows a first example of a standard data sheet for a UHF/SHF tag ('end product').

> **Note**
> The values in the table are only examples given for guidance, but they show the boxes that need to be ticked – with a clear indication of the measurement conditions, of course!

Table 20.1

			Minimum	Typical	Maximum
Environment in which measurements are made			Air, free space		
Tag operating temperature range			-40 to $+65\,^\circ$C		
Conforms to communication protocol			ISO 18000-6 B		

	Symbol	Unit	Minimum	Typical	Maximum
Range of operating frequencies		MHz	860	–	960
Nominal tag tuning frequency or frequencies	f_0	MHz	900	910	920
Bandwidth	B_p	MHz	–	100	–
Quality factor	Q		–	10	–
Carrier modulation index	M_i	%	90	–	100
Radiation pattern					
Electrical fields					
E_{thres_read}		$V\,m^{-1}$	–	–	0.7
E_{thres_write}		$V\,m^{-1}$	–	–	1
E_{max} (absolute maximum rating)		$V\,m^{-1}$	–	–	22
Maximum tag junction temperature for E_{max}	$T_{j\ max}$	$^\circ$C	–	–	85
Total thermal resistance of tag	$R_{th\ tag}$	$^\circ C\,W^{-1}$	–	–	120
Guaranteed minimum variation of radar cross-section	ΔRCS	cm^2	70	–	–
For $E_{min\ rcs}$		$V\,m^{-1}$	0.7	–	–
For $E_{max\ rcs}$		$V\,m^{-1}$	–	–	22
Rise time of the field E	t_r	ms	0.5	–	0.9
Maximum tag reset time	$t_{reset\ max}$	ms	–	–	1.2

Table 20.2 provides a second example – which is strange at first sight – of a standard data sheet for a UHF/SHF tag operating in the 860–960 MHz band.

Table 20.2

			Minimum	Typical	Maximum
Environment in which measurements are made			Air, free space		
Range of operating temperature			-40 to $+65\,^\circ$C		
Conforms to communication protocol			ISO 18000-6 B		

	Symbol	Unit	Minimum	Typical	Maximum
Range of operating frequencies		MHz	860	–	960
Nominal tuning frequency of the tag	f_0	MHz	1100	1110	1120
Bandwidth	B_p	MHz	–	110	–
Quality factor	Q		–	10	–
etc.					

> **Note**
> Here again, the values shown in the table are only examples provided for guidance, but they do represent the boxes that need to be ticked.

This is not as odd as it may seem. These are indeed the characteristics of the tag in air, in free space, as clearly stated in this data sheet. However, what is not stated, but must be deduced, is that, if this tag is to operate in the standard RFID band from 860 to 960 MHz, it must be placed in an environment that detunes the tag's tuning frequency by about 200 MHz and resets it appropriately so that it can operate correctly in the frequency band concerned (e.g. by placing the tag behind a vehicle windscreen; see Chapter 7).

At least this has the merit of being clear to everybody.

Conclusions

So we arrive at the end of this book, having introduced the basic physical operating principles of contactless RFID devices for UHF and SHF, the scope of their applications and the information environments in which they are located. As you will have realized, this is a complicated field, because most of the parameters involved combine with and overlap each other, and it has been my aim to guide you through the maze in the clearest possible way.

I hope that I have answered most of your questions and demystified most of the problems involved in moving from theory to practice in these applications. I also trust that you will now find it easier to deal with the specification and implementation of projects or applications. If so, I will have succeeded.

Admittedly, many areas have had to be omitted. I have not dealt in any detail with the physical design of antennas or encryption methods, because these are matters for specialists. If you are interested in these subjects, they are covered in many specialist texts that are currently available.

I have also avoided any detailed description of the communication protocols (frames, requests, collision management, etc.) of the ISO 18000-6 family. This is not due to any reluctance on my part, but rather to the fact that, sooner or later, you will be obliged to peruse these documents in detail to extract the last morsel of information. There would not have been room for such an approach in this book. For example, Amendment 1 of ISO 18000-6, dealing with Part C – EPC C1 G2 – contains 128 pages in which each data bit is important, and it took five days of intensive work by a six-person group at an ISO working meeting to re-read, check and agree on the content of each of them! The same is true of application-based comparisons between barcode labels and contactless labels, or between types of data organization and databases associated with the processing of the contents of tags, etc. After all, we must leave a few topics for future editions, or for other more specialized writers on these subjects. Each to his own!

After this brief digression, I must point out that this whole field is currently evolving rapidly and there will inevitably be many new developments in the years to come, given the major industrial and economic interests involved. I will certainly return to these matters at the appropriate time.

RFID at Ultra and Super High Frequencies: Theory and Application   Dominique Paret
© 2009 John Wiley & Sons, Ltd

The Future

Before taking leave of you, having described many possible applications in this book, I must mention that I often meet people who ask, 'And what about the future?' Once again, therefore, I have consulted my favourite crystal ball and come up with the following prognostications.

Briefly, in addition to contactless smart cards, immobilizers, labels, etc., as described above, workers in this field are looking towards applications in the medium term, including intelligent microsystems with on-board microcontrollers, remotely powered and capable of controlling remote sensors of all types, embedded in materials (for mechanical stress measurements for concrete pillars, intelligent airbag systems, etc.), using new families of MEMS currently under development. These devices generally demand a rather large amount of computing power, and therefore have a somewhat higher power consumption and require technological advances to resolve the problems of remotely powering systems via contactless RF links.

Without giving away too many secrets, I can reveal that many working groups at the ISO have already spent much time examining the possible structure of protocols between remote sensors – for measuring temperature, pressure, acceleration, pH (the measurement of the concentration of H^+ ions in a body, indicating its acid–base balance; for pure water, $pH = 7$) and every other measurable physical quantity – and contactless tag architecture, with the aim of standardizing the communication interface between the RF part (the 'air interface') and the sensor, with a view to developing families of interoperable systems. The ISO has also completed a great deal of work on communication protocols for the geolocation of objects and/or persons fitted with tags, in order to track them with the aid of GPS devices (applications include the geographical monitoring of containers carried on trucks and the movement of elderly people in hospitals, etc.).

Moreover, it is clear that our brave new world of the 21st century will be based, like it or not, on communication in all its forms, and one of the main contributors to this development will be contactless radio links.

Already, the world is highly dependent on GPS, GSM, WAP, Bluetooth, WiFi, WiMax, WireLAN, e-commerce, intelligent labelling, the use of smart cards for transaction certification and security protection, etc., and all these systems operate with RF links, whether at LF, HF, UHF or SHF. The age of radio has returned with a vengeance, and so have its analogue-based physical principles, problems of interference and jamming, careless or fraudulent eavesdropping and all the complexities of local and international standards that constantly bedevil us. If I have put a special emphasis on bit coding, types of modulation, shapes of spectra, and the like, it is because all these aspects will have to be standardized – especially in view of globalization – to enable every user to operate without disturbing his neighbour, while ensuring the best possible return on investment.

The development of contactless technology, RFID and mobile RFID, NFC including RFID, will therefore be dependent on harmonious global standardization, which has been discussed for each area of application in turn. This challenge will not be easy to meet and will take time to resolve. However, this is the price that the end user – you and I – must pay for the resulting benefits in everyday life. Therefore, as I have already implied in this book, once again the future lies before you! Welcome to the 'intelligence environment' and 'ontology'! While awaiting these future developments, and hoping that you have enjoyed reading this book as much as I have enjoyed writing it, I look forward to meeting you again for yet more extraordinary adventures!

Goodbye for now. However, if you are perplexed about any of these matters, you are welcome to contact the author at: dp-consulting@orange.fr.

Useful Addresses, Component Manufacturers and Further Reading

Standardization and Similar Authorities

ETSI
European Telecommunication Standards Institute
Route des Lucioles – Valbonne – 06921 Sofia Antipolis Cedex
Telephone: + 33 (0) 4 92 94 42 00   Fax: + 33 (0) 4 93 65 47 16

ISO
International Organization for Standardization (ISO)
1, ch. de la Voie-Creuse
Case postale 56
CH-1211 Geneva 20, Switzerland
Telephone: + 41 22 749 01 11   Fax: + 41 22 733 34 30

Component Manufacturers

Numerous component manufacturers (NXP/Philips Semiconductors, Electro Marin, Alien, Impinj, Infineon, TI, Motorola, ST Microelectronics, Temic, MicroChip, etc.) have been mentioned in this book. Instead of filling a dozen pages with their addresses, fax numbers, etc., I advise readers to visit the websites of these large companies, which are generally highly detailed and will give you the latest news about RFID components.

If you still need more, search under 'RFID' on Google for a vast number of useful, but mostly commercial, pages.

Further Reading

RFID – Identification Radiofréquence et Cartes à Puce Sans Contact, by D. Paret, Dunod, 2nd edition, 2001.

RFID at Ultra and Super High Frequencies: Theory and Application   Dominique Paret
© 2009 John Wiley & Sons, Ltd

RFID – Applications en Identification Radiofréquence et Cartes à Puce sans Contact, by D. Paret, Dunod, 2003.

RFID – Handbuch, by Klaus Finkenzeller, Hanser, ISBN 3-446-19376-6. A general overview covering most contactless applications.

RFID – Handbook, by Klaus Finkenzeller, John Wiley and Sons, Ltd, ISBN 0-471-98851-0 (translation of the German text).

RF/ID – Radio Frequency Identification – Application 2000, by J. D. Gerdeman, Research Triangle Consultants – RTC Inc., PO Box 12031, Research Triangle Park, NC 27709, ISBN 1-883872-01-4.

Index